B PHYSICS AT

HADRON MACHINES

Related Titles from AIP Conference Proceedings

B PHYSICS AT HADRON MACHINES

9th International Conference on *B* Physics at Hadron Machines
Beauty 2003

Pittsburgh, Pennsylvania　　　*14 –18 October 2003*

EDITORS
Manfred Paulini
Carnegie Mellon University
Pittsburgh, Pennsylvania

Samim Erhan
University of California at Los Angeles
Los Angeles, California

SPONSORING ORGANIZATIONS
U.S. Department of Energy
U.S. National Science Foundation
Carnegie Mellon University

Editors:

Manfred Paulini
Department of Physics
Carnegie Mellon University
Pittsburgh, PA 15213
USA

E-mail: paulini@cmu.edu

Samim Erhan
Department of Physics
University of California at Los Angeles
Los Angeles, CA 90095
USA

E-mail: erhan@physics.ucla.edu

L.C. Catalog Card No. 2004110782
ISBN 0-7354-0203-5
ISSN 0094-243X

Printed in the United States of America

CONTENTS

Preface ix
Committees xi

RECENT RESULTS

Recent CLEO Results 3
    K. M. Ecklund (*Representing the CLEO Collaboration*)
Recent Results from DØ 7
    V. Jain (*Representing the DØ Collaboration*)
Recent Results from CDF 11
    R. Harr (*Representing the CDF Collaboration*)
Measurement of Time Dependent *CP* Violation in $B^0 \to \phi K_s^0$ 16
    M. Krishnamurthy (*Representing the BaBar Collaboration*)

CP VIOLATION

Time Dependent *CP* Violation at Belle 23
    K. Miyabayashi (*Representing the Belle Collaboration*)
CP Violation Prospects at the Tevatron 30
    P. Maksimović (*Representing the CDF and DØ Collaborations*)
Learning the Weak Phase γ from *B* Decays 35
    J. L. Rosner
Towards α (ϕ_2) and γ (ϕ_3) at the *B* Factories 42
    A. Watson (*Representing the BaBar Collaboration*)
CP Violation Beyond the Standard Model 50
    D. London

B PRODUCTION AND SPECTROSCPY

Measurement of the $b\bar{b}$ Cross Section at HERA-*B* 57
    M. zur Nedden (*Representing the HERA-B Collaboration*)
Hadronic Beauty Production at HERA 63
    A. B. Meyer (*Representing the H1 and ZEUS Collaborations*)
Heavy Flavor Production at the Tevatron 67
    C. Chen (*Representing the CDF and DØ Collaborations*)
B Reconstruction and Spectroscopy at DØ 72
    E. von Toerne (*Representing the DØ Collaboration*)

CHARM PHYSICS

Review of Recent Results in Charm Physics 79
    J. Engelfried
The CLEO-c Research Program 82
    D. Asner (*Representing the CLEO Collaboration*)
Charmed-Strange Mesons Experimental Results 87
    J. Wang
Charm Physics at Belle 92
    K. Trabelsi (*Representing the Belle Collaboration*)

DETECTORS, HARDWARE, AND COMPUTING

The Status of the ATLAS Inner Detector 99
    H.-G. Moser (*Representing the ATLAS Collaboration*)

Status of the Compact Muon Solenoid Detector at the Large Hadron Collider................104
    H. Neal (*Representing the CMS Collaboration*)
BTeV Detector Update................109
    D. Cinabro (*Representing the BTeV Collaboration*)
LHCb: Reoptimized Detector and Tracking Performance................114
    G. Raven (*Representing the LHCb Collaboration*)
Status of the LHCb RICH and Hadron Particle Identification................119
    M. Adinolfi (*Representing the LHCb Collaboration*)
LHCb Calorimeters and Muon System - Lepton Identification................123
    F. Machefert (*Representing the LHCb Collaboration*)
ATLAS Software for *B* Physics................127
    F. Parodi (*Representing the ATLAS Collaboration*)
Grid Computing in High Energy Physics................131
    P. Avery

$|V_{ucb}|$, $|V_{ub}|$ AND FACTORIZATION

Factorization and the Soft-Collinear Effective Theory: Color-Suppressed Decays................141
    S. Mantry, D. Pirjol, and I. W. Stewart
Semileptonic Branching Ratios and Moments................146
    G. Della Ricca
Present Status of Our Knowledge of $|V_{cb}|$................151
    M. Artuso
The Status of $|V_{ub}|$................156
    L. Gibbons (*Representing the CLEO Collaboration*)

B LIFETIMES AND MIXING

***B* Lifetime Measurements at the Tevatron**................167
    D. Zieminska (*Representing the CDF and DØ Collaborations*)
***B* Tagging and Mixing at the Tevatron**................172
    T. Miao (*Representing the CDF and DØ Collaborations*)

RARE DECAYS

Theory of Radiative and Rare *B* Decays................181
    G. Isidori
Radiative and EW Penguin *B* Decays with the Belle Detector................188
    T. Ziegler (*Representing the Belle Collaboration*)
Spectroscopy and Rare Decays at CDF................193
    C.-J. S. Lin (*Representing the CDF Collaboration*)

B TRIGGER AT FUTURE EXPERIMENTS

The ATLAS *B* Physics Trigger................201
    S. George (*Representing the ATLAS Collaboration*)
Online Event Selection at the CMS Experiment................207
    M. Konecki (*Representing the CMS Collaboration*)
The LHCb Trigger System: Implementation & Performance................214
    O. Callot (*Representing the LHCb Collaboration*)
The BTeV Trigger System................218
    M. H. L. S. Wang (*Representing the BTeV Collaboration*)

B PHYSICS AT FUTURE HADRON MACHINES

B Physics at LHC with the CMS Detector..225
    N. Marinelli (*Representing the CMS Collaboration*)
LHCb Physics Performance 231
    U. Uwer (*Representing the LHCb Collaboration*)
Physics Reach of the BTeV Experiment 236
    B. Cox (*Representing the BTeV Collaboration*)
ATLAS *B* Physics Performance Update 242
    P. Eerola (*Representing the ATLAS Collaboration*)

THE FUTURE OF FLAVOR PHYSICS

Flavor Prospects...251
    E. Witten
Footprints of New Physics in the *B* System 255
    Y. Grossman
Ultra-Rare *B* Decays 263
    B. Grinstein
B Physics in the LHC Era 269
    R. N. Cahn

PANEL DISCUSSION

Models and Methods: Can Theory Meet the *B* Physics Challenge?...............................277
    J. N. Butler
Models and Methods: Can Theory Meet the *B* Physics Challenge?...............................279
    L. Wolfenstein
Models and Methods: Can Theory Meet the *B* Physics Challenge?...............................280
    U. Nierste
Models and Methods: Can Theory Meet the *B* Physics Challenge?...............................283
    D. London

List of Participants 285
Program...287
Photos 293
Author Index 297

PREFACE

This volume of AIP Conference Proceedings contains the write-up's of the talks presented at the 9th International Conference on B Physics at Hadron Machines, Beauty 2003, held at Carnegie Mellon University in Pittsburgh, Pennsylvania, USA, from Tuesday October 14, to Saturday October 18, 2003. This conference was a successor of the previous one held in June 2002 at Santiago de Compostela in Galicia, Spain. It continues the successful series of earlier ones which began in 1993 in the Liblice Castle, near Prague, Czech Republic, followed by Le Mont St. Michel, France (1994), Oxford, United Kingdom (1995), Rome, Italy (1996), Los Angeles, U.S.A. (1997), Bled, Slovenia (1999) and Maagan, Israel (2000).

The purpose of this conference is to review recent results in the field of B physics and CP violation and to explore the experimental reach of current and future generations of B physics experiments at hadron machines. The conference is open to all experimental and theoretical physicists interested in this field. The conference consisted of plenary talks with no parallel sessions. The site that served as webpage for the conference can be found at URL: http://www-hep.phys.cmu.edu/beauty2003/

The conference program has been planned by the International Advisory Committee under the Chair of Peter Schlein assisted by the Local Organizing Committee. An adequate and timely formulation of the program was assured matching the continuous evolution in the field of B physics at hadron machines. The program covered the full range of B physics activities worldwide, including recent results from the e^+e^- B factories. The B physics program at running and future experiments at hadron machines as well as R&D programs in detector and trigger technologies were reviewed. It has always been the scope of the workshop organization to combine both the fundamental physics insights and the key technical developments which allow its experimental unveiling. While earlier conferences in this series highlighted the progress in research and development of the various experimental apparatus and detector components, the focus of the conference has shifted to include recent physics results from running experiments. In particular, the e^+e^- machines at SLAC and KEK truly initiated a new era of studying CP violation in B decays and constraining the unitarity of the Cabibbo-Kobayashi-Maskawa quark mixing matrix. But also the collider experiments at the Fermilab Tevatron showed first detailed results. In addition, excellent theoretical reviews greatly enhanced the discussion of experimental results.

There are in fact three main developments covered in these proceedings which focused the attention of the Beauty 2003 workshop:

1. Continued impressive state of the art results from Belle and BaBar on the CKM origin of CP violation, along with other impressive achievements of their experimental programs.

2. First results from the Tevatron experiments CDF and DØ together with recent advancements from the recently approved BTeV experiment.

3. Reports on the significant progress in preparation of the LHC experiments, including B physics studies and related technical developments, such as triggering and GRID computing.

The conference concluded with a panel discussion on "Models and Methods: Can Theory Meet the B Physics Challenge?" organized and chaired by Joel Butler, Fermilab.

We do not want to end this introduction without mentioning the pleasant aspects of the social program of the conference which delighted the participants with a number of enjoyable events. Among them was an excursion to Fallingwater, a home designed by the famed architect Frank Lloyd Wright and built over a waterfall in the Laurel Highlands, south-east of Pittsburgh, as well as a Conference Banquet at the Carnegie Museum of Natural History with a reception in the museum's Hall of Dinosaurs and a dinner in the Hall of Architecture.

The next Beauty Conference on B Physics at Hadron Machines will be held in May 2005 in Perugia, Italy, hosted by INFN Firenze.

Samim Erhan
Manfred Paulini
Editors

International Advisory Committee

J. Butler (FNAL)
R. Cashmore (CERN)
D. Denegri (Saclay)
N. Ellis (CERN)
S. Erhan (UCLA)
H. Evans (Columbia)
F. Ferroni (Rome)
M. Gronau (Haifa)
J.D. Hansen (Copenhagen)
N. Harnew (Oxford)
K. Kinoshita (Cincinnati)
N. Lockyer (Pennsylvania)
H. Newman (Caltech)
J. Rosner (Chicago)
Y. Rozen (Haifa)
P. Schlein (UCLA), Chair
P. Sphicas (CERN)
S. Stone (Syracuse)

Local Organizing Committee

R. Briere (CMU)
T. Ferguson (CMU)
F. Gilman (CMU)
R. Holman (CMU)
A. Leibovich (Pitt)
L.-F. Li (CMU)
M. Paulini (CMU), Chair
I. Rothstein (CMU)
J. Russ (CMU)
P. Shepard (Pitt)
H. Vogel (CMU)
L. Wolfenstein (CMU)

RECENT RESULTS

Recent CLEO Results

Karl M. Ecklund

(Representing the CLEO Collaboration)

Laboratory of Elementary-Particle Physics, Cornell University Ithaca, NY 14853, U.S.A.

Abstract. I report B physics results from the CLEO collaboration, highlighting measurements of the Cabibbo-Kobayashi-Maskawa matrix elements $|V_{cb}|$ and $|V_{ub}|$. I report a recent measurement of $|V_{ub}|$ through study of the q^2 dependence of $\bar{B} \to \pi \ell \bar{\nu}$ and $\bar{B} \to \rho \ell \bar{\nu}$. I also describe new measurements of the inclusive semileptonic branching fraction $\mathscr{B}(\bar{B} \to X e \bar{\nu})$ and of moments of the hadronic invariant mass spectrum in $\bar{B} \to X \ell \bar{\nu}$, with impact on $|V_{cb}|$.

1. INTRODUCTION

CLEO's recent measurements of the Cabibbo-Kobayashi-Maskawa matrix elements $|V_{ub}|$ and $|V_{cb}|$ are still competitive in the era of B Factory statistics because these measurements are systematically and theoretically limited, and CLEO's well-understood detector and analysis techniques bring added value to the world knowledge of these couplings in semileptonic B decays. Along with $|V_{td}|$ and $|V_{ts}|$, inferred from measurements of B mixing, $|V_{ub}|$ and $|V_{cb}|$ form an important part of the B CP puzzle: the sides of the Unitarity Triangle. Combined with measurements of the angles from the CP violating phases, will the Unitarity Triangle hold together or indicate the presence of new physics? Precise measurements of $|V_{ub}|$ and $|V_{cb}|$ are required for this test.

2. EXCLUSIVE $|V_{ub}|$ MEASUREMENT

Recently CLEO measured $|V_{ub}|$ in the exclusive modes $\bar{B} \to [\pi/\rho/\omega/\eta]\ell\bar{\nu}$ [1]. The neutrino is reconstructed from the missing energy and momentum of the event, taking advantage of CLEO's large solid angle (95%). Tracks reconstructed from hits in the drift chamber and silicon are combined with neutral showers in the calorimeter to form missing energy and momentum estimates. Considerable effort is made to remove spurious tracks and showers from hadronic interactions, in order to give the best estimate of the neutrino energy and momentum. When the neutrino candidate is combined with a lepton and light meson candidate, energy and momentum conservation leads to signal peaks in $\Delta E = E - E_{\text{beam}}$ and the B candidate invariant mass $M_{m\ell\nu}$, with $S/B \approx 1$. We perform a simultaneous maximum likelihood fit in ΔE and $M_{m\ell\nu}$ to seven sub-modes:

TABLE 1. Branching fractions for exclusive $b \to u\ell\bar{\nu}$ modes. In order, the errors given are statistical, systematic, due to the π form factor, and due to the ρ form factors. For $\eta\ell\bar{\nu}$, the last error is due to the η form factor.

Mode	BF $\times 10^4$
$B^0 \to \pi^- \ell^+ \nu$	$1.33 \pm 0.18 \pm 0.11 \pm 0.01 \pm 0.07$
$B^0 \to \rho^- \ell^+ \nu$	$2.17 \pm 0.34 \, ^{+0.47}_{-0.54} \pm 0.01 \pm 0.41$
$B^+ \to \eta \ell^+ \nu$	$0.84 \pm 0.31 \pm 0.10 \pm 0.09$

$\pi^{\pm}$, π^0, $\rho^{\pm}$, ρ^0, $\omega/\eta \to \pi^+\pi^-\pi^0$, and $\eta \to \gamma\gamma$. In the fit, we use isospin symmetry to constrain the semileptonic widths $\Gamma^{SL}(\pi^{\pm}) = 2\Gamma^{SL}(\pi^0)$ and $\Gamma^{SL}(\rho^{\pm}) = 2\Gamma^{SL}(\rho^0) \approx 2\Gamma^{SL}(\omega)$, where the final approximate equality is inspired by constituent quark symmetry. We find clear signals for π and ρ/ω and a 3.2 sigma significance for $\eta\ell\bar{\nu}$. The branching fractions are given in Table 1.

Signals for π (Fig. 1a) and ρ are extracted separately in three q^2 bins. The differential decay rate for $\pi\ell\bar{\nu}$,

$$\frac{d\Gamma}{dq^2} = \frac{G_F^2}{24\pi^3} |V_{ub}|^2 p_\pi^3 \left| f_+(q^2) \right|^2,$$

includes a form factor f_+ which encodes the hadronic physics for the $B \to \pi$ transition. For $\rho\ell\bar{\nu}$, which has a vector meson in the final state, there are two additional form factors. Given form factors from theory, we extract $|V_{ub}|$ from a fit to $d\Gamma/dq^2$ (Fig. 1b). Combining the $\bar{B} \to \pi\ell\bar{\nu}$ and $\bar{B} \to \rho\ell\bar{\nu}$ results, we find

$$|V_{ub}| = (3.17 \pm 0.17|_{\text{st}} \, ^{+0.16}_{-0.17}|_{\text{sy}} \, ^{+0.53}_{-0.39}|_{\text{th}} \pm 0.03|_{\text{ff}}) \times 10^{-3},$$

where the quoted errors are, in order, statistical, experimental systematic, and theoretical from form factor normalization and shape. This result uses form factors from Lattice QCD ($q^2 > 16$ GeV^2) and light cone sum rules

CP722, *B Physics at Hadron Machines*, edited by M. Paulini and S. Erhan
© 2004 American Institute of Physics 0-7354-0203-5/04/$22.00

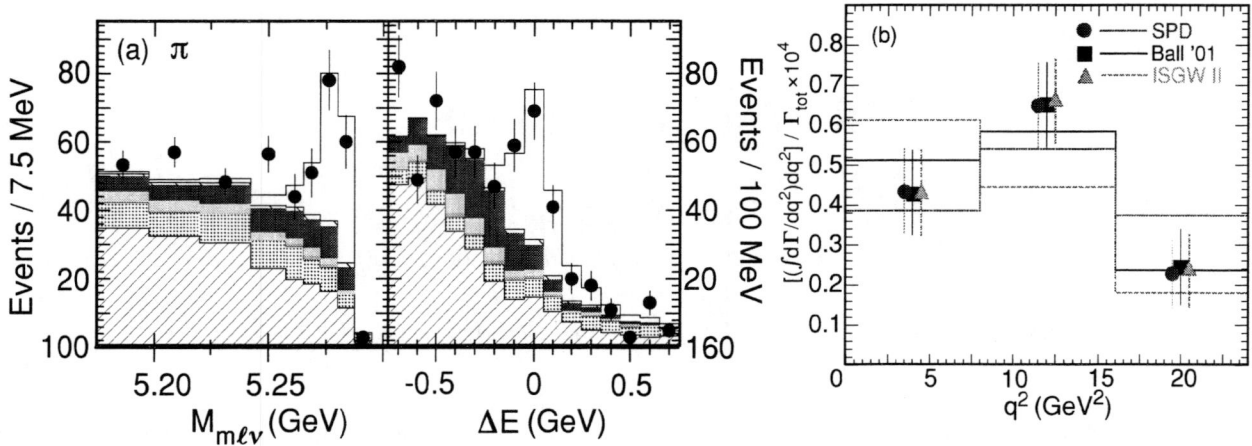

FIGURE 1. Exclusive $B \to \pi \ell \bar{\nu}$: (a) projections of ML fit to $M_{m\ell\nu}$ and ΔE and (b) fit to $d\Gamma/dq^2$.

$(q^2 > 16 \text{ GeV}^2)$ where each are most reliable. In a test of $\bar{B} \to \pi \ell \bar{\nu}$ form factors, ISGW2 [2] is disfavored (Fig. 1b).

By binning in q^2, this analysis has relaxed the theoretical assumption on the shape of the form factor made in earlier analyses. The theoretical uncertainty on the form factor normalization currently limits the precision of the $|V_{ub}|$ extraction. In the future, unquenched Lattice QCD calculations [3] can improve the $\bar{B} \to \pi \ell \bar{\nu}$ form factor in a limited region of q^2. We find good agreement between measurements of $|V_{ub}|$ using inclusive techniques [4–6] and other exclusive measurements [7, 8].

3. INCLUSIVE $|V_{cb}|$ MEASUREMENT

A measurement of $|V_{cb}|$ is possible using the inclusive semileptonic decay rate. The experimental inputs are the branching fraction for $\bar{B} \to X_c \ell \bar{\nu}$ and the B lifetime. The inclusive decay rate $\Gamma_c^{SL} = \gamma_c |V_{cb}|^2$, where γ_c comes from theory.

Within the framework of heavy quark effective theory (HQET) [9], the inclusive semileptonic decay rate is expanded in a double series in α_s^n and $1/M^n$, where M is the heavy quark mass. Hadronic effects enter both in the perturbative expansion and as expansion parameters, defined to be matrix elements of non-perturbative QCD operators. At $\mathcal{O}(1/M^2)$ there are two parameters: λ_1, which is proportional to the kinetic energy of the b quark in the B meson, and λ_2, which comes from the chromomagnetic operator. An additional parameter $\bar{\Lambda}$ relates the B meson mass to the b quark mass. From the B-B^* mass difference, we use $\lambda_2 = 0.128 \pm 0.010 \text{ GeV}^2$. The other parameters can be estimated (*e.g.* in quark models) but they can also be measured using spectral moments in inclusive B decay. Moments, *e.g.*, of the lepton energy spectrum, are

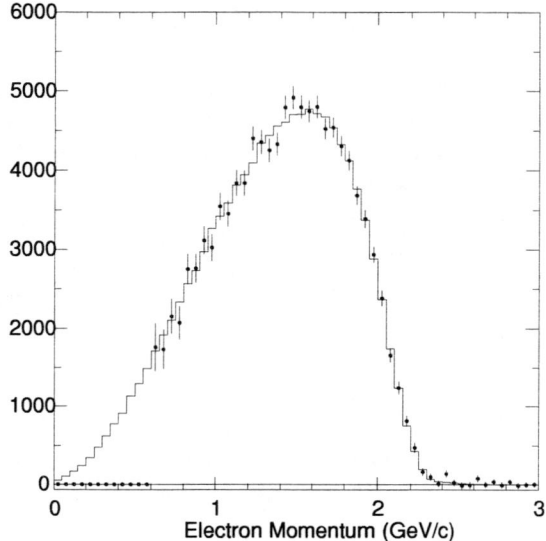

FIGURE 2. Unfolded primary $\bar{B} \to X e \bar{\nu}$ electron spectrum using a high momentum lepton tag. The line shows a fit to exclusive $\bar{B} \to X e \bar{\nu}$ decays, used to extrapolate below the cut at 600 MeV/c (preliminary).

also computed in HQET, allowing extraction of λ_1 and $\bar{\Lambda}$ from two or more spectral measurements.

3.1. Inclusive Semileptonic Branching Fraction

CLEO has a new preliminary measurement of the inclusive semileptonic branching fraction using a high-momentum ($p > 1.5$ GeV/c) lepton tag. The analysis is an update of Ref. [10], where the lepton tag identi-

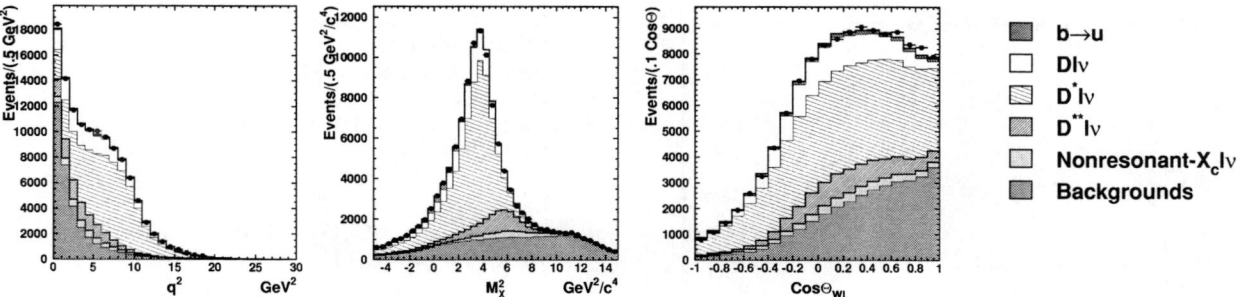

FIGURE 3. Projections of likelihood fit to differential decay rate. For the three independent kinematic variables, we have chosen q^2, M_X^2, and $\cos\theta_{W\ell}$, the helicity angle of the virtual W decay.

fies a sample of B decays with high purity (98%). Additional electrons may come from the decay chain of the same B or from the decay of the other B meson in the event ($e^+e^- \to \Upsilon(4S) \to B\bar{B}$). Secondary electrons ($b \to c \to e$) and primary electrons are separated using kinematic and charge correlations, with a known correction from B^0-$\bar{B}^0$ mixing. The new semileptonic branching fraction is $(10.88 \pm 0.08 \pm 0.33)\%$, in agreement with measurements from LEP and B factory data [11]. The spectrum of electrons above 600 MeV/c is also obtained (Fig. 2), from which spectral moments will be measured.

3.2. Extraction of $|V_{cb}|$ Using the Heavy Quark Expansion

Using CLEO's new inclusive branching fraction of $(10.8 \pm 0.3)\%$, subtracting a 1% relative contribution from $b \to u\ell\bar{\nu}$, and the PDG average B^0 and B^+ lifetimes [12], we find the semileptonic decay rate $\Gamma_c^{SL} = (0.44 \pm 0.02) \times 10^{-10}$ MeV. Using the HQET expansion for the decay rate and HQET parameters $\bar{\Lambda}$ and λ_1 from CLEO measurements of the moments of the $B \to X_s\gamma$ photon spectrum [13] and $\bar{B} \to X\ell\bar{\nu}$ hadronic mass spectrum [14], we obtain $|V_{cb}| = 0.0411 \pm 0.0005|_{\bar{\Lambda},\lambda_1} \pm 0.0007|_{\Gamma} \pm 0.0009|_{HQE}$. The overall precision of 3% is limited by theoretical uncertainties from the unknown $\mathcal{O}(1/M^3)$ heavy quark expansion parameters. There is an unquantifiable error from the parton-hadron duality assumption of the inclusive approach.

3.3. Spectral Moments as a Test of HQE

It is essential to test the heavy quark expansion and the assumption of parton-hadron duality implicit in inclusive determinations of $|V_{cb}|$. Previously CLEO published an analysis of the lepton energy spectral moments in $\bar{B} \to X_c\ell\bar{\nu}$ with a cut at 1.5 GeV [15], showing good

agreement with HQET expectations and the hadronic mass and $B \to X_s\gamma$ photon spectrum analyses. At Lepton Photon 2003, CLEO presented new preliminary results from an analysis of the hadronic invariant mass spectrum in $\bar{B} \to X_c\ell\bar{\nu}$ [16].

Like the exclusive $|V_{ub}|$ measurement, we use the neutrino reconstruction technique to estimate the neutrino energy and momentum. Combined with a lepton (electron or muon) candidate with $p > 1.0$ GeV/c, we can reconstruct the hadronic invariant mass recoiling against the lepton and neutrino:

$$M_X^2 = M_B^2 + q^2 - 2E_{\text{beam}}(E_\ell + E_\nu) + 2|\vec{p}_B||\vec{q}|\cos\theta_{Bq},$$

where $q^2(\vec{q})$ is the lepton-neutrino (virtual W) invariant mass squared (momentum). Only the last term in the exact equation for M_X^2 is unknown; fortunately, because it is small in the $\Upsilon(4S)$ rest frame, the approximation from neglecting this term is adequate.

We fit the three-dimensional differential decay rate to contributions from $\bar{B} \to D\ell\bar{\nu}$, $\bar{B} \to D^*\ell\bar{\nu}$, $\bar{B} \to D^{**}\ell\bar{\nu}$, $\bar{B} \to X_c\ell\bar{\nu}$, and $\bar{B} \to X_u\ell\bar{\nu}$ (Fig. 3), allowing contributions from fake lepton, $e^+e^- \to q\bar{q}$, and $b \to c \to \ell$ backgrounds, which are estimated using data and Monte Carlo simulation. From the fit results, we extract the first and second moments of the hadronic recoil mass. The first moment $\langle M_X^2 - \bar{M}_D^2 \rangle$ is computed as a function of the minimum lepton energy cut $E_\ell^{\min}$, varied between 1.0 and 1.5 GeV (Table 2). The first moment is plotted versus the lepton energy cut in Fig. 4, overlaid with results from a similar analysis of *BABAR* [17] and expectations from HQET given the CLEO measurement of $\langle E_\gamma \rangle$ in $B \to X_s\gamma$. The dotted lines show the range of uncertainty from $1/M^3$ terms in the heavy quark expansion.

There is good agreement with theory and between experiments, which use complementary techniques. *BABAR* uses a reconstructed B tag with smaller systematic but larger statistical uncertainty, while without the B tag, CLEO has higher efficiency but larger backgrounds and attendant systematic uncertainties, notably from $e^+e^- \to q\bar{q}$.

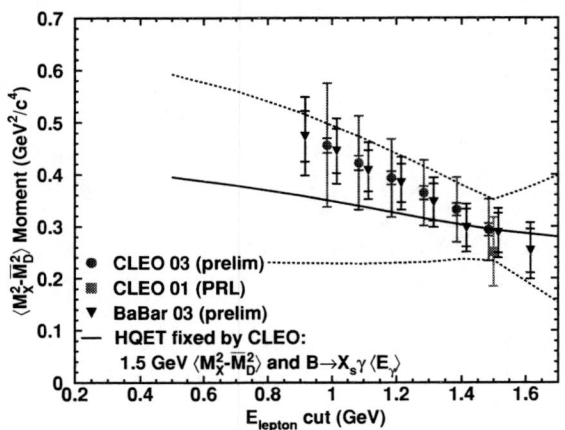

FIGURE 4. Hadronic recoil mass moments versus the lepton energy cut. For both CLEO and *BABAR*, the inner (outer) error bars give the statistical (total) uncertainty. N.B., there is substantial correlation among the points as the lepton energy cut is varied.

TABLE 2. Hadronic recoil mass moments versus the lepton energy cut. The errors on the entries in the table are the statistical, detector systematics, and model dependence, respectively.

E_ℓ^{min} Cut (GeV)	$\langle M_X^2 - \bar{M}_D^2 \rangle$ [(GeV/c^2)^2]
1.0	$0.456 \pm 0.014 \pm 0.045 \pm 0.109$
1.1	$0.422 \pm 0.014 \pm 0.031 \pm 0.084$
1.2	$0.393 \pm 0.013 \pm 0.027 \pm 0.069$
1.3	$0.364 \pm 0.013 \pm 0.030 \pm 0.054$
1.4	$0.332 \pm 0.012 \pm 0.027 \pm 0.055$
1.5	$0.293 \pm 0.012 \pm 0.033 \pm 0.048$

4. CONCLUSION

CLEO's direct contributions to beauty physics are nearing an end, but recent measurements of CKM matrix elements $|V_{cb}|$ and $|V_{ub}|$ are still among the best available. In the future, CLEO's contribution to flavor physics and the test of the CKM paradigm for *CP* violation will come from measurements at charm threshold [18].

REFERENCES

1. Athar, S. B., et al., [CLEO Collaboration], *Phys. Rev.* **D68**, 072003 (2003).
2. Scora, D., and Isgur, N., *Phys. Rev.* **D52**, 2783–2812 (1995).
3. El-Khadra, A., in these proceedings (2004).
4. Bornheim, A., et al., [CLEO Collaboration], *Phys. Rev. Lett.* **88**, 231803 (2002).
5. Aubert, B., et al., [BaBar Collaboration], *Phys. Rev. Lett.* **92**, 071802 (2004).
6. Kakuno, H., et al., [Belle Collaboration], *Phys. Rev. Lett.* **92**, 101801 (2004).
7. Behrens, B. H., et al., [CLEO Collaboration], *Phys. Rev.* **D61**, 052001 (2000).
8. Aubert, B., et al., [BaBar Collaboration], *Phys. Rev. Lett.* **90**, 181801 (2003).
9. Manohar, A. V., and Wise, M. B., *Heavy quark physics*, Cambridge University Press, 2000, chap. 4.
10. Barish, B., et al., [CLEO Collaboration], *Phys. Rev. Lett.* **76**, 1570–1574 (1996).
11. Heavy Flavor Averaging Group, Updates of Semileptonic Results (2003), for summer conferences.
12. Hagiwara, K., et al., [Particle Data Group Collaboration], *Phys. Rev.* **D66**, 010001 (2002), URL: http://pdg.lbl.gov.
13. Chen, S., et al., [CLEO Collaboration], *Phys. Rev. Lett.* **87**, 251807 (2001).
14. Cronin-Hennessy, D., et al., [CLEO Collaboration], *Phys. Rev. Lett.* **87**, 251808 (2001).
15. Mahmood, A. H., et al., *Phys. Rev.* **D67**, 072001 (2003).
16. Huang, G. S., et al., [CLEO Collaboration], arXiv:hep-ex/0307081 (2003).
17. Aubert, B., et al., [BaBar Collaboration], arXiv:hep-ex/0307046 (2003).
18. Asner, D., in these proceedings (2004).

Recent Results from DØ

Vivek Jain

(Representing the DØ Collaboration)

Brookhaven National Laboratory, Department of Physics, Upton, NY 11973, U.S.A.

Abstract. We discuss recent B physics results from DØ. The measurements presented here correspond to an integrated luminosity of ~ 115 pb^{-1} of data collected at the Tevatron, between April 2002 and June 2003, at a center of mass energy of the $p\bar{p}$ collision of 1.96 TeV.

1. INTRODUCTION

The B physics program at DØ, is designed to be complementary to the program at the B factories at SLAC and KEK and includes studies of B_s^0 oscillations, quarkonia (J/ψ, χ_c, Υ), searches for rare decays such as $B_s^0 \rightarrow \mu^+\mu^-$, B spectroscopy, *e.g.*, B_{sJ}^*, lifetimes of the different B hadrons, search for the lifetime difference in CP eigenstates of B_s^0 mesons, study of beauty baryons, B_c properties, b production cross-section, *etc.*

One of the more important topics in B physics is the search for B_s^0-$\bar{B}_s^0$ mixing. Global fits to the unitarity triangle, assuming that the Standard Model is correct, indicate that the 95% CL interval for the mixing parameter Δm_s is (14.2-28.1) ps^{-1} [1]. The current limit[1] is $\Delta m_s > 14.9$ ps^{-1} at the 95% CL [1]. A measured value of Δm_s much larger than the upper limits given here could pose a problem for the Standard Model.

2. DØ DETECTOR

For the current run of the Tevatron (Run II), the DØ detector went through a major upgrade. The inner tracking system was completely revamped. The detector now includes a Silicon Tracker, surrounded by Scintillating Fiber Tracker, both of which are enclosed in a 2 Tesla solenoidal magnetic field. Pre-shower counters are located before the calorimeter to aid in electron and photon identification. The muon system has been improved, *e.g.*, more shielding was added to reduce beam background.

The DØ detector has excellent tracking and lepton acceptance. Tracks with pseudo-rapidity (η) as large as

---

[1] The limit is derived by combining limits from 13 different measurements

2.5-3.0 ($\theta \approx 10°$) and transverse momentum (p_T) as low as 180 MeV/c are reconstructed.

The muon system can identify muons within $|\eta| < 2.0$. The minimum p_T of the reconstructed muons varies as a function of η. In most of the results presented here, we required muons to have $p_T > 2$ GeV/c.

Low p_T electron identification is currently limited to $p_T > 2$ GeV/c and $|\eta| < 1.1$. However, we are working to improve both the momentum and η coverage.

A Silicon based (hardware) trigger is being commissioned which will allow triggering on long-lived particles, such as the daughters of charm and beauty hadrons. We expect to start including this trigger in the online system after the current Tevatron shutdown ends in Mid-November. We are currently making impact parameter cuts at Level 3 (software trigger). The hardware trigger will allow us to make these cuts at an earlier stage.

3. DATA SAMPLE

The results presented here are based on data collected between April 2002 and June 2003. The dataset corresponds to an integrated luminosity of about 115 pb^{-1}. We collected data with a dimuon as well as a single muon trigger. To reduce the data rate, a luminosity dependent prescale was applied to the single muon trigger (the prescale was 1 for instantaneous $\mathscr{L} < 20 \cdot 10^{30}$ cm^{-2}s^{-1}). In both these triggers, the requirement that muons have hits in all layers of the muon system implies that they have total momentum ≥ 3 GeV/c.

4. RESULTS

B_s^0 oscillations are one of three processes where the flavour of the initial state changes by two units. The others being $K\bar{K}$ and $B^0\bar{B}^0$ mixing. Studies of $K\bar{K}$ mixing yield results on indirect and direct CP violation, whereas the (unexpected) large value of $B^0\bar{B}^0$ mixing implied that the top quark was much heavier than expectations. If our current understanding of the Standard Model and the CKM matrix are correct, then B_s^0 oscillations should be in the allowed region, and any deviation could be a sign of new physics. If the mixing parameter, Δm_s, is very large, then the difference in the widths of the CP eigenstates of the B_s^0 may be detectable.

The significance of a B mixing measurement can be expressed as,

$$\text{Sig.} = \sqrt{\frac{N\varepsilon\mathscr{D}^2}{2}}\exp^{-(\Delta m_s \times \sigma_t)^2/2}\sqrt{\frac{S}{S+B}}, \quad (1)$$

where N is the number of reconstructed B_s^0 events, $\varepsilon\mathscr{D}^2$ is a measure of how well we know the flavour of the B_s^0 at production, σ_t is the proper time resolution, and $S/(S+B)$ expresses the cleanliness of the signal.

To study B_s^0 oscillations, we therefore need three ingredients, (a) final state reconstruction, (b) ability to measure B decay lengths, and (c) flavour tagging of the B both at production and decay[2].

4.1. Final State Reconstruction

We can reconstruct B_s^0 mesons in both hadronic and semileptonic final states. The advantage of the former is that since it is a fully reconstructed decay, the proper time resolution is much better, whereas the branching fraction for the latter is much larger ($\approx 10\%$ vs. 0.5%).

In Fig. 1, we show the $B_s^0 \to D_s\mu X$ signal for $\int\mathscr{L} \approx 6.2$ pb^{-1}. The blue shaded histogram represents the wrong sign combinations. By comparison, we expect to see about 1.0 $B_s^0 \to D_s\pi$ events/pb^{-1}.

In Fig. 2, we present the inclusive $B \to D^{*-}\mu X$ signal for $\int\mathscr{L} \approx 6.2$ pb^{-1}. The blue line represents the wrong sign combinations. We can use this mode to study B^0 mixing and for measuring the inclusive B lifetime.

We will also reconstruct hadronic B_s^0 decays, e.g., $B_s^0 \to D_s\pi$, as well as semi-electronic final states. Details about B hadron reconstruction at DØ can be found in E. von Toerne's contribution [2] to these proceedings.

4.2. Lifetime Measurement

In Fig. 3, we present a measurement of the inclusive B lifetime using $B \to D^0\mu X$ decays.

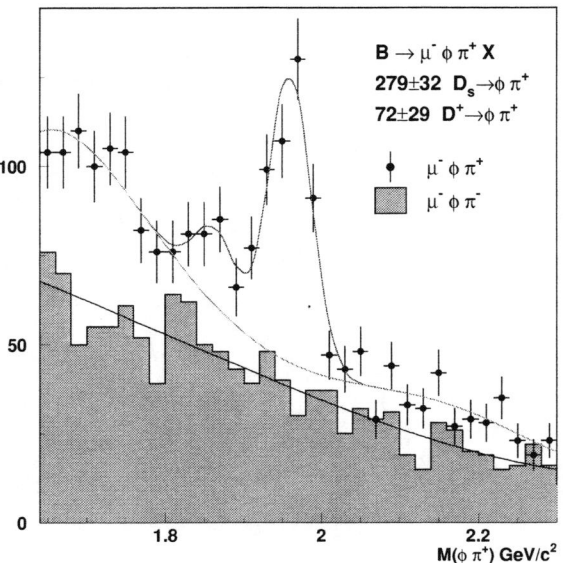

FIGURE 1. B_s^0 semileptonic yield.

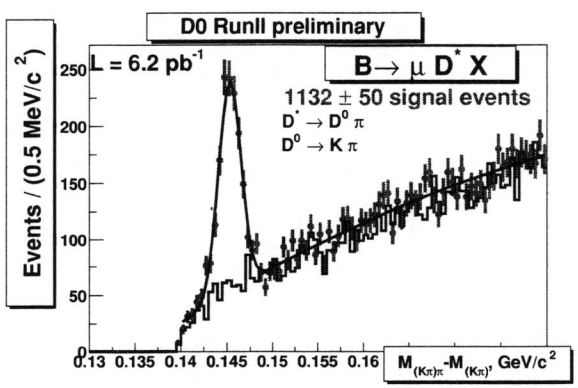

FIGURE 2. $B \to D^{*-}\mu X$.

The background shape is obtained from the D^0 mass sidebands. Both sidebands and signal regions are fit with a combination of a Gaussian function to represent zero-lifetime background and exponentials (convoluted with Gaussian resolutions functions) to represent a non-zero lifetime (the convolution takes resolution smearing into account).

From the fit, we measure the inclusive B lifetime to be 438 ± 25(stat.) μm, which agrees with the world average, 472 ± 2 μm [3]. We have also measured lifetimes for other B hadrons, and details can be found in Daria Zieminska's contribution [4] to these proceedings.

---

[2] The latest results can be found at http://www-d0.fnal.gov/Run2Physics/ckm/

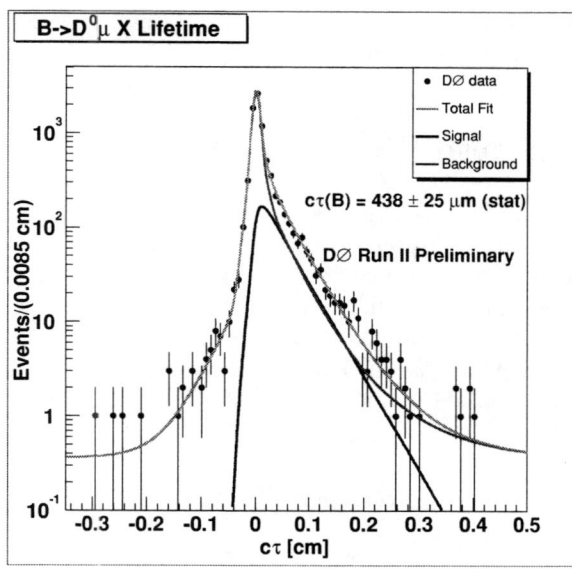

FIGURE 3. Lifetime for $B \to D^0 \mu X$.

4.3. Flavour Tagging

To be able to do a mixing measurement, one needs to know the flavour of the B hadron at the time of production and decay. By using flavour-specific decays, one can easily tag the flavour at the time of decay. Tagging the B flavour at production is more work. At DØ, we plan to use the following techniques:

- Soft Lepton Tagging: The sign of the lepton produced in the semileptonic decay of the other B in the event is used to tag the flavour of the B meson of interest. We then make the assumption that the flavour of the decay B is opposite to that of the tag B. There will be some contamination due to the fact that the other B meson can mix, or that we pick up the tag lepton from a charm semileptonic decay. This method has low efficiency, but very high tagging power. We also plan to use electrons.

- Jet Charge Tagging: We take all tracks opposite to the decay B and form a track jet. The assumption is that these tracks are produced in the fragmentation of the other b quark, as well as in the decay of the other B hadron. This method has high efficiency, but has poorer tagging power.

- Same Side Tagging: In this technique, we identify tracks produced in the fragmentation of the b quark which gives rise to the decay B. In addition, the decay B can come from a resonance, e.g., $B^{**+} \to B^0 \pi^+$, and one can use such pions to identify the flavour of the decay B at the time of production. This method has high efficiency, but has poorer tagging power.

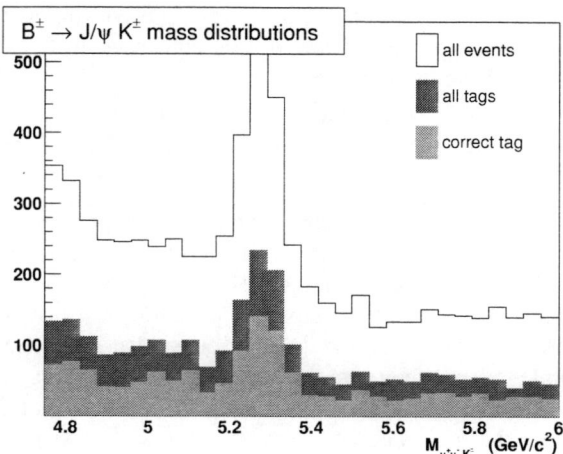

FIGURE 4. Fits for Jet Charge tagging results.

TABLE 1. Summary of flavour tagging results.

	ε	Dilution ($\mathscr{D}$)	$\varepsilon \mathscr{D}^2$
Soft Muon	5%	57%	$1.6 \pm 1.1\%$
Jet Charge	47%	27%	$3.3 \pm 1.7\%$
Same Side	79%	26%	$5.5 \pm 2.0\%$

We benchmarked these techniques using a sample of $B^+ \to J/\psi K^+$ decays. Since the B^+ does not oscillate, it provides a good testing ground.

In Fig. 4, we show the B^+ peak for all events, events with a (jet-charge) tag, and events with the correct tag. We use the fits to these peaks to determine the efficiency (ε) and tagging power (Dilution) of the various techniques. $\varepsilon = (N_R + N_W)/N_{all}$, and Dilution $= (N_R - N_W)/(N_R + N_W)$, where N_{all}, N_R and N_W are the number of all events, correctly tagged and wrongly tagged events, respectively.

The results for the various techniques are summarized in Table 1. We are in the process of checking these results on a sample of B_s^0 and B^0 events.

4.4. B_s^0 Mixing

We expect to observe the following four classes of events:

- $B_s^0 \to D_s \mu X$ collected with the single muon trigger. The flavour at production will be tagged using all three techniques discussed above.

- $B_s^0 \to D_s \pi$ collected with the single muon trigger. In this case, we can use the trigger muon as the flavour tag. The tagging power will be very high in this case.

- $B_s^0 \to D_s \mu X$ collected with the dimuon trigger. The flavour at production will be tagged using the second muon in the event.
- $B_s^0 \to D_s e X$ collected with the single muon trigger. The flavour at production will be tagged using the trigger muon.

The semileptonic events have poorer proper time resolution compared to the hadronic mode, but much larger statistics. We are in the process of improving our estimate of the proper time resolution for both hadronic and semileptonic events. In addition, we can combine all modes to get a better measurement or a better limit.

If we assume a resolution of ≈ 150 fs, we project that using the first class of events (with $\int \mathcal{L} \approx 500$ pb^{-1}), we can make a 3σ measurement if $\Delta m_s \approx 12$ ps^{-1} or a 1.5σ measurement if $\Delta m_s \approx 15$ ps^{-1}. The third and fourth class of events have a very similar reach in Δm_s.

For hadronic events (with the same luminosity), if the resolution is ≈ 110 fs, the corresponding numbers are a 3σ measurement if $\Delta m_s \approx 12$ ps^{-1} or a 2.2σ measurement if $\Delta m_s \approx 15$ ps^{-1}.

4.5. Quarkonia

At the Tevatron, J/ψ can be produced either promptly (direct as well as indirect), *i.e.*, $p\bar{p} \to J/\psi$ or $\chi_c \to J/\psi$) or as a product in B decays. Υ by contrast are only produced promptly. The study of quarkonia sheds light on the strong interaction, especially non-perturbative QCD[3].

In Run I of the Tevatron, the production cross-section of direct J/ψ was about 50 times larger than the Color Singlet Model.

With the new dataset, we plan to update the results on the J/ψ production cross-section as a function of p_T and η, as well as study polarization effects. In addition, we are working to measure the absolute cross-section and polarization of the Υ states.

5. CONCLUSIONS

The DØ detector has started to produce results in the field of B physics. We currently have ≈ 220 pb^{-1} of data on tape, and expect to get up to ≈ 500 pb^{-1} by the end of 2004.

As a stepping stone to B_s^0 mixing, we plan to measure Δm_d to benchmark our analyses techniques. In addition, we are pursuing a vigorous program, which includes

measurements of B lifetimes, rare decays, studies of quarkonia, beauty baryons, B_c, *etc*.

ACKNOWLEDGMENTS

I would like to acknowledge members of the B group at DØ for assisting me with this talk, especially Brad Abbott, Christos Leonidopoulos and Rick Van Kooten.

I would also like to thank the organizers for a very well organized and stimulating conference.

This work was supported by the U.S. Department of Energy under Contract No. DE-AC02-98CH10886.

REFERENCES

1. Battaglia, M., et al., arXiv:hep-ph/0304132 (2003).
2. von Toerne, E., in these proceedings (2004).
3. Hagiwara, K., et al., [Particle Data Group Collaboration], *Phys. Rev.* **D66**, 010001 (2002).
4. Zieminska, D., in these proceedings (2004).

---

[3] An excellent review, covering both theory and experiment, can be found at http://hep.physics.indiana.edu/~zieminsk/talks.html. Select the Trento workshop talk.

Recent Results from CDF

Robert Harr

(Representing the CDF Collaboration)

Wayne State University, Detroit, Michigan 48202, U.S.A.

Abstract. We report on recent heavy quark results from CDF in Run II. We focus on a selection of mature analyses that demonstrate the capabilities of the experiment to extract interesting physics from the data. A few of the results presented have already been submitted for publication and papers are being prepared for most of the others.

1. INTRODUCTION

Having more than 200 pb^{-1} of physics data in hand and more on the way, CDF's B physics program is in full swing. The triggers for B physics were enhanced in the upgrade for Run II, adding the ability to trigger on J/ψ's down to $p_T = 0$, and the ability to trigger on high p_T tracks with significant impact parameter to the beam spot (displaced tracks). Now, in addition to the topics accessible with leptonic or semi-leptonic triggers, analyses that require fully hadronic charm and beauty decays are possible with data selected by the displaced track trigger. Early exploitation of this data resulted in the first three Run II physics papers, a measurement of the D_s^+-D^+ mass difference [1], measurements of charm production cross sections [2], and an improved limit on the rare decay $D^0 \to \mu^+\mu^-$ [3].

2. *B* HADRON MASSES AND LIFETIMES

Measurements of B hadron masses and lifetimes are a prelude to CP measurements in B decays. We use exclusive J/ψ decay modes which give us good signal statistics with little background: $B^0 \to J/\psi K^{0*}$, $B^+ \to J/\psi K^+$, $B_s^0 \to J/\psi \phi$, and $\Lambda_b \to J/\psi \Lambda$. The results are summarized in Table 1. The B_s^0 (Fig. 1) and Λ_b (Fig. 2) measurements are the world's best. The B^0 and B^+ mass measurements are competitive with the best measurements available.

Lifetime measurements were performed for the same set of decays. The proper time distributions for B_s^0 and Λ_b are shown in Figs. 3 and 4, respectively, and the extracted lifetimes are reported in Table 1. These results are competitive with the LEP results, and will improve

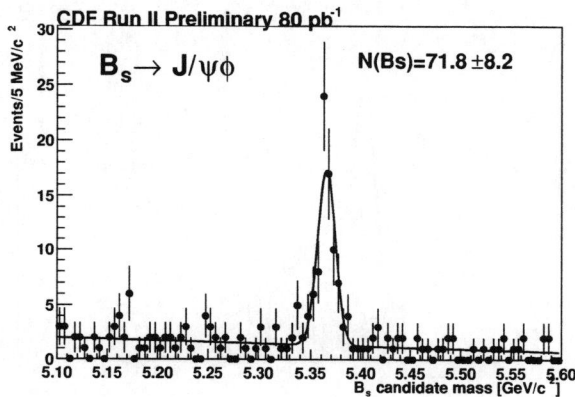

FIGURE 1. Invariant mass distribution for $J/\psi \phi$ combinations where $J/\psi \to \mu^+\mu^-$ is used to trigger the event, and $\phi \to K^+K^-$.

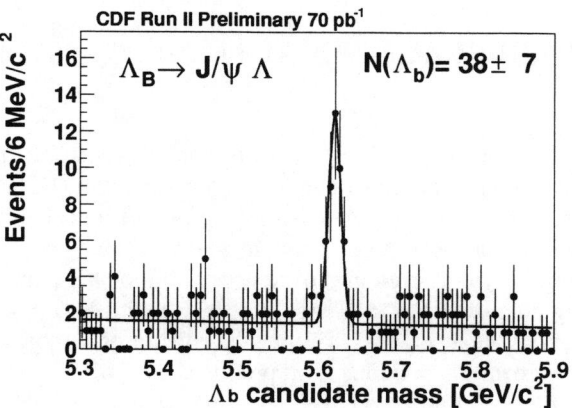

FIGURE 2. Invariant mass distribution for $J/\psi \Lambda$ combinations where $J/\psi \to \mu^+\mu^-$ is used to trigger the event, and $\Lambda \to p\pi^-$.

CP722, *B Physics at Hadron Machines,* edited by M. Paulini and S. Erhan

TABLE 1. Summary of CDF measurements of B hadron masses and lifetimes. The first error is statistical and the second systematic. The B^0 and B^+ lifetime results agree within errors with the precise measurements of BaBar and Belle. The amount of data used varies by analysis.

	Mass [MeV/c^2]	Lifetime [ps]
B^0	$5280.30 \pm 0.92 \pm 0.96$	$1.51 \pm 0.06 \pm 0.02$
B^+	$5279.32 \pm 0.68 \pm 0.94$	$1.63 \pm 0.05 \pm 0.04$
B_s^0	$5365.50 \pm 1.29 \pm 0.94$	$1.33 \pm 0.14 \pm 0.02$
Λ_b	$5620.4 \pm 1.6 \pm 1.2$	$1.25 \pm 0.26 \pm 0.10$

with additional data. The results for the B^0 and B^+ lifetimes are consistent with the precise measurements of BaBar and Belle. Additional details about lifetime measurements were presented by D. Zieminska [4].

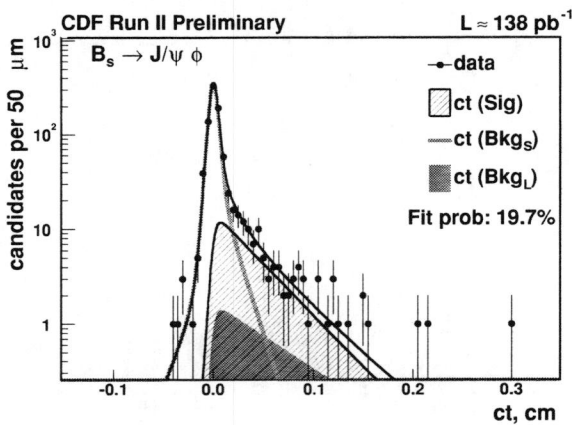

FIGURE 3. Proper time distribution for $J/\psi \phi$ combinations where $J/\psi \rightarrow \mu^+\mu^-$ is used to trigger the event, and $\phi \rightarrow K^+K^-$.

3. THE RARE DECAYS $B_s^0 \rightarrow \mu^+\mu^-$ AND $D^0 \rightarrow \mu^+\mu^-$

Using data selected with the dimuon trigger, CDF has searched for the flavor-changing neutral current (FCNC) decay $B_s^0 \rightarrow \mu^+\mu^-$. After applying optimized selection criteria, one event remained in the B_s^0 search window (Fig. 5), yielding an improved upper limit on the branching fraction of 9.5×10^{-7} (1.2×10^{-6}) at the 90% (95%) confidence level. This is more than a factor of 2 improvement over the previous limit produced by CDF in Run I.

An upper limit on the branching fraction of $B^0 \rightarrow \mu^+\mu^-$ is derived simultaneously yielding values of 2.5×10^{-7} and 3.1×10^{-7} at the 90% and 95% confidence levels, respectively.

Data selected with the displaced track trigger were used to improve the limit on the branching fraction of

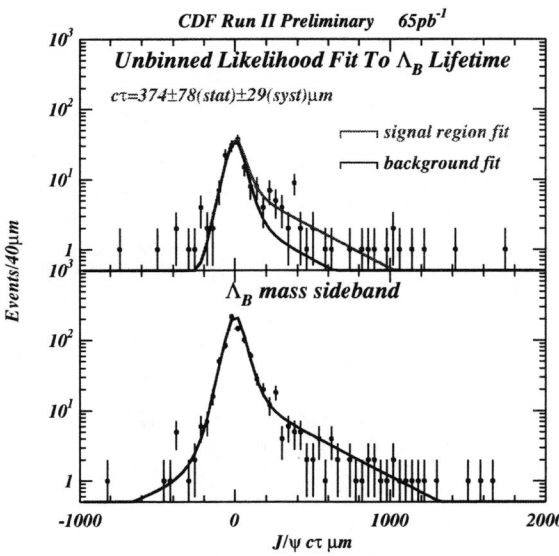

FIGURE 4. Proper time distribution for $J/\psi \Lambda$ combinations where $J/\psi \rightarrow \mu^+\mu^-$ is used to trigger the event, and $\Lambda \rightarrow p\pi^-$.

the FCNC decay $D^0 \rightarrow \mu^+\mu^-$. This search begins by reconstructing a clean sample of the kinematically similar $D^0 \rightarrow \pi^+\pi^-$ decays (Fig. 6), then using muon identification to select $D^0 \rightarrow \mu^+\mu^-$ candidates. The $D^0 \rightarrow \pi^+\pi^-$ decays serve also as normalization. A new upper limit of 2.5×10^{-6} at the 90% confidence level is derived from zero candidates in the search window, and is almost a factor of two better than the previous limit [3]. Additional details about rare decay searches at CDF are available in the talk of C.J. Lin [5].

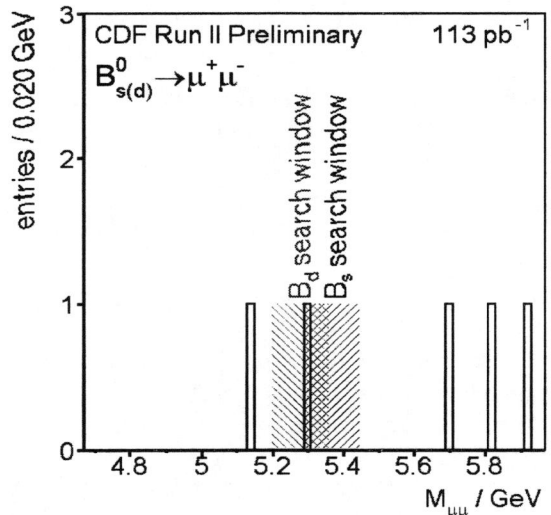

FIGURE 5. The invariant mass distribution for $\mu^+\mu^-$ combinations in the B^0 and B_s^0 mass regions.

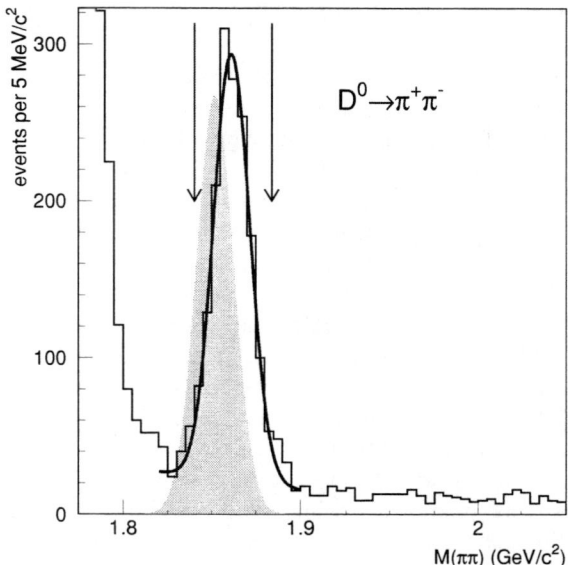

FIGURE 6. The invariant mass distribution for the D^* tagged $D^0 \to \pi^+\pi^-$ candidates. To derive the limit on $D^0 \to \mu^+\mu^-$, these events were searched for cases where both tracks penetrate into the muon detector. The shaded Gaussian shows how the invariant mass is shifted when $D^0 \to \pi^+\pi^-$ is reconstructed as $D^0 \to \mu^+\mu^-$.

4. OBSERVATION OF $X(3872)$

In the summer of 2003, Belle reported the observation of a new state of narrow width in B decays [6]. It is dubbed the $X(3872)$. The state is observed in the decay chain $B \to XK \to J/\psi\,\pi^+\pi^-K$ with a mass close to 3872 MeV/c^2. At the Tevatron, this state could be produced in B decays, or directly in $\bar{p}p$ collisions. Using $J/\psi \to \mu^+\mu^-$ triggered data, CDF observed a signal in the $J/\psi\,\pi^+\pi^-$ final state (Fig. 7) [7].

The Belle collaboration reported that the decay favors high $\pi^+\pi^-$ invariant mass. To verify this observation, we looked at the events with $\pi^+\pi^-$ invariant mass greater than 500 MeV/c^2, and see a number of signal events consistent with the case of no invariant mass requirement (Fig. 8), but sitting on about half as many background events. The significance of the signal is large, in excess of ten standard deviations.

The nature of this state is a bit of a mystery. The leading alternatives are (1) a $\bar{c}c$ state, probably the $1\,^3D_2$, or (2) a D^*D molecule. The second alternative is possible as the observed mass is very near the D^*D threshold. Determining the quantum numbers of $X(3872)$ should settle this issue, and the large sample seen at CDF – almost 20 times the number seen at Belle – can play a role.

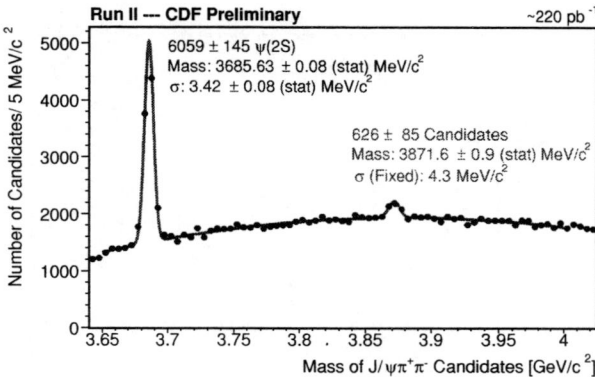

FIGURE 7. Mass distribution for $J/\psi\,\pi^+\pi^-$ combinations.

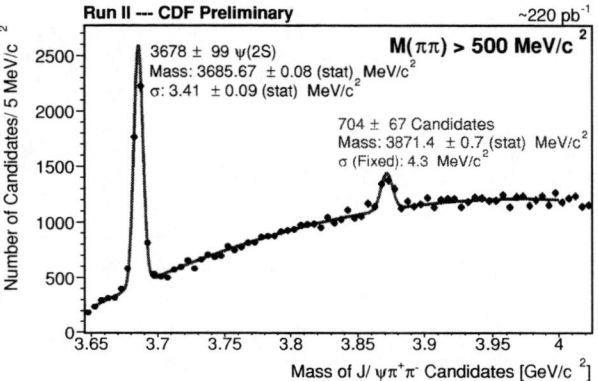

FIGURE 8. Invariant mass distribution of $J/\psi\,\pi^+\pi^-$ combinations for candidates where the $\pi^+\pi^-$ invariant mass exceeds 500 MeV/c^2.

5. HADRONIC B DECAYS

The displaced track trigger selects a sample enriched in heavy quark decays. The trigger requires an event to have two oppositely-charged tracks of $p_T > 2.0$ GeV/c with large impact parameter to the beam line, between 120 μm and 1000 μm. Further requirements are imposed to reject backgrounds and keep the overall trigger rate within a suitable range. Fully reconstructible decays are important for CP and mixing studies where good proper time resolution is needed. The signals reported here represent our initial attempts to reconstruct fully hadronic decays. We are still learning how to fully exploit the capabilities of the detector to extract physics signals in these data.

5.1. $B^0/B_s^0 \to h^+h^-$ Decays

The decays $B^0 \to \pi^+\pi^-$, $B^0 \to K^\pm\pi^\mp$, $B_s^0 \to K^\pm\pi^\mp$, and $B_s^0 \to K^+K^-$ are potential modes for CP measure-

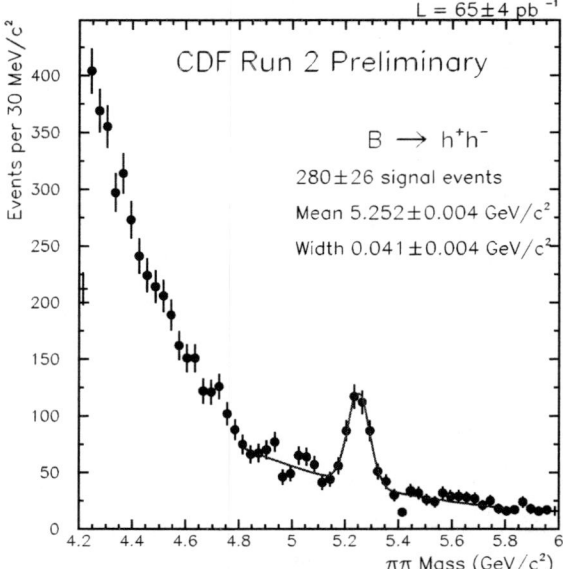

FIGURE 9. The $\pi^+\pi^-$ invariant mass distribution for events selected by the displaced track trigger.

ments by CDF. Although the branching fractions are small, the final state has two high-p_T tracks that are relatively efficient for the displaced-track trigger. However, the invariant mass distributions overlap such that kinematics alone is insufficient to differentiate between the possible mass assignments, as shown in Fig. 9. We statistically separate the modes with a combination of invariant mass, decay kinematics, and dE/dx particle identification.

Based on the derived fractions, we find the ratio of branching ratios

$$\frac{\mathscr{B}(B^0 \to \pi^+\pi^-)}{\mathscr{B}(B^0 \to K^\pm\pi^\mp)} = 0.26 \pm 0.15 \pm 0.055 \,,$$

and the CP asymmetry

$$
\begin{aligned}
\mathscr{A}_{CP} &= \frac{N(\bar{B}^0 \to K^-\pi^+) - N(B^0 \to K^+\pi^-)}{N(\bar{B}^0 \to K^-\pi^+) + N(B^0 \to K^+\pi^-)} \\
&= 0.02 \pm 0.15 \pm 0.017.
\end{aligned}
$$

We report the first observation of $B_s^0 \to K^+K^-$, and measured the ratio of branching ratios

$$\frac{\mathscr{B}(B_s^0 \to K^+K^-)}{\mathscr{B}(B^0 \to \pi^+\pi^-)} = 2.71 \pm 0.73 \pm 0.35 \pm 0.81 \,,$$

where the systematic error on f_s/f_d is included separately.

5.2. $B^0 \to D^-\pi^+$ and $B_s^0 \to D_s^-\pi^+$ Decays

Modes with more final state particles present different reconstruction challenges, and more complex backgrounds that can affect the signal extraction. The modes $B^0 \to D^-\pi^+$ and $B_s^0 \to D_s^-\pi^+$ serve as good examples. $B_s^0 \to D_s^-\pi^+$ with its flavor-tagged final state is one of the modes that will be used to observe B_s^0 mixing.

The $D^-\pi^+$ and $D_s^-\pi^+$ invariant mass distributions from data are shown in Figs. 10 and 12. The distributions have a complex, multi-peaked structure below the B^0 or B_s^0 mass, respectively. Since these data are selected by the displaced-track trigger, most of the minimum bias combinatoric background is removed and what remains comes primarily from heavy quark decays. By simulating a variety of related B meson decays with particle mis-assignments or missing neutrals, we are able to reproduce the peaking backgrounds (Figs. 11 and 13). Allowing the overall normalization of each contribution to vary, we fitted the data distributions for signal, peaking B meson backgrounds, and a smooth background contribution, as shown by the red, dashed curves in Figs. 10 and 12.

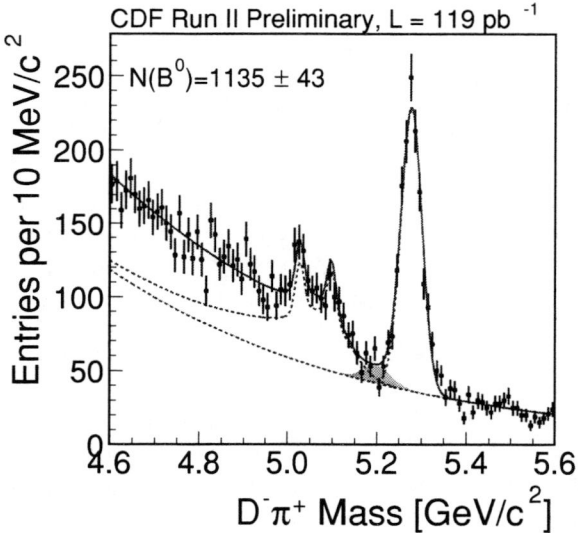

FIGURE 10. The $D^-\pi^+$ invariant mass distribution for events selected by the displaced track trigger.

6. SUMMARY

Run II of the Fermilab Tevatron is in its third year. The CDF experiment is running smoothly, and has collected more than 200 pb^{-1} of physics data being exploited for a wide variety of new results. The CDF collaboration has produced its first three Run II publications: the single best measurement of the mass difference $m(D_s^+)$ −

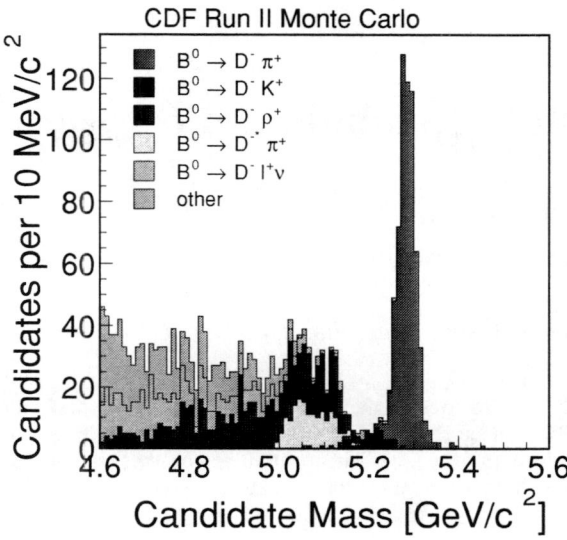

FIGURE 11. The $D^-\pi^+$ invariant mass distribution for Monte Carlo simulation of a number of background modes.

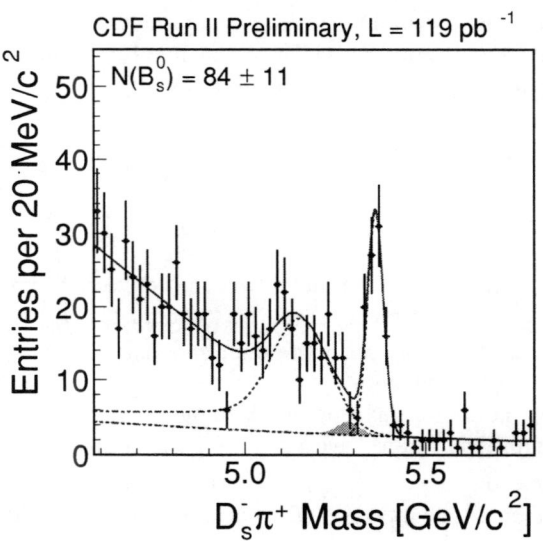

FIGURE 12. The $D_s^-\pi^+$ invariant mass distribution for events selected by the displaced track trigger.

$m(D^+)$ [1]; the first measurement of the charm cross sections in high energy hadron collisions [2]; and a new upper limit on the FCNC decay $D^0 \rightarrow \mu^+\mu^-$ [3]. Papers are in preparation for many other results: observation of a new state $X(3872)$ seen in the $J/\psi\pi^+\pi^-$ final state; new mass measurements of B hadrons; an improved limit on the FCNC decay $B_s^0 \rightarrow \mu^+\mu^-$; the ratio of branching fractions and relative production for $B_s^0 \rightarrow D_s^-\pi^+$ and $B^0 \rightarrow D^-\pi^+$; and the branching fraction of $\Lambda_b \rightarrow \Lambda_c\pi$. Measurements of CP violation, and discovery of B_s^0 mix-

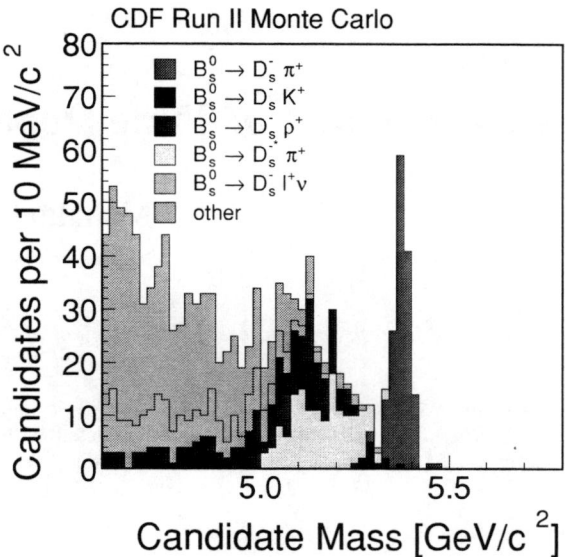

FIGURE 13. The $D_s^-\pi^+$ invariant mass distribution for Monte Carlo simulation of a number of background modes.

ing are major goals of Run II. While these measurements require the high statistics of the full Run II data, work is already underway to develop the tools and techniques for these challenging analyses: B hadron lifetime measurements; and extracting signals in fully-reconstructed, flavor-tagged B_s^0 modes for mixing measurements. Discussions of tagging studies and prospects for mixing and CP violation measurements at CDF were presented by other speakers at the conference [8, 9].

ACKNOWLEDGMENTS

This work was supported by the U.S. Department of Energy.

REFERENCES

1. Acosta, D., et al., [CDF Collaboration], *Phys. Rev.* **D68**, 072004 (2003).
2. Acosta, D., et al., [CDF Collaboration], *Phys. Rev. Lett.* **91**, 241804 (2003).
3. Acosta, D., et al., [CDF Collaboration], *Phys. Rev.* **D68**, 091101 (2003).
4. Zieminska, D., in these proceedings (2004).
5. Lin, C., in these proceedings (2004).
6. Abe, K., et al., [Belle Collaboration], arXiv:hep-ex/0308029 (2003).
7. Acosta, D., et al., [CDF Collaboration], arXiv:hep-ex/0312021 (2003).
8. Miao, T., in these proceedings (2004).
9. Maksimovic, P., in these proceedings (2004).

Measurement of Time Dependent *CP* Violation in $B^0 \to \phi K_S^0$

Mahalaxmi Krishnamurthy

(Representing the BaBar Collaboration)

University of Tennessee, Department of Physics and Astronomy, Knoxville, TN 37996, U.S.A.

Abstract. We present a measurement of the time dependent *CP* asymmetries in the decay $B^0 \to \phi K_S^0$ using a sample of approximately 120 million $B\bar{B}$ pairs collected with the *BABAR* detector at the PEP-II asymmetric-energy B meson factory at SLAC. The values of the *CP*-violating parameters derived from a time-dependent fit to the ϕK_S^0 sample are $C_{\phi K} = -0.38 \pm 0.37 \pm 0.12$ and $S_{\phi K} = 0.45 \pm 0.43 \pm 0.07$, where the first uncertainty is statistical and the second systematic.

Decays of B mesons into charmless hadronic final states with a ϕ meson are dominated by $b \to s\bar{s}s$ gluonic penguins (Fig. 1), with a possible small contribution from electroweak penguins, while other Standard Model (SM) amplitudes are strongly suppressed [1–4]. In this model, *CP* violation arises from a single complex phase in the Cabibbo-Kobayashi-Maskawa (CKM) quark-mixing matrix [5, 6]. Since many scenarios of physics beyond the SM introduce additional diagrams with heavy particles in the penguin loops and new *CP*-violating phases [7, 8], comparison of *CP*-violating observables with the SM expectations is a sensitive probe for new physics. In the SM, neglecting CKM-suppressed contributions, the time dependent *CP*-violating asymmetry in decays $B^0 \to \phi K_S^0$ and $B^0 \to J/\psi K_S^0$ is proportional to the same parameter $\sin 2\beta$, where the latter decay is dominated by non-loop diagrams.

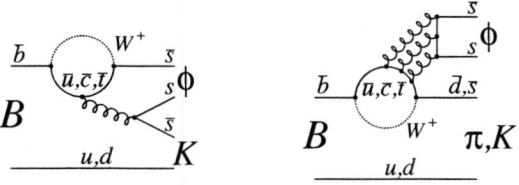

FIGURE 1. Examples of quark-level diagrams for $B \to \phi K$. Left: internal penguin. Right: flavor-singlet penguin.

This analysis is based on about 120 million $B\bar{B}$ pairs collected at SLAC with the *BABAR* detector [9] at the PEP-II asymmetric-energy e^+e^- storage ring operating at an energy of the $\Upsilon(4S)$ resonance.

In the time dependent *CP* asymmetry measurement, each candidate event consists of a fully reconstructed neutral B meson, B_{CP}, decaying into ϕK_S^0, and a partially reconstructed recoil B meson, B_{tag}. We examine B_{tag}

for evidence that it decayed either as B^0 or $\bar{B}^0$ (flavor tag). The decay-time distribution of B decays to a *CP* eigenstate with a B^0 or $\bar{B}^0$ tag depends on both the B^0-$\bar{B}^0$ oscillation amplitude and the amplitudes of B^0 and $\bar{B}^0$ decays to this final state. The decay rate $f_+(f_-)$ when the tagging meson is a $B^0(\bar{B}^0)$ is given by

$$
f_\pm(\Delta t) = \frac{e^{-|\Delta t|/\tau_{B^0}}}{4\tau_{B^0}} [\, 1 \pm S_{\phi K} \sin(\Delta m_d \Delta t) \\
\mp C_{\phi K} \cos(\Delta m_d \Delta t)],
\tag{1}
$$

where $\Delta t = t_{CP} - t_{\text{tag}}$ is the difference between the proper decay times of the B_{CP} and B_{tag} candidates, τ_{B^0} is the neutral B meson mean lifetime, and Δm_d is the B^0-$\bar{B}^0$ oscillation frequency.

In the SM, decays that proceed purely via the $b \to s\bar{s}s$ penguin transitions have $S_{\phi K} = -\eta_f \sin 2\beta$ and $C_{\phi K} = 0$, where $\beta \equiv \arg\left[-V_{cd}V_{cb}^*/V_{td}V_{tb}^*\right]$ (V_{ik} is the CKM matrix element for quarks i,k) and η_f is the *CP* eigenvalue ($\eta_f = -1$ for ϕK_S^0). Thus, the time dependent *CP*-violating asymmetry is

$$
\mathcal{A}_{CP}(\Delta t) \equiv \frac{f_+(\Delta t) - f_-(\Delta t)}{f_+(\Delta t) + f_-(\Delta t)} \\
= S_{\phi K} \sin(\Delta m_d \Delta t).
\tag{2}
$$

Additional decay diagrams with non-zero weak and strong phases can move the *CP* parameters away from these values (direct *CP* violation). Measurements of $\sin 2\beta$ in tree-dominated charmonium–neutral-kaon final states have been reported by *BABAR* [10] and Belle [11, 12].

We fully reconstruct B meson candidates in the decay modes ϕK_S^0 and ϕK^+, with $\phi \to K^+ K^-$, $K_S^0 \to \pi^+\pi^-$. For the K^+ track and the charged-track daughters of the

CP722, *B Physics at Hadron Machines*, edited by M. Paulini and S. Erhan

ϕ, we require at least 12 measured drift-chamber hits and a minimal transverse momentum p_T of 0.1 GeV/c. The tracks must originate within 1.5 cm in xy and ± 10 cm along the beam (z) axis to the nominal beam spot. Looser criteria are applied to tracks belonging to $K_S^0 \to \pi^+\pi^-$ and B_{tag}. We combine pairs of oppositely charged tracks originating from a common vertex to form K_S^0 and ϕ candidates. A $K_S^0 \to \pi^+\pi^-$ candidate is accepted on the basis of requirements on the two-pion invariant mass (within 20 MeV/c^2 of the nominal K_S^0 mass [13]), the flight-length (ℓ) significance ($\ell/\sigma_\ell > 3$), and the angle between the line connecting the ϕ and K_S^0 decay vertices and the K_S^0 momentum (< 0.045 rad).

The kaon tracks are distinguished from pions and protons using dE/dx information from the drift chamber in conjunction with dE/dx information from the silicon vertex tracker, SVT, for $p_T < 0.7$ GeV/c, or with the measured Cherenkov angle and the number of photons recorded by the internally reflecting Cherenkov detector, DIRC, when $p_T > 0.7$ GeV/c. For the kaon tracks that are used to reconstruct the ϕ meson, we require the two-kaon invariant mass to be within 16 MeV/c^2 of the nominal ϕ mass [13].

For an extended unbinned maximum likelihood (ML) fit, we parameterize the distributions of kinematic and topological variables for signal and background events in terms of probability density functions (PDF's) [14]. The background arises primarily from random combinations of tracks produced in the quark-antiquark continuum. It does not show any CP asymmetry. Background from other B decays is established with Monte Carlo simulations and side-band data. The shapes of event variable distributions are obtained from signal and background Monte Carlo samples and data control samples.

Each B candidate is characterized by the energy difference $\Delta E = (q_\Upsilon \cdot q_B/\sqrt{s}) - \sqrt{s}/2$ and the beam-energy–substituted mass $m_{\text{ES}} = [(s/2 + \vec{p}_\Upsilon \cdot \vec{p}_B)^2/E_\Upsilon^2 - \vec{p}_B^2]^{1/2}$ [9]. Here, q_Υ and q_B are four-momenta of the $\Upsilon(4S)$ and the B candidate, $s \equiv (q_\Upsilon)^2$ is the square of the center-of-mass (CM) energy, $\vec{p}_\Upsilon$ and $\vec{p}_B$ are the three-momenta of the $\Upsilon(4S)$ and the B in the laboratory frame, and $E_\Upsilon \equiv q_\Upsilon^0$ is the energy of the $\Upsilon(4S)$ in the laboratory frame. For signal events, ΔE peaks at zero and m_{ES} peaks at the nominal B mass. Our selection requires $|\Delta E| < 0.2$ GeV and $m_{\text{ES}} > 5.2$ GeV/c^2. We also use the ϕ helicity angle θ_H, which is defined as the angle between the directions of the K^+ and the parent B in the ϕ rest frame. The $\cos\theta_H$ distribution is a quadratic function for pseudo-scalar-vector B decay modes and is nearly uniform for the combinatorial background.

In continuum events, particles appear bundled into jets, which can be parameterized with several variables computed in the CM frame. We use the angle θ_T between the thrust axis of the B candidate and the thrust axis of the other charged and neutral particles [9]. We require the angle θ_T to satisfy $|\cos\theta_T| < 0.9$. Other quantities that characterize the event topology are the CM angle θ_B between the B momentum and the beam axis and the sum of the momenta p_i of the other charged and neutral particles in the event weighted with the Legendre polynomials $L_n(\theta_i)$, $n = 0, 2$, where θ_i is the angle between the momentum of particle i and the thrust axis of the B candidate. We combine these variables into a Fisher discriminant $\mathcal{F}$ [15]. Contamination from other B decays are negligible, as demonstrated from Monte Carlo (MC) simulation studies. Possible K^+K^- S-wave contributions from B decays are suppressed by their uniform distribution of $\cos\theta_H$. Resonant behavior ($f_0(980)$ or $a_0(980)$) is not observed in the invariant mass spectrum below the ϕ mass.

The flavor of B_{tag} is determined using a multivariate tagging algorithm [10]. The information is broken down into five hierarchical and mutually exclusive categories. The tagging performance is measured from fully reconstructed neutral B decays into the $D^{(*)-}X^+$ ($X^+ = \pi^+, \rho^+, a_1^+$) and $J/\psi K^{*0}$ ($K^{*0} \to K^+\pi^-$) flavor eigenstates (B_{flav} sample). The tagging efficiency is about 66%, the effective efficiency $Q = \sum_c \varepsilon_c(1 - 2w_c)^2$ with ε_c and w_c the efficiency and the mistag probability per tag category c, respectively, is $Q = (28.7 \pm 0.7)\%$. We apply identical tagging procedures to the B_{flav} and ϕK_S^0 samples, which enables us to measure the tagging efficiencies and mistag rates applicable to the ϕK_S^0 events with the much larger B_{flav} sample.

The time difference Δt is obtained from the measured distance between the positions of the B_{CP} and B_{tag} decay vertices along the beam axis (z-axis) and the known boost of the e^+e^- system. For B_{CP}, we achieve an average z-vertex position resolution of better than 60 μm as determined from Monte Carlo, which compares well to the resolution obtained for the final state $J/\psi K_S^0$ [10]. For 98% of candidates with a reconstructed vertex, the average Δz (Δt) resolution is 180 μm (1.1 ps). We require $|\Delta t| < 20$ ps and $\sigma_{\Delta t} < 3$ ps, where $\sigma_{\Delta t}$ is the error on Δt calculated for each event. As the Δt resolution is dominated by the tagging vertex in the event, we can characterize the resolution with the much larger B_{flav} sample, which we fit simultaneously with the CP samples. The amplitudes for the B_{CP} asymmetries and for the B_{flav} flavor oscillations are reduced by the same factor due to wrong tags. Both distributions are convolved with a common Δt resolution function and backgrounds are accounted for by adding terms to the likelihood, incorporated with different assumptions about their Δt evolution and convolved with a separate resolution function [16].

The likelihood for candidate j in the B flavor category c is obtained by summing the product of event yield $N_{i,c}$ and probability $\mathcal{P}_{i,c}$ over signal and background hypotheses i. The total likelihood $\mathscr{L}$ is given by

$$\mathscr{L} = \frac{1}{N!} \exp\left(-\sum_i N_{i,c}\right) \prod_j \left[\sum_i N_{i,c} \mathscr{P}_{i,c}(\vec{x}_j; \vec{\alpha}_i)\right] \quad (3)$$

with N being the total number of events entering the fit.

In the time dependent CP fit, the probabilities $\mathscr{P}_{i,c}$ are products of PDF's for each of the independent variables $\vec{x}_j = \{m_{ES}, \Delta E, \mathscr{F}, \cos\theta_H, \Delta t\}$ in the CP samples and $\{m_{ES}, \Delta E, \Delta t\}$ in the B_{flav} sample. The $\vec{\alpha}_i$ are the parameters of the distributions in $\vec{x}_j$, which are either floated in the fit, or fixed to values derived from signal Monte Carlo, data control channels, on-resonance sidebands. Event yields are extracted in an initial fit without tagging and Δt information, and are fixed in the final CP fit. The total sample consists of 86,200 B_{flav} and 2138 ϕK_S^0. We find 70 ± 9 ϕK_S^0 signal candidates where the signal yield agrees well with our determination of the branching fraction $B^0 \to \phi K^0$ [17]. Fig. 2 shows the m_{ES} distributions of ϕK_S^0 events (where a requirement on the ΔE distribution to be between ± 3 standard deviations around zero is made).

We then determine the CP parameters $S_{\phi K}$ and $C_{\phi K}$ along with additional 38 free parameters: the relative efficiency per tagging category (4 parameters), the average mistag fraction and the difference between B^0 and $\bar{B}^0$ mistags for each tagging category (8 parameters), the signal Δt resolution (9), and time dependence (6), Δt resolution (3) and mistag fractions (8) for the background. We fix τ_{B^0} and Δm_d to the world averages [13]. The determination of the mistag fractions and Δt resolution parameters is dominated by the high-statistics B_{flav} sample. The fit was established with a parameterized simulation of a large number of data-seized experiments and full detector simulated events for the different signal and background samples. We also apply the CP fit to the control channel ϕK^+ where one expects $S_{\phi K} = C_{\phi K} = 0$. We measure $S_{\phi K} = 0.23 \pm 0.24$ and $C_{\phi K} = -0.14 \pm 0.18$ with statistical errors, only.

The simultaneous fit to the ϕK_S^0 and flavor decay modes yields:

$$
\begin{aligned}
S_{\phi K} &= 0.45 \pm 0.43(\text{stat}) \pm 0.07(\text{syst}) \\
C_{\phi K} &= -0.38 \pm 0.37(\text{stat}) \pm 0.12(\text{syst})
\end{aligned}
$$

Fig. 3 shows the Δt distributions of the B^0- and the $\bar{B}^0$-tagged subsets.

We consider systematic uncertainties in the CP coefficients ($\sigma_{S_{\phi K}}, \sigma_{C_{\phi K}}$) due to the event-yield determination (0.01, 0.05), the parameterization of PDF's for the event yield in signal and background (0.01, 0.04), background composition and CP asymmetry of the background, including an estimate of the error due to potential S-wave contamination (0.03, 0.05), the assumed parameterization

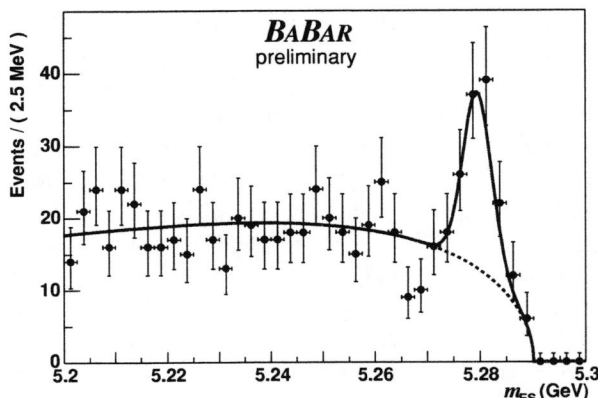

FIGURE 2. Distribution of the event variable m_{ES} for the ϕK_S^0 final state, after a requirement on the ΔE distribution to be between ± 3 standard deviations around zero is made. The solid line represents the fit for the event yield, the dotted line shows the total background.

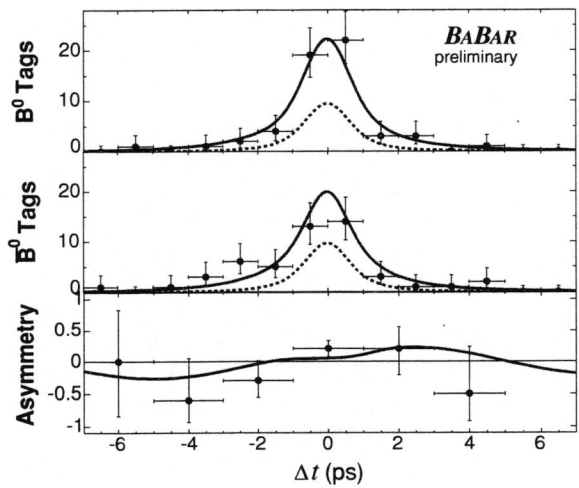

FIGURE 3. The top two plots show the Δt distributions of B^0- and $\bar{B}^0$-tagged events, respectively. The solid lines refer to the fit for all events; the dashed lines correspond to the background. The bottom plot shows the raw asymmetry.

of the Δt resolution function (0.03, 0.02), the m_{ES} background parameterization (0.02, 0.05), a possible difference in the efficiency for B^0 and $\bar{B}^0$ (0.00, 0.02), the fixed values for Δm_d and τ_B (0.00, 0.01), the bias in the coefficients due to the fit procedure, (0.04) 0.05) and the uncertainties in the SVT alignment and beamspot position (0.02,0.02). These are all included in the uncertainty without making corrections to the final results. We estimate errors due to the effect of doubly CKM-suppressed decays [18] to be (0.01, 0.03).

In summary, we have measured the time dependent CP asymmetries in the B meson final state ϕK_S^0. The values for ($S_{\phi K}$, $C_{\phi K}$) agree within one standard deviation with the ones measured in the charmonium–neutral-kaon channels [10–12]; the central value of $S_{\phi K}$ is also consistent with no *CP* asymmetry at approximately the 1σ level.

ACKNOWLEDGMENTS

I would like to thank the organizers of the Beauty 2003 conference for their hospitality. This work is supported by DOE and NSF (USA), NSERC (Canada), IHEP (China), CEA and CNRS-IN2P3 (France), BMBF and DFG (Germany), INFN (Italy), FOM (The Netherlands), NFR (Norway), MIST (Russia), and PPARC (United Kingdom). Individuals have received support from the A. P. Sloan Foundation, Research Corporation, and Alexander von Humboldt Foundation.

REFERENCES

1. Deshpande, N. G., and Trampetic, J., *Phys. Rev.* **D41**, 895 (1990).
2. Deshpande, N. G., and He, X.-G., *Phys. Lett.* **B336**, 471–476 (1994).
3. Fleischer, R., *Z. Phys.* **C62**, 81–90 (1994).
4. Grossman, Y., Ligeti, Z., Nir, Y., and Quinn, H., *Phys. Rev.* **D68**, 015004 (2003).
5. Cabibbo, N., *Phys. Rev. Lett.* **10**, 531–532 (1963).
6. Kobayashi, M., and Maskawa, T., *Prog. Theor. Phys.* **49**, 652–657 (1973).
7. Grossman, Y., and Worah, M. P., *Phys. Lett.* **B395**, 241–249 (1997).
8. Fleischer, R., *Int. J. Mod. Phys.* **A12**, 2459–2522 (1997).
9. Aubert, B., et al., [BaBar Collaboration], *Nucl. Instrum. Meth.* **A479**, 1–116 (2002).
10. Aubert, B., et al., [BaBar Collaboration], *Phys. Rev. Lett.* **89**, 201802 (2002).
11. Abe, K., et al., [Belle Collaboration], *Phys. Rev.* **D66**, 071102 (2002).
12. Miyabayashi, K., in these proceedings (2004).
13. Hagiwara, K., et al., [Particle Data Group Collaboration], *Phys. Rev.* **D66**, 010001 (2002).
14. Aubert, B., et al., [BaBar Collaboration], *Phys. Rev. Lett.* **87**, 151801 (2001).
15. Fisher, R. A., *Annals Eugen.* **7**, 179–188 (1936).
16. Aubert, B., et al., [BaBar Collaboration], *Phys. Rev.* **D66**, 032003 (2002).
17. Aubert, B., et al., [BaBar Collaboration], *Phys. Rev.* **D69**, 011102 (2004).
18. Long, O., Baak, M., Cahn, R. N., and Kirkby, D., *Phys. Rev.* **D68**, 034010 (2003).

CP VIOLATION

Time Dependent *CP* Violation at Belle

Kenkichi Miyabayashi

(Representing the Belle Collaboration)

Nara Women's University, Department of Physics, Kita-Uoya-Nishi-machi, Nara, 630-8506, Japan

Abstract. We report measurements of time-dependent *CP* violation in the neutral *B* meson system using a data sample corresponding to 140 fb^{-1} collected by the Belle detector at the KEKB e^+e^- collider. One *B* meson decaying into a *CP* eigenstate is fully reconstructed and the flavor of the accompanying *B* meson is identified by its decay products. The *CP* violation parameters are determined from the distribution of proper time intervals between the two *B* decays. Here, we cover a precise measurement of the Standard Model *CP* violation parameter $\sin(2\phi_1)$, as well as *CP* asymmetry measurements in neutral *B* meson decays into $J/\psi \ \pi^0$ and $\pi^+ \ \pi^-$ decays.

1. INTRODUCTION

In the Standard Model (SM), the Kobayashi-Maskawa (KM) quark-mixing matrix [1] has an irreducible complex phase that gives rise to *CP* violation in weak interactions. In particular, the SM predicts large *CP*-violating asymmetries in the time-dependent rates of B^0 and $\bar{B}^0$ decays into a common *CP* eigenstate f_{CP} [2, 3]. In the decay chain $\Upsilon(4S) \to B^0 \bar{B}^0 \to f_{CP} f_{\text{tag}}$, where one of the *B* mesons decays at time t_{CP} to a final state f_{CP} and the other decays at time t_{tag} to a final state f_{tag} that distinguishes between B^0 and $\bar{B}^0$, the decay rate has a time dependence given by [4]

$$\mathscr{P}(\Delta t) = \frac{e^{-|\Delta_t|/\tau_{B^0}}}{4\tau_{B^0}} \left\{ 1 + q \cdot \left[\mathscr{S}_{f_{CP}} \sin(\Delta m_d \Delta t) \right. \right.$$
$$\left. \left. + \mathscr{A}_{f_{CP}} \cos(\Delta m_d \Delta t) \right] \right\},$$

where τ_{B^0} is the B^0 lifetime, Δm_d is the mass difference between the two B^0 mass eigenstates, $\Delta t = t_{CP} - t_{\text{tag}}$, and the *b* flavor charge $q = +1 \ (-1)$ when the tagging *B* meson is a $B^0 \ (\bar{B}^0)$. The *CP*-violating parameters $\mathscr{S}_{f_{CP}}$ and $\mathscr{A}_{f_{CP}}$ are given by

$$\mathscr{S}_{f_{CP}} \equiv \frac{2\mathscr{I}m(\lambda)}{|\lambda|^2 + 1}, \qquad \mathscr{A}_{f_{CP}} \equiv \frac{|\lambda|^2 - 1}{|\lambda|^2 + 1}, \qquad (1)$$

where λ is a complex parameter that depends on both the B^0-$\bar{B}^0$ mixing and the amplitudes for B^0 and $\bar{B}^0$ decay to f_{CP}. To a good approximation in the SM, $|\lambda|$ is equal to the absolute value of the ratio of the $\bar{B}^0 \to f_{CP}$ to $B^0 \to f_{CP}$ decay amplitudes.

By choosing an appropriate f_{CP}, we can access or determine the aimed KM angle[1] or can probe potential new physics. In this report, first of all, we describe the experimental apparatus and the data sample used for *CP* violation measurements. Then, after explaining the analysis procedure, each measurement is reported.

2. EXPERIMENTAL APPARATUS AND DATA SAMPLE

The results presented here are based on a data sample of 140 fb^{-1} (corresponding to $15.2 \times 10^7 \ B\bar{B}$ pairs) collected at the $\Upsilon(4S)$ resonance with the Belle detector [5] at the KEKB asymmetric-energy e^+e^- (3.5 on 8 GeV) collider [6]. For the $B^0 \to \pi^+\pi^-$ mode, we used 78 fb^{-1} (corresponding to $85 \times 10^6 \ B\bar{B}$ pairs).

At KEKB, the $\Upsilon(4S)$ is produced with a boost of $\beta\gamma = 0.425$ nearly along the electron beamline (z). Since the B^0 and $\bar{B}^0$ mesons are nearly at rest in the $\Upsilon(4S)$ center-of-mass system (cms), Δt can be determined from the displacement in z between the f_{CP} and f_{tag} decay vertices: $\Delta t \simeq (z_{CP} - z_{\text{tag}})/\beta\gamma c \equiv \Delta z/\beta\gamma c$.

The Belle detector is a large-solid-angle magnetic spectrometer that consists of a three-layer silicon vertex detector (SVD), a 50-layer central drift chamber (CDC), an array of aerogel threshold Cherenkov counters (ACC), a barrel-like arrangement of time-of-flight scintillation counters (ToF), and an electromagnetic calorimeter comprised of CsI(Tl) crystals (ECL) located inside a super-

---

[1] $\phi_1 (= \beta) [-V_{cd}V_{cb}^*/V_{td}V_{tb}^*]$ and $\phi_2 (= \alpha) [-V_{td}V_{tb}^*/V_{ud}V_{ub}^*]$.

CP722, *B Physics at Hadron Machines*, edited by M. Paulini and S. Erhan
© 2004 American Institute of Physics 0-7354-0203-5/04/$22.00

conducting solenoid coil that provides a 1.5 T magnetic field. An iron flux-return located outside of the coil is instrumented to detect K_L^0 mesons and to identify muons (KLM). The detector is described in detail elsewhere [5].

3. ANALYSIS PROCEDURE

In order to measure time-dependent CP violation parameters, we have to carry out the following steps: (1) reconstruction of B decays into f_{CP}, (2) flavor tagging, (3) vertexing to measure Δt and (4) unbinned maximum likelihood fit.

The B candidate selection is carried out using two observables in the rest frame of the $\Upsilon(4S)$ (cms): the beam-energy constrained mass $M_{bc} \equiv \sqrt{E_{beam}^2 - (\Sigma \vec{p}_i)^2}$ and the energy difference $\Delta E \equiv \Sigma E_i - E_{beam}$, where $E_{beam} = \sqrt{s}/2$ is the cms beam energy, and $\vec{p}_i$ and E_i are the cms three-momenta and energies of the B meson decay products.

Charged leptons, kaons, pions and Λ baryons that are not associated with the reconstructed $B^0 \rightarrow f_{CP}$ decay are used to identify the b flavor of the accompanying B meson, denoted by f_{tag}. Based on the measured properties of these tracks, two parameters, q and r, are assigned to each event. The first, q, has the discrete value $+1$ (-1) when the tag-side B meson is more likely to be a B^0 ($\bar{B}^0$). The parameter r is an event-by-event MC determined flavor-tagging dilution factor that ranges from $r = 0$ for no flavor discrimination to $r = 1$ for an unambiguous flavor assignment. It is used only to sort data into six intervals of r, according to the estimated flavor purity. The wrong-tag probabilities for each of these intervals, w_l ($l = 1,6$), which are used in the final fit, are determined directly from the data samples of B^0 decays to exclusively reconstructed self-tagging channels. The difference of the wrong-tag fractions between B^0 and $\bar{B}^0$ are also determined from the same samples and implemented as Δw_l ($l = 1,6$) into the final fit. We obtain w_l and Δw_l using time-dependent B^0-$\bar{B}^0$ mixing. The wrong tag fractions for each r interval are given elsewhere [7].

The decay vertices of B^0 mesons are reconstructed using tracks that have enough SVD hits: *i.e.* both z and r-ϕ hits in at least one SVD layer and at least one additional layer with a z-hit, where the r-ϕ plane is perpendicular to the z-(beams)-axis. Each vertex position is required to be consistent with the IP profile, which is determined run-by-run and smeared in the r-ϕ plane by 21 μm to account for the B meson decay length. With these requirements, we are able to determine a vertex even in the case where only one track has enough associated SVD hits. The vertex position for the $B^0 \rightarrow f_{CP}$ decay is reconstructed using lepton tracks from the J/ψ in $\sin(2\phi_1)$ and $B^0 \rightarrow J/\psi \pi^0$ case. In the $B^0 \rightarrow \pi^+ \pi^-$

case, the candidate charged pion pair is used. The algorithm for the f_{tag} vertex reconstruction is chosen to minimize the effect of long-lived particles, secondary vertices from charmed hadrons and a small fraction of poorly reconstructed tracks [8]. From all the charged tracks with associated SVD hits except those used for $B^0 \rightarrow f_{CP}$ reconstruction, we select tracks with a position error in the z-direction of less than 500 μm, and with an impact parameter with respect to the f_{CP} vertex of less than 500 μm. Track pairs with opposite charges are removed if they have an invariant mass within ± 15 MeV/c^2 of the nominal K_S^0 mass. If the reduced χ^2 associated with the f_{tag} vertex exceeds 20, the track making the largest χ^2 contribution is removed and the vertex is refitted. This procedure is repeated until an acceptable reduced χ^2 is obtained.

We determine $\mathcal{S}_{f_{CP}}$ and $\mathcal{A}_{f_{CP}}$ for each mode by performing an unbinned maximum-likelihood fit to the observed Δt distribution. The probability density function (PDF) expected for the signal distribution is given by

$$
\begin{aligned}
\mathcal{P}_{sig}(\Delta t) &= \\
&= \frac{e^{-|\Delta t|/\tau_{B^0}}}{4\tau_{B^0}} \Big\{ 1 - q\Delta w_l + q(1-2w_l) \cdot \\
&\quad \cdot \Big[\mathcal{S}_{f_{CP}} \sin(\Delta m_d \Delta t) + \mathcal{A}_{f_{CP}} \cos(\Delta m_d \Delta t) \Big] \Big\}
\end{aligned}
$$

to account for the effect of incorrect flavor assignment. The distribution is convolved with the proper-time interval resolution function $R_{sig}(\Delta t)$, which takes into account the finite vertex resolution. $R_{sig}(\Delta t)$ is formed by convolving four components: the detector resolutions for z_{CP} and z_{tag}, the shift in the z_{tag}-vertex position due to secondary tracks originating from charmed particle decays, and the kinematic approximation that the B mesons are at rest in the cms [8]. A small component of broad outliers in the Δz distribution, caused by mis-reconstruction, is represented by a Gaussian function $P_{ol}(\Delta t)$. We determine the following likelihood value for each event:

$$
\begin{aligned}
P_i(\Delta t_i; & \mathcal{S}_{f_{CP}}, \mathcal{A}_{f_{CP}}) \\
= & (1 - f_{ol}) \int_{-\infty}^{\infty} \Big[f_{sig} \mathcal{P}_{sig}(\Delta t') R_{sig}(\Delta t_i - \Delta t') \\
& + (1 - f_{bkg}) \mathcal{P}_{bkg}(\Delta t') R_{bkg}(\Delta t_i - \Delta t') \Big] d(\Delta t') \\
& + f_{ol} P_{ol}(\Delta t_i) \quad\quad\quad (2)
\end{aligned}
$$

where f_{ol} is the outlier fraction and f_{sig} is the signal probability. $\mathcal{P}_{bkg}(\Delta t)$ is the PDF for the background events. Depending on f_{CP}, we split the backgrounds into several categories if needed. The resolution functions are modeled by a sum of two Gaussians. We fix τ_{B^0} and Δm_d

at their world-average values.[2] The only free parameters in the final fit are $\mathscr{S}_{f_{CP}}$ and $\mathscr{A}_{f_{CP}}$, which are determined by maximizing the likelihood function

$$\mathscr{L} = \prod_i P_i(\Delta t_i; \mathscr{S}_{f_{CP}}, \mathscr{A}_{f_{CP}}), \qquad (3)$$

where the product is over all events.

4. MEASUREMENT OF *CP* VIOLATION PARAMETER $\sin(2\phi_1)$

If B^0 and $\bar{B}^0$ decays into f_{CP} are dominated by the $b \to c\bar{c}s$ process, the SM predicts $\mathscr{S}_{f_{CP}} = \sin(2\phi_1)$ and $\mathscr{A}_{f_{CP}} = 0$ with negligible corrections from strong interactions [2, 3]. Changes exist in the analysis with respect to our earlier result [10]. We apply a new proper-time interval resolution function that reduces systematic uncertainties in $\sin(2\phi_1)$ and also in Δm_d and in the lifetime measurements (τ_{B^0}, τ_{B+}). We introduce b flavor-dependent wrong-tag fractions to accommodate possible differences between B^0 and $\bar{B}^0$ decays. We also adopt a multi-parameter fit to the flavor-specific control samples to obtain the resolution parameters and wrong-tag fractions simultaneously. There are other improvements in the estimation of background components, which become possible with increased statistics.

We reconstruct B^0 decays to the following *CP* eigenstates[3]: $J/\psi K_S^0$, $\psi(2S)K_S^0$, $\chi_{c1}K_S^0$, $\eta_c K_S^0$ for $\xi_f = -1$ and $J/\psi K_L^0$ for $\xi_f = +1$, where $\xi_f = +1(-1)$ corresponding to *CP*-even (-odd) final states. We also use $B^0 \to J/\psi K^{*0}$ decays where $K^{*0} \to K_S^0 \pi^0$. Here, the final state is a mixture of even and odd *CP*, depending on the relative orbital angular momentum of the J/ψ and K^{*0}. We find that the final state is primarily $\xi_f = +1$; the $\xi_f = -1$ fraction is $0.19 \pm 0.02(\text{stat}) \pm 0.03(\text{syst})$ [11].

The reconstruction and selection criteria for all f_{CP} channels used in the measurement are described in detail elsewhere [12, 13]. J/ψ and $\psi(2S)$ mesons are reconstructed via their decays to $\ell^+\ell^-$ ($\ell = \mu, e$). The $\psi(2S)$ is also reconstructed via $J/\psi \pi^+ \pi^-$, and the χ_{c1} via $J/\psi \gamma$. The η_c is detected in the $K_S^0 K^- \pi^+$, $K^+ K^- \pi^0$, and $p\bar{p}$ modes. For the $J/\psi K_S^0$ mode, we use $K_S^0 \to \pi^+\pi^-$ and $\pi^0\pi^0$ decays; for the other modes, we only use $K_S^0 \to$

---

[2] For measurements of $\sin 2\phi_1$ and *CP* violation in $B^0 \to J/\psi \pi^0$, in the likelihood function, we use the following B^0-$\bar{B}^0$ mixing parameter and B meson lifetimes; $\Delta m_d = 0.502 \pm 0.007$ ps^{-1}, $\tau_{B^0} = 1.537 \pm 0.015$ ps and $\tau_{B+} = 1.671 \pm 0.018$ ps which can be found in the 2003 Review of Particle Physics (web-edition) [9]. For the *CP* violation measurement in the $B^0 \to \pi^+\pi^-$ decay, we use the values in [9].

[3] Throughout this paper, the inclusion of the charge conjugate decay mode is implied unless otherwise stated.

TABLE 1. The numbers of reconstructed $B \to f_{CP}$ candidates after flavor tagging and vertex reconstruction, N_{events}, and the estimated signal purity in the signal region for each f_{CP} mode. ξ_f denotes the *CP* eigenvalue. J/ψ mesons are reconstructed in $J/\psi \to \mu^+\mu^-$ or e^+e^- decays. Candidate K_S^0 mesons are reconstructed as $K_S^0 \to \pi^+\pi^-$ decays unless otherwise written explicitly.

Mode	ξ_f	N_{events}	Purity
$J/\psi K_S^0 (\pi^+\pi^-)$	-1	1997	0.98
$J/\psi K_S^0 (\pi^0\pi^0)$	-1	288	0.82
$\psi(2S)(\ell^+\ell^-)K_S^0$	-1	145	0.93
$\psi(2S)(J/\psi\pi^+\pi^-)K_S^0$	-1	163	0.88
$\chi_{c1}(J/\psi\gamma)K_S^0$	-1	101	0.96
$\eta_c(K_S^0 K^\pm \pi^\mp)K_S^0$	-1	123	0.72
$\eta_c(K^+K^-\pi^0)K_S^0$	-1	74	0.70
$\eta_c(p\bar{p})K_S^0$	-1	20	0.90
$J/\psi K^{*0}(K_S^0\pi^0)$	+1*	174	0.93
$J/\psi K_L^0$	+1	2332	0.60

* (81%)

$\pi^+\pi^-$. For reconstructed $B \to f_{CP}$ candidates other than $J/\psi K_L^0$, we identify B decays using ΔE and M_{bc}. Candidate $B^0 \to J/\psi K_L^0$ decays are selected by requiring ECL and/or KLM hit patterns that are consistent with the presence of a shower induced by a K_L^0 meson. The centroid of the shower is required to be within a 45° cone centered on the K_L^0 direction inferred from two-body decay kinematics and the measured four-momentum of the J/ψ. This constraints p_B^{cms}, the reconstructed momentum of the B meson in the cms frame, to lie in the range $0.20 < p_B^{\text{cms}} < 0.45$ GeV/c.

Figure 1 shows the observed Δt distributions for the $q\xi_f = +1$ and $q\xi_f = -1$ event samples (top), the asymmetry between two samples with $0 < r \leq 0.5$ (middle) and with $0.5 < r \leq 1.0$ (bottom). The asymmetry in the region $0.5 < r \leq 1.0$, where wrong-tag fractions are small as shown in Ref. [7], clearly demonstrates large *CP* violation.

We determine $\sin(2\phi_1)$ from an unbinned maximum likelihood fit to the observed Δt distributions. In the fit, we float $\sin(2\phi_1) = -\mathscr{S}_{f_{CP}}/\xi_f$ as free parameter and set $\mathscr{A}_{f_{CP}} = 0$. The signal fraction f_{sig} is calculated as a function of p_B^{cms} for $J/\psi K_L^0$ and of ΔE and M_{bc} for other modes. The result of the fit is

$$\sin(2\phi_1) = 0.733 \pm 0.057(\text{stat}) \pm 0.028(\text{syst}).$$

In order to test the assumption of $\mathscr{A}_{f_{CP}} = 0$, we also performed a fit with $\mathscr{A}_{CP} \equiv -\xi_f \text{Im}\lambda/|\lambda|$ and $|\lambda|$ as free parameters, keeping everything else the same:

$$|\lambda| = 1.007 \pm 0.041(\text{stat}),$$
$$\mathscr{A}_{CP} = 0.733 \pm 0.057(\text{stat}),$$

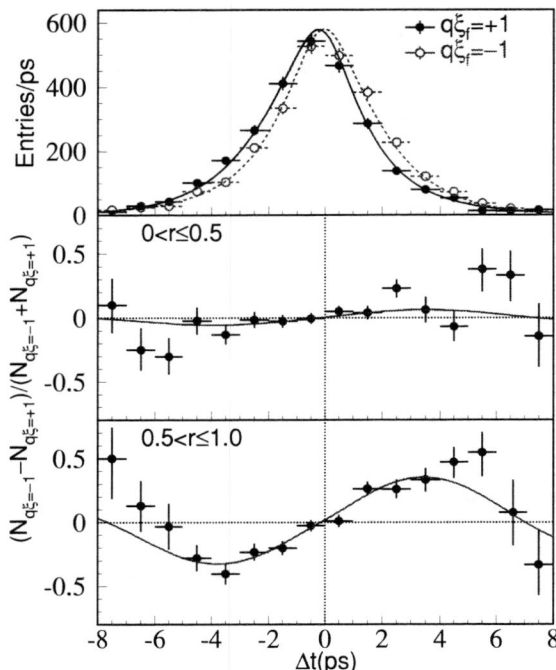

FIGURE 1. Measurement of $\sin(2\phi_1)$ using 140 fb^{-1} collected by the Belle detector. (a) Proper time difference for each flavor; $q = +1$ and -1 denotes B^0 and $\bar{B}^0$, respectively. ξ_f is the *CP* eigenvalue of the final states. (b) The raw asymmetry for the poorly flavor-tagged sample. (c) The raw asymmetry for the well-tagged sample.

for all *CP* modes combined. This result is consistent with the assumption used in our analysis. The reader can find more details in Ref. [7].

5. TIME-DEPENDENT *CP* VIOLATION IN $B^0 \to J/\psi \pi^0$

The SM predicts $\mathscr{S}_{f_{CP}} = -\xi_f \sin(2\phi_1)$, and $\mathscr{A}_{f_{CP}} = 0$ (or equivalently $|\lambda| = 1$) for both $b \to c\bar{c}s$ and the leading contributions to $b \to c\bar{c}d$ (*e.g.* the tree diagram). Hence, for $f_{CP} = J/\psi \pi^0$, which is a *CP*-even final state, $\mathscr{S}_{J/\psi\pi^0}$ becomes $-\sin(2\phi_1)$ if the tree diagram dominates. If penguin contributions or other contributions are substantial, a precision measurement of the time-dependent *CP* asymmetry in $b \to c\bar{c}d$ may reveal values for $\mathscr{S}_{J/\psi\pi^0}$ and $\mathscr{A}_{J/\psi\pi^0}$ that differ from what is expected.

A study of the *CP* asymmetry in $B^0 \to J/\psi \pi^0$ decays has been reported by the BaBar collaboration [14]. Here we report a measurement of time-dependent *CP*-violating parameters in $B^0 \to J/\psi \pi^0$ decays using the higher statistics data accumulated by the Belle detector [15].

J/ψ mesons are reconstructed in the same way as in

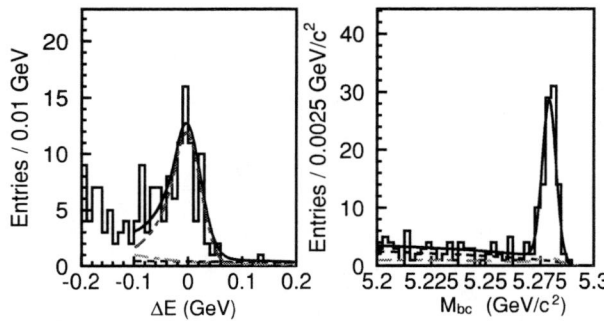

FIGURE 2. The ΔE(left) and $M_{\rm bc}$(right) distributions for $B^0 \to J/\psi \pi^0$ candidates. The superimposed curves show fitted contributions from signal (red two-dot-dash, plotted only in ΔE distribution), $B \to J/\psi X$ background (green dot-dash), combinatorial background (blue dash) and the sum of all the contributions (black solid). See text for further details.

the $J/\psi K_S^0$ reconstruction, requiring both lepton tracks to be positively identified as such. π^0's are formed from photon pairs that have an invariant mass in the range 0.118 to 0.15 GeV/c^2. The π^0 momentum is obtained by applying a mass constrained fit.

Candidate events are selected by requiring $5.270 < M_{\rm bc} < 5.290$ GeV/c^2 and $-0.10 < \Delta E < 0.05$ GeV. The number of reconstructed $B^0 \to J/\psi \pi^0$ candidates is 103. The ΔE and $M_{\rm bc}$ distributions for the candidate events are shown in Fig. 2. To assign $f_{\rm sig}$, we determine event distribution functions in the ΔE-$M_{\rm bc}$ plane for both signal and background. In this analysis, backgrounds are studied using a large sample of MC events along with events outside of the signal region. We split the backgrounds into two categories, one being B decays having a J/ψ ($B \to J/\psi X$) and the other being combinatorial background to which random combinations of particles in $B\bar{B}$ decays and continuum events contribute. The purity of the signal is estimated to be $(86\pm10)\%$

After flavor tagging and vertexing, 91 candidates are used to the final fit which results in the *CP* violation parameters

$$\mathscr{S}_{J/\psi\pi^0} = -0.72 \pm 0.42(\text{stat}) \pm 0.08(\text{syst}),$$
$$\mathscr{A}_{J/\psi\pi^0} = -0.01 \pm 0.29(\text{stat}) \pm 0.07(\text{syst}). \quad (4)$$

Figure 3 shows the raw asymmetry in each Δt bin without background subtraction. The curve shows the result of the unbinned-maximum likelihood fit to the Δt distribution, $\mathscr{S}_{J/\psi\pi^0} \sin(\Delta m_d \Delta t) + \mathscr{A}_{J/\psi\pi^0} \cos(\Delta m_d \Delta t)$.

The resulting *CP* violation parameters are consistent with those obtained for $B^0 \to J/\psi K_S^0$ and other decays governed by $b \to c\bar{c}s$ transition and suggest that penguin and other contributions to this decay mode are not large.

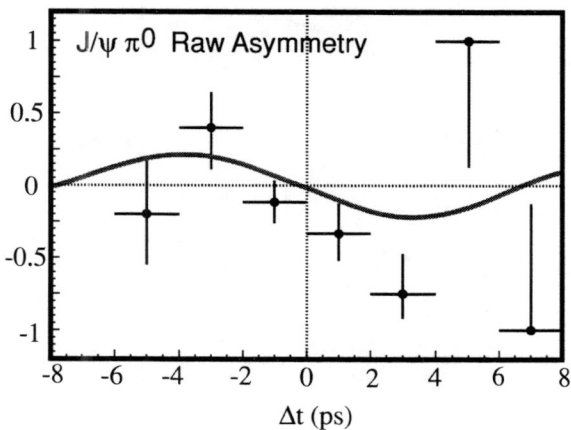

FIGURE 3. The Δt asymmetry. The curve shows the result of the unbinned-maximum likelihood fit.

6. TIME-DEPENDENT *CP* VIOLATION IN $B^0 \to \pi^+ \pi^-$

A measurement of time-dependent *CP*-violating asymmetries in the mode $B^0 \to \pi^+ \pi^-$ is sensitive to direct *CP* violation and the CKM angle ϕ_2. If the decay proceeded only via a $b \to u$ tree amplitude, we would have $\mathscr{S}_{\pi\pi} = \sin(2\phi_2)$ and $\mathscr{A}_{\pi\pi} = 0$, or equivalently $|\lambda| = 1$. The situation is complicated by the possibility of significant contributions from gluonic $b \to d$ penguin amplitudes that have a different weak phase and additional strong phases [16–21]. As a result, $\mathscr{S}_{\pi\pi}$ may not be equal to $\sin(2\phi_2)$ and direct *CP* violation, $\mathscr{A}_{\pi\pi} \neq 0$, may occur.

The $B^0 \to \pi^+ \pi^-$ event selection is described in detail elsewhere [22]. We use oppositely charged track pairs that are positively identified as pions according to the likelihood ratio for a particle to be a $K^{\pm}$ meson, KID = $\mathscr{L}(K)/[\mathscr{L}(K)+\mathscr{L}(\pi)]$, which is based on the combined information from the ACC and the CDC dE/dx measurements. Here, we use KID<0.4 as the default requirement for the selection of pions.

Candidate B mesons are reconstructed using ΔE and M_{bc}. The signal region is defined as 5.271 GeV/c^2 < M_{bc} < 5.287 GeV/c^2 and $|\Delta E|$ < 0.057 GeV, corresponding to $\pm 3\sigma$ from the central values. In order to suppress background from the $e^+e^- \to q\bar{q}$ continuum ($q = u$, d, s, c), we form signal and background likelihood functions, $\mathscr{L}_S$ and $\mathscr{L}_{BG}$, from two variables. One is a Fisher discriminant determined from six modified Fox-Wolfram moments [22–24]; the other is the cms B flight direction with respect to the z-axis. We reduce the continuum background by imposing requirements on the likelihood ratio $LR = \mathscr{L}_S/(\mathscr{L}_S + \mathscr{L}_{BG})$ for candidate events.

In this analysis, we optimize the expected sensitivity by including additional candidate events with a lower signal likelihood ratio. The requirements on LR vary for

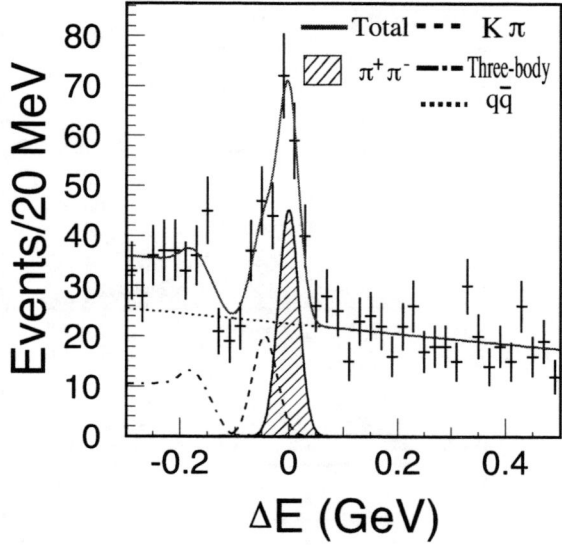

FIGURE 4. ΔE distribution in the M_{bc} signal region for $B^0 \to \pi^+ \pi^-$ candidates with LR > 0.825. The sum of the signal and background functions is shown as a solid curve. The hatched area represents the $\pi^+ \pi^-$ component, the dashed curve represents the $K^+ \pi^-$ component, the dotted curve represents the continuum background, and the dot-dashed curve represents the charmless three-body B decay background component.

different values of the tagging dilution factor r, since the separation of continuum background from the B signal varies with r; there are 12 distinct regions in the LR-r plane.

Figure 4 shows the ΔE distributions for the $B^0 \to \pi^+ \pi^-$ candidates that are in the M_{bc} signal region with LR > 0.825, after flavor tagging and vertex reconstruction. The $B^0 \to \pi^+ \pi^-$ signal yield for LR > 0.825 is extracted by fitting the ΔE distribution with a Gaussian signal function plus contributions from misidentified $B^0 \to K^+ \pi^-$ events, three-body B decays, and continuum background. The fit yields 106^{+16}_{-15} $\pi^+\pi^-$ events, 41^{+10}_{-9} $K^+\pi^-$ events and 128^{+5}_{-6} continuum events in the signal region. The errors do not include systematic uncertainties unless otherwise stated.

The unbinned maximum likelihood fit to the 760 $B^0 \to \pi^+\pi^-$ candidates (391 B^0- and 369 $\bar{B}^0$-tags), containing 163^{+24}_{-23} $\pi^+\pi^-$ signal events, yields $\mathscr{A}_{\pi\pi} = +0.77$ and $\mathscr{S}_{\pi\pi} = -1.23$. In Figs. 5(a) and (b), we show the raw, unweighted Δt distributions for the 148 B^0- and 127 $\bar{B}^0$-tagged events with LR > 0.825. The fit curves use $\mathscr{A}_{\pi\pi}$ and $\mathscr{S}_{\pi\pi}$ values that are obtained from all of the LR-r regions. The background-subtracted Δt distributions are shown in Fig. 5(c). Figure 5(d) shows the background-subtracted *CP* asymmetry between the B^0- and $\bar{B}^0$-tagged events as a function of Δt. The result of the fit is superimposed and shown by the solid curve.

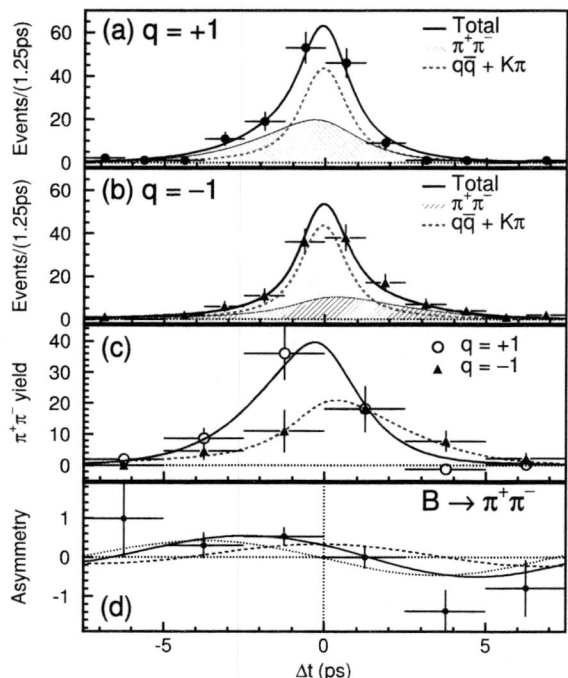

FIGURE 5. The raw, unweighted Δt distributions for the 275 $B^0 \to \pi^+\pi^-$ candidates with $LR > 0.825$ in the signal region: (a) 148 candidates with $q = +1$, *i.e.* the tag side is identified as B^0; (b) 127 candidates with $q = -1$; (c) $B^0 \to \pi^+\pi^-$ yields after background subtraction. The errors are statistical only and do not include the error on the background subtraction; (d) the CP asymmetry for $B^0 \to \pi^+\pi^-$ after background subtraction. In Figs. (a) through (c), the curves show the results of the unbinned maximum likelihood fit to the Δt distributions of the 760 $B^0 \to \pi^+\pi^-$ candidates. In Fig. (d), the solid curve shows the resultant CP asymmetry, while the dashed (dotted) curve is the contribution from the cosine (sine) term.

The statistical errors on $\mathscr{A}_{\pi\pi}$ and $\mathscr{S}_{\pi\pi}$ are estimated by a frequentist approach, where we use MC pseudo-experiments to determine acceptance regions. We quote the rms values of the MC $\mathscr{A}_{\pi\pi}$ and $\mathscr{S}_{\pi\pi}$ distributions as the statistical errors of our measurement. The systematic error on $\mathscr{A}_{\pi\pi}f$ is primarily due to uncertainties in the background fractions and vertexing. For $\mathscr{S}_{\pi\pi}$, the background fractions and a possible fit bias near the physical boundary are the two leading components. We obtain

$$\mathscr{A}_{\pi\pi} = +0.77 \pm 0.27(\text{stat}) \pm 0.08(\text{syst}),$$
$$\mathscr{S}_{\pi\pi} = -1.23 \pm 0.41(\text{stat})^{+0.08}_{-0.07}(\text{syst}).$$

We use the Feldman-Cousins frequentist approach [25] to determine the statistical significance of our measurement. In order to form confidence intervals, we use the $\mathscr{A}_{\pi\pi}$ and $\mathscr{S}_{\pi\pi}$ distributions of the fits to MC pseudo-experiments for various input values of $\mathscr{A}_{\pi\pi}$ and $\mathscr{S}_{\pi\pi}$. Figure 6 shows the resulting two-dimensional confidence regions in the $\mathscr{A}_{\pi\pi}$ vs. $\mathscr{S}_{\pi\pi}$ plane. The case

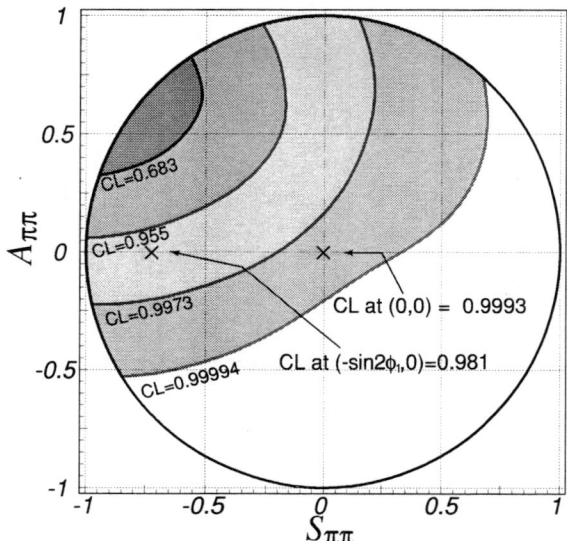

FIGURE 6. Confidence regions for $\mathscr{A}_{\pi\pi}$ and $\mathscr{S}_{\pi\pi}$.

that CP symmetry is conserved, $\mathscr{A}_{\pi\pi} = \mathscr{S}_{\pi\pi} = 0$, is ruled out at the 99.93% confidence level (CL), equivalent to $3.4\,\sigma$ significance assuming Gaussian errors [26].

7. CONCLUSION AND PROSPECT

Since the first e^+e^- collision in 1999, the KEKB luminosity has been improving. A peak luminosity of $10^{34}\,\text{cm}^{-2}\,\text{s}^{-1}$ as well as an integrated luminosity record of 608 pb^{-1} were recently achieved.

Measurements of $\sin(2\phi_1)$ and CP violation in $B^0 \to J/\psi\,\pi^0$ and $B^0 \to \pi^+\pi^-$ decays were successfully performed. During shutdown in 2003 summer, the new SVD having four layers of silicon strip detector with the smaller radius beampipe in the interaction region were installed. The data acquisition system was also partially upgraded to reach a higher bandwidth. By the next summer an additional ~ 100 fb^{-1} of data is expected.

ACKNOWLEDGMENTS

We wish to thank the KEKB accelerator group for the excellent operation of the KEKB accelerator. We acknowledge support from the Ministry of Education, Culture, Sports, Science, and Technology of Japan and the Japan Society for the Promotion of Science; the Australian Research Council and the Australian Department of Industry, Science and Resources; the National Science Foundation of China under contract No. 10175071; the Department of Science and Technology of India; the BK21 program of the Ministry of Education of Korea and the

CHEP SRC program of the Korea Science and Engineering Foundation; the Polish State Committee for Scientific Research under contract No. 2P03B 01324; the Ministry of Science and Technology of the Russian Federation; the Ministry of Education, Science and Sport of the Republic of Slovenia; the National Science Council and the Ministry of Education of Taiwan; and the U.S. Department of Energy.

REFERENCES

1. Kobayashi, M., and Maskawa, T., *Prog. Theor. Phys.* **49**, 652–657 (1973).
2. Carter, A. B., and Sanda, A. I., *Phys. Rev.* **D23**, 1567 (1981).
3. Bigi, I. I. Y., and Sanda, A. I., *Nucl. Phys.* **B193**, 85 (1981).
4. Bigi, I. I. Y., Khoze, V. A., Uraltsev, N. G., and Sanda, A. I., *Adv. Ser. Direct. High Energy Phys.* **3**, 175–248 (1989), a general review of the formalism is given here.
5. Abashian, A., et al., [Belle Collaboration], *Nucl. Instrum. Meth.* **A479**, 117–232 (2002).
6. Kurokawa, S., and Kikutani, E., *Nucl. Instrum. Meth.* **A499**, 1–7 (2003).
7. Abe, K., et al., [Belle Collaboration], arXiv:hep-ex/0308036 (2003).
8. Tajima, H., et al., arXiv:hep-ex/0301026 (2003).
9. Hagiwara, K., et al., [Particle Data Group Collaboration], *Phys. Rev.* **D66**, 010001 (2002), URL: http://pdg.lbl.gov.
10. Abe, K., et al., [Belle Collaboration], *Phys. Rev.* **D66**, 071102 (2002).
11. Abe, K., et al., [Belle Collaboration], *Phys. Lett.* **B538**, 11–20 (2002).
12. Abe, K., et al., [Belle Collaboration], *Phys. Rev. Lett.* **87**, 091802 (2001).
13. Abe, K., et al., [Belle Collaboration], *Phys. Rev.* **D66**, 032007 (2002).
14. Aubert, B., et al., [BaBar Collaboration], *Phys. Rev. Lett.* **91**, 061802 (2003).
15. Abe, K., et al., [Belle Collaboration], arXiv:hep-ex/0308053 (2003).
16. Gronau, M., *Phys. Rev. Lett.* **63**, 1451 (1989).
17. London, D., and Peccei, R. D., *Phys. Lett.* **B223**, 257 (1989).
18. Beneke, M., Buchalla, G., Neubert, M., and Sachrajda, C. T., *Nucl. Phys.* **B606**, 245–321 (2001).
19. Keum, Y. Y., Li, H.-N., and Sanda, A. I., *Phys. Rev.* **D63**, 054008 (2001).
20. Ciuchini, M., Franco, E., Martinelli, G., Pierini, M., and Silvestrini, L., *Phys. Lett.* **B515**, 33–41 (2001).
21. Gronau, M., and Rosner, J. L., *Phys. Rev.* **D65**, 013004 (2002).
22. Casey, B. C. K., et al., [Belle Collaboration], *Phys. Rev.* **D66**, 092002 (2002).
23. Fox, G. C., and Wolfram, S., *Phys. Rev. Lett.* **41**, 1581 (1978).
24. Abe, K., et al., [Belle Collaboration], *Phys. Lett.* **B511**, 151–158 (2001).
25. Feldman, G. J., and Cousins, R. D., *Phys. Rev.* **D57**, 3873–3889 (1998).
26. Abe, K., et al., [Belle Collaboration], *Phys. Rev.* **D68**, 012001 (2003).

CP Violation Prospects at the Tevatron

Petar Maksimović

(Representing the CDF and DØ Collaborations)

The Johns Hopkins University, Department of Physics and Astronomy, Baltimore, MD 21218, U.S.A.

Abstract. CDF and DØ are caught between the idealistic projections from the mid-1990's based on the Run I results and their TDR's on one side, and realistic extrapolations from the recent Run II measurements on the other. In this paper, we report only on the latter, and describe recent progress in other areas.

1. INTRODUCTION

The hadronic environment poses several challenges for *B* physics. Tevatron events are "messy", due to the presence of tracks from fragmentation, underlying event – hadronized remnants of p and $\bar{p}$, and pile-up event – another collision somewhere else along the beam axis, in addition to the $b\bar{b}$ pair. Furthermore, in detectors such as DØ [1] and CDF [2], if one b quark is central ($|\eta| < 1$) then the other one is also central only 20-40% of the time. The reconstruction of the *B* hadron on the 'opposite side' is improved with a better detector coverage; this is the reason why DØ does better than CDF in certain aspects of the *B* physics program such as opposite side flavor tagging.

However, these downsides are offset by a very large $b\bar{b}$ production cross section ($\sim 100~\mu$b to produce one b quark with $p_T > 6$ GeV/c within $|\eta| < 1$), and a variety of b hadron species that the b quark hadronizes into. Nevertheless, the rate of the soft QCD interactions is larger than the $b\bar{b}$ production rate by three orders of magnitude; the trigger system thus plays a critical role in formulating a *B* physics program at a hadron collider experiment. In the past decade, both CDF and DØ developed fast electronics which can analyze data from the silicon detectors in real time and trigger on events that contain displaced tracks. Such a system has been operational at CDF since the commencement of Run II, and is put in place at DØ in the fall of 2003. For this reason, at this point only CDF has physics results from the fully hadronic *B* hadron decays, although that is expected to change in the near future as the new data streams in. CDF also has a faster DAQ system than DØ, allowing for a higher Level 1 Accept trigger rate, and thereby providing a slight edge in certain hadronic *B* decay modes.

The measurements of *CP* violation in the *B* sector depend on three quantities: the number of events, the resolution on the proper time of the *b* hadron decay, and the effective flavor tagging efficiency, $\varepsilon\mathscr{D}^2$. All three are discussed in detail in these proceedings, alongside with projections for the measurement of Δm_s [3]. In here, we merely summarize the relevant information.

There are two scenarios used in extrapolating the integrated luminosity into the future. The 'base' luminosity is more conservative and assumes only improvements proven to be work. The 'design' luminosity is more optimistic and relies on the use of electron cooling. Through FY2007, the Tevatron is expected to deliver 2.1 fb^{-1} in the base scenario, and 3.8 fb^{-1} in the design scenario. Through FY2009, including the use of the recycler, the total integrated luminosity options are increased to 4.4 fb^{-1} in the base and to 8.6 fb^{-1} in the design scenario. We consider 2 fb^{-1} for our projections, since it is not clear how effective the *B* triggers presently deployed at CDF and DØ would work in the high-luminosity regime.

In terms of flavor tagging, DØ is doing better because of a larger coverage, especially regarding "opposite side" flavor tagging. CDF's implementations of kaon tagging rely on the use of the Time-of-Flight detector and can be deployed on both the 'same side' and the "opposite side", but are still not ready. In the remainder of this paper, we assume CDF's total effective tagging efficiency to be $\varepsilon\mathscr{D}^2 = 5\%$. As a comparison, the sum of preliminary $\varepsilon\mathscr{D}^2$ values for the taggers DØ reported so far is $(10.4 \pm 3.4)\%$.

Finally, the resolution on the proper time of *b* hadron decays is 67 fs for CDF, and 110 fs for DØ. However, with the inclusion of the fine-pitch innermost layer of the silicon detector, "Layer 00", and with the addition of an event-by-event primary vertex location instead of the run-averaged beam line, CDF's resolution on the proper time will go down to 50 fs.

CP722, *B Physics at Hadron Machines*, edited by M. Paulini and S. Erhan
© 2004 American Institute of Physics 0-7354-0203-5/04/$22.00

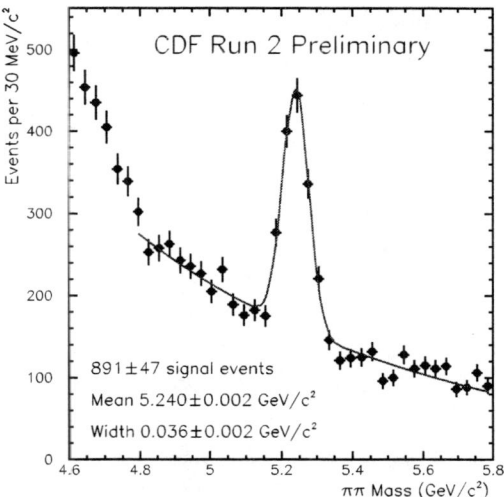

FIGURE 1. Invariant mass of h^+h^- candidates using the pion mass hypothesis for both tracks, corresponding to approximately 190 pb^{-1}. The first 65 pb^{-1} are used in the measurement.

Furthermore, there are other improvements that will enhance the yields. In the fall of 2003, the trigger system of the DØ detector has undergone a significant upgrade in which the two track-based trigger systems have been added. CDF is also in the process of upgrading various components of the trigger and the DAQ system, which will be completed in 2004 and 2005. Moreover, the innermost layer of CDF's silicon detector, known as "Layer 00," is beginning to be used for physics.

2. *CP* VIOLATION IN $B \rightarrow h^+h^-$ DECAYS

The CDF's two-track-trigger collects $B \rightarrow h^+h^-$ decays. This sample is a mixture of $B^0 \rightarrow K\pi$, $B^0 \rightarrow \pi\pi$, $B^0_s \rightarrow KK$ and $B^0_s \rightarrow K\pi$ decay modes. A sum of all four components is shown in Fig. 1. In order to measure the composition of this sample, one needs to rely on some kind of particle identification in order to tell kaons from pions. Unfortunately, the K-π separation of the Time-of-Flight system is negligible in this momentum regime (*i.e.*, above 2 GeV/c), and the dE/dx separation between kaons and pions is only about 1.2 standard deviations. Therefore, separating components event-by-event is not possible; one can only measure the contribution of each decay mode on a statistical basis.

For each particle, a probability density function is formed that includes the following: dE/dx information, calibrated on $D^{*+} \rightarrow D^0\pi^+$ candidates where $D^0 \rightarrow$

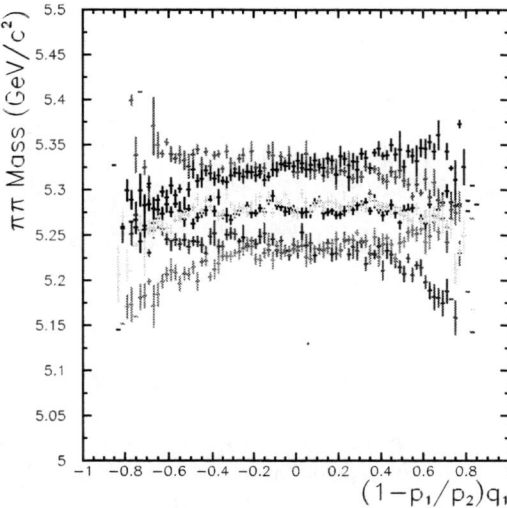

FIGURE 2. Invariant mass distribution of h^+h^- candidates versus $(1 - (p_1/p_2))q_1$ (where $p_1 < p_2$).

TABLE 1. Yields for the signatures comprising the $B \rightarrow h^+h^-$ sample obtained from the combined unbinned likelihood fit.

Decay mode	Yield in 65 pb^{-1}
$B^0 \rightarrow K^+\pi^-$	148 ± 17 (stat) ± 17 (syst)
$B^0 \rightarrow \pi^+\pi^-$	39 ± 14 (stat) ± 17 (syst)
$B^0_s \rightarrow K^+K^-$	90 ± 17 (stat) ± 17 (syst)
$B^0_s \rightarrow K^-\pi^+$	3 ± 11 (stat) ± 17 (syst)

$K^-\pi^+$; the invariant mass of the two-track pair, calculated using the pion hypothesis for both tracks; and a kinematic quantity $(1 - (p_1/p_2))q_1$, where p_i are the momenta of the two tracks, and q_1 is the charge of the one with the lower momentum. This expression combines the kinematic and charge information. The correlation between it and $m(hh)$ from the Monte Carlo simulation is given in Fig. 2 and shows that a probability density function based on both is sensitive to the difference between $K^+\pi^-$ from π^+K^-. This allows the extraction of direct *CP* violation information from an unbinned likelihood fit to $B \rightarrow hh$ data.

The result of the fit to the first 65 pb^{-1} of data is shown in Table 1. This is the first observation of $B^0_s \rightarrow K^+K^-$, and results in:

$$\frac{\mathscr{B}(B^0_s \rightarrow K^{\pm}K^{\mp})}{\mathscr{B}(B^0 \rightarrow K^{\pm}\pi^{\mp})} = 2.71 \pm 1.15$$

where the error includes the systematic uncertainty on f_s/f_d ratio. The null measurement of $B^0_s \rightarrow K^-\pi^+$ is also interesting from the theoretical standpoint. A fit for the direct *CP* violation asymmetry in $B^0 \rightarrow K^+\pi^-$ decays

TABLE 2. Projected yields for the signatures comprising the $B \to h^+ h^-$ sample for 2 fb^{-1} and 3.5 fb^{-1}.

Decay mode	Yield in 2 fb^{-1}	Yield in 3.5 fb^{-1}
$B^0 \to K^+ \pi^-$	6700	11,725
$B^0 \to \pi^+ \pi^-$	1770	3097
$B_s^0 \to K^+ K^-$	4040	7070
$B_s^0 \to K^- \pi^+$	1070	1870

yields:

$$\mathscr{A}_{CP}(B^0 \to K^+ \pi^-) = 0.02 \pm 0.15 \text{ (stat)} \pm 0.02 \text{ (syst)}$$

The systematic uncertainty on $\mathscr{A}_{CP}(B^0 \to K^+ \pi^-)$ is already comparable to the measurements at the B factories.

We now turn to the projections for the yields and CP violation in $B^0 \to \pi^+ \pi^-$ and $B_s^0 \to K^+ K^-$. Table 2 gives the projected yields for the four decay modes comprising the $B \to h^+ h^-$ sample for 2 fb^{-1} and 3.5 fb^{-1}. The yield for $B_s^0 \to K^- \pi^+$ comes from theoretical prediction for the branching ratio; the other three are obtained by scaling our current yield by the increase in luminosity and by the expected increase in trigger efficiency (in place since June 2003). These predictions do not take into account other upgrades of the CDF trigger and DAQ system that are currently in progress.

CP asymmetry in both $B^0 \to \pi^+ \pi^-$ and $B_s^0 \to K^+ K^-$ decays (measuring $\sin 2\alpha$ and $\sin 2\gamma$, respectively) will be extracted from a simultaneous fit to the $B \to h^+ h^-$ sample which also includes proper time of the B candidate decay and its production flavor, following a method outlined by Fleischer [4]. Experimentally, a very different time dependence of $\mathscr{A}_{CP}(B^0)$ (Eq. (1)) and $\mathscr{A}_{CP}(B_s^0)$ (Eq. (2)) helps further separate $B^0 \to \pi^+ \pi^-$ and $B_s^0 \to K^+ K^-$ decays:

$$\mathscr{A}_{CP}(B^0) = A_{CP}^{\text{dir}} \cos \Delta m_d t + A_{CP}^{\text{mix}} \sin \Delta m_d t \quad (1)$$

$$\mathscr{A}_{CP}(B_s^0) = A_{CP}^{\text{dir}} \cos \Delta m_s t + A_{CP}^{\text{mix}} \sin \Delta m_s t \quad (2)$$

In the case of $B^0 \to \pi^+ \pi^-$, the displaced track trigger favors A_{CP}^{mix} since there is a bias toward longer proper times which affects the cosine term much more than the sine. The effective flavor tagging efficiency assumed for both B^0 and B_s^0 mesons is $\varepsilon \mathscr{D}^2 = 5\%$. The expected errors on $\mathscr{A}_{CP}(B^0)$ as functions of the integrated luminosity are shown in Fig. 3. Both errors are a factor of four below earlier expectations [5].

CP asymmetry in $B_s^0 \to K^+ K^-$ is given in Fig. 4. In this case the displaced track trigger bias affects both terms of Eq. (2) equally due to a very high value of Δm_s, and thus only one curve is shown for both A_{CP}^{dir} and A_{CP}^{mix}. However, Δm_s is unknown, so we consider $x_s \equiv \Delta m_s / \tau_{B_s^0}$ equal to 20, 30 and 50.

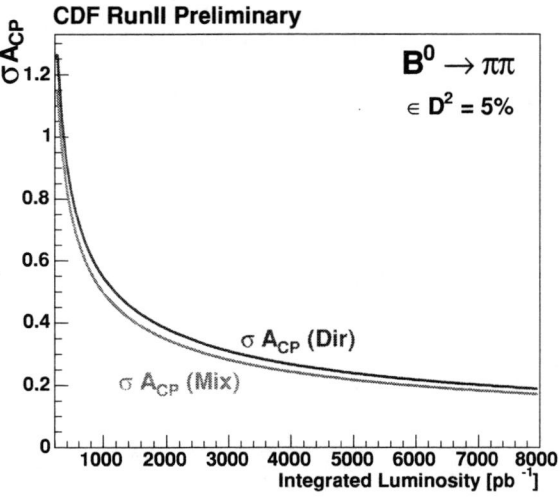

FIGURE 3. Projected error on $\mathscr{A}_{CP}$ in $B^0 \to \pi^+ \pi^-$ as a function of the integrated luminosity. The upper curve is the expected error on A_{CP}^{dir}, and the lower curve is the expected error on A_{CP}^{mix}.

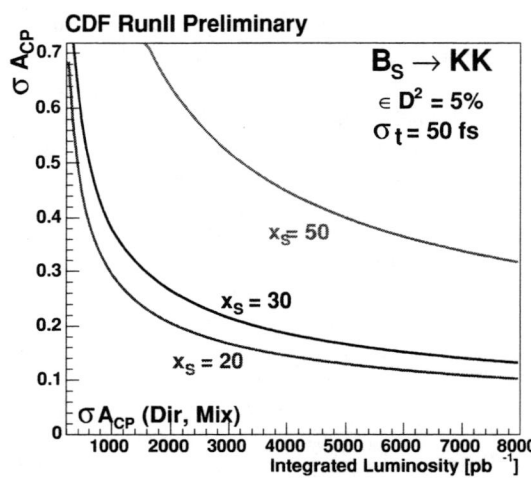

FIGURE 4. Projected error on $\mathscr{A}_{CP}$ in $B_s^0 \to K^+ K^-$ as a function of the integrated luminosity.

3. CP VIOLATION IN $B \to DK$ DECAYS

Several methods use the $B \to DK$ mode, mostly to measure γ [6, 7]. They all rely on having a large sample of $B \to D\pi$ first, and then reconstruct the Cabibbo suppressed $B \to DK$ decay by using particle id to separate kaons from pions. CDF has already reconstructed $B^+ \to \bar{D}^0 \pi^+$ (shown in Fig. 5), $B^0 \to \bar{D}^{(*)-} \pi^+$ and $B_s^0 \to D_s^- \pi^+$ (shown in Fig. 6) with very good signal to noise. However, in CDF the separation of $B \to DK$ from $B \to D\pi$ has not been attempted yet, which is why at this time we cannot provide estimates of the error on γ, but only

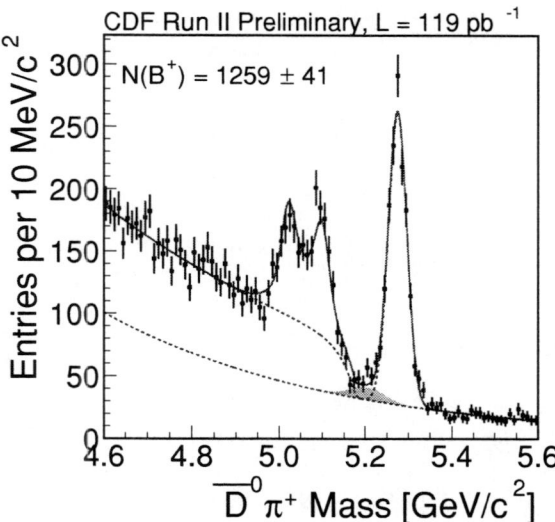

FIGURE 5. Invariant mass of $B^+ \to \bar{D}^0 \pi^+$.

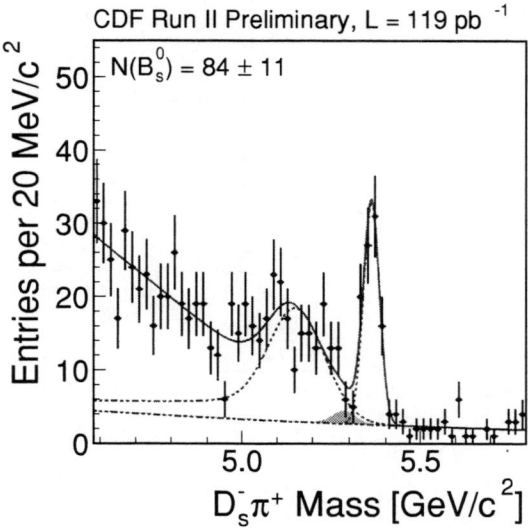

FIGURE 6. Invariant mass of $B_s^0 \to D_s^- K^+$.

quote expected rates for various $B \to D\pi$ and $B \to DK$ channels. Such rates are given in Table 3. The yields for $B \to D\pi$ decay modes are extrapolated from the events observed so far. These yields are then multiplied by

$$\frac{\mathscr{B}(B^+ \to \bar{D}^0 K^+)}{\mathscr{B}(B^+ \to \bar{D}^0 \pi^+)} = (8.31 \pm 0.35 \pm 0.20)\% \sim 8\%$$

(an average of the BaBar and Belle measurements) to obtain the expected yields for $B \to DK$. To set the scale, the expected contribution from $B \to DK$ decays to the invariant mass distributions of $B \to D\pi$ candidates is shown as a shaded Gaussian centered at ~ 5.2 in Fig. 5 and ~ 5.3 in Fig. 6.

4. *CP* VIOLATION IN $B_s^0 \to J/\psi\phi$

CP violation studies of the decay mode $B_s^0 \to J/\psi\phi$, followed by $J/\psi \to \mu^+\mu^-$ and $\phi \to K^+K^-$ have received a lot of attention in recent years [5]. After separating the sample of $B_s^0 \to J/\psi\phi$ decays into a *CP* even and a *CP* odd component using the transversity analysis, one can use this sample to measure $\Delta\Gamma_s/\Gamma_s$. An almost orthogonal approach entails studying the time-dependent *CP* asymmetry and fitting it with $\mathscr{A}_{CP} D \sin\Delta m_s t$. In the Standard Model, $\mathscr{A}_{CP}$ measures $arg(-V_{ts}V_{tb}^*/V_{cs}V_{cb}^*)$ which is doubly Cabibbo-suppressed and thus very small. Observing $\mathscr{A}_{CP}$ significantly different from zero in this decay mode would be a sign of new physics.

Due to di-muon triggers, this decay chain can be reconstructed fairly easily by both CDF and DØ. DØ has an advantage of a better muon coverage, and it has observed a higher yield: 133 ± 17 candidates in 114 pb^{-1},

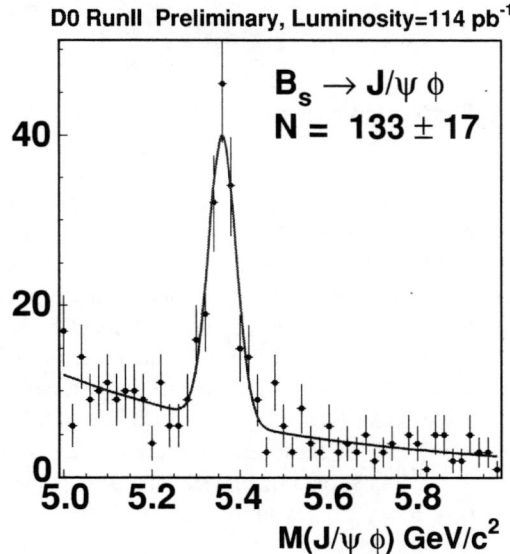

FIGURE 7. Invariant mass of $B_s^0 \to J/\psi\phi$ candidates.

while CDF has 120 ± 13 events in 140 pb^{-1}. The yields trivially scale with luminosity. We expect that both experiments will be able to perform well in this area.

5. DIRECT *CP* VIOLATION

Decays with flavor specific final states can be used for searches for direct *CP* violation. These measurements are easy for the experiments in a hadronic environment since they do not require flavor tagging. Several classes of de-

TABLE 3. Projected yields for the $B \to D\pi$ and $B \to DK$ decay modes for 2 fb^{-1} and 3.5 fb^{-1}.

Decay mode	Yield in 2 fb^{-1}	Yield in 3.5 fb^{-1}
$B^+ \to \bar{D}^0\pi^+, \bar{D}^0 \to K^+\pi^-$	48,000	84,000
$B^+ \to \bar{D}^0 K^+, \bar{D}^0 \to K^+\pi^-$	3990	6980
$B^+ \to \bar{D}^0 K^+, (\bar{D}^0 \to KK + \bar{D}^0 \to \pi\pi)$	520	910
$B_s^0 \to D_s^-\pi^+, D_s^- \to \phi\pi^-$	3200	5600
$B_s^0 \to D_s^- K^+, D_s^- \to \phi\pi^-$	256	448

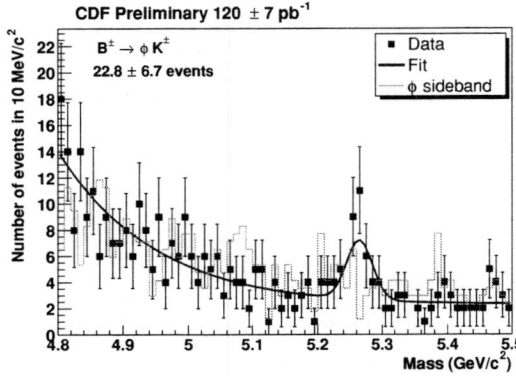

FIGURE 8. Invariant mass of $B^+ \to \phi K^+$ candidates.

cays are of particular interest. Various $B \to \phi X$ decays ($B^+ \to \phi K^+$, $B^0 \to \phi K^{*0}$, $B^+ \to \phi K^{*+}$, $B^+ \to \phi\phi K^+$, and $B_s^0 \to \phi\phi$) are intriguing because of an apparent disagreement between the measurements of $\sin 2\beta$ in $B^0 \to \phi K_s^0$ and $B^0 \to J/\psi K_s^0$ which by now has reached $\sim 2.5\sigma$ level [8, 9]. Furthermore, the angular analyses of of these channels provide several quantities simultaneously sensitive to new physics [10]. Both CDF and DØ will be able to reconstruct all of these decay modes. CDF's signal for $B^+ \to \phi K^+$ is shown in Fig. 8. The CDF trigger upgrades, both those already in progress and the ones being discussed (such as a dedicated ϕ trigger), may increase the yield of all ϕX signatures by up to a factor of two.

Another category of decay modes interesting from the perspective of CP violation are the b baryons. $\Lambda_b \to pK$ and $\Lambda_b \to p\pi$ are expected to exhibit CP violation in the Standard Model. $\Lambda_b \to \phi\Lambda$ is interesting for the reasons similar to $B \to \phi X$ decays [10].

However, many of these decays have not been studied extensively from the theoretical stand-point, and more work will be needed in order to draw conclusions pertinent to physics beyond the Standard Model.

The displaced track trigger is a windfall for charm physics. In the past year, CDF has output several results that have surpassed measurements from experiments like E791, FOCUS, and CLEO that have traditionally dominated the field of charm physics. For example, the mea-

surements of the direct CP asymmetries

$$\mathscr{A}_{CP}(D^0 \to K^+K^-) = (2.0 \pm 1.2 \pm 0.6)\% \quad (3)$$
$$\mathscr{A}_{CP}(D^0 \to \pi^+\pi^-) = (1.0 \pm 1.3 \pm 0.6)\% \quad (4)$$

based on 120 pb^{-1} of CDF data are now the best in the world.

The errors from Eqs. (3) and (4) trivially scale down with the integrated luminosity. For example, for 2 fb^{-1} we expect $\sigma(\mathscr{A}_{CP}(D^0 \to K^+K^-)) = 0.3\%$ and $\sigma(\mathscr{A}_{CP}(D^0 \to \pi^+\pi^-)) = 0.4\%$. The measurement of $\mathscr{A}_{CP}$ in $D^+ \to \pi^+\pi^-\pi^+$ is now in progress at CDF, and the expected error in 2 fb^{-1} is 0.2%.

6. CONCLUSION

CDF and DØ are taking data ready for CP violation studies. The understanding of tracking and of most low-level analysis components (such as dE/dx) is excellent in both detectors. CDF and DØ are focused on CP violation in B_s^0 decays and b baryons, as well as the study of B decay modes with low branching ratios. Direct CP violation measurements will be performed in all flavor-specific final states. Finally, both collaborations will strive to fully exploit the possibilities in the charm sector.

REFERENCES

1. Abachi, S., et al., [DØ Collaboration], The DØ Upgrade: The Detector and its Physics, Tech. Rep. FERMILAB-PUB-96-357-E (1996).
2. Blair, R., et al., [CDF Collaboration], The CDF II Detector: Technical Design Report, Tech. Rep. FERMILAB-PUB-96-390-E (1996).
3. Miao, T., in these proceedings (2004).
4. Fleischer, R., *Phys. Lett.* **B459**, 306–320 (1999).
5. Anikeev, K., et al., arXiv:hep-ph/0201071 (2001).
6. Gronau, M., and Wyler, D., *Phys. Lett.* **B265**, 172–176 (1991).
7. Aleksan, R., Kayser, B., and London, D., arXiv:hep-ph/9312338 (1993).
8. Higuchi, T., in these proceedings (2004).
9. Mahalaxmi, K., in these proceedings (2004).
10. London, D., in these proceedings (2004).

Learning the Weak Phase γ from B Decays

Jonathan L. Rosner

Laboratory for Elementary-Particle Physics, Cornell University, Ithaca, NY 14850, U.S.A.[1]

Abstract. The current status of some methods to determine the weak phase γ of the Cabibbo-Kobayashi-Maskawa (CKM) matrix element V_{ub}^* using B decays is discussed, and comments are made on accuracy achievable in the next few years.

1. INTRODUCTION

The observed CP violation in K and B decays can be interpreted in terms of phases of elements of the Cabibbo-Kobayashi-Maskawa (CKM) matrix. While $\beta = \mathrm{Arg}(V_{td}^*)$ is well-determined from the CP asymmetry in $B^0 \to J/\psi K_S^0$, current information on $\gamma = \mathrm{Arg}(V_{ub}^*)$ is much less precise, with $39° < \gamma < 80°$ at 95% confidence level (CL) [1]. In order to learn γ one must generally separate strong and weak phases from one another in two-body B decays. We describe several areas in which progress in this work has been accomplished, and what improvements lie ahead. Some additional details are noted in an earlier review [2].

In Section 2 we compare the determination of β from $B^0 \to J/\psi K_S^0$ with the more difficult determination of $\alpha = \pi - \beta - \gamma$ from $B^0 \to \pi^+ \pi^-$. We then discuss some uses of information from various decay modes of $B \to K\pi$ in Sec. 3. One obtains useful constraints on γ with some assumptions about SU(3) flavor symmetry from the decays $B \to VP$ (Sec. 4) and $B \to PP$ (Sec. 5), where V and P denote light vector and pseudoscalar mesons. The decays $B \to D_{CP}K$, where D_{CP} denotes a CP eigenstate of a neutral charmed meson, also provide useful constraints (Sec. 6). We summarize in Sec. 7.

2. β FROM $B^0 \to J/\psi K_S^0$ VERSUS α FROM $B^0 \to \pi^+ \pi^-$

The unitarity of the CKM matrix is conveniently expressed in terms of the triangle of Fig. 1. Here, for example, $1 - \bar{\rho} - i\bar{\eta} = -V_{tb}^* V_{td}/V_{cb}^* V_{cd}$. (See Ref. [2] for other definitions.)

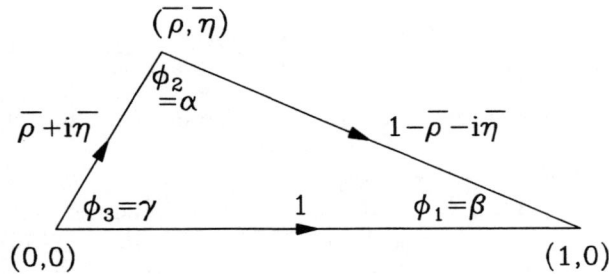

FIGURE 1. The unitarity triangle. Two conventions for its angles are shown.

2.1. $B^0 \to J/\psi K_S^0$

The CP asymmetry in the decay $B^0 \to J/\psi K_S^0$ is simple to analyze because there is only one main subprocess $\bar{b} \to \bar{c} c \bar{s}$. The direct decay (with zero weak phase) interferes with $B^0 \to \bar{B}^0$ mixing (with weak phase $e^{-2i\beta}$). The most recent BaBar and Belle measurements, when averaged, provide $\sin 2\beta = 0.736 \pm 0.049$ [3] without much ambiguity.

2.2. $B^0 \to \pi^+ \pi^-$

Here there are two types of amplitude, "T" (tree) and "P" (penguin), contributing to the decay. (For a discussion of amplitudes within flavor SU(3) see Refs. [4] and [5].) Different weak and strong phases can complicate the analysis. If one had only a tree amplitude, the direct amplitude $A(B^0 \to \pi^+ \pi^-) \sim e^{i\gamma}$ would interfere with $A(B^0 \to \bar{B}^0 \to \pi^+ \pi^-) \sim e^{-2i\beta} e^{-i\gamma}$ to provide a measure of the relative phase $2(\beta + \gamma) = 2\pi - 2\alpha$. So, in the absence of the penguin contribution, one would measure α. One seeks an estimate of $|P/T|$ or observables not requiring this ratio.

---

[1] On leave from Enrico Fermi Institute and Department of Physics, University of Chicago, 5640 S. Ellis Avenue, Chicago, IL 60637.

The neutral B mass eigenstates may be written

$$B_L^0 = p|B^0\rangle + q|\bar{B}^0\rangle \; ; \quad B_H^0 = p|B^0\rangle - q|\bar{B}^0\rangle \; , \quad (1)$$

where

$$q/p = e^{-2i\beta} \; , \quad \lambda \equiv (q/p)(\bar{A}/A) \; , \quad (2)$$

$$A \equiv A(B \to f) \; , \quad \bar{A} \equiv A(\bar{B} \to \bar{f}) \; . \quad (3)$$

Observables in the time-dependence of $\left\{ \begin{matrix} B^0 \\ \bar{B}^0 \end{matrix} \right\}_{t=0} \to$

$\pi^+\pi^-$ (or any other final state) are:

$$\Gamma(t) \sim e^{-\Gamma|t|}[1 \mp S \sin\Delta mt \mp A \cos\Delta mt] \; , \quad t \equiv t_{\text{decay}} - t_{\text{tag}} \; , \quad (4)$$

with $S = 2\text{Im}\lambda/(1 + |\lambda|^2)$ and

$$A = \frac{|\lambda|^2 - 1}{|\lambda|^2 + 1} = A_{CP} = \frac{\Gamma(\bar{B} \to \bar{f}) - \Gamma(B \to f)}{\Gamma(\bar{B} \to \bar{f}) + \Gamma(B \to f)} \; . \quad (5)$$

The experimental data [6, 7] on these asymmetries in $B \to \pi^+\pi^-$ are shown in Table 1.

With no penguin contributions, $S_{\pi\pi} = \sin 2\alpha < 0$ would favor $\alpha > 90°$. With a penguin-to-tree ratio $|P/T| \simeq 0.3$ estimated from $B \to K\pi$ using flavor SU(3) symmetry, one finds instead the parametric dependence of the time-dependent asymmetries on α and a relative strong phase δ [8, 9] as shown in Fig. 2 [10].

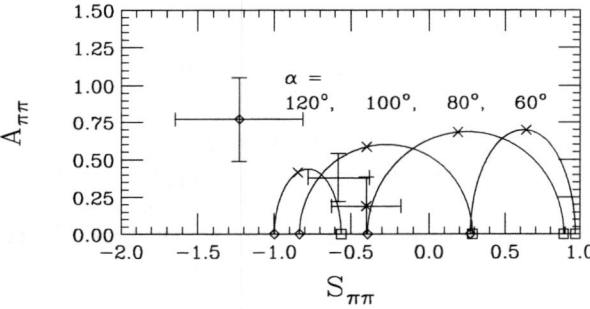

FIGURE 2. Curves describing behavior of $S_{\pi\pi}$ and $A_{\pi\pi}$ as the relative strong phase δ between penguin and tree amplitudes is varied from 0° (diamonds) through 90° (crosses) to 180° (squares). The curves are labeled by values of α. Plotted points: BaBar (cross), Belle (diamond), average (no symbol).

Unless δ is near $\pi/2$, curves for different values of α intersect at the same values of $S_{\pi\pi}$ and $A_{\pi\pi}$. A quantity which is useful in resolving this discrete ambiguity is $R_{\pi\pi} = \frac{\Gamma(B^0 \to \pi^+\pi^-)}{\Gamma(B^0 \to \pi^+\pi^-)_{\text{tree}}}$. In Ref. [11] it was found that $R_{\pi\pi} = 0.87^{+0.11}_{-0.28}$, which slightly favors larger strong phases and hence larger values of α for given $(S_{\pi\pi}, A_{\pi\pi})$. A related analysis has appeared recently in Ref. [12]. Information on $B_s^0 \to K^+K^-$ may be combined with that on $B \to \pi^+\pi^-$ with the help of flavor SU(3) to separate out penguin and tree contributions [13]. The time-dependence of $B_s^0(t) \to K^+K^-$ provides a complementary method [14].

3. INFORMATION FROM $B \to K\pi$

A great deal of information can be obtained from $B \to K\pi$ decay rates averaged over CP, supplemented with measurements of direct CP asymmetries. One probes in this manner tree-penguin interference in various processes. Denoting amplitudes with $|\Delta S| = 1$ by primed quantities, several comparisons can be made:

- $B^0 \to K^+\pi^-$ ($T' + P'$) vs. $B^+ \to K^0\pi^+$ (P') [15–18];
- $B^+ \to K^+\pi^0$ ($T' + P' + C'$) vs. $B^+ \to K^0\pi^+$ (P') [18–21];
- $B^0 \to K^0\pi^0$ vs. other modes [18, 22–26].

The data which are used in these analyses are summarized in Table 2.

In all these comparisons it is helpful to use flavor SU(3) (often only U-spin, i.e., $s \leftrightarrow d$). We give the example of $B^0 \to K^+\pi^-$ in detail. The tree amplitude for this process is $T' \sim V_{us}V_{ub}^*$, with weak phase γ, while the penguin amplitude is $P' \sim V_{ts}V_{tb}^*$ with weak phase π. We denote the penguin-tree relative strong phase by δ and define $r \equiv |T'/P'|$. Then we may write

$$A(B^0 \to K^+\pi^-) = |P'|[1 - re^{i(\gamma+\delta)}] \; , \quad (6)$$

$$A(\bar{B}^0 \to K^-\pi^+) = |P'|[1 - re^{i(-\gamma+\delta)}] \; , \quad (7)$$

$$A(B^+ \to K^0\pi^+) = A(B^- \to \bar{K}^0\pi^-) = -|P'| \; , \quad (8)$$

where the last two amplitudes are expected to be equal in the approximation that small annihilation amplitudes are neglected. A test for this assumption is the absence of a CP asymmetry in $B^+ \to K^0\pi^+$ (or in $B^+ \to \bar{K}^0K^+$, where it would be bigger [27]).

One now forms the ratio

$$R \equiv \frac{\Gamma(B^0 \to K^+\pi^-) + \Gamma(\bar{B}^0 \to K^-\pi^+)}{2\Gamma(B^+ \to K^0\pi^+)}$$

$$= 1 - 2r\cos\gamma\cos\delta + r^2 \; . \quad (9)$$

Fleischer and Mannel [15] pointed out that $R \geq \sin^2\gamma$ for any r, δ so if $1 > R$ one can get a useful bound. However, if one uses

$$RA_{CP} = -2r\sin\gamma\sin\delta \quad (10)$$

as well and eliminates δ one can get a more powerful constraint, illustrated in Fig. 3.

At the 1σ level, $R < 1$, leading to an upper bound $\gamma < 80°$ which happens to coincide with that of Ref. [15]. We have used $R = 0.898 \pm 0.071$ and $A_{CP} = -0.095 \pm 0.029$ based on recent averages [7] of CLEO, BaBar, and Belle data, and $r = |T'/P'| = 0.142^{+0.024}_{-0.012}$. The most conservative bound arises for the smallest $|A_{CP}|$ and largest r. The allowed region lies between the curves $A_{CP} = 0$ and $|A_{CP}| = 0.124$ (1σ). In order to estimate the tree amplitude and $r = |T'/P'|$ we have used factorization

TABLE 1. Time-dependent asymmetries in $B^0 \to \pi^+\pi^-$.

Observable	BaBar	Belle	Average
$S_{\pi\pi}$	$-0.40\pm0.22\pm0.33$	$-1.23\pm0.41^{+0.08}_{-0.07}$	-0.58 ± 0.20
$A_{\pi\pi}$	$0.19\pm0.19\pm0.05$	$0.77\pm0.27\pm0.08$	0.38 ± 0.16

TABLE 2. Branching ratios and CP asymmetries for $B \to K\pi$ decays [7].

Decay mode	Amplitude	$\mathscr{B}$ (units of 10^{-6})	A_{CP}
$B^+ \to K^0\pi^+$	P'	21.78 ± 1.40	0.016 ± 0.057
$B^+ \to K^+\pi^0$	$-(P'+C'+T')/\sqrt{2}$	12.53 ± 1.04	0.00 ± 0.12
$B^0 \to K^+\pi^-$	$-(T'+P')$	18.16 ± 0.79	-0.095 ± 0.029
$B^0 \to K^0\pi^0$	$(P'-C')/\sqrt{2}$	11.68 ± 1.42	0.03 ± 0.37

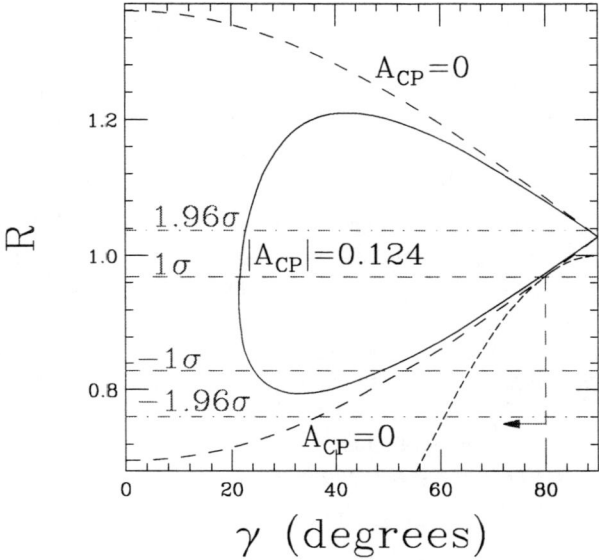

$$\gamma \ (\text{degrees})$$

FIGURE 3. Behavior of R for $r=0.166$ and $A_{CP}=0$ (dashed curves) or $|A_{CP}|=0.124$ (solid curve) as a function of the weak phase γ. Horizontal dashed lines denote $\pm1\sigma$ experimental limits on R, while dot-dashed lines denote 95% CL ($\pm1.96\sigma$) limits. The short-dashed curve denotes the Fleischer-Mannel bound $\sin^2\gamma \leq R$.

in $B^+ \to \pi^-\ell^+\nu_\ell$ at low q^2 [11] and $\left|\frac{T'}{T}\right| = \frac{f_K}{f_\pi}\left|\frac{V_{us}}{V_{ud}}\right| \simeq$ $(1.22)(0.23) = 0.28$. One could use processes in which T dominates, such as $B^0 \to \pi^+\pi^-$ or $B^+ \to \pi^+\pi^0$, but these are contaminated by contributions from P and C, respectively.

In such an approach one always must question the validity of SU(3) flavor symmetry. SU(3) breaking is taken into account in the ratio of tree amplitudes, but no breaking is taken in other amplitudes, since we do not assume factorization for C or P and therefore cannot account for the breaking merely via ratios of decay constants. We have assumed the same relative tree-penguin strong phases for $|\Delta S|=1$ and $\Delta S=0$ amplitudes. Tests

of these assumptions will be available once one observes penguin-dominated $B \to K\bar{K}$ decays and charmless B^0_s decays; there are also numerous relations implied between CP-violating rate differences [28, 29].

The process $B^+ \to K^+\pi^0$ also provides constraints on γ. The deviation of the ratio

$$R_c \equiv \frac{\Gamma(B^+ \to K^+\pi^0) + \Gamma(B^- \to K^-\pi^0)}{\Gamma(B^+ \to K^0\pi^+)} = 1.15\pm0.12 \tag{11}$$

from 1, when combined with $A_{CP} = 0.00\pm0.12$, $r_c = |(T'+C')/P'| = 0.195\pm0.016$ and an estimate of the electroweak penguin (EWP) $\delta_{EW} \equiv |P'_{EW}|/|T'+C'| = 0.65\pm0.15$, leads to a 1σ lower bound $\gamma > 40°$. Details may be found in Refs. [2, 18–21]. The most conservative bound arises for smallest A_{CP}, largest r_c, and largest $|P'_{EW}|$, and is shown in Fig. 4.

Another ratio

$$R_n \equiv \frac{\bar{\Gamma}(B^0 \to K^+\pi^-)}{2\bar{\Gamma}(B^0 \to K^0\pi^0)}$$

$$= \left|\frac{p'+t'}{p'-c'}\right|^2 = 0.78\pm0.10 \tag{12}$$

involves the decay $B^0 \to K^0\pi^0$. Here the bar denotes CP-averaged decay widths, while small letters denote amplitudes which include EWP contributions. This ratio should be same to leading order in $|t'/p'|$ and $|c'/p'|$ as

$$R_c = \left|\frac{p'+t'+c'}{p'}\right|^2, \tag{13}$$

but the two ratios differ by 2.4σ. Possibilities (see, e.g., Refs. [18, 30]) include (1) new physics, e.g., in the EWP amplitude, and (2) an underestimate of the π^0 detection efficiency in all experiments, leading to an overestimate of any branching ratio involving a π^0. The latter possibility can be taken into account by considering the ratio $(R_n R_c)^{1/2} = 0.96\pm0.08$, in which the π^0 efficiency cancels. As shown in Fig. 5, this ratio leads only to the conservative bound $\gamma \leq 88°$.

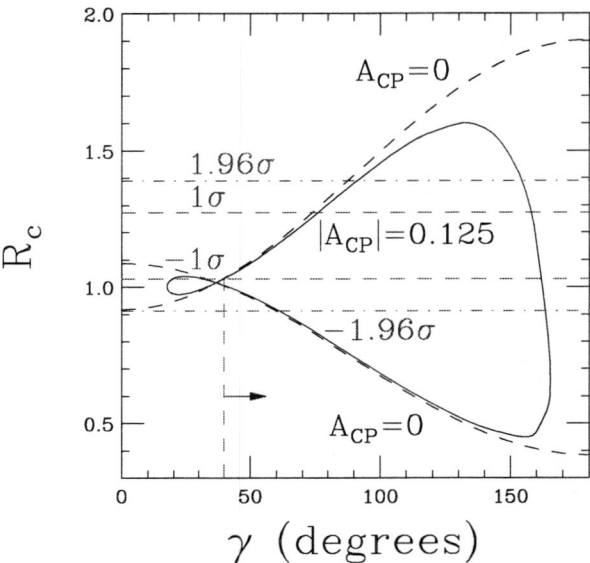

FIGURE 4. Behavior of R_c for $r_c = 0.21$ (1σ upper limit) and $A_{CP}(K^+\pi^0) = 0$ (dashed curves) or $|A_{CP}(K^+\pi^0)| = 0.125$ (solid curve) as a function of the weak phase γ. Horizontal dashed lines denote $\pm 1\sigma$ experimental limits on R_c, while dot-dashed lines denote 95% CL ($\pm 1.96\sigma$) limits. We have taken $\delta_{EW} = 0.80$ (its 1σ upper limit), which leads to the most conservative bound on γ.

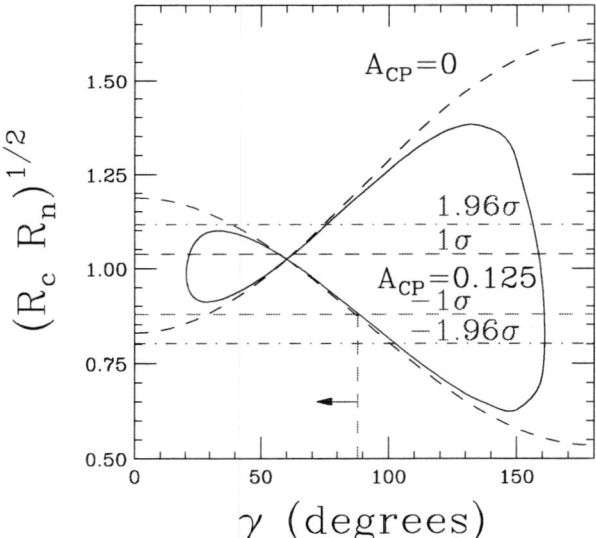

FIGURE 5. Behavior of $(R_c R_n)^{1/2}$ for $r_c = 0.18$ (1σ lower limit) and $A_{CP}(K^+\pi^0) = 0$ (dashed curves) or $|A_{CP}(K^+\pi^0)| = 0.125$ (solid curve) as a function of the weak phase γ. Horizontal dashed lines denote $\pm 1\sigma$ experimental limits on $(R_c R_n)^{1/2}$, while dot-dashed lines denote 95% CL ($\pm 1.96\sigma$) limits. Upper branches of curves correspond to $\cos\delta_c(\cos\gamma - \delta_{EW}) < 0$, while lower branches correspond to $\cos\delta_c(\cos\gamma - \delta_{EW}) > 0$. Here we have taken $\delta_{EW} = 0.50$ (its 1σ lower limit), which leads to the most conservative bound on γ.

4. INFORMATION FROM $B \to VP$

Although the decays $B \to VP$ are characterized by more amplitudes than $B \to PP$ (since the final particles do not belong to the same flavor-symmetry multiplet), data have become so abundant that useful global fits can be performed [31, 32]. We label amplitudes by the meson (pseudoscalar P or vector V) containing the spectator quark. Some features of the fit of Ref. [32] are:

- $|t_P/t| \simeq f_\rho/f_\pi$, where t is the tree amplitude in $B \to PP$ decays, as would be expected for a weak current producing a charged meson.
- Penguin amplitudes satisfy $p'_V \simeq -p'_P$, as proposed long ago by Lipkin [33–35].
- Small CP asymmetries in many processes imply small strong phases.
- The time-dependent asymmetries in $B \to \rho\pi$ are crucial in resolving discrete ambiguities, as has also been found in the QCD factorization approach of Refs. [24–26].

Three local χ^2 minima are found: $\gamma = (26 \pm 5)°$, $(63 \pm 6)°$ (a range compatible with fits [1] to CKM parameters), and $(162^{+5}_{-6})°$ (incompatible with $\beta \simeq 24°$ since $\alpha + \beta + \gamma = \pi$). At 95% CL the solution compatible with CKM fits implies $51° \leq \gamma \leq 73°$ and small strong phases in accord with the expectations of QCD factorization [26]. Some predictions for as-yet-unseen decay modes are shown in Table 3.

In this fit the free parameters were:

- $p'_{P,V}$ (penguin amplitudes); their relative phase ϕ;
- $t_{P,V}$ (tree amplitudes); their strong phases $\delta_{P,V}$ with respect to $p_{P,V}$;
- Color-suppressed $c_{P,V}$ taken real with respect to $t_{P,V}$;
- Electroweak penguins $P'_{EW(P,V)}$ taken real with respect to $p'_{P,V}$;
- The weak phase γ.

One thus has 12 parameters (11 if p'_V/p'_P is assumed to be real, and 10 if we assume $p'_V/p'_P = -1$) to fit 34 data points. The resulting dependence of $\chi^2_{\min}$ on γ is shown in Fig. 6.

The relative phases of amplitudes are specified by CP-averaged decay rates as well as by CP asymmetries. They are illustrated in Fig. 7.

The p'_P amplitude in this diagram is taken to be real and positive. The weak phases of $t_{V,P}$ and $\bar{t}_{V,P}$ are included. There is a small relative phase between t_V and t_P, as expected in QCD factorization [26]. The relative phases of p'_V and p'_P are such that they contribute constructively to $B \to K^*\eta$, as anticipated by Lipkin [33–35]. From these phases one expects constructive tree-

TABLE 3. Predictions of the favored fit of Ref. [32] for some as-yet-unseen $B \to VP$ decays.

As yet unseen decay mode	Predicted $\mathscr{B}$ Units of 10^{-6}	Present limit	Comments
$B^+ \to \bar{K}^{*0} K^+$	0.50 ± 0.05	< 5.3	Pure p_P
$B^+ \to K^{*+} \pi^0$	$15.0^{+3.3}_{-2.8}$	< 31	EWP enhancement
$B^+ \to \rho^+ K^0$	12.6 ± 1.6	< 48	Pure p'_V
$B^0 \to \rho^0 K^0$	$7.2^{+2.1}_{-1.9}$	< 12.4	EWP enhancement

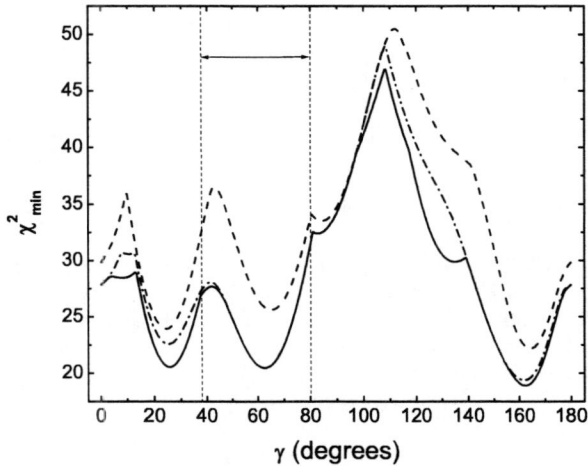

FIGURE 6. $(\chi^2)_{\min}$, obtained by minimizing over all remaining fit parameters, as a function of the weak phase γ. Dashed curve: $p'_V/p'_P = -1$ (24 d.o.f.); dash-dotted curve: p'_V/p'_P real (23 d.o.f.); solid curve: p'_V/p'_P complex (22 d.o.f.). Vertical dashed lines show the limits $39° \leq \gamma \leq 80°$ from fits [1] to CKM parameters.

penguin interference in the CP-averaged rate for $B^0 \to K^{*+} \pi^-$ and destructive interference in $B^0 \to K^+ \rho^-$.

5. $B \to PP$ DECAYS WITH η AND η'

A global fit to $B \to PP$ decays under the same assumptions as the fit to $B \to VP$ decays is still in progress [7]. However, $B \to PP$ decays with η and η' have been analyzed using flavor symmetry in Ref. [36]. It is found that the large CP asymmetry in $B^+ \to \pi^+ \eta$ reported by the BaBar Collaboration [37] implies a comparable A_{CP} in $B^+ \to \pi^+ \eta'$. We have

$$A_{CP}(\pi^+ \eta) = -\frac{0.91 \sin \alpha \sin \delta}{1 - 0.91 \cos \alpha \cos \delta} = -0.51 \pm 0.19 \ , \tag{14}$$

$$\begin{aligned} \bar{\mathscr{B}}(\pi^+ \eta) &= 4.95 \times 10^{-6} (1 - 0.91 \cos \alpha \cos \delta) \\ &= (4.12 \pm 0.85) \times 10^{-6} \ , \end{aligned} \tag{15}$$

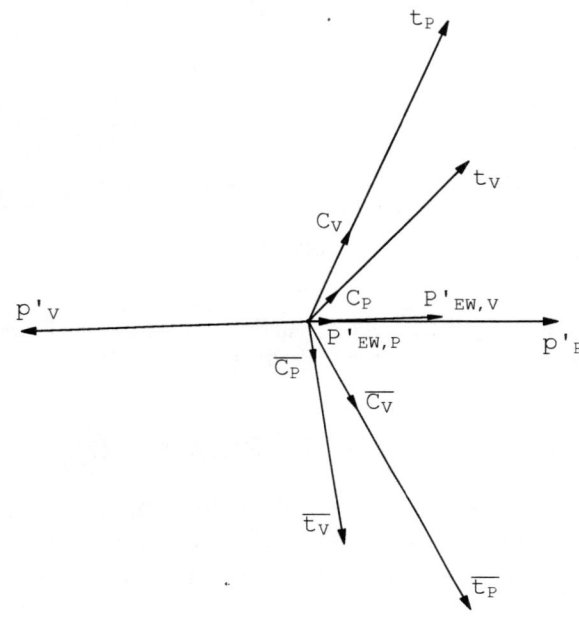

FIGURE 7. Magnitudes and phases of dominant invariant amplitudes in solution with $\gamma \simeq 63°$ and complex p'_V/p'_P [32].

leading to the predictions

$$A_{CP}(\pi^+ \eta') = -\frac{\sin \alpha \sin \delta}{1 - \cos \alpha \cos \delta} \simeq -0.57 \ , \tag{16}$$

$$\begin{aligned} \bar{\mathscr{B}}(\pi^+ \eta') &= 3.35 \times 10^{-6} (1 - \cos \alpha \cos \delta) \\ &\simeq 2.7 \times 10^{-6} \ . \end{aligned} \tag{17}$$

Tree and penguin amplitudes are of comparable magnitude in both these processes, leading to the possibility of large CP asymmetries which appears to be realized in the data. The scatter of predictions for $B^+ \to \pi^+ \eta'$ is shown in Fig. 8. The central values are based on $(\alpha, \delta) \simeq (78°, 28°)$ with the discrete ambiguities $(\alpha \leftrightarrow \delta)$ or $\alpha \to \pi - \alpha$, $\delta \to \pi - \delta$.

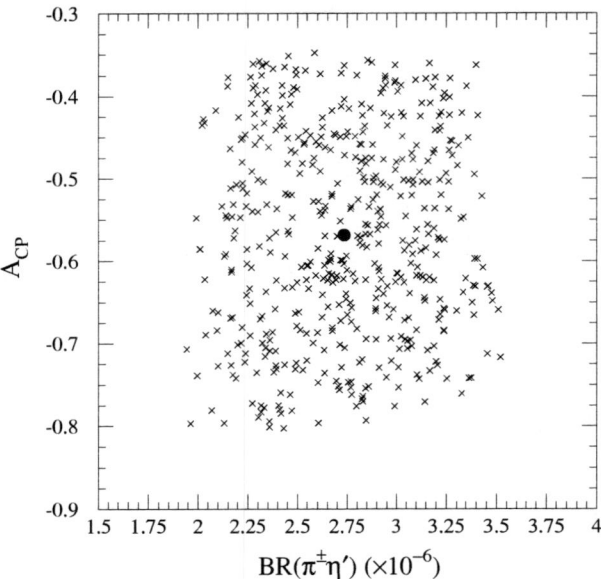

FIGURE 8. Predicted values of the averaged branching ratio and direct CP asymmetry for the decays $B^{\pm} \to \pi^{\pm}\eta'$.

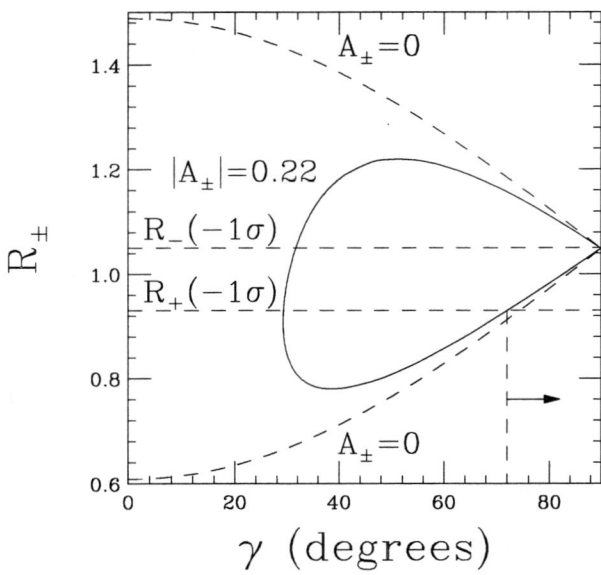

FIGURE 9. Behavior of $R_{\pm}$ for $A_{\pm} = 0$ (dashed curves) or $|A_{\pm}| = 0.22$ (solid curve) as a function of the weak phase γ. Horizontal dashed lines denote -1σ experimental limits on $R_{\pm}$. We have taken parameters (including $r = 0.22$) which lead to the most conservative bound on γ.

6. CONSTRAINTS FROM $B \to D_{\text{CP}}K$

We present the discussion of M. Gronau [10]. One wishes to compare $B^{\pm} \to D_{CP}K^{\pm}$ with the CKM-favored process $B^- \to D^0K^-$, thereby probing the quark subprocesses

$$\frac{A(\bar{b} \to \bar{u}c\bar{s})}{A(\bar{b} \to \bar{c}u\bar{s})} = re^{i(\gamma+\delta)}, \quad \frac{A(b \to u\bar{c}s)}{A(b \to c\bar{u}s)} = re^{i(-\gamma+\delta)}.$$

$$(18)$$

One then considers the ratios

$$R_{\pm} \equiv \frac{\Gamma(D^0_{CP=\pm}K^-) + \Gamma(D^0_{CP=\pm}K^+)}{\Gamma(D^0K^-)}$$

$$= 1 + r^2 \pm 2r\cos\gamma\cos\delta \quad (19)$$

and the CP asymmetries

$$A_{\pm} \equiv \frac{\Gamma(D^0_{CP=\pm}K^-) - \Gamma(D^0_{CP=\pm}K^+)}{\Gamma(D^0_{CP=\pm}K^-) + \Gamma(D^0_{CP=\pm}K^+)}$$

$$= \pm 2r\sin\gamma\sin\delta/R_{\pm}. \quad (20)$$

The relevant data are summarized in Table 4. The ratio $\Gamma(B^- \to D^0K^-)/\Gamma(B^- \to D^0\pi^-)$ was evaluated [10] by taking the average of CLEO [38], Belle [39], and BaBar [40] values.

We take $(R_+ + R_-)/2 = 1 + r^2$ so $r \geq 0.22$ at 1σ. The average $|A_{\pm}|$ is $0.11 \pm 0.11 \leq 0.22$ at 1σ; the most conservative bound is obtained for smallest r and largest $|A_{\pm}|$ and at 1σ is $\gamma \geq 72°$, as shown in Fig. 9. Note that one R must be below $1 + r^2$ while the other must be above it.

7. SUMMARY

A number of promising bounds on γ stemming from various B decays have been mentioned. So far all are statistics-limited. At 1σ we have found

- R ($K^+\pi^-$ vs. $K^0\pi^+$) gives $\gamma \lesssim 80°$;
- R_c ($K^+\pi^0$ vs. $K^0\pi^+$) gives $\gamma \gtrsim 40°$;
- R_n ($K^+\pi^-$ vs. $K^0\pi^0$) should equal R_c; $(R_cR_n)^{1/2}$ gives $\gamma \lesssim 88°$;
- $B \to D_{CP}K$ decays give $\gamma \gtrsim 72°$.

A flavor-SU(3) analysis of $B \to VP$ decays favors $\gamma = (63 \pm 6)°$, or $51° \leq \gamma \leq 73°$ at 95% CL. Several as-yet-unseen decay modes are predicted, such as $B^+ \to \rho^+K^0$ and $B^+ \to K^{*+}\pi^0$. SU(3) relations among rate differences remain to be tested. We predict $2.0 \leq \mathscr{B}(\pi^+\eta')/10^{-6} \leq 3.5$, $-0.34 \geq A_{CP}(\pi^+\eta') \geq -0.80$. A global $B \to PP$ analysis, still in progress, is complicated by possible $B \to K\pi$ inconsistencies or new physics in (e.g.) $B^0 \to K^0\pi^0$.

The future of most such γ determinations remains for now in experimentalists' hands, as one can see from Figs. 3-5 and 9. Uncertainties in SU(3) breaking are probably already the limiting factor on the error in γ from Fig. 6, and better estimates will require flavor SU(3) tests at levels of $\mathscr{B} \simeq 1/2 \times 10^{-6}$. We have noted (see, e.g., [16]) that measurements of rate ratios in $B \to K\pi$ can ultimately pinpoint γ to within about $10°$. The required accuracies in R, R_c, and R_n to achieve this goal can be es-

TABLE 4. Ratios $R_\pm$ and CP asymmetries $A_\pm$ for $B \to D_{CP}K$ decays.

	R_+	A_+	R_-	A_-
Belle [39]	1.12 ± 0.24	0.06 ± 0.19	1.30 ± 0.25	-0.19 ± 0.18
BaBar [40]	1.06 ± 0.21	0.07 ± 0.18		
Average	1.09 ± 0.16	0.065 ± 0.132	1.30 ± 0.25	-0.19 ± 0.18

timated from Figs. 3-5. For example, knowing $(R_c R_n)^{1/2}$ to within 0.05 would pin down γ to within $10°$ if this ratio lies in the most sensitive range of Fig. 5.

A complementary approach to the flavor-SU(3) method is the QCD factorization formalism of Refs. [24–26]. It predicts small strong phases (as found in our analysis) and deals directly with flavor-SU(3) breaking; however, it involves some unknown form factors and meson wave functions and appears to underestimate the magnitude of $B \to VP$ penguin amplitudes.

ACKNOWLEDGMENTS

I thank Cheng-Wei Chiang, Michael Gronau, Zumin Luo, Matthias Neubert, and Denis Suprun for enjoyable collaborations, and Maury Tigner for extending the hospitality of the Laboratory for Elementary-Particle Physics. This work was supported in part by the United States Department of Energy through Grant No. DE FG02 90ER40560 and by the John Simon Guggenheim Memorial Foundation.

REFERENCES

1. Hocker, A., Lacker, H., Laplace, S., and Le Diberder, F., *Eur. Phys. J.* **C21**, 225–259 (2001).
2. Rosner, J. L., *AIP Conf. Proc.* **689**, 150–174 (2003).
3. Browder, T. E., *Int. J. Mod. Phys.* **A19**, 965–974 (2004).
4. Gronau, M., Hernandez, O. F., London, D., and Rosner, J. L., *Phys. Rev.* **D50**, 4529–4543 (1994).
5. Gronau, M., Hernandez, O. F., London, D., and Rosner, J. L., *Phys. Rev.* **D52**, 6374–6382 (1995).
6. Jawahery, H. (Presented at Lepton-Photon 2003 Symposium).
7. Chiang, C.-W., Gronau, M., Rosner, J. L., and Suprun, D. A., arXiv:hep-ph/0404073 (2004), see also Heavy Flavor Averaging Group, Lepton-Photon 2003 branching ratios and CP asymmetries, at URL http://www.slac.stanford.edu/xorg/hfag/rare/.
8. Gronau, M., and Rosner, J. L., *Phys. Rev.* **D65**, 093012 (2002).
9. Gronau, M., and Rosner, J. L., *Phys. Rev.* **D66**, 053003 (2002).
10. Gronau, M., arXiv:hep-ph/0306308 (2003), invited talk at Flavor Physics and *CP* Violation (FPCP 2003), Paris, France, May 2003.
11. Luo, Z., and Rosner, J. L., *Phys. Rev.* **D68**, 074010 (2003).
12. Buchalla, G., and Safir, A. S., arXiv:hep-ph/0310218 (2003).
13. Fleischer, R., *Phys. Lett.* **B459**, 306–320 (1999).
14. Gronau, M., and Rosner, J. L., *Phys. Rev.* **D65**, 113008 (2002).
15. Fleischer, R., and Mannel, T., *Phys. Rev.* **D57**, 2752–2759 (1998).
16. Gronau, M., and Rosner, J. L., *Phys. Rev.* **D57**, 6843–6850 (1998).
17. Gronau, M., and Rosner, J. L., *Phys. Rev.* **D65**, 013004 (2002).
18. Gronau, M., and Rosner, J. L., *Phys. Lett.* **B572**, 43–49 (2003).
19. Neubert, M., and Rosner, J. L., *Phys. Lett.* **B441**, 403–409 (1998).
20. Neubert, M., and Rosner, J. L., *Phys. Rev. Lett.* **81**, 5076–5079 (1998).
21. Neubert, M., *JHEP* **02**, 014 (1999).
22. Buras, A. J., and Fleischer, R., *Eur. Phys. J.* **C11**, 93–109 (1999).
23. Buras, A. J., and Fleischer, R., *Eur. Phys. J.* **C16**, 97–104 (2000).
24. Beneke, M., Buchalla, G., Neubert, M., and Sachrajda, C. T., *Nucl. Phys.* **B606**, 245–321 (2001).
25. Beneke, M., and Neubert, M., *Nucl. Phys.* **B651**, 225–248 (2003).
26. Beneke, M., and Neubert, M., *Nucl. Phys.* **B675**, 333–415 (2003).
27. Falk, A. F., Kagan, A. L., Nir, Y., and Petrov, A. A., *Phys. Rev.* **D57**, 4290–4300 (1998).
28. Deshpande, N. G., and He, X.-G., *Phys. Rev. Lett.* **75**, 1703–1706 (1995).
29. Gronau, M., *Phys. Lett.* **B492**, 297–302 (2000).
30. Grossman, Y., *Int. J. Mod. Phys.* **A19**, 907–917 (2004).
31. Aleksan, R., Giraud, P. F., Morenas, V., Pene, O., and Safir, A. S., *Phys. Rev.* **D67**, 094019 (2003).
32. Chiang, C.-W., Gronau, M., Luo, Z., Rosner, J. L., and Suprun, D. A., *Phys. Rev.* **D69**, 034001 (2004).
33. Lipkin, H. J., *Phys. Lett.* **B254**, 247–252 (1991).
34. Lipkin, H. J., *Phys. Lett.* **B415**, 186–192 (1997).
35. Lipkin, H. J., *Phys. Lett.* **B433**, 117–124 (1998).
36. Chiang, C.-W., Gronau, M., and Rosner, J. L., *Phys. Rev.* **D68**, 074012 (2003).
37. Aubert, B., et al., [BaBar Collaboration], arXiv:hep-ex/0303039 (2003).
38. Bornheim, A., et al., [CLEO Collaboration], *Phys. Rev.* **D68**, 052002 (2003).
39. Swain, S. K., et al., [Belle Collaboration], *Phys. Rev.* **D68**, 051101 (2003).
40. Aubert, B., et al., [BaBar Collaboration], *Phys. Rev. Lett.* **92**, 202002 (2004).

Towards $\alpha\,(\phi_2)$ and $\gamma\,(\phi_3)$ at the B Factories

Alan Watson

(Representing the BaBar Collaboration)

School of Physics and Astronomy, The University of Birmingham, Birmingham B15 2TT, U.K.

Abstract. The main aim of the BaBar and Belle experiments is to test the Standard Model explanation of CP violation in the B system through measurements of the angles of the Unitarity Triangle. One angle (β/ϕ_1) is already well measured by the two collaborations. This paper describes some of the recent results from the two experiments from channels which are sensitive to the remaining angles.

1. INTRODUCTION

The primary objective of the BaBar and Belle experiments is to test whether the Standard Model's single source of CP violation, the single non-trivial phase of the CKM matrix, can provide a full description of CP violation in B meson decays. This proposition is to be tested by over-constraining the so-called "Unitarity Triangle" through multiple measurements of all three of its angles and by requiring both, that measurements made using different processes are consistent with each other and that the three angles sum to 180°.

The two collaborations use different naming conventions for these angles. The angles BaBar calls α, β and γ are, respectively, called ϕ_2, ϕ_1 and ϕ_3 by Belle. In this paper the author will use the BaBar convention.

Indirect constraints on the Unitarity Triangle may be obtained from measurements of B^0 and B_s^0 mixing, $|V_{ub}/V_{cb}|$, and CP violation in K^0 decays. A global fit to these results [1] produces a single solution and the 95% confidence limits [2]:

$$15.5° \;<\; \beta \;<\; 26.4°,$$
$$77° \;<\; \alpha \;<\; 120°,$$
$$39° \;<\; \gamma \;<\; 80°.$$

The angle β has been well measured by both collaborations, with an average value of $\sin(2\beta) = 0.734 \pm 0.055$ [3] in good agreement with the Standard Model fit. The remaining angles pose much greater experimental challenges. In this paper we describe the status of some of the measurements which may be used to constrain these angles.

2. THE EXPERIMENTS

The BaBar [4] and Belle [5] detectors both have tracking systems comprising silicon vertex trackers and drift chambers within a 1.5 T solenoidal field. Also within their superconducting coils are particle identification systems and electromagnetic calorimeters comprised of CsI(Tl) crystals, while outside the coils the return yokes of both experiments are instrumented to detect muons and neutral hadrons. For the analyses described here the particle identification systems are particularly important. Both experiments use energy loss (dE/dx) within the tracking detectors. Surrounding the BaBar drift chamber is an internally reflecting ring-imaging Cerenkov detector, while Belle uses an array of aerogel threshold Cerenkov detectors and a barrel-like arrangement of time-of-flight scintillators. Both systems are capable of providing 2.5 σ separation between pions and kaons up to the maximum momentum of charged hadrons from two-body B decays (around 4 GeV/c).

The BaBar and Belle experiments are sited at the PEP-II [6] and KEK-B [7] asymmetric e^+e^- colliders. In these machines the inequality in the energies of the e^- beam and the e^+ beam (9 GeV vs. 3.1 GeV in PEP-II, 8 GeV vs. 3.5 GeV in KEK-B) provides a boost to the $\Upsilon(4S)$ meson which allows the decay time difference between the two B mesons to be measured. Measurements of CP-violating asymmetries in neutral B decays are made as a function of the decay time difference between the decay of interest and the other B decay, whose characteristics (charge vs. momentum of tracks, presence of high-momentum charged kaons or leptons) are used to identify the charge of the b quark or antiquark at the time of its decay ("tagging"), and hence the type ($B^0/\bar{B}^0$) of the CP-measuring B meson at that time.

CP722, *B Physics at Hadron Machines*, edited by M. Paulini and S. Erhan

3. MEASUREMENT OF α/ϕ_2

3.1. Principles

The CP angle α can in principle be measured from $b \to u\bar{u}d$ decays, such as $B^0 \to \pi^+\pi^-$. In the absence of Penguin contributions, interference between the direct decay and mixing followed by decay to the same final state will produce a time-dependent asymmetry:

$$A_{\pi\pi}(B^0/\bar{B}^0) = S_{\pi\pi}\sin(\Delta m_d \Delta t) - C_{\pi\pi}\cos(\Delta m_d \Delta t), \quad (1)$$

where Δm_d is the mass difference between the two B^0 mass eigenstates and Δt the time difference between the $\bar{B}^0$ and B^0 decays. The coefficients of the sine and cosine terms are given by:

$$S_{\pi\pi} = \frac{2\mathrm{Im}\lambda_{\pi\pi}}{1+|\lambda_{\pi\pi}|^2} \quad \text{and} \quad C_{\pi\pi} = \frac{1-|\lambda_{\pi\pi}|^2}{1+|\lambda_{\pi\pi}|^2}. \quad (2)$$

If only the $b \to u\bar{u}d$ amplitude contributes to this decay $|\lambda| = 1$, giving $C_{\pi\pi} = 0$ and $S_{\pi\pi} = \sin(2\alpha)$. In the presence of other amplitudes (*i.e.* penguins), $C_{\pi\pi}$ will in general differ from 0 (direct CP violation), while $S_{\pi\pi}$ will no longer measure $\sin(2\alpha)$ but rather an "effective alpha" (α_{eff}) which differs from the true α by an unknown amount $\Delta\alpha$. Information from other B decay modes is then needed in order to constrain or measure the penguin-induced shift $\Delta\alpha$.

3.2. $B \to \pi\pi$ Decays

Both collaborations study branching fractions and asymmetries using similar sized data samples: 81 fb^{-1} for BaBar and 78 fb^{-1} for Belle. Larger data samples of 113 fb^{-1} and 140 fb^{-1}, respectively, are used to search for $B^0/\bar{B}^0 \to \pi^0\pi^0$.

The main background to charmless B decay measurements comes from combinatorics in light-flavoured (u, d, s, c) events, which make up the majority of hadronic events even at the $\Upsilon(4S)$ resonance. To suppress these, both collaborations use similar methods. Two discriminating variables are used to select candidates with the center-of-mass energy and momentum expected for B mesons. The first, $\Delta E = E^*_{beam} - E^*_B$ compares the expected and measured center-of-mass energies for the B candidate, and should peak at zero. The position of the peak is shifted if the wrong mass hypothesis is assumed for any of the particles (e.g. if a kaon is treated as a pion), and so it can discriminate between $\pi\pi$ and πK final states. The second, $m_{ES} = \sqrt{E^{*2}_{beam} - p^{*2}_B}$, should peak at the B mass, but is sensitive only to the measured momenta. Light flavour continuum events give smooth

TABLE 1. Measured $B \to \pi\pi$ branching fractions ($\times 10^{-6}$).

Mode	BaBar	Belle	Average
$\pi^+\pi^-$	$4.7 \pm 0.6 \pm 0.2$	$4.4 \pm 0.6 \pm 0.3$	4.6 ± 0.4
$\pi^\pm\pi^0$	$5.5 \,^{+1.0}_{-0.9} \pm 0.6$	$5.0 \pm 1.2 \pm 0.5$	5.3 ± 0.6
$\pi^0\pi^0$	$2.1 \pm 0.6 \pm 0.3$	$1.7 \pm 0.6 \pm 0.3$	1.9 ± 0.5

TABLE 2. Results of time-dependent $B^0 \to \pi^+\pi^-$ asymmetry fits.

	$C_{\pi\pi}$	$S_{\pi\pi}$
BaBar	$-0.19 \pm 0.19 \pm 0.05$	$-0.40 \pm 0.22 \pm 0.03$
Belle	$-0.77 \pm 0.27 \pm 0.08$	$-1.23 \pm 0.41 \,^{+0.08}_{-0.07}$
Average*	-0.38 ± 0.16	-0.58 ± 0.20

* Heavy Flavour Averaging Group [14]. The averaging has a confidence level of only 0.047

distributions in both variables with no peaking in the signal region. In addition, event shape variables are used to discriminate between the "spherical" $B\bar{B}$ events and the more "jet-like" continuum.

The precise methods do differ between the collaborations. BaBar makes loose cuts on these variables to produce a signal-enhanced sample, then performs a multi-variant fit to m_{ES}, ΔE and the event shape variables to estimate the signal yield. The Cerenkov angle θ_c is included in the fit, so that yields for $B \to \pi\pi$ and $B \to K\pi$ are estimated simultaneously. Belle instead makes tighter cuts on event shape, m_{ES} and particle identification likelihoods to obtain a higher purity sample of a single final state, then fits the ΔE distribution to estimate signal yield. The branching fractions measured by the two collaborations are summarized in Table 1 [8–11].

The CP asymmetry amplitudes are estimated by unbinned maximum likelihood fits to the Δt distributions for the B^0- and $\bar{B}^0$-tagged events, as well as to m_{ES}, ΔE and, in BaBar's analysis, event shape and θ_c distributions. Extensive Monte Carlo tests were performed to validate the fitting procedures. The results are summarized in Table 2 [12, 13]. In addition both experiments have found charge asymmetries in $B^\pm \to \pi^\pm\pi^0$ decays to be consistent with zero.

Interpreting the observed asymmetries is more challenging. Grossman and Quinn [15] proposed a simple limit on the magnitude of the effect to be determined from the branching fraction ratio

$$\sin^2(\alpha_{eff} - \alpha) < \frac{\mathscr{B}(B^0/\bar{B}^0 \to \pi^0\pi^0)}{\mathscr{B}(B^\pm \to \pi^\pm\pi^0)}. \quad (3)$$

Unfortunately, the relatively large branching fraction for $B \to \pi^0\pi^0$ means that this only yields $|\Delta\alpha| < 48°$ at 90% CL. Isospin analysis [16] as yet yields no significant constraints, and will require $\mathscr{O}(\text{ab}^{-1})$ of data to achieve a

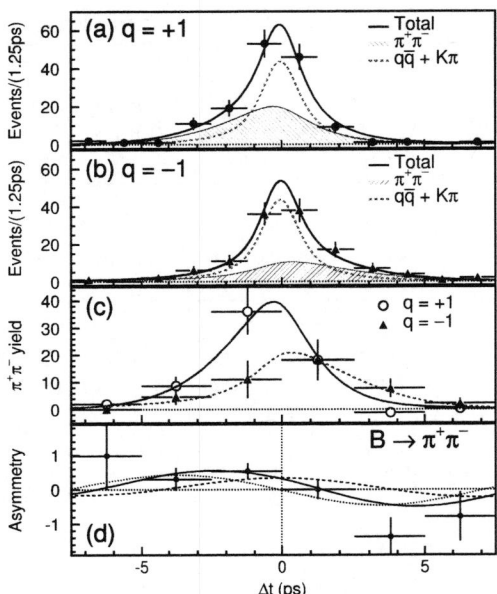

FIGURE 1. Belle's decay-time distributions for B^0 and $\bar{B}^0$ decays to $\pi^+\pi^-$ and the resulting charge asymmetry.

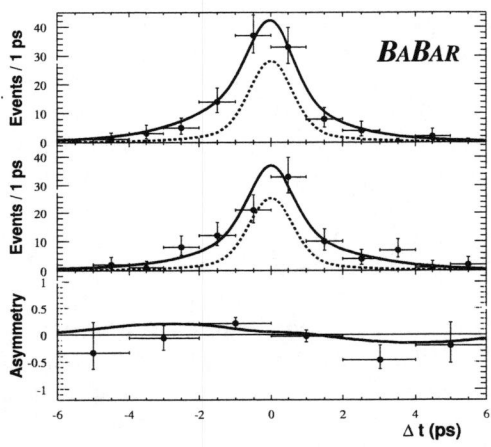

FIGURE 2. BaBar's decay-time distributions for B^0 (top) and $\bar{B}^0$ (middle) decays to $\pi^+\pi^-$ and the resulting charge asymmetry.

precision which would test our theoretical understanding. More significant constraints on α are obtained by the Heavy Flavour Averaging Group [14], by identifying the penguin contribution to $\pi^+\pi^-$ with the enhanced penguins of $K^+\pi^-$ or with $B^+ \to K^0\pi^+$. Tighter constraints yet are obtained if QCD factorization estimates of the penguin-to-tree amplitude ratio are used. The constraints are summarized in Fig. 3.

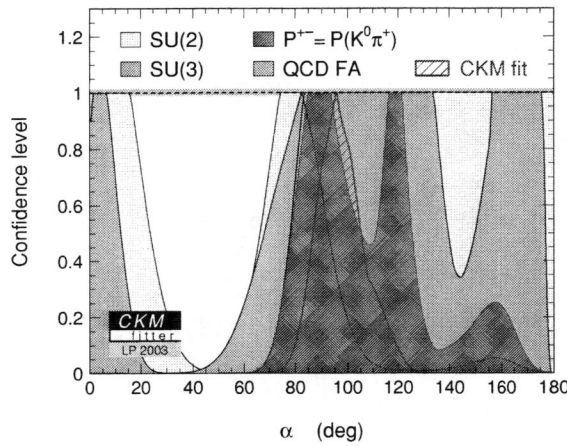

FIGURE 3. Constraints on α from $B \to \pi\pi$ results, with additional information and assumptions as described in the text, compared with the expectation from the CKM fit.

3.3. $B \to \rho\pi$ Decays

$B \to \rho\pi$ is more complex, because the final state is not a CP eigenstate. In the long term it offers the promise that a time-dependent Dalitz plot analysis, including interference between the different ρ bands, should allow the tree and penguin amplitudes to be measured and α to be determined unambiguously. However, with the present statistics only a quasi-two-body analysis is possible.

The decay probabilities for B^0 and $\bar{B}^0$ to $\rho^\pm\pi^\mp$ are given by:

$$f^{\rho^\pm h^\mp}_{B^0}(\Delta t) = (1 \pm A_{CP}(\rho h))e^{-|\Delta t|/\tau} \times$$
$$[1 + (S_{\rho h} \pm \Delta S_{\rho h})\sin(\Delta m\Delta t) - (C_{\rho h} \pm \Delta C_{\rho h})\cos(\Delta m\Delta t)]$$

$$f^{\rho^\pm h^\mp}_{\bar{B}^0}(\Delta t) = (1 \pm A_{CP}(\rho h))e^{-|\Delta t|/\tau} \times$$
$$[1 - (S_{\rho h} \pm \Delta S_{\rho h})\sin(\Delta m\Delta t) + (C_{\rho h} \pm \Delta C_{\rho h})\cos(\Delta m\Delta t)]$$

where the C and S terms are analogous to those in $\pi\pi$ decays, A_{CP} is the overall charge asymmetry between $B^0/\bar{B}^0 \to \rho^+\pi^-$ and $B^0/\bar{B}^0 \to \rho^-\pi^+$ (an example of direct CP violation), while ΔS and ΔC are CP-conserving dilution factors due to strong phase differences between the amplitudes contributing to these decays and to any asymmetry between $\Gamma(B^0 \to \rho^+\pi^-) + \Gamma(\bar{B}^0 \to \rho^-\pi^+)$ and $\Gamma(B^0 \to \rho^-\pi^+) + \Gamma(\bar{B}^0 \to \rho^+\pi^-)$.

Both collaborations see clear signals for $B \to \rho^\pm\pi^\mp$ [17, 18] and $\rho^0\pi^\pm$ [19, 20]. BaBar also measures branching fractions and asymmetries for $\rho^\pm\pi^0$, where there are 2 π^0s in the final state. There is some disagreement between the two experiments' measurements of $B \to \rho^0\pi^0$, where Belle [18] measures a branching fraction more than twice of BaBar's limit. However, the

TABLE 3. Measured $B \to \rho\pi$ branching fractions ($\times 10^{-6}$).

Mode	BaBar	Belle
$\rho^{\pm}\pi^{\mp}$	$22.6 \pm 1.8 \pm 2.2$	$29.1 \, {}^{+5.0}_{-4.9} \pm 4.0$
$\rho^{\pm}\pi^{0}$	$11.0 \pm 1.9 \pm 1.9$	
$\rho^{0}\pi^{\pm}$	$9.3 \pm 1.0 \pm 0.8$	$8.0 \, {}^{+2.3}_{-2.0} \pm 0.7$
$\rho^{0}\pi^{0}$	< 2.5 @ 90% CL	$6.0 \, {}^{+2.9}_{-2.3} \pm 1.2$
	$(0.9 \pm 0.7 \pm 0.4)$	

TABLE 4. Asymmetry parameters and dilution factors from $B \to \rho^{\pm}\pi^{\mp}$ analyses.

Parameter		Value
$A_{CP}^{\rho\pi}$	(Belle)	$-0.38 \, {}^{+0.19}_{-0.21} \, {}^{+0.04}_{-0.05}$
$A_{CP}^{\rho\pi}$	(BaBar)	$-0.114 \pm 0.062 \pm 0.027$
$A_{CP}^{\rho K}$		$-0.18 \pm 0.12 \pm 0.08$
$S_{\rho\pi}$		$-0.13 \pm 0.18 \pm 0.04$
$C_{\rho\pi}$		$+0.35 \pm 0.13 \pm 0.03$
$\Delta S_{\rho\pi}$		$+0.33 \pm 0.18 \pm 0.03$
$\Delta C_{\rho\pi}$		$+0.20 \pm 0.13 \pm 0.05$

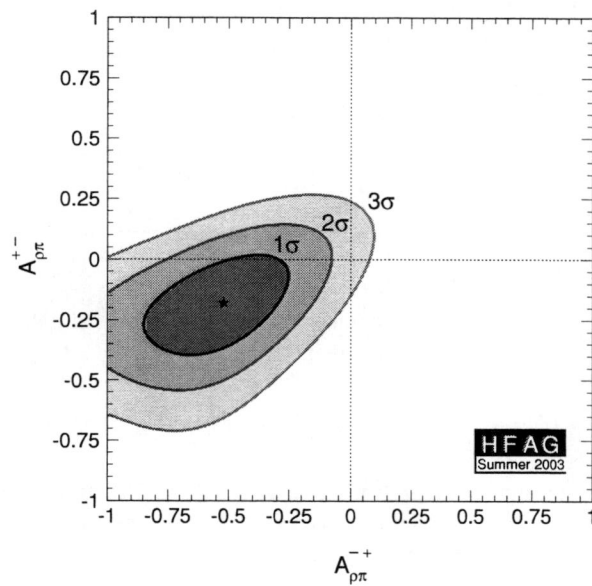

FIGURE 4. Probability contours for A_{+-} vs. A_{-+} [21].

difference in the central values measured by the two experiments has a statistical significance only slightly $> 2\sigma$. The branching fractions are summarized in Table 3.

BaBar has fitted the Δt distributions for B^0 and $\bar{B}^0$ tagged $\rho^{\pm}\pi^{\mp}$ candidates to extract the five parameters described above. BaBar also fitted the charge asymmetry for $B \to \rho^{\pm}K^{\mp}$ events. Belle has directly measured the A_{CP} parameter by counting unambiguous $\rho^{+}\pi^{-}$ and $\rho^{-}\pi^{+}$ events. These results are summarized in Table 4. Measurements of the charge asymmetries for $\rho^{\pm}\pi^{0}$ and $\rho^{0}\pi^{\pm}$ by BaBar were both consistent with 0 at the level of 1.5σ.

At present, no significant constraints on α have been made from these results. BaBar has re-expressed their measurements in terms of the charge asymmetries for the two pairs of CP-conjugate decays $\bar{B}^0 \to \rho^{+}\pi^{-}$, $B^0 \to \rho^{-}\pi^{+}$ and $\bar{B}^0 \to \rho^{-}\pi^{+}$, $B^0 \to \rho^{+}\pi^{-}$. They find:

$$A_{+-} = \frac{N(\bar{B}^0 \to \rho^{+}\pi^{-}) - N(B^0 \to \rho^{-}\pi^{+})}{N(\bar{B}^0 \to \rho^{+}\pi^{-}) + N(B^0 \to \rho^{-}\pi^{+})}$$
$$= -0.52 \, {}^{+0.17}_{-0.19} \pm 0.07,$$

$$A_{-+} = \frac{N(\bar{B}^0 \to \rho^{-}\pi^{+}) - N(B^0 \to \rho^{+}\pi^{-})}{N(\bar{B}^0 \to \rho^{-}\pi^{+}) + N(B^0 \to \rho^{+}\pi^{-})}$$
$$= -0.18 \pm 0.13 \pm 0.05.$$

These results provide evidence for direct CP violation in $B \to \rho^{\pm}\pi^{\mp}$ decays, with the confidence level for $A_{+-} = A_{-+} = 0$ being only 1.5%.

3.4. $B \to \rho\rho$ Decays

The particular feature of $B \to \rho\rho$ is the polarization of the two ρ mesons, which must be either both longitudinal or both transverse. The former will give CP-even eigenstates, while the latter will contain a mixture of CP-even and CP-odd terms. In general, an angular analysis is needed to separate the two states. However, the longitudinal term is expected to dominate, with

$$f_L = \frac{\Gamma_L}{\Gamma_L + \Gamma_T} \approx 1 - O(m_\rho^2/m_B^2). \quad (4)$$

This considerably simplifies the CP analysis. Experimentally the longitudinal and transverse fractions are determined by fitting the angular distributions and correlations between the ρ decays. This is important for branching fraction measurements, since the reconstruction efficiencies depend on the π^0 momenta, and thus on the ρ decay angles and hence the longitudinal fraction.

Both experiments have observed and measured branching fractions and longitudinal polarization fractions for $B^{\pm} \to \rho^{\pm}\rho^{0}$ [22, 23]. BaBar has also measured those for $B^0 \to \rho^{+}\rho^{-}$ [24]. In both channels the expectation of longitudinal dominance is verified. These results are summarized in Table 5.

Belle measures the $B^{\pm} \to \rho^{\pm}\rho^{0}$ charge asymmetry directly by separately estimating signals for the two charge states. They find $N(\rho^{+}\rho^{0}) = 29.3 \pm 9.1$, and $N(\rho^{-}\rho^{0}) = 29.3 \pm 9.5$, giving a charge asymmetry $A_{CP}(\rho^{\pm}\rho^{0}) = 0.00 \pm 0.22 \pm 0.03$.

TABLE 5. Measured $B \to \rho\rho$ branching fractions ($\times 10^{-6}$) and longitudinal polarization fractions.

Mode	Branching Ratios	f_L
$\rho^{\pm}\rho^0$ BaBar	$22.5 \pm 5.7 \pm 5.8$	$0.97\ ^{+0.03}_{-0.07} \pm 0.04$
$\rho^{\pm}\rho^0$ Belle	$31.7 \pm 7.1\ ^{+3.8}_{-6.7}$	$0.95 \pm 0.11 \pm 0.02$
$\rho^+\rho^-$ BaBar	$27\ ^{+7}_{-6}\ ^{+5}_{-7}$	$0.99\ ^{+0.01}_{-0.07} \pm 0.03$
$\rho^0\rho^0$ BaBar	< 2.1 (90% CL)	

BaBar fits the Δt distributions for longitudinally-polarized $\rho^+\rho^-$ events, using the form:

$$f_{B/\bar{B}}(\Delta t) = \frac{1}{4\tau} e^{-|\Delta t|/\tau}[1 \pm S_L \sin(\Delta m \Delta t) \mp C_L \cos(\Delta m \Delta t)]. \quad (5)$$

A simultaneous fit for yield, longitudinal fraction and CP asymmetries yields $f_L = 0.98 \pm 0.03$ and asymmetries $C_L = -0.21 \pm 0.25 \pm 0.11$ and $S_L = -0.37 \pm 0.36 \pm 0.17$, consistent with zero.

Although the current measurements do not yield very strong constraints on α, this channel does offer the prospect of a relatively straightforward measurement in the near future. As with $\pi^+\pi^-$ there is a simple relation between S_L and α_{eff}:

$$S_L = \sqrt{1 - C_L^2} \sin(\alpha_{eff}). \quad (6)$$

As for $\pi\pi$, $|\alpha_{eff} - \alpha|$ may be constrained by a Grossman-Quinn bound:

$$\sin^2(\alpha_{eff} - \alpha) < \frac{\mathscr{B}(B^0/\bar{B}^0 \to \rho^0\rho^0)}{\mathscr{B}(B^{\pm} \to \rho^{\pm}\rho^0)}. \quad (7)$$

Using BaBar's limit $\mathscr{B}(B^0 \to \rho^0\rho^0) < 2.1 \times 10^{-6}$, and conservatively taking this as the limit on the longitudinal decay mode, we find at 90% confidence level that $\sin^2(\alpha_{eff} - \alpha) < 19°$. This is already more than a factor of two better than the corresponding limit for $\pi\pi$, and may improve with further statistics if the branching fraction is significantly lower than this limit. Thus this channel offers the possibility of a simple, reasonably-accurate measurement of α in the coming years.

4. MEASUREMENT OF γ/ϕ_3

4.1. Principles

The third angle of the Unitarity Triangle is that between $V_{cd}V_{cb}^*$ and $V_{ud}V_{ub}^*$. Constraints on it may therefore be obtained from channels where there is interference between $b \to c$ and $b \to u$ transitions. The resulting asymmetries are correspondingly small. On the positive side, in contrast to the α-sensitive channels described above,

these measurements do not suffer from penguin pollution.

4.2. $B^{\pm} \to D^0 K^{\pm}$ Decays

This method measures the weak phase difference between the Cabibbo-allowed decay $B^- \to D^0 K^-$ (a $b \to c\bar{u}s$ transition) and the suppressed decay $B^- \to \bar{D}^0 K^-$ (a $b \to u\bar{c}s$ transition), which is equal to γ. These two processes may interfere if the D^0 and $\bar{D}^0$ decay to the same final state.

One method is to study channels where the $D^0/\bar{D}^0$ decay to CP eigenstates, either even (e.g. $\pi\pi$, KK) or odd (e.g. $K_S^0\pi^0$, $K_S^0\phi$). Measurements of the rates and charge asymmetries in the even and odd decays should allow the extraction of γ. Both experiments are actively pursuing such measurements, but at present, statistics are too limited to constrain γ.

Belle [25] has recently presented results from another method, based on $D^0/\bar{D}^0$ decays to a common 3-body state $K_S^0\pi^+\pi^-$. Interference between the $b \to c$ and $b \to u$ channels may be observed by fitting the $K_S^0\pi\pi$ Dalitz plots from B^- and B^+ decays. The Dalitz plot amplitudes are given by:

$$A(B^- \to K_S^0\pi^+\pi^-K^-) = f(m_-^2, m_+^2) + re^{i(\delta-\gamma)}f(m_+^2, m_-^2) \quad (8)$$

$$A(B^+ \to K_S^0\pi^+\pi^-K^+) = f(m_+^2, m_-^2) + re^{i(\delta-\gamma)}f(m_-^2, m_+^2) \quad (9)$$

where $m_{\pm}^2$ is the mass-squared of the $K_S^0\pi^{\pm}$ combination, r is the ratio of the amplitudes for the Cabibbo-suppressed and Cabibbo-allowed decays and δ is the strong phase difference between the two. A simultaneous fit to both Dalitz plots allows the parameters r, γ and δ to be extracted.

With sufficient statistics a binned fit to the Dalitz plot would allow a model-independent measurement. At present, in order to maximize the statistical power of the method, a model is chosen for the Dalitz plot amplitude for $D^0 \to K_S^0\pi^+\pi^-$, $f(m^2(K_S^0\pi^-), m^2(K_S^0\pi^+))$. This model was obtained by fitting a set of resonant amplitudes, plus a non-resonant term, to 57,800 $D^{*+} \to D^0\pi^+ \to (K_S^0\pi^+\pi^-)\pi^+$ events reconstructed in 78 fb^{-1} of Belle data. Monte Carlo studies were used to estimate the effect of using different D^0 decay models on the measurement of γ, and a systematic uncertainty $\Delta\gamma = 10°$ was assigned for the model-dependence.

140 fb^{-1} of data were used in this analysis, yielding 107 ± 12 signal events over a background of 33 ± 3. The fit yields two mirrored solutions for γ and δ, each 180° apart in both angles. Concentrating on the solution closest to the Standard Model fit result, Belle finds $\gamma = 95°\ ^{+25°}_{-20°} \pm 13° \pm 10°_{model}$. The 90% confidence limits

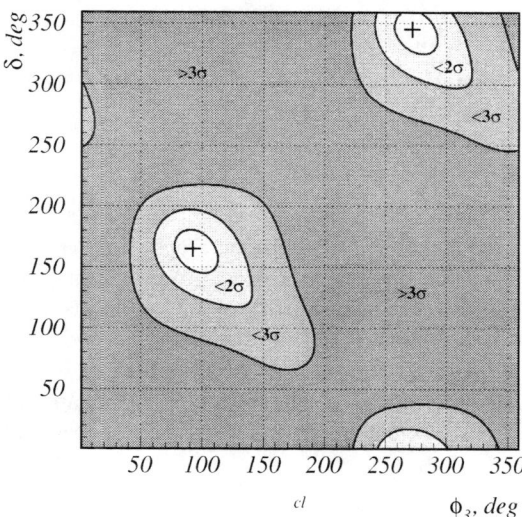

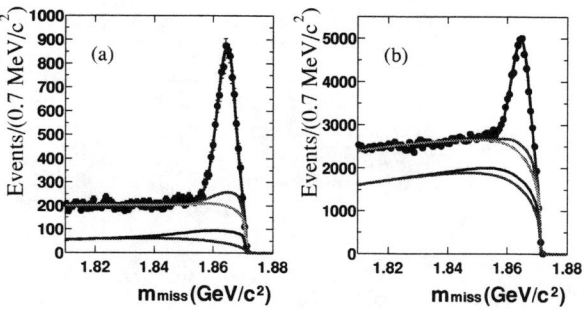

FIGURE 6. Missing mass signals from partially-reconstructed $D^*\pi$ decays, for lepton-tagged (left) and kaon-tagged (right) events.

FIGURE 5. Confidence levels in the $\phi_3(\gamma) - \delta$ plane from the Dalitz plot analysis.

obtained are $61° < \gamma < 142°$. The results are summarized in Fig. 5.

It is estimated that this method could reach 10° statistical precision with 1 ab^{-1} of data.

4.3. $B^0 \to D^{(*)\pm}\pi^{\mp}$ Decays

In this channel, the interference between the direct $B^0 \to D^{(*)-}\pi^+$ decay, via a $\bar{b} \to \bar{c}d\bar{u}$ transition, and mixing followed by the decay $\bar{B}^0 \to D^{(*)\pm}\pi^{\mp}$ via $b \to u\bar{c}d$ is sensitive to $\sin(2\beta + \gamma)$. However, although there is no penguin pollution, there is an unknown strong phase difference between the two contributing amplitudes which must be measured. The time-evolution of the decay is described by:

$$P_{B^0}(D^{\mp}\pi^{\pm}) \propto e^{-\Gamma|\Delta t|}(1 \pm C\cos(\Delta m_d \Delta t) + S^{\mp}\sin(\Delta m_d \Delta t)) \tag{10}$$

$$P_{\bar{B}^0}(D^{\mp}\pi^{\pm}) \propto Ne^{-\Gamma|\Delta t|}(1 \mp C\cos(\Delta m_d \Delta t) - S^{\mp}\sin(\Delta m_d \Delta t)) \tag{11}$$

and similar equations for $D^*\pi$, where

$$C = \frac{1-r^2}{1+r^2} \approx \quad \text{and} \quad S^{\mp} = \frac{2r}{1+r^2}\sin(2\beta + \gamma \pm \delta) \tag{12}$$

and $r = |V_{ub}^*V_{cd}/V_{cb}V^*ud| \approx 0.02$. Thus by measuring the time-dependent asymmetries for both B and $\bar{B}$ tagged events, it is possible to separate δ from $2\beta + \gamma$, allowing $2\beta + \gamma$ to be determined (with a 4-fold ambiguity).

Two analysis methods have been used, based on full and partial reconstruction. Both collaborations have presented analyses based on fully-reconstructed D and D^* decays [26, 27]. These methods provide high purity but relatively limited statistics. BaBar, analyzing 81 fb^{-1} of data, find (5207 ± 87) $D\pi$ candidates (85% purity) and (4746 ± 78) $D^*\pi$ (94% purity), while Belle in 140 fb^{-1} find 8375 $D\pi$ (88% purity) and 7556 $D^*\pi$ (95% purity). Fits to the Δt distributions for different signal and tag charge combinations were then used to measure the CP asymmetries.

The results of the fit are summarized in Table 6. Note that Belle presents their results in a different form, as $2r\sin(2\beta + \gamma \pm \delta)$. For the purposes of comparison, these have been transformed trigonometrically to the same form used by BaBar.

BaBar [28] has also presented an analysis based on partial reconstruction of the decay $B^0 \to D^{*\mp}\pi^{\pm}$. This decay produces two distinctive pions, a "fast" π from the B decay and a "slow" π from the D^* decay. Beam constraints then allow the "missing mass" to be reconstructed, which is required to be consistent with the mass of the unreconstructed D^0. Combinatorial backgrounds are much higher in this method, and also depend on the flavour tag. High momentum leptons are a good signature of B decays. Lepton-tagged events thus have a much higher B purity than kaon tagged events, as can be seen in Fig. 6. In 81 fb^{-1}, BaBar reconstructs 6406 ± 129 lepton-tagged $D^*\pi$ and 25157 ± 323 kaon-tagged events. A simultaneous time-dependent fit to the lepton- and kaon-tagged samples yields

$$2r\sin(2\beta + \gamma)\cos(\delta) = -0.063 \pm 0.024 \pm 0.014. \tag{13}$$

BaBar has interpreted their results in terms of $\sin(2\beta + \gamma)$ by minimizing a χ^2 between the various measurements and the values of $\sin(2\beta + \gamma)$ and $\delta_{D\pi}/\delta_{D^*\pi}$, assuming values of r_D and r_{D^*} obtained from studies of $B^0 \to D_s^{(*)}\pi$, with an additional 30% theoretical uncertainty added. Table 7 summarizes the limits obtained from the fully- and partially-reconstructed analyses and from combining the two analyses. This

TABLE 6. *CP* asymmetry measurements using $B^0 \to D^{(*)}\pi$.

	BaBar	Belle
$2r_{D^*\pi}\sin(2\beta+\gamma)\cos(\delta_{D^*\pi})$	$-0.068 \pm 0.038 \pm 0.021$	$0.063 \pm 0.041 \pm 0.016 \pm 0.036$
$2r_{D^*\pi}\sin(2\beta+\gamma)\sin(\delta_{D^*\pi})$	$0.031 \pm 0.070 \pm 0.035$	$0.030 \pm 0.041 \pm 0.016 \pm 0.036$
$2r_{D\pi}\sin(2\beta+\gamma)\cos(\delta_{D\pi})$	$-0.022 \pm 0.038 \pm 0.021$	$0.058 \pm 0.038 \pm 0.013 \pm 0.036$
$2r_{D\pi}\sin(2\beta+\gamma)\sin(\delta_{D\pi})$	$0.025 \pm 0.068 \pm 0.035$	$0.036 \pm 0.068 \pm 0.013 \pm 0.036$

TABLE 7. Limits on $|\sin(2\beta+\gamma)|$ from BaBar's $B^0 \to D^{(*)}\pi$ measurements.

	68% CL	90% CL
Fully reconstructed	> 0.69	-
Partially reconstructed	> 0.88	> 0.75
Combined	> 0.89	> 0.76

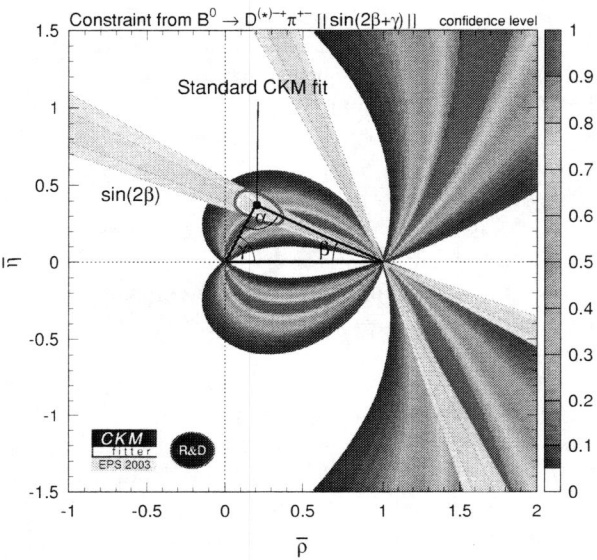

FIGURE 7. Probability contours in the $\rho - \eta$ plane from $D^{(*)}\pi$ constraints on $\sin(2\beta+\gamma)$.

combination excludes $|\sin(2\beta+\gamma)| = 0$ with a confidence of 99.5%.

The results are in good agreement with the solution favoured by the Standard Model fit, as shown in Fig. 7 [2].

5. SUMMARY AND PROSPECTS

The analyses described here are all in their early stages. Larger data samples in the coming years will increase the statistical significance of the measurements, allow better control of systematic uncertainties, and permit more sophisticated analyses to be made. Other channels, not described here, may bring additional constraints. If the predictions of the QCD factorization approach are validated, it will be possible to extract stronger constraints from these measurements. At present, such constraints, as exist, are all consistent with the Standard Model solution. However, the analyses described here will be statistically-limited for several years to come, and significant improvements in precision are to be expected in the future.

ACKNOWLEDGMENTS

The author would like to thank the BaBar and Belle collaborations for their assistance in compiling this summary, the PEP-II and KEKB accelerator groups, the many bodies who fund the two experiments, in particular PPARC who fund the author's work, the Heavy Flavour Averaging Group for providing a very valuable resource, and the organizing committee for their work organizing an highly enjoyable and informative conference.

REFERENCES

1. Hocker, A., Lacker, H., Laplace, S., and Le Diberder, F., *Eur. Phys. J.* **C21**, 225–259 (2001).
2. CKMFitter working group, September 2003 results (2003), URL: http://www.slac.stanford.edu/~laplace/ckmfitter/ckm_results_s%ummer03.html.
3. Heavy Flavour Averaging Group, results for Summer 2003 (2003), URL: http://www.slac.stanford.edu/xorg/hfag/triangle/summer2003/in%dex.html.
4. Aubert, B., et al., [BaBar Collaboration], *Nucl. Instrum. Meth.* **A479**, 1–116 (2002).
5. Abashian, A., et al., *Nucl. Instrum. Meth.* **A479**, 117–232 (2002).
6. PEP-II Conceptual Design Report (1993), SLAC-0418.
7. Kurokawa, S., and Kikutani, E., *Nucl. Instrum. Meth.* **A499**, 1–7 (2003).
8. Aubert, B., et al., [BaBar Collaboration], *Phys. Rev. Lett.* **89**, 281802 (2002).
9. Aubert, B., et al., [BaBar Collaboration], *Phys. Rev. Lett.* **91**, 021801 (2003).
10. Chao, Y., et al., [Belle Collaboration], arXiv:hep-ex/0311061 (2003), submitted to *Phys. Rev. D Rapid Communications*.

11. Abe, K., et al., [Belle Collaboration], *Phys. Rev. Lett.* **91**, 261801 (2003).

12. Abe, K., et al., [Belle Collaboration], *Phys. Rev.* **D68**, 012001 (2003).

13. Aubert, B., et al., [BaBar Collaboration], Updates to h^+h^- asymmetry measurements for Summer 2003 (preliminary) (2003), URL: `https://oraweb.slac.stanford.edu:8080/pls/slacquery/BABAR_DOC%UMENTS.DetailedIndex?P_BP_ID=3592`.

14. Heavy Flavour Averaging Group, α_{eff} constraints from $\pi\pi$ (2003), URL: `http://www.slac.stanford.edu/xorg/hfag/triangle/summer2003/in%dex.shtml#alphaeff_pipi`.

15. Grossman, Y., and Quinn, H. R., *Phys. Rev.* **D58**, 017504 (1998).

16. Gronau, M., and London, D., *Phys. Rev. Lett.* **65**, 3381–3384 (1990).

17. Aubert, B., et al., [BaBar Collaboration], *Phys. Rev. Lett.* **91**, 201802 (2003).

18. Abe, K., et al., [Belle Collaboration], arXiv:hep-ex/0307077 (2003), contribution to EPS HEP conference 2003.

19. Aubert, B., et al., [BaBar Collaboration], arXiv:hep-ex/0311049 (2003), submitted to *Phys. Rev. Lett.*

20. Gordon, A., et al., [Belle Collaboration], *Phys. Lett.* **B542**, 183–192 (2002).

21. Heavy Flavour Averaging Group (2003), URL: `http://www.slac.stanford.edu/xorg/hfag/triangle/summer2003/in%dex.shtml#alphaeff_rhopi`.

22. Zhang, J., et al., [Belle Collaboration], *Phys. Rev. Lett.* **91**, 221801 (2003).

23. Aubert, B., et al., [BaBar Collaboration], *Phys. Rev. Lett.* **91**, 171802 (2003).

24. Aubert, B., et al., [BaBar Collaboration], arXiv:hep-ex/0308024 (2003), contribution to Lepton-Photon 2003 conference.

25. Abe, K., et al., [Belle Collaboration], arXiv:hep-ex/0308043 (2003), contribution to Lepton-Photon 2003 conference.

26. Aubert, B., et al., [BaBar Collaboration], arXiv:hep-ex/0308018 (2003), contribution to Lepton-Photon 2003 conference.

27. Abe, K., et al., [Belle Collaboration], arXiv:hep-ex/0308048 (2003), contribution to Lepton-Photon 2003 conference.

28. Aubert, B., et al., [BaBar Collaboration], arXiv:hep-ex/0307036 (2003), contribution to EPS HEP conference 2003.

CP Violation Beyond the Standard Model

David London

Physics Department, McGill University, 3600 University St., Montréal QC, Canada H3A 2T8
and
Lab. René J.-A. Lévesque, Université de Montréal, C.P. 6128, succ. centre-ville, Montréal, QC, Canada H3C 3J7

Abstract. I review CP-violating signals of physics beyond the Standard Model in the B system. I examine the prospects for finding these effects at future colliders, with an emphasis on hadron machines.

1. INTRODUCTION

I have been asked to review the various CP-violating signals for physics beyond the Standard Model (SM) in the B system, with a particular emphasis on future hadron colliders. Now, in any discussion of this type, one has to consider the following question: will this new physics (NP) be discovered directly or not? If the assumption is that the NP will not be observed directly at hadron colliders, then the aim of measuring CP violation in the B system is to find evidence for NP. This "discovery signal" study is model-independent. That is, if some signal is seen, one will know that NP is present, but one will not know what kind of NP it is. On the other hand, if one assumes that this NP can be produced directly at hadron colliders, then its discovery will also probably reveal its identity, though not the details of its properties. In this case, the study of B physics is still useful – it will furnish "diagnostic tests" of this NP. The point is that the future study of CP violation in the B system is important, though what we will learn depends on what is discovered (or not) through other measurements. In particular, it is essential to consider both possibilities – that the NP is discovered directly, or not – in any discussion of CP violation signals of NP at future colliders [1]. In this talk I will attempt to address both of these scenarios.

If new physics exists, it can affect the B system in many different ways:

1. It can lead to new effects in B_s^0-$\bar{B}_s^0$ mixing or the $b \to s$ penguin amplitude, *i.e.* in the $b \to s$ flavour-changing neutral current (FCNC).
2. It can enter the $b \to d$ FCNC, *i.e.* B^0-$\bar{B}^0$ mixing or the $b \to d$ penguin.
3. It can affect tree-level decays, such as $b \to c\bar{q}q'$, $b \to u\bar{q}q'$, though this is less favoured theoretically.

Of course, any particular NP model may contain all of these effects, and all three classes of signals should be considered.

2. A SIGN OF NEW PHYSICS?

As is well known, there is a hint of a discrepancy in CP violation in $B^0(t) \to \phi K_S^0$ – the Belle measurement of β from this mode disagrees with that obtained from $B^0(t) \to J/\psi K_S^0$, though there is no disagreement in the BaBar measurement [2]. If this discrepancy is confirmed, it would point to new physics in the $\bar{b} \to \bar{s}s\bar{s}$ penguin amplitude, *i.e.* in the $b \to s$ FCNC. Many models of NP have been proposed to explain this effect: Z- or Z'-mediated FCNC's, non-minimal supersymmetry (SUSY), SUSY with R-parity violation, left-right symmetric models, anomalous t-quark couplings, *etc.* [3]. If this effect is confirmed, we will want to distinguish among these models, either through other B physics measurements, or through direct searches at hadron colliders.

This measurement raises an interesting question: is only the $\bar{b} \to \bar{s}s\bar{s}$ decay affected, or are all $b \to s$ FCNC amplitudes affected? For example, is there sizable NP in B_s^0-$\bar{B}_s^0$ mixing? This question can be answered by making measurements of a variety of B decays.

One key task of hadron colliders is the measurement of B_s^0-$\bar{B}_s^0$ mixing. This is of great interest in any case, but the potential discrepancy in $B^0(t) \to \phi K_S^0$ only serves to emphasize its importance. In order to make this measurement, it will be necessary to resolve oscillations in the B_s system. Once it has been demonstrated that this is possible, one can turn to CP tests involving B_s^0 mesons.

Even if new physics is discovered directly, one cannot test the CP nature of the NP couplings to ordinary particles – this is the domain of B physics. Hadron colliders will make several important CP measurements involving

CP722, *B Physics at Hadron Machines,* edited by M. Paulini and S. Erhan

B_s^0 mesons:

- Indirect CP violation in $B_s^0(t) \to D_s^+ D_s^-$, $J/\psi \phi$, $J/\psi \eta'$, etc. This measures the phase of B_s^0-$\bar{B}_s^0$ mixing, which is $\simeq 0$ in the SM.

- The measurement of $\mathscr{A}_{CP}^{mix}(B_s^0(t) \to D_s^{\pm} K^{\mp})$ probes γ in the SM [4]. In fact, this might possibly be the first direct measurement of this CP phase. Its value can be compared to that obtained from $\mathscr{A}_{CP}(B^{\pm} \to DK^{\pm})$ at B factories [5].

- Mixing-induced CP asymmetry in $\mathscr{A}_{CP}^{mix}(B_s^0(t) \to \phi\phi)$. This decay is analogous to $B^0(t) \to \phi K_s^0$. Here, one will need to perform an angular analysis, discussed in more detail below. Within the SM, this CP asymmetry is expected to be $\simeq 0$.

In all cases, any discrepancy with the SM prediction points specifically to new physics, with new phases, in B_s^0-$\bar{B}_s^0$ mixing and/or the $b \to s$ penguin. (Note that not all models of NP predict new phases. For example, in the minimal supersymmetric SM with minimal flavour violation, there are no new phases – the couplings of all SUSY particles track the CKM matrix.)

As an aside, suppose that the phase of B_s^0-$\bar{B}_s^0$ mixing is measured in, say, $\mathscr{A}_{CP}^{mix}(B_s^0(t) \to \psi\phi)$. The CKM phase $\chi \sim 2\text{-}5\%$ is extracted. Within the SM [6],

$$\sin(\chi) = \left| \frac{V_{us}}{V_{ud}} \right|^2 \frac{\sin(\beta)\sin(\gamma - \chi)}{\sin(\beta + \gamma)} . \qquad (1)$$

A discrepancy in this relation points to the presence of NP, though we can't pinpoint precisely where it enters.

3. DIRECT CP VIOLATION

Other good tests for new physics involve direct CP asymmetries. Like any CP-violating signal, direct CP violation can only come about when there are two interfering amplitudes. In this case, such CP violation corresponds to a difference in the rate for a B decay process and its CP-conjugate. If a given decay has only a single amplitude, then the direct CP asymmetry must vanish.

There are many decays which are dominated by a single amplitude in the SM. Examples of these include $B \to J/\psi K$ and ϕK, $B^0 \to D_s^+ D^-$, $B_s^0 \to D_s^+ D_s^-$, $B_c^+ \to J/\psi \pi^+$, etc. If a direct CP asymmetry is measured in any of these modes, it implies the presence of NP in a penguin or tree amplitude. Note that many models of NP affect $b \to s$ or $b \to d$ penguin amplitudes; fewer affect tree amplitudes. A complete study of direct CP asymmetries will probe various NP models. If NP has already been found, this is a good way to study the new couplings.

One particularly useful decay is $B^+ \to \pi^+ K^0$. In the SM, we have $|A(B^+ \to \pi^+ K^0)| \simeq |A(B^- \to \pi^- \bar{K}^0)|$.

Thus, any direct CP violation implies new physics, specifically in the $b \to s$ penguin. In this case the transition $\bar{b} \to \bar{s}d\bar{d}$ is affected. (Note that there is also a hint of NP in $B \to K\pi$ [2]. This is a good way of testing for this NP.)

4. TRIPLE PRODUCTS

One potential weakness of direct CP asymmetries is that

$$\mathscr{A}_{CP}^{dir} \propto \sin(\phi)\sin(\delta) , \qquad (2)$$

where ϕ and δ are, respectively, the weak and strong phase differences between the SM and NP amplitudes. Thus, if $\delta = 0$, $\mathscr{A}_{CP}^{dir} = 0$, even if there is a NP contribution. This possibility can be addressed by measuring in addition triple-product correlations (TP's).

Triple product correlations take the form $\vec{v}_1 \cdot (\vec{v}_2 \times \vec{v}_3)$, where each v_i is a spin or momentum. TP's are odd under time reversal (T) and hence, by the CPT theorem, also constitute potential signals of CP violation. One can establish the presence of a nonzero TP by measuring a nonzero value of the asymmetry

$$A_T \equiv \frac{\Gamma(\vec{v}_1 \cdot (\vec{v}_2 \times \vec{v}_3) > 0) - \Gamma(\vec{v}_1 \cdot (\vec{v}_2 \times \vec{v}_3) < 0)}{\Gamma(\vec{v}_1 \cdot (\vec{v}_2 \times \vec{v}_3) > 0) + \Gamma(\vec{v}_1 \cdot (\vec{v}_2 \times \vec{v}_3) < 0)} , \qquad (3)$$

where Γ is the decay rate for the process in question.

The most obvious place to search for triple products is in the decay $B \to V_1 V_2$, where both V_1 and V_2 are vector mesons. In this case, the TP takes the form $\vec{\varepsilon}_1^{*T} \times \vec{\varepsilon}_2^{*T} \cdot \hat{p}$, where $\vec{p}$ is the momentum of one of the final vector mesons, and $\vec{\varepsilon}_1$ and $\vec{\varepsilon}_2$ are the polarizations of V_1 and V_2. Note that TP's can be obtained by performing an angular analysis of the $B \to V_1 V_2$ decay. However, as seen from Eq. (3), a full angular analysis is not necessary.

Now, because triple products are odd under T, they can be faked by strong phases. That is, one can obtain a TP signal even if the weak phases are zero. In order to obtain a true CP-violating signal, one has to compare the TP in $B \to V_1 V_2$ with that in $\bar{B} \to \bar{V}_1 \bar{V}_2$. The CP-violating TP is found by *adding* the two T-odd asymmetries [7]:

$$\mathscr{A}_T \equiv \frac{1}{2}(A_T + \bar{A}_T) . \qquad (4)$$

Thus, neither tagging nor time dependence is necessary to measure TP's. One can in principle combine measurements of charged and neutral B decays [8].

The main point is that the CP-violating TP asymmetry of Eq. (4) takes the form

$$\mathscr{A}_T \propto \sin(\phi)\cos(\delta) . \qquad (5)$$

That is, unlike $\mathscr{A}_{CP}^{dir}$ [Eq. (2)], the triple product does not vanish if $\delta = 0$. Thus, TP's are complementary to direct *CP* asymmetries. In order to completely test for the presence of NP, it is necessary to measure both direct *CP* violation and triple products.

This then begs the question: which $B \to V_1 V_2$ decays are expected to yield large TP's in the SM? Interestingly, the answer is *none* [7, 9–11]! It is straightforward to see how this comes about.

As noted above, all *CP*-violating effects require the interference of two amplitudes, with different weak phases. Thus, there can be no triple products in decays which in the SM are dominated by a single decay amplitude.

Now consider other $B \to V_1 V_2$ decays. Within factorization, the amplitude can be written

$$\sum_{\mathcal{O},\mathcal{O}'} \left\{ \langle V_1 | \mathcal{O} | 0 \rangle \langle V_2 | \mathcal{O}' | B \rangle + \langle V_2 | \mathcal{O} | 0 \rangle \langle V_1 | \mathcal{O}' | B \rangle \right\},$$

$$(6)$$

where $\mathcal{O}$ and $\mathcal{O}'$ are SM operators. The key point is that TP's are a *kinematical CP*-violating effect [12]. That is, to produce a TP in a given decay, both of the above amplitudes must be present, with a relative weak phase.

For example, consider the decay $B^0 \to D^{*+} D^{*-}$. There is a tree amplitude, proportional to $V_{cb}^* V_{cd}$, and a penguin amplitude, proportional to $V_{tb}^* V_{td}$. Given that there are two amplitudes with a relative weak phase, one would guess that a *CP*-violating triple product would be produced. However, this is not the case. In fact, both amplitudes contribute to $\langle D^{*+} | \mathcal{O} | 0 \rangle \langle D^{*-} | \mathcal{O}' | B \rangle$; there is no $\langle D^{*-} | \mathcal{O} | 0 \rangle \langle D^{*+} | \mathcal{O}' | B \rangle$. (That is, in the SM one has only $\bar{b} \to \bar{c}$ transitions; $\bar{b} \to c$ transitions do not occur.) Thus, despite the presence of two amplitudes in this decay, no TP is produced, at least within factorization.

Using the above argument, we note that there are three classes of $B \to V_1 V_2$ decays in the SM, all of which are expected to have zero or small triple products:

1. Decays governed by a single weak decay amplitude, such as $B \to J/\psi K^*$, $B_s^0 \to \phi\phi$, $B_s^0 \to D_s^* D_s^*$, $B_c^+ \to J/\psi \rho^+$, *etc.* Because there is only one amplitude, there can be no *CP*-violating effects, including TP's. This is model-independent.

2. Color-allowed decays with two weak decay amplitudes, such as $\bar{B}^0 \to D^{*+} D^{*-}$, $\bar{B}_s^0 \to D_s^{*+} D_s^{*-}$, $\bar{B}_s^0 \to K^{*+} K^{*-}$, $B_c^- \to \bar{D}^{*0} \rho^-$, $B_c^- \to \bar{D}^{*0} K^{*-}$, *etc.* The two amplitudes are usually a tree and a penguin diagram, though it is possible to have two penguin contributions. As argued above, both decay amplitudes contribute to the same kinematical amplitude, so that all TP's vanish. Since the decays are colour-allowed, non-factorizable corrections are expected to be small, so that the prediction of tiny TP's is robust.

3. Color-suppressed decays with two weak amplitudes, such as $B^- \to \rho^0 K^{*-}$, $\bar{B}_s^0 \to \phi K^*$, $B_c^- \to J/\psi D^{*-}$, *etc.* Once again, within factorization,

both amplitudes contribute to the same kinematical amplitude, so that the TP's vanish. However, non-factorizable effects may be large in colour-suppressed decays. We have tried to be conservative in our estimates of such effects, and still find tiny TP's for such decays [11]. This conclusion is clearly model-dependent. (In any case, the branching ratios for such decays are very small.)

The fact that all TP's in $B \to V_1 V_2$ are expected to vanish or be very small in the SM makes this an excellent class of measurements to search for new physics. In the SM, all couplings to the *b*-quark [*i.e.* the operators $\mathcal{O}'$ in Eq. (6)] are left-handed. Within factorization, the discovery of a large TP in a $B \to V_1 V_2$ decay would point to new physics with large couplings to the right-handed *b*-quark [10, 11]. Many new-physics models, though not all, have such couplings. As an example, supersymmetry with R-parity-violating couplings can explain the apparent discrepancy in $B^0(t) \to \phi K_s^0$. Such a model of NP will also contribute to $B \to \phi K^*$ decays, leading to TP's. In the SM, such TP's vanish; with this type of NP, one can get TP asymmetries as large as 15-20% [11]! The upshot is that triple products are excellent diagnostic tests for new physics. Some NP models predict large TP's, so that null measurements can strongly constrain (or eliminate) such models.

5. TIME-DEPENDENT ANGULAR ANALYSIS

Consider now a $V_1 V_2$ state which in the SM is dominated by a single amplitude. Suppose that there is a new-physics amplitude, with a different weak phase, contributing to this decay. Above, I have argued that one can detect such an amplitude by looking for both direct *CP* violation and triple products. However, much more information can be obtained if a time-dependent angular analysis of the corresponding $B^0(t) \to V_1 V_2$ decay can be performed [8].

The time-dependent decay rate for $B^0(t) \to V_1 V_2$ is given by

$$\Gamma(B^0(t) \to V_1 V_2) = e^{-\Gamma t} \sum_{\lambda \leq \sigma} \left(\Lambda_{\lambda\sigma} + \Sigma_{\lambda\sigma} \cos(\Delta M t) - \rho_{\lambda\sigma} \sin(\Delta M t) \right) g_\lambda g_\sigma. \quad (7)$$

In the above, the helicity indices λ and σ take the values $\{0, \|, \perp\}$, and the g_λ are known functions of the kinematic angles. For a given helicity λ, $\Lambda_{\lambda\lambda}$ essentially measures the total rate, while $\Sigma_{\lambda\lambda}$ and $\rho_{\lambda\lambda}$ represent the direct and indirect *CP* asymmetries, respectively. The quantity $\Lambda_{\perp i}$ ($i = \{0, \|\}$) is simply the triple product discussed earlier.

Now, there are 18 observables in this decay rate. However, if there is no new physics, there are only 6 theoretical parameters. This implies that there are 12 relations among the observables. They are:

$$\Sigma_{\lambda\lambda} = \Lambda_{\perp i} = \Sigma_{\|0} = 0 \,,$$

$$\rho_{ii}/\Lambda_{ii} = -\rho_{\perp\perp}/\Lambda_{\perp\perp} = \rho_{\|0}/\Lambda_{\|0} \,,$$

$$2\Lambda_{\|0}\Lambda_{\perp\perp}\left(\Lambda_{\lambda\lambda}^2 - \rho_{\lambda\lambda}^2\right) =$$

$$\left[\Lambda_{\lambda\lambda}^2\rho_{\perp 0}\rho_{\perp\|} + \Sigma_{\perp 0}\Sigma_{\perp\|}\left(\Lambda_{\lambda\lambda}^2 - \rho_{\lambda\lambda}^2\right)\right] \,,$$

$$\rho_{\perp i}^2\Lambda_{\perp\perp}^2 = \left(\Lambda_{\perp\perp}^2 - \rho_{\perp\perp}^2\right)\left(4\Lambda_{\perp\perp}\Lambda_{ii} - \Sigma_{\perp i}^2\right) \,. \quad (8)$$

The violation of any of these relations will be a smoking-gun signal of NP. If the NP conspires to make direct *CP* violation and TP's small, it can still be detected through one of these other signals [8]. Thus, if a time-dependent angular analysis can be performed, there are many more ways to search for new physics.

However, even more can be done! Suppose that some signal for new physics is found. In this case, it is straightforward to show that there are more theoretical parameters than independent observables, so that one cannot solve for the NP parameters. However, because the expressions relating the observables to the theoretical parameters are nonlinear, one can actually put a *lower bound* on the NP parameters [8]. This is extremely important, as it allows us to get direct information on the NP through measurements in the *B* system.

6. α FROM $B^0 \to K^{(*)}\bar{K}^{(*)}$ DECAYS

I now turn to the extraction of α from $B^0_{d,s} \to K^{(*)}\bar{K}^{(*)}$ decays [13]. Consider first $B^0 \to K^0\bar{K}^0$, which is a pure $b \to d$ penguin:

$$
\begin{aligned}
A(B^0 \to K^0\bar{K}^0) &= P_u V_{ub}^* V_{ud} + P_c V_{cb}^* V_{cd} + P_t V_{tb}^* V_{td} \\
&= \mathscr{P}_{uc} e^{i\gamma} e^{i\delta_{uc}} + \mathscr{P}_{tc} e^{-i\beta} e^{i\delta_{tc}} \,, \quad (9)
\end{aligned}
$$

where $\mathscr{P}_{uc} \equiv |(P_u - P_c)V_{ub}^* V_{ud}|$, $\mathscr{P}_{tc} \equiv |(P_t - P_c)V_{tb}^* V_{td}|$, and I have explicitly written out the strong phases δ_{uc} and δ_{tc}, as well as the weak phases β and γ. By measuring $B^0(t) \to K^0\bar{K}^0$, one can extract 3 observables – the total rate, and the direct and indirect *CP* asymmetries. However, these depend on the 4 unknowns $\mathscr{P}_{uc}$, $\mathscr{P}_{tc}$, $\Delta \equiv \delta_{uc} - \delta_{tc}$ and α, so that *CP* phase information cannot be obtained.

Now consider a second pure $b \to d$ penguin decay of the form $B^0 \to K^*\bar{K}^*$, where K^* represents any excited neutral kaon. This decay can be treated completely analogously to $B^0 \to K^0\bar{K}^0$, with unprimed parameters and observables being replaced by ones with tildes. The measurement of the time-dependent rate again allows one to

extract 3 observables, which depend on the 4 unknowns $\widetilde{\mathscr{P}}_{uc}$, $\widetilde{\mathscr{P}}_{tc}$, $\widetilde{\Delta}$ and α. Again, there are more observables than unknowns, so that one cannot extract α. However, one can combine measurements from the two decays to write

$$\frac{\mathscr{P}_{tc}^2}{\widetilde{\mathscr{P}}_{tc}^2} = f(\alpha, \text{observables}) \,. \quad (10)$$

Note that the CKM matrix elements $|V_{tb}^* V_{td}|$ cancel in this ratio. From this ratio, we see that we could solve for α if we knew the value of $\mathscr{P}_{tc}/\widetilde{\mathscr{P}}_{tc}$.

This information can be obtained by considering $B^0_s \to K^{(*)}\bar{K}^{(*)}$ decays. Consider the decay $B^0_s \to K^0\bar{K}^0$, which is a pure $b \to s$ penguin:

$$
\begin{aligned}
A(B^0_s \to K^0\bar{K}^0) &= P'_u V_{ub}^* V_{us} + P'_c V_{cb}^* V_{cs} + P'_t V_{tb}^* V_{ts} \\
&= \mathscr{P}'_{uc} e^{i\gamma} e^{i\delta'_{uc}} + \mathscr{P}'_{tc} e^{i\delta'_{tc}} \\
&\simeq \mathscr{P}'_{tc} e^{i\delta'_{tc}} \,. \quad (11)
\end{aligned}
$$

Here $\mathscr{P}'_{uc} \equiv |(P'_u - P'_c)V_{ub}^* V_{us}|$ and $\mathscr{P}'_{tc} \equiv |(P'_t - P'_c)V_{tb}^* V_{ts}|$. However, since $|V_{ub}^* V_{us}/V_{tb}^* V_{ts}| \simeq 2\%$, the u-quark piece $\mathscr{P}'_{uc}$ is negligible compared to $\mathscr{P}'_{tc}$, leading to the last line above. Therefore the measurement of $\mathscr{B}(B^0_s \to K^0\bar{K}^0)$ gives $\mathscr{P}'_{tc}$. Similarly, the measurement of $\mathscr{B}(B^0_s \to K^*\bar{K}^*)$ gives $\widetilde{\mathscr{P}}'_{tc}$.

The key point is that, in the flavour SU(3) limit, we have $\mathscr{P}'_{tc}/\widetilde{\mathscr{P}}'_{tc} = \mathscr{P}_{tc}/\widetilde{\mathscr{P}}_{tc}$. Thus, using Eq. (10), one can obtain α. Now, uncertainties due to SU(3) breaking are typically $\sim 25\%$. However, these leading-order effects *cancel* in the double ratio $(\mathscr{P}'_{tc}/\widetilde{\mathscr{P}}'_{tc})/(\mathscr{P}_{tc}/\widetilde{\mathscr{P}}_{tc})$. One is left with only second-order SU(3)-breaking effects, *i.e.* a theoretical error of at most 5% [13].

One can compare the value of α extracted with this method with that obtained elsewhere (*e.g.* in $B \to \pi\pi$ or $B \to \rho\pi$). A discrepancy would point to new physics in the $b \to d$ or $b \to s$ penguin.

It is useful to list some experimental considerations. First, B^0_s mesons are involved, and all branching ratios are $\sim 10^{-6}$. Second, the $K^{(*)}$ and $\bar{K}^{(*)}$ mesons are detected through their decays to charged π's and K's only, requiring good K/π separation. Finally, no π^0 detection needed. All in all, this method is particularly appropriate to hadron colliders.

7. TRIPLE PRODUCTS IN Λ_b DECAYS

Another class of decays which can only be studied at hadron colliders are those involving Λ_b baryons. Consider the decays $\Lambda_b \to F_1 P$ and $F_1 V$, where F_1 is a fermion (p, Λ, ...), P is a pseudoscalar (K^-, η, ...), and V is a vector (K^{*-}, ϕ, ...). In these decays, triple products are possible [14]. In $\Lambda_b \to F_1 P$, only one TP is possible: $\vec{p}_{F_1} \cdot (\vec{s}_{F_1} \times \vec{s}_{\Lambda_b})$, where $\vec{s}_i$ is the spin of particle i. On the other hand, since the decay $\Lambda_b \to F_1 V$ involves three spins and one final momentum, four TP's are possible.

As in $B \to V_1 V_2$ decays, within factorization we require a right-handed coupling to b-quarks to generate a TP. For certain $F_1 P$ final states, one can "grow" a sizable right-handed current due to the Fierz transformations of some of the SM operators. However, for $F_1 V$ final states, there are no such right-handed currents. Thus, all TP's are expected to vanish in the SM for $\Lambda_b \to F_1 V$ decays.

We find that $\mathscr{A}_T^{pK} = -18\%$, but the TP's for all other fermion-pseudoscalar final states ($pK^{*-}, \Lambda\eta, \Lambda\eta', \Lambda\phi$) are small, at most $O(1\%)$. Once again, the fact that almost all TP's are expected to be small implies that this is a good place to look for new physics. In fact, one can use TP's in Λ_b decays as a diagnostic tool for NP [15].

8. RADIATIVE DECAYS

Finally, I consider radiative decays of B mesons. The inclusive partial rate asymmetries can be calculated reliably in the SM [16]:

$$\mathscr{A}_{CP}^{dir}(b \to s\gamma) = 0.5\%,$$
$$\mathscr{A}_{CP}^{dir}(b \to d\gamma) = -10\%. \quad (12)$$

If measurements of these asymmetries are found to differ from their SM values, this will indicate the presence of new physics. Indeed, large deviations are possible in several models of NP [17].

Exclusive partial rate asymmetries in $B \to K^*\gamma$ and $B \to \rho\gamma$ are not known as well – there are important bound-state corrections [18]. However, if significant deviations from the values calculated for inclusive decays are found in exclusive decays, this probably points to new physics.

One can also consider mixing-induced CP asymmetries (e.g. $B^0(t) \to \rho\gamma$, $B_s^0(t) \to \phi\gamma$). In the SM the photon polarization is opposite for B and $\bar{B}$ decays, so that no interference is possible. That is, $\mathscr{A}_{CP}^{mix}(b \to s\gamma, b \to d\gamma) \simeq 0$ in the SM. However, one can get a significant $\mathscr{A}_{CP}^{mix}$ in certain models of NP (e.g. left-right symmetric models, SUSY, models with exotic fermions) [19].

9. CONCLUSION

In summary, there are numerous signals of new physics in B and Λ_b decays. Furthermore, there are many ways of determining which types of NP might be responsible for these signals. If the NP is discovered directly, the measurement of CP violation in the B system can be used to probe its couplings. Thus, the study of B processes is complementary to direct searches for NP. Hadron collid-

ers have a significant role to play in the discovery of NP, as well as in its identification.

ACKNOWLEDGMENTS

This work was financially supported by NSERC of Canada.

REFERENCES

1. See also Cahn, R., in these proceedings.
2. For a discussion, see Grossman, Y., in these proceedings.
3. A partial list of references includes Hiller, G., *Phys. Rev.* **D66**, 071502 (2002); Datta, A., *Phys. Rev.* **D66**, 071702 (2002); Ciuchini, M., and Silvestrini, L., *Phys. Rev. Lett.* **89**, 231802 (2002); Raidal, M., *Phys. Rev. Lett.* **89**, 231803 (2002); Dutta, B., Kim, C. S., and Oh, S., *Phys. Rev. Lett.* **90**, 011801 (2003); Lee, J. P., and Lee, K. Y., *Eur. Phys. Jour.* **C29**, 373 (2003); Khalil, S., and Kou, E., *Phys. Rev.* **D67**, 055009 (2003); Chiang, C. W., and Rosner, J. L., *Phys. Rev.* **D68**, 014007 (2003); Giri, A. K., and Mohanta, R., *Phys. Rev.* **D68**, 014020 (2003); Barger, V., Chiang, C. W., Langacker, P., and Lee, H. S., hep-ph/0310073.
4. Aleksan, R., Dunietz, I., and Kayser, B., *Z. Phys.* **C54**, 653 (1992).
5. For a review of *CP* violation in $B \to DK$ decays, see Beneke, M., Ciuchini, M., Faccini, R., Gardner, S., London, D., Sakai, Y., and Soni, A., eConf **C0304052**, WG401 (2003).
6. Aleksan, R., Kayser, B., and London, D., *Phys. Rev. Lett.* **73**, 18 (1994); Silva, J. P., and Wolfenstein, L., *Phys. Rev.* **D55**, 5331 (1997).
7. Valencia, G., *Phys. Rev.* **D39**, 3339 (1989).
8. London, D., Sinha, N., and Sinha, R., hep-ph/0304230.
9. Kramer, G., and Palmer, W. F., *Phys. Rev.* **D45**, 193 (1992), *Phys. Lett.* **B279**, 181 (1992), *Phys. Rev.* **D46**, 2969 (1992); Kramer, G., Palmer, W. F., and Mannel, T., *Z. Phys.* **C55**, 497 (1992); Kramer, G., Palmer, W. F., and Simma, H., *Nucl. Phys.* **B428**, 77 (1994); Kamal, A. N., and Luo, C. W., *Phys. Lett.* **B388** 633 (1996).
10. Atwood, D., and Soni, A., *Phys. Rev. Lett.* **81**, 3324 (1998), *Phys. Rev.* **D59**, 013007 (1999).
11. Datta, A., and London, D., hep-ph/0303159.
12. Kayser, B., *Nucl. Phys. Proc. Suppl.* **13**, 487 (1990).
13. Datta, A., and London, D., *Phys. Lett.* **B533**, 65 (2002).
14. Bensalem, W., Datta, A., and London, D., *Phys. Lett.* **B538**, 309 (2002).
15. Bensalem, W., Datta, A., and London, D., *Phys. Rev.* **D66**, 094004 (2002).
16. For a review, see Hurth, T., Lunghi, E., and Porod, W., hep-ph/0310282, and references therein.
17. For example, see Kagan, A. L., and Neubert, M., *Phys. Rev.* **D58**, 094012 (1998); Kiers, K., Soni, A., and Wu, G. H., *Phys. Rev.* **D62**, 116004 (2000).
18. Greub, C., Simma, H., and Wyler, D., *Nucl. Phys.* **B434**, 39 (1995) [Erratum-ibid. **B444**, 447 (1995)].
19. Atwood, D., Gronau, M., and Soni, A., *Phys. Rev. Lett.* **79**, 185 (1997).

B PRODUCTION AND SPECTROSCOPY

Measurement of the $b\bar{b}$ Cross Section at HERA-B

Martin zur Nedden

(Representing the HERA-B Collaboration, DESY Hamburg)

Humboldt Universität zu Berlin, Institut für Physik, Germany .

Abstract. The $b\bar{b}$ production cross section has been measured in collisions of 920 GeV protons off a nuclei target using the HERA-B detector. The identification of $b\bar{b}$ events was done via inclusive bottom quark decays into J/ψ by exploiting the longitudinal separation of the $J/\psi \to \ell^+\ell^-$ decay vertices from the primary proton nucleus interaction, where both the $\mu^+\mu^-$ and the e^+e^- decay channels have been reconstructed. The first measurement, using data collected during a short period in 2000, yields a cross section in the combined analysis of $\sigma(b\bar{b}) = 32^{+14}_{-12}(\text{stat})^{+6}_{-7}(\text{sys})$ nb/nucleon. On the most recent data samples taken in 2002/03, a much more accurate measurement with the uncertainties reduced to 10% is expected.

1. INTRODUCTION

Despite the good theoretical description of open charm and charmonium production in various experimental environments, the open beauty production description is still not satisfactory at least for higher center of mass energies. While the behavior of differential cross sections as a function of kinematical variables is usually reproduced quite well, the absolute values are still in disagreement by a factor of 2 to 7 [1]. This has been observed by several experiments: in $\gamma\gamma$ reactions at LEP (OPAL [2] and L3 [3]), in $p\bar{p}$ interactions at CDF [4], and in ep scattering at HERA (H1 [5, 6] and ZEUS [7]). Although new calculations for CDF of the theoretical cross section [8] reduced the discrepancy, the theoretical predictions contain still large uncertainties and the measured cross sections and QCD NLO expectations are not yet in good agreement [9]. This emphasizes the need of precise measurement of the $b\bar{b}$ production cross sections as an important input to perturbative QCD calculations.

For the measurement of $\sigma(b\bar{b})$ in the kinematical region of $\sqrt{s} \approx 40$ GeV the recent situation is not satisfactory. Only two experimental measurements exist for this region those of the E789 [10] and E771 [11] collaborations (see. Tab. 1). Their results show large uncertainties and poor compatibility. A new accurate measurement is therefore highly desirable and would be a sensitive test for QCD predictions.

2. THE HERA-B EXPERIMENT

The HERA-B [12, 13] experiment (Fig. 1) has been designed to identify B mesons decays in a dense hadronic

TABLE 1. Other $\sigma(b\bar{b})$ Measurements.

Exp.	E_p [GeV]	$\sigma(J/\psi)$ [nb/nucleon]	Channel
E789	800	$5.7 \pm 1.5 \pm 1.3$	$b \to J/\psi(\mu^+\mu^-)X$
E771	800	$43^{+27}_{-17} \pm 7$	$\mu(\text{s.l.}) / b(\bar{b})$ dec.
Theory prediction: $9 \leq \sigma(b\bar{b}) \leq 55$ nb/nucleon			

environment with a large geometrical coverage. The B mesons are produced by interaction of the proton in the halo of the 920 GeV HERA beam with different target wires, which can be used individually or simultaneously. Since there are various target materials available, the A dependence of heavy quark production can be measured. The events from different wires can be separated easily by applying accurate spatial separation cuts.

The $b\bar{b}$ production cross section $\sigma^A_{b\bar{b}}$ on a nucleus of atomic number A is obtained from the inclusive reaction

$$pA \to b\bar{b}X \text{ with } b\bar{b} \to J/\psi X' \to (e^+e^-/\mu^+\mu^-)X' \quad (1)$$

B hadron decays into a J/ψ are distinguished from the large prompt J/ψ background by exploiting the long B lifetime in a detached vertex analysis. HERA-B is able to reconstruct the J/ψ either in the $\mu^+\mu^-X$ or in the e^+e^-X decay channel. The availability of both channels is important to increase the statistics and to cross check the results.

In the data taking period of 2002/03 starting in October 2002 HERA-B was routinely running and able to collect $164 \cdot 10^6$ events triggered by a dilepton J/ψ trigger. The data were taken using a trigger configuration, selecting the events by the requirement of two pretrigger (PT) seeds originating either from the muon pretrig-

CP722, *B Physics at Hadron Machines*, edited by M. Paulini and S. Erhan
© 2004 American Institute of Physics 0-7354-0203-5/04/$22.00

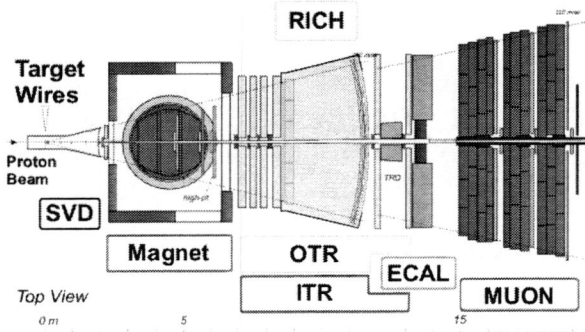

FIGURE 1. HERA-*B* is a fixed target spectrometer at the HERA proton beam at $E_p = 920$ GeV.

ger or the pretrigger of the electromagnetic calorimeter (ECAL). The PT information was then used as seeds for the first level trigger (FLT), where a single track candidate was searched. The PT and FLT information was finally passed to the second level trigger (SLT) where the FLT track had to be confirmed and combined with a second track. On the SLT in addition a vertex constraint was applied. Using this trigger configuration, HERA-*B* was able to collect up to 1400 fully reconstructed J/ψ's per hour. With the achieved event yield and an interaction rate of 5 MHz we collected about 300,000 J/ψ's approximately equally distributed to both decay channels $J/\psi \rightarrow e^+e^-$ and $J/\psi \rightarrow \mu^+\mu^-$. Furthermore, for other charmonium states a quite large statistics could be achieved: $N(\chi_c) \approx 20,000$, $N(\psi') \approx 5,000$. In the *B* sector the number of candidates was increased by more than an order of magnitude with respect to the year 2000: $N(b \rightarrow J/\psi X) \approx 180$, $N(\Upsilon(1S)) \approx 60$.

3. MEASUREMENT

In order to minimize the systematic errors and to remove the dependence on the absolute luminosity determination, the $\sigma(b\bar{b})$ cross section is measured relative to the known prompt J/ψ cross section $\sigma_{J/\psi}^A$ [14, 15]. Within the HERA-*B* acceptance, the first fixed-target experiment covering the negative x_f region ($x_F \in [-0.35, 0.15]$, $x_F = \frac{p_L^{cms}}{(p_L^{cms})_{max}}$), the cross section can be written as as:

$$\Delta\sigma_{b\bar{b}} = \Delta\sigma_r \cdot \frac{n_B}{n_P} \cdot \frac{1}{\varepsilon_R \cdot \varepsilon_B^{\Delta z} \cdot \mathscr{B}(b\bar{b} \rightarrow J/\psi X)} \quad (2)$$

where n_B/n_P are the observed b and prompt J/ψ events, ε_R the relative efficiency ($\varepsilon_R = \varepsilon_B^{J/\psi}/\varepsilon_P^{tot}$, $\varepsilon_B^{tot} = \varepsilon_B^{J/\psi} \cdot \varepsilon_B^{\Delta z}$) of B to prompt J/ψ efficiency ratio (trigger + reconstruction + selection), $\mathscr{B}(b\bar{b} \rightarrow J/\psi) = (2.32 \pm 0.20)\%$ is the branching ratio measured at LEP [16], $\Delta\sigma_r$ is the

TABLE 2. Reference J/ψ prompt cross sections of other experiments.

Exp.	Target	E_p [GeV]	$\sigma(J/\psi)$ [nb/nucleon]	α
E789	Au	800	$442 \pm 2 \pm 88$	0.9 ± 0.02
E771	Si	800	$375 \pm 4 \pm 30$	0.92 ± 0.008

reference (prompt J/ψ) cross section re-weighted to the HERA-*B* acceptance and finally $\Delta\sigma_{b\bar{b}}$ the measured cross section in HERA-*B* acceptance.

The reference prompt J/ψ production cross section per nucleon $\sigma(pN \rightarrow J/\psi X)$ was previously measured by two fixed target experiments (E789 and E771, see Tab. 2, [14, 15]). After correcting for the most recent measurement of the nuclear dependence A^α using $\alpha = 0.955 \pm 0.005$ (E866, [17]) and scaling to HERA-*B* energies [18] a reference cross section of $\sigma(pN \rightarrow J/\psi) = (357 \pm 8 \pm 27)$ nb/nucleon is obtained. Within the acceptance of HERA-*B* only a fraction of $f_P = (77 \pm 1)\%$ [14] of the prompt J/ψ and $f_B = (90.6 \pm 0.5)\%$ of the $b\bar{b}$ events can be measured. Therefore, the reference prompt J/ψ cross section to be used writes as:

$$\Delta\sigma_r = f_P \cdot \sigma(pN \rightarrow J/\psi) \cdot A^\alpha \quad (3)$$

Since no nuclear suppression has been observed in D Meson production [19] and a similar behavior is expected in the b channel, a nuclear dependence of $\alpha = 1.0$ is assumed for the $b\bar{b}$ production cross section [20] in this analysis, i.e. $\sigma_{b\bar{b}}^A = \sigma(pN \rightarrow b\bar{b}) \cdot A$.

4. MONTE CARLO MODELS

A Monte Carlo (MC) simulation is used to determine the efficiencies of Eq. 2 and to describe the background contribution of the prompt J/ψ events to the $b \rightarrow J/\psi X$ decay channel. The simulation of heavy flavour quark (Q) production in $pA \rightarrow Q\bar{Q}X$ interactions and the heavy quark hadronization in the nuclear environment is done using the PYTHIA 5.7 event generator [21]. For the remaining part of the process (X) as light quark production inside the nucleus, secondary interactions and pA inelastic interactions, the FRITIOF 7.02 [22] package is used.

The $b\bar{b}$ events generated by PYTHIA are weighted using a model with various contributions. First, the generated b quark kinematics (x_F, p_T) are given by a model by M. Mangano [23, 24] based on the most recent next-to-next-to-leading-logarithmic (NNLL) MRST parton density functions [25] with a b quark mass of $m_b = 4.75$ GeV/c^2 and a QCD renormalization scale of $\mu_R = \sqrt{m_b^2 + p_T^2}$. Furthermore, the intrinsic transverse momenta k_T of the colliding partons are

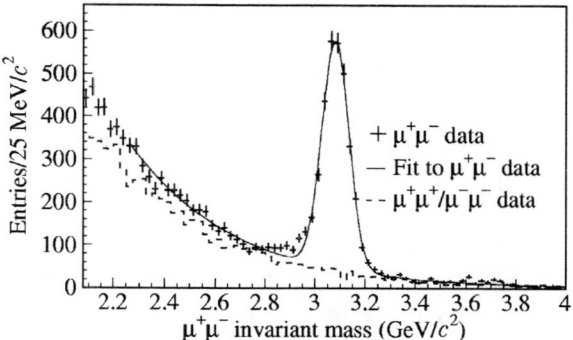

FIGURE 2. The prompt $J/\psi \rightarrow \mu^+\mu^-$ signal of the data taken in 2000. The signal has been optimized to obtain a higher significance $\Sigma = \sqrt{S/(S+B)}$. The fit (solid line) assumes a Gaussian in the signal region and describes the background with an exponential.

smeared with a Gaussian distribution leading to $\langle k_T^2 \rangle = 0.5$ GeV$^2/c^2$ [26]. Finally, the b fragmentation to hadrons is described by a Peterson function using a parameter of $\varepsilon_{Peterson} = 0.006$ [27].

To obtain a realistic description of the kinematical distribution of the prompt J/ψ's, the generated events are re-weighted to match the known prompt J/ψ differential cross sections measured in proton-gold collisions [14]. These results were obtained in the positive x_F ($x_f > -0.05$) region while HERA-B has mainly access to the negative x_f region. MC studies based on the Color Octet Model [28] of charmonium production show a symmetric X_F distribution of the prompt J/ψ decays. Therefore the experimental parameterization of [14] was used to extrapolate to the full x_F space.

5. PROMPT J/ψ SELECTION

Since the number of prompt J/ψ decays n_P is used as a normalization factor of the $b\bar{b}$ cross section measurement, n_P will be determined before applying the detached vertex cuts. The selection of the $J/\psi \rightarrow \ell^+\ell^-$ differs between the μ and e channel due to the different shapes of the background. To select the $\mu^+\mu^-$ decays, a dimuon vertex is required in addition to muon identification cuts in the Muon and RICH systems. The final selection used for the analysis based on the data taken in 2000 is presented in Fig. 2. The selection of the e events is more difficult. Aside from the dilepton vertex two additional identifications cuts were applied:

- Electrons are fully absorbed in the ECAL, therefore the ratio of the cluster energy E associated with the track and the track momenta p should be around 1. The E/p spectrum has to be compatible with

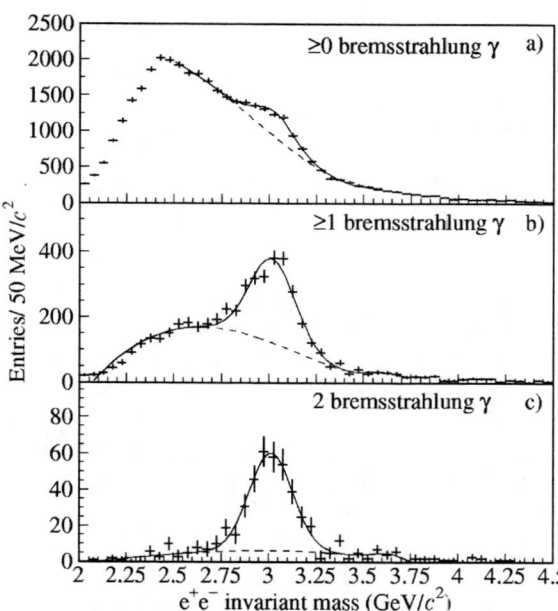

FIGURE 3. The prompt $J/\psi \rightarrow e^+e^-$ signal of the data taking period of 2000, with an E/p cut at 1 σ and 0 (a), 1 (b) or 2 (c) BR requirements in addition. The signal is described by a Gaussian and the background shape by a polynomial.

a Gaussian distribution of mean 1.00 and width $\sigma \approx 9\%$.

- Bremsstrahlung (BR) photons that are emitted by electrons traveling through the layers of material before the magnet keeping the original direction of the lepton. To purify the electron signal, a photon cluster in the ECAL could be required in conjunction with the electron.

In Fig. 3 the invariant mass distributions for the data taken in 2000 are presented. In addition to the requirement that E/p is within $\pm 3\sigma$ it shows, how the significance of the signal improves by requiring 1 or 2 BR clusters.

6. DETACHED ANALYSIS

To identify b hadrons from the decay chain in Eq. 1 their decay length Δz, defined as the distance along the beam axis between the J/ψ decay vertex and the closest target wire, is used (Fig. 4). Additionally, cuts on the minimum impact parameter of both leptons are applied. For the selection in the μ channel in 2000 the cuts were set for the decay length to $|\Delta z| > 0.4$ cm and $|\Delta z| > 7.5 \cdot \sigma_Z$, where σ_Z is the uncertainty of the secondary vertex position. For the impact parameter to the primary vertex the cut was set at $I_P > 160$ μm. In case of the e channel the cuts were chosen to $\Delta z > 0.5$ cm, the

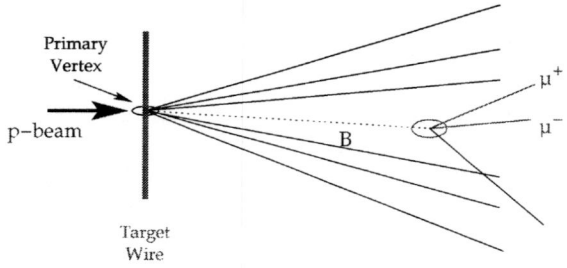

FIGURE 4. Average decay length of B mesons at HERA-B: $DL \approx 8$ mm $\gg \sigma_{\Delta z}$.

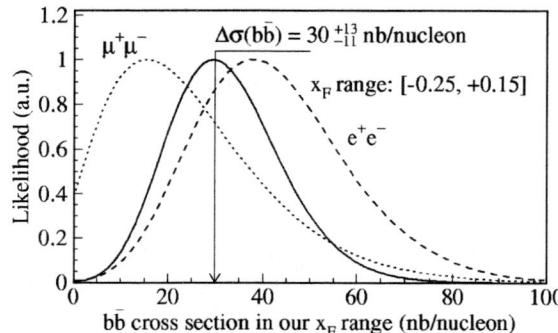

FIGURE 5. The combined fit for the 2000 measurement.

minimal distance in the z-plane, z_w, of the wire to any other track to $z_W > 250\ \mu$m or for the impact parameter to the wire to $I_W > 200\ \mu$m.

6.1. The 2000 $\sigma(b\bar{b})$ Measurement

To extract the number of detached J/ψ's N_B, an unbinned likelihood fit was performed on the invariant mass spectra of the selected detached data. The target materials used in 2000 were mainly carbon (77%, $A = 12$) and titanium (23%, $A = 48$). Overall, in the e channel a number of $n_B = 8.6^{+3.9}_{-3.2}$ detached J/ψ's were found, compared to the number of $n_P = 5710 \pm 380$ of prompt J/ψ's. Due to the good knowledge of the particle identification efficiencies, no BR tag requirement was necessary and n_P was determined applying a $\pm 3\ \sigma$ cut in E/p. For the μ channel $n_B = 1.9^{+2.2}_{-1.5}$ and $n_P = 2880 \pm 60$ was extracted.

To get the maximum information on the $b\bar{b}$ production cross section a combined fit to the e^+e^- and $\mu^+\mu^-$ data samples was performed (Fig. 5) delivering the final result in the HERA-B x_F range of $\Delta\sigma(b\bar{b}) = 30^{+13}_{-11}$ nb/nucleon. Extrapolating to the full x_F range a value of [29]

$$\sigma(b\bar{b}) = 32^{+14}_{-12}(\text{stat})^{+6}_{-7}(\text{sys})\ \text{nb/nucleon} \qquad (4)$$

was found. The main sources of systematic uncertainties originate (for μ/e) from the reference cross section (11%), the b production and decay model (8%), the trigger simulation (5%) and the description of the background shape (11/20%) and fluctuations (11/23%).

6.2. New Measurement (2002/03)

The search for detached J/ψ vertices was performed on the new data of 2002/03 on 40% of the full statistics. Using this amount of data a number of $n^\mu_B = 38 \pm 7$ (Fig. 6) detached J/ψ's were found in the μ channel and $n^e_B = 31 \pm 9$ (Fig. 7) within the e channel. Extrapolating

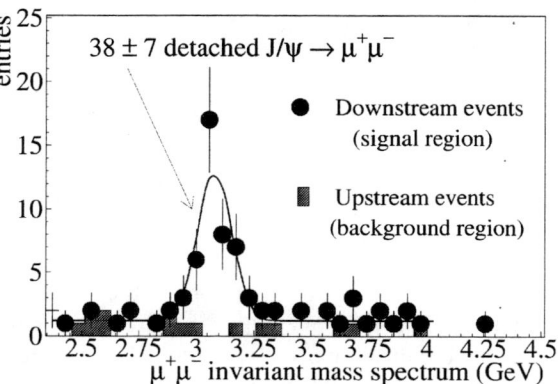

FIGURE 6. Detached Signal of 40% of the data of 2002/03 in the $\mu^+\mu^-$ channel.

these numbers to the full statistics, we expect a total number of detached events in 2002/03 of $n^\mu_B \approx 95 \pm 11$ and $n^e_B \approx 80 \pm 13$. This corresponds to an improvement of the statistical error from $\approx 40\%$ (2000) to $\approx 10\%$ (2002/03) and a reduction of the systematic uncertainties from 24% (33%) down to 16% (18%) for $\mu(e)$.

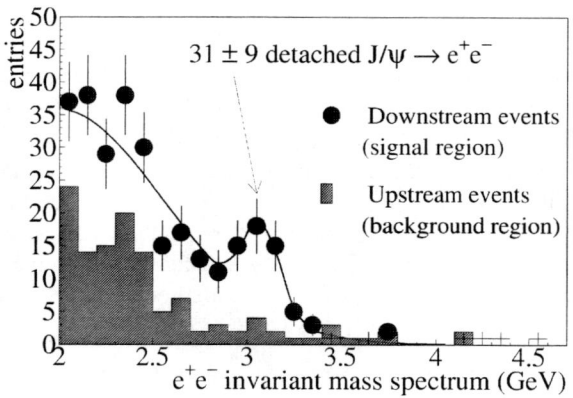

FIGURE 7. Detached Signal of 40% of the data of 2002/03 in the e^+e^- channel.

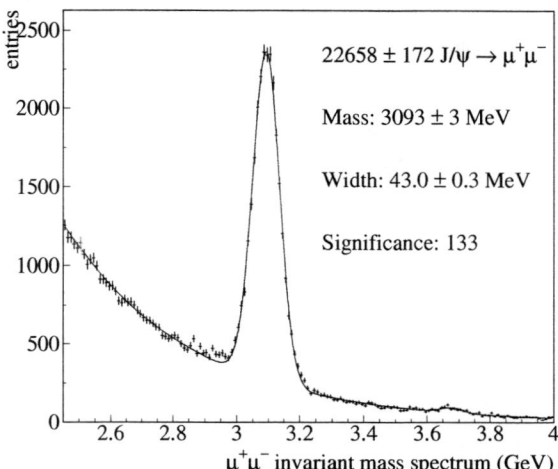

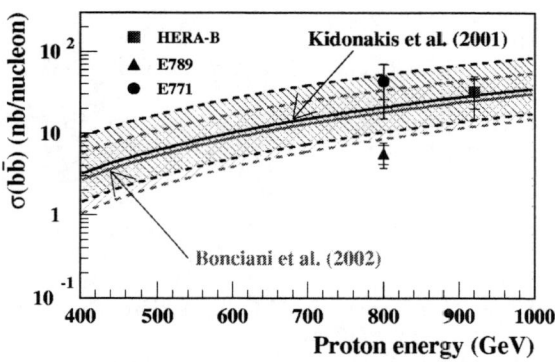

FIGURE 9. Comparison of the HERA-*B* (2000) $\sigma(b\bar{b})$ measurement value with other experiments and the theoretical predictions.

FIGURE 8. The prompt $J/\psi \rightarrow \mu^+\mu^-$ signal of 16 % of the data taken in 2002/03. The signal has been optimized to obtain a higher significance $\Sigma = \sqrt{S/(S+B)}$.

A first complete analysis of the $b\bar{b}$ production cross section was done on the sample shown in Fig. 8 corresponding to 16 % of the full $\mu^+\mu^-$ data sample of 2002/03. The selection cuts for the detached J/ψ's were set to $|\Delta z| > 0.4$ cm, $|\Delta z| > 12 \cdot \sigma_z$ and $I_P > 160 \, \mu$m. Out of a total amount of $n_P = 22,658 \pm 172$ prompt events a number of $n_B = 11^{+4.4}_{-3.7}$ events within in the detached signal was extracted. Since this sample was taken using a carbon wire, a value of $\Delta\sigma_r^C = (245 \pm 6 \pm 19)$nb/nucleon was calculated for the corresponding reference prompt J/ψ cross section.

The efficiency evaluations and the detached cut optimizations are under evaluation. The cuts are optimized by maximizing the significance function $\Sigma = S/\sqrt{S+BG}$, where S stands for signal events and BG for the corresponding background. Using B events from MC for the signal events, the background is described using various models including real data samples and MC events. Once the cuts are optimized on this subsample, they will be used in a blind analysis for the whole data sample to avoid any biased cut selection. Preliminary evaluation provide a measurement of the production cross section $\sigma(b\bar{b})$ on the 2002/03 data in full agreement with the previous result (Eq. 4).

7. CONCLUSIONS

The measurement using the data of 2000 shows good agreement with QCD calculations beyond NLO [30, 31] and the existing experimental results (Fig. 9) [10, 11, 29]. The measurement with new data (2002/03), dealing with ~ 30 times more statistics, can be used to reduce the the-

oretical uncertainties, originating mainly from uncertainties on the mass of the b quark ($m_b \in [4.5, 5.0]$ GeV/c^2).

To analyze the full data sample, the cut optimization will be done on the presented 16 % of the data. Afterwards, the fixed cuts will be used for a blind analysis on the full sample. For the whole available data sample, ~ 95 detached J/ψ's in the μ channel and ~ 80 in the e channel are expected, enabling us to improve the statistical and systematic errors of the first measurement in 2000. From the updated results, an input on b quark mass and on QCD calculations is expected. A preliminary evaluation of the $b\bar{b}$ production cross sections with reduced statistics is fully compatible with the measurement of 2000.

Searches for exclusive B decays $B^\pm \rightarrow J/\psi K^\pm$, $B^0 \rightarrow J/\psi K^\pm \pi^\mp$ and $B^0 \rightarrow J/\psi K_S^0$ are in progress, as well as searches for double semileptonic decays $b\bar{b} \rightarrow \mu\mu X$, $b \rightarrow \mu\nu X$. These analyses will provide alternative $\sigma_{b\bar{b}}$ measurements and allow to cross check the inclusive cross section measurement.

REFERENCES

1. Salam, G. P., *Acta Phys. Polon.* **B33**, 2791–2812 (2002).
2. OPAL Collaboration, Tech. Rep. OPAL Physics Note PN 455 (2001).
3. Acciarri, M., et al., [L3 Collaboration], *Phys. Lett.* **B503**, 10–20 (2001).
4. Acosta, D., et al., [CDF Collaboration], *Phys. Rev.* **D65**, 052005 (2002).
5. Adloff, C., et al., [H1 Collaboration], *Phys. Lett.* **B467**, 156–164 (1999).
6. Adloff, C., et al., [H1 Collaboration], *Phys. Lett.* **B518**, 331–332 (2001).
7. Breitweg, J., et al., [ZEUS Collaboration], *Eur. Phys. J.* **C18**, 625–637 (2001).
8. Nason, P., arXiv:hep-ph/0301003 (2003).
9. Frixione, S., arXiv:hep-ph/0111368 (2001).

10. Jansen, D. M., et al., *Phys. Rev. Lett.* **74**, 3118–3121 (1995).

11. Alexopoulos, T., et al., [E771 Collaboration], *Phys. Rev. Lett.* **82**, 41–44 (1999).

12. E. Hartouni, et al., (HERA-B Collaboration), HERA-B: An Experiment to Study *CP* Violation in the *B* System using an Internal Target at the HERA Proton Ring. Design Report, Tech. Rep. DESY-PRC-95-01 (1995).

13. HERA-B Collaboration, HERA-B: Report on Status and Prospects. Executive Summary, Tech. Rep. DESY-PRC-00-04 (2000).

14. Schub, M. H., et al., [E789 Collaboration], *Phys. Rev.* **D52**, 1307–1315 (1995).

15. Alexopoulos, T., et al., [E771 Collaboration], *Phys. Rev.* **D55**, 3927–3932 (1997).

16. Groom, D. E., et al., [Particle Data Group Collaboration], *Eur. Phys. J.* **C15**, 1–878 (2000).

17. Leitch, M. J., et al., [FNAL E866/NuSea Collaboration], *Phys. Rev. Lett.* **84**, 3256–3260 (2000).

18. Alexopoulos, T., et al., [E771 Collaboration], *Phys. Lett.* **B374**, 271–276 (1996).

19. Leitch, M. J., et al., [E789 Collaboration], *Phys. Rev. Lett.* **72**, 2542–2545 (1994).

20. Vogt, R., [Hard Probe Collaboration], *Int. J. Mod. Phys.* **E12**, 211–270 (2003).

21. Sjostrand, T., *Comput. Phys. Commun.* **82**, 74–90 (1994).

22. Pi, H., *Comput. Phys. Commun.* **71**, 173–192 (1992).

23. Mangano, M. L., Nason, P., and Ridolfi, G., *Nucl. Phys.* **B373**, 295–345 (1992).

24. Nason, P., Dawson, S., and Ellis, R. K., *Nucl. Phys.* **B327**, 49–92 (1989).

25. Martin, A. D., Roberts, R. G., Stirling, W. J., and Thorne, R. S., *Phys. Lett.* **B531**, 216–224 (2002).

26. Chay, J.-g., Ellis, S. D., and Stirling, W. J., *Phys. Rev.* **D45**, 46–54 (1992).

27. Peterson, C., Schlatter, D., Schmitt, I., and Zerwas, P. M., *Phys. Rev.* **D27**, 105 (1983).

28. Cho, P. L., and Leibovich, A. K., *Phys. Rev.* **D53**, 150–162 (1996).

29. Abt, I., et al., [HERA-B Collaboration], *Eur. Phys. J.* **C26**, 345–355 (2003).

30. Kidonakis, N., Laenen, E., Moch, S., and Vogt, R., *Phys. Rev.* **D64**, 114001 (2001).

31. Bonciani, R., Catani, S., Mangano, M. L., and Nason, P., *Nucl. Phys.* **B529**, 424–450 (1998).

Hadronic Beauty Production at HERA

Andreas B. Meyer

(Representing the H1 and ZEUS Collaborations)

Hamburg University, Luruper Chaussee 149, 22761 Hamburg, Germany

Abstract. Recent results on beauty production in electron proton collisions at 300-318 GeV center-of-mass energies are presented. The cross sections measured by the H1 and ZEUS experiments at HERA are compared with perturbative QCD calculations.

1. INTRODUCTION

The measurement of heavy flavour processes at the HERA ep-collider is a powerful means of exploring the dynamics of the strong interactions described by QCD. Perturbative QCD (pQCD) has been seen to successfully describe inclusive production cross sections wherever Q^2, the negative 4-momentum transfer squared, and/or the transverse momenta of the final state are sufficiently large to provide an energy scale for perturbative expansions in orders of the strong coupling constant α_S. In heavy flavour production, the largeness of the heavy quark mass provides an alternative energy scale for perturbative calculations, and it has been shown that pQCD provides a reasonable description of charm production. In contrast, the first measurements of beauty production at HERA [1–3] revealed significant excesses of data over pQCD predictions by factors of ~ 3, in line with similar findings in $p\bar{p}$ and $\gamma\gamma$ collisions at the Tevatron [4–7] and at LEP [8]. In this paper, recent HERA beauty measurements are presented, based on the H1 and ZEUS data recorded between 1995 and 2000.

2. THEORY

In the Standard Model, the dominant mechanism for beauty production in ep-collisions at HERA is the direct boson gluon fusion process as shown in Fig. 1a). Further contributions are expected from resolved photon processes as illustrated in Fig. 1b). All these processes are directly sensitive to the gluon density in the proton. The kinematics at HERA spans a large range in the photon virtuality Q^2 from photoproduction, where $Q^2 < 1$ GeV/c^2 to deep inelastic scattering (DIS), where Q^2 can reach values much higher than the squared b quark mass. It is expected that resolved photon processes are

suppressed towards larger Q^2.

The large mass of the beauty quark provides a hard scale, which renders a small coupling constant α_s in the hard subprocess. This facilitates the calculation of cross sections by means of pQCD. One distinguishes the following two approaches:

- In the so-called massive or fixed order scheme [9], u, d and s are the only active flavours in the proton, and charm and beauty are dynamically produced in the hard scattering. This approach should work well for the kinematic region $p_T \leq m_b$, where p_T is the transverse momentum of the outgoing b quark. Calculation tools up to next-to-leading order (α_s^2) (NLO) are available in the form of Monte Carlo integration programs [10, 11].

- At higher transverse momenta the so called massless or re-summed approach [12–16] should be applicable, where charm and beauty are regarded as active flavours (massless partons) in the proton and in the photon and fragmented only after the hard process into massive quarks. This ansatz incorporates excitation processes, such as the two right most diagrams shown in Fig. 1b).

3. EXPERIMENTAL TECHNIQUE

The total beauty production cross-section at HERA is two orders of magnitude smaller than that of charm. High p_T muons from semileptonic b decays are used to identify the beauty events. In the single track tag analyses, the selected muon is required to be associated with a jet. To separate beauty events from charm and light quark background, two observables are used which exploit the large mass and the long lifetime of the b quark, respectively. First, the transverse momentum p_T^{rel} of the muon

CP722, *B Physics at Hadron Machines*, edited by M. Paulini and S. Erhan

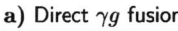

a) Direct γg fusion   b) Resolved γ processes

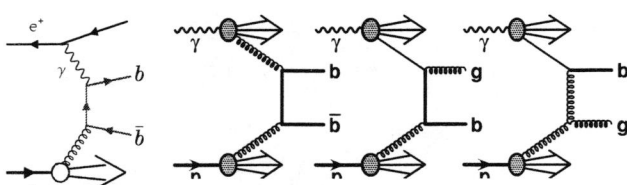

FIGURE 1. Beauty production processes in leading order pQCD.

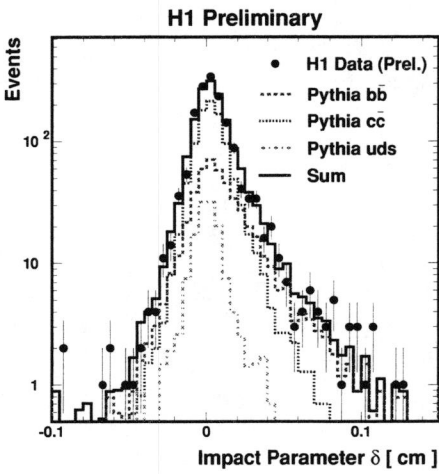

FIGURE 2. Impact parameter distribution as measured by H1 [17]. The lines show the decomposition obtained from a fit of the data into simulated contributions from b (dashed), c (dotted) and light-quark (dashed-dotted) events using the PYTHIA MC generator [18, 19].

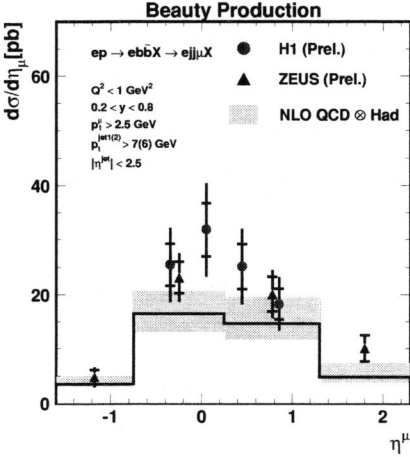

FIGURE 3. Differential beauty cross sections as a function of pseudorapidity of the muon.

with respect to the axis of the associated jet: for muons from b decays this exhibits a much harder spectrum than for the other sources. Second, the impact parameter δ of the muon track with respect to the primary event vertex: for muons from b decays this takes larger values as compared to the other sources. This variable is only used by the H1 experiment, where a central silicon vertex detector provides the necessary track resolution (Fig. 2).

The relative contributions of beauty and background in the data are determined from likelihood fits to the above observables.

In the double track tag measurements a D^{*+} meson is required in addition to the muon. It is reconstructed in the decay channel $D^{*+} \to D^0 \pi_s^+ \to K^- \pi^+ \pi_s^+$. The angular and charge correlation between the muon and the D^{*+} meson are exploited to identify charm and beauty contributions.

4. RESULTS

New preliminary H1 results [17] are available for beauty production in photoproduction, exploiting a three times larger data sample than for the first results [1]. The measurements are based on the muon single track tag technique described in the previous section. Figure 3 shows the single differential cross-sections for $ep \to eb\bar{b}X \to e\,\text{jet}\,\text{jet}\,\mu$ as a function of the muon pseudorapidity. The H1 data are in agreement with the ZEUS data presented last year [20]. The data points are also compared to a NLO calculation in the massive scheme based on FMNR [10]. For the calculation, fragmentation is performed using the Peterson function [21]. The errors of the theory prediction are dominated by the uncertainties of the renormalization scale and the b quark mass. *All* the data points in Fig. 3 lie above the theory calculation. However, data and calculation agree at the two sigma level. Figure 4 shows the ratio of the measured cross sections and the NLO calculation as a function of the muon transverse momentum. At the lowest momenta, the theory undershoots the data by a factor of two.

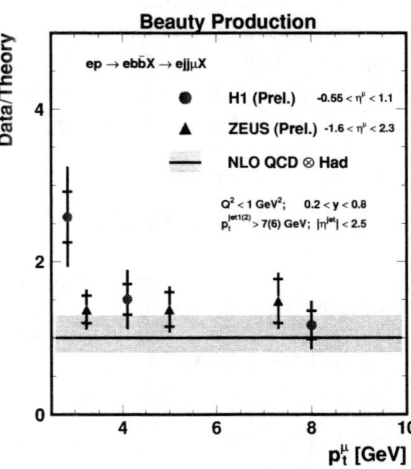

FIGURE 4. Ratio of data and NLO calculation as a function of transverse momentum of the muon.

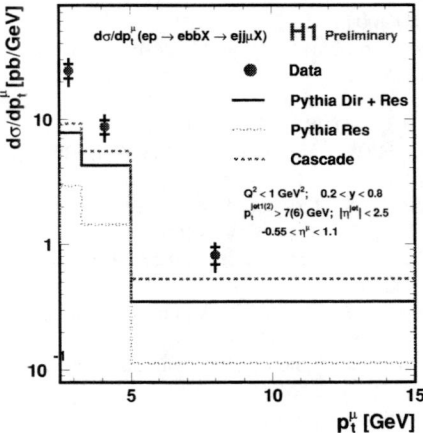

FIGURE 5. Differential beauty cross-sections as a function of transverse momentum of the muon.

The other data points are also above the theory but at a level of $\leq 50\%$. Figure 5 depicts the measured H1 cross-sections in bins of the muon transverse momentum compared to predictions from the leading order Monte Carlo programs PYTHIA [18, 19] and CASCADE [22, 23]. PYTHIA is run in the massless mode and simulates direct and resolved photon processes. Higher orders are approximated in PYTHIA and CASCADE by gluon showers generated using the DGLAP and CCFM evolution prescriptions, respectively. With the settings used, CASCADE and PYTHIA clearly fall below the data.

The contribution from resolved photon processes is investigated by the observable $x_\gamma =$

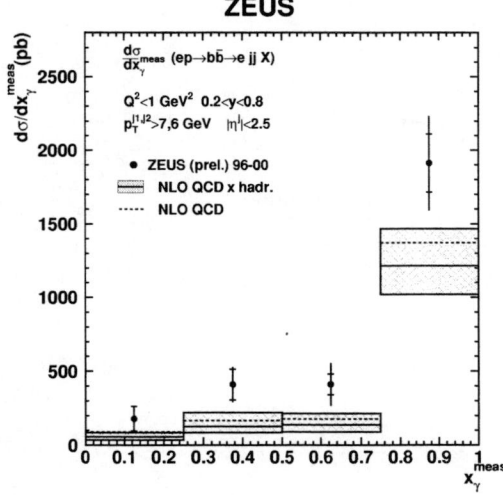

FIGURE 6. Differential beauty cross-section data from ZEUS [20] as a function of x_γ (see text) in comparison with a NLO calculation (band) [10].

$\sum_{j1,j2}(E_T^j e^{-\eta^j})/(2yE_e)$. x_γ can be interpreted in the proton rest frame as the fractional photon momentum of the parton (gluon or b quark) that undergoes the hard scattering process. Figure 6 shows the ZEUS data. Resolved processes, reconstructed at $x_\gamma \leq 0.75$, have a significant contribution to the total cross section of about 30%. The NLO calculation describes the shape of the data, but falls low in normalization.

ZEUS presented last year the first differential cross-section measurements [24] in the deep inelastic scattering regime. Figure 7 shows the cross sections in bins of the photon virtuality Q^2, compared with a NLO calculation in the massive scheme based on [11]. In contrast to the photoproduction case discussed above, data and theory agree well.

ZEUS has presented at this conference a first double track measurement in the deep inelastic scattering regime using D^{*+} mesons and muons [25]. The measured cross-section is shown in Fig. 8. It lies above the NLO calculation in the massive scheme based on HVQDIS [11]. Also shown in Fig. 8 is the corresponding ZEUS result in photoproduction [26] which is also above an NLO calculation based on FMNR [10]. The ZEUS $D^*\mu$ photoproduction results agree with those of the H1 experiment [27].

5. CONCLUSIONS AND OUTLOOK

Recent measurements of beauty production in ep-collisions have been presented. Almost all measured cross-sections are above the QCD calculations which are performed here in the massive scheme in NLO. How-

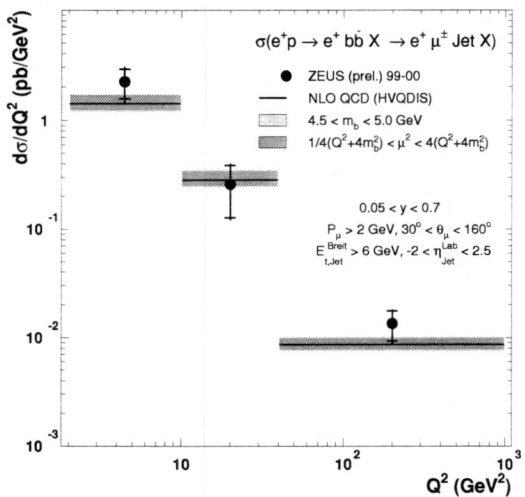

FIGURE 7. Differential beauty cross sections in deep inelastic scattering as a function of the photon virtuality Q^2.

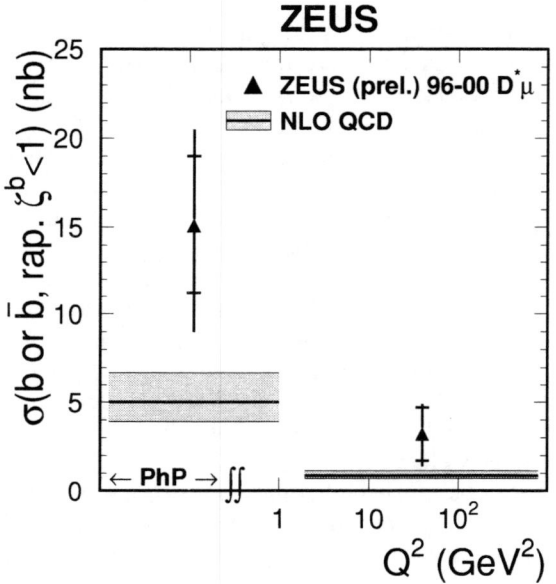

FIGURE 8. $D^{*+}\mu$ cross-sections in photoproduction (left) and DIS (right).

ever, in contrast to some of the earlier measurements, the excesses have now decreased to a level, at which the data are compatible with the theory predictions within errors. In the new H1 photoproduction measurement [17], it is recognized that a large part of the significant excess observed in the corresponding older measurements [1, 2] was due to the fact that the data cross-sections were extrapolated outside their visible range using a leading order Monte Carlo simulation.

The upgraded HERA II collider with a factor of five larger luminosity and the improved H1 and ZEUS detectors will allow beauty measurements in ep-collisions to be made with much higher precision and in an extended kinematic phase-space, further testing our understanding of the physics of heavy flavour production.

REFERENCES

1. Adloff, C., et al., [H1 Collaboration], *Phys. Lett.* **B467**, 156–164 (1999).
2. Adloff, C., et al., [H1 Collaboration] (2000), Contributed paper to ICHEP 2000, Osaka.
3. Breitweg, J., et al., [ZEUS Collaboration], *Eur. Phys. J.* **C18**, 625–637 (2001).
4. Abe, F., et al., [CDF Collaboration], *Phys. Rev. Lett.* **71**, 2396–2400 (1993).
5. Abachi, S., et al., [DØ Collaboration], *Phys. Rev. Lett.* **74**, 3548–3552 (1995).
6. Abachi, S., et al., [DØ Collaboration], *Phys. Lett.* **B370**, 239–248 (1996).
7. Abe, F., et al., [CDF Collaboration], *Phys. Rev.* **D53**, 1051–1065 (1996).
8. Acciarri, M., et al., [L3 Collaboration], *Phys. Lett.* **B503**, 10–20 (2001).
9. Frixione, S., Nason, P., and Ridolfi, G., *Nucl. Phys.* **B454**, 3–24 (1995).
10. Frixione, S., Mangano, M. L., Nason, P., and Ridolfi, G., *Phys. Lett.* **B348**, 633–645 (1995).
11. Harris, B. W., and Smith, J., *Phys. Rev.* **D57**, 2806–2812 (1998).
12. Kniehl, B. A., Kramer, M., Kramer, G., and Spira, M., *Phys. Lett.* **B356**, 539–545 (1995).
13. Kniehl, B. A., Kramer, G., and Spira, M., *Z. Phys.* **C76**, 689–700 (1997).
14. Binnewies, J., Kniehl, B. A., and Kramer, G., *Z. Phys.* **C76**, 677–688 (1997).
15. Binnewies, J., Kniehl, B. A., and Kramer, G., *Phys. Rev.* **D58**, 014014 (1998).
16. Cacciari, M., and Greco, M., *Phys. Rev.* **D55**, 7134–7143 (1997).
17. Adloff, C., et al., [H1 Collaboration] (2003), Contributed paper no. 117 to EPS03, Aachen.
18. Sjostrand, T., *Comput. Phys. Commun.* **82**, 74–90 (1994).
19. Norrbin, E., and Sjostrand, T., *Eur. Phys. J.* **C17**, 137–161 (2000).
20. Chekanov, S., et al., [ZEUS Collaboration] (2002), Contributed paper no. 785 to ICHEP 2002, Amsterdam.
21. Peterson, C., Schlatter, D., Schmitt, I., and Zerwas, P. M., *Phys. Rev.* **D27**, 105 (1983).
22. Jung, H., and Salam, G. P., *Eur. Phys. J.* **C19**, 351–360 (2001).
23. Jung, H., *Comput. Phys. Commun.* **143**, 100–111 (2002).
24. Chekanov, S., et al., [ZEUS Collaboration] (2002), Contributed paper no. 783 to ICHEP 2002, Amsterdam.
25. Chekanov, S., et al., [ZEUS Collaboration] (2003), Contributed paper no. 675 to EPS03, Aachen.
26. Chekanov, S., et al., [ZEUS Collaboration] (2002), Contributed paper no. 784 to ICHEP 2002, Amsterdam.
27. Adloff, C., et al., [H1 Collaboration] (2002), Contributed paper no. 1016 to ICHEP 2002, Amsterdam.

Heavy Flavor Production at the Tevatron

Chunhui Chen

(Representing the CDF and DØ Collaborations)

Department of Physics and Astronomy, University of Pennsylvania, Philadelphia, PA, U.S.A.

Abstract. Using a subset of the current Run II data, CDF and DØ have performed several measurements of heavy flavor production. In this paper, we present a new measurement of prompt charm meson production by CDF. We also report the latest CDF II measurements of inclusive J/ψ production and b production without a requirement on the minimum transverse momentum of the J/ψ and b quark. These are the first measurements of the total inclusive J/ψ and b quark cross section in the central rapidity region at a hadron collider. Results of the J/ψ cross section as a function of rapidity, and b-jet production cross section measured by DØ are also reviewed.

1. INTRODUCTION

Measurements of the production cross section of heavy flavor quarks (c and b quarks) in $p\bar{p}$ collisions provide an opportunity to test predictions based on Quantum Chromodynamics (QCD). No only is QCD one of the four fundamental forces of nature, but also many searches of new physics beyond the Standard Model require a good understanding of QCD backgrounds.

The bottom quark production cross section has been measured by both the CDF and DØ experiments [1, 2] at the Fermilab Tevatron in $p\bar{p}$ collision at $\sqrt{s} = 1.8$ TeV, and found to be about three times larger than the next-to-leading order (NLO) QCD calculation [3, 4]. Since then, several theoretical explanations have been proposed to solve the disagreement, such as large contributions from NNLO QCD processes, possible contributions from "new physics" [5], *etc.* Recently, a more accurate description of b quark fragmentation has reduced the discrepancy to a factor 1.7 [6]. Nevertheless, further experimental measurements on the heavy flavor production are essential to shed light on the remaining clouds of this long standing issue.

Since the end of Run I in 1996, the Tevatron has undergone a major upgrade. The center-of-mass energy of the $p\bar{p}$ collision was increased from $\sqrt{s} = 1.8$ TeV to $\sqrt{s} = 1.96$ TeV. The large luminosity enhancement dramatically increases the discovery reach and moves the experimental program into a regime of precision hadron collider physics. At the same time, both the CDF [7] and DØ [8] experimental collaborations significantly upgraded their detectors to enrich the physics capabilities, especially for the heavy flavor physics program. The Tevatron Run II data taking period started in April 2002.

So far, the accelerator has delivered about 300 pb^{-1} integrated luminosity and both experiments collected more than 200 pb^{-1} physics quality data. In this paper, we will give a brief summary of the latest results on heavy flavor production from the Tevatron Run II.

2. PROMPT CHARM MESON PRODUCTION CROSS SECTION

In order to help understanding the discrepancy between the measured Run I bottom quark cross section measurement and the theoretical calculation, one can repeat the previous measurement with better control of experimental uncertainties. An alternative approach is to examine the production rate of another heavy quark. Charm meson production cross sections have not been measured in $p\bar{p}$ collisions and may help to solve this disagreement. For the CDF II upgrade, one of the crucial improvements with respect to CDF I is the implementation of a Silicon Vertex Trigger (SVT) [9], which allows CDF to trigger on displaced tracks originating from long-lived charm and bottom hadrons. With this trigger, a large amount of fully reconstructed charm mesons has been collected, opening a new window for heavy flavor production studies at hadron colliders.

Using just (5.8 ± 0.3) pb^{-1} of data, CDF performed a measurement of the prompt charm meson production cross section [10]. The charm mesons are reconstructed in the following decay modes: $D^0 \to K^-\pi^+$, $D^{*+} \to D^0\pi^+$ with $D^0 \to K^-\pi^+$, $D^+ \to K^-\pi^+\pi^+$, $D_s^+ \to \phi\pi^+$ with $\phi \to K^+K^-$, and their charge conjugate, as shown in Fig. 1. We separate charm directly produced in the $p\bar{p}$ interaction (prompt charm) from charm originating from

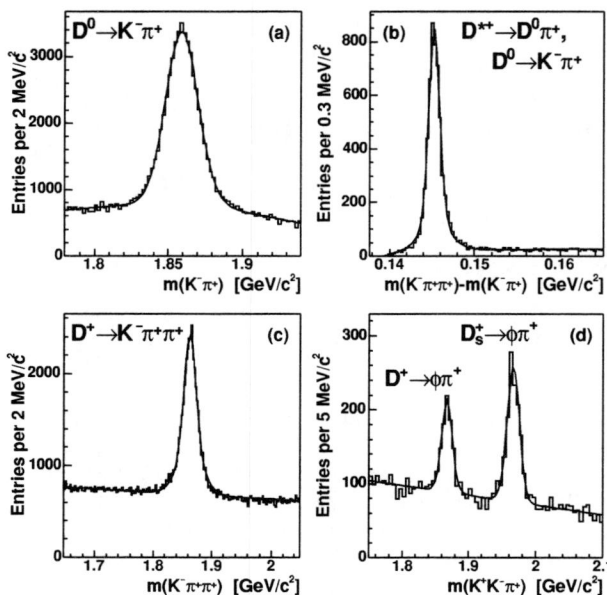

FIGURE 1. Charm signals summed over all p_T bins: (a) invariant mass distribution of $D^0 \to K^- \pi^+$ candidates; (b) mass difference distribution of $D^{*+} \to D^0 \pi^+$ candidates; (c) invariant mass distribution of $D^+ \to K^- \pi^+ \pi^+$ candidates; (d) invariant mass distribution of $D^+ \to \phi \pi^+$ and $D_s^+ \to \phi \pi^+$ candidates.

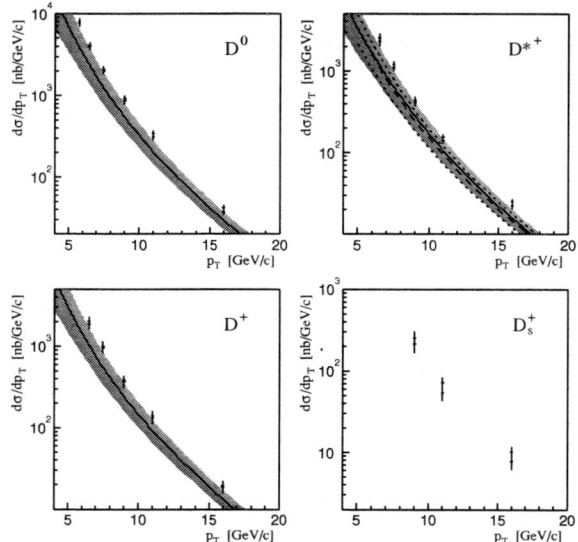

FIGURE 2. The measured differential cross sections for $|y| \leq 1$, shown by the points. The inner bars represent the statistical uncertainties; the outer bars are the quadratic sums of the statistical and systematic uncertainties. The solid curves are the theoretical predictions from Cacciari and Nason [11], with the uncertainties indicated by the shaded bands. The dashed curve shown with the D^{*+} cross section is the theoretical prediction from Kniehl [12]; the dotted lines indicate the uncertainty. No prediction is available yet for D_s^+ production.

B meson decays (secondary charm) using the impact parameter of the charm candidate with respect to the primary interaction point. A prompt charm meson points to the beamline and its impact parameter is zero. A secondary charm meson candidate does not necessarily point back to the primary vertex, as it has a rather wide impact parameter distribution. The shape of the impact parameter distribution of secondary charm is obtained from a Monte Carlo simulation. The detector impact parameter resolution is modeled from a sample of $K_S^0 \to \pi^+ \pi^-$ decays that satisfy the same trigger requirement. The prompt charm fractions are measured as a function of charm meson p_T. Averaged over all p_T bins, $(86.6 \pm 0.4)\%$ of the D^0 mesons, $(88.1 \pm 1.1)\%$ of D^{*+}, $(89.1 \pm 0.4)\%$ of D^+, and $(77.3 \pm 3.8)\%$ of D_s^+ are promptly produced (statistical uncertainties only).

The measured prompt charm meson integrated cross sections are found to be
$\sigma(D^0, p_T \geq 5.5 \text{ GeV}/c, |y| \leq 1) = 13.3 \pm 0.2 \pm 1.5 \,\mu\text{b}$,
$\sigma(D^{*+}, p_T \geq 6.0 \text{ GeV}/c, |y| \leq 1) = 5.2 \pm 0.1 \pm 0.8 \,\mu\text{b}$,
$\sigma(D^+, p_T \geq 6.0 \text{ GeV}/c, |y| \leq 1) = 4.3 \pm 0.1 \pm 0.7 \,\mu\text{b}$ and
$\sigma(D_s^+, p_T \geq 8.0 \text{ GeV}/c, |y| \leq 1) = 0.75 \pm 0.05 \pm 0.22 \,\mu\text{b}$,
where the first uncertainty is statistical and the second systematic. The measured differential cross sections are compared to two recent calculations [11, 12], as shown in Figs. 2 and 3. They are higher than the theoretical predictions by about 100% at low p_T and 50% at high p_T.

However, they are compatible within uncertainties. The same models also underestimate B meson production at $\sqrt{s} = 1.8$ TeV by similar factors [2, 6, 13].

3. INCLUSIVE J/ψ PRODUCTION CROSS SECTION

The CDF II detector has improved the dimuon trigger system with a lower threshold of 1.5 GeV/c. This allows us to trigger on $J/\psi \to \mu^- \mu^+$ events with transverse momentum all the way down to $p_T(J/\psi) = 0$ GeV/c. Using about 39.7 pb^{-1} of data, CDF reconstructed 299800 ± 800 J/ψ candidates (statistical uncertainty only), as shown in Fig. 4. A new measurement of the total inclusive J/ψ cross section in the central rapidity region $|y| \leq 0.6$ has been carried out. This is the first measurement of the total inclusive J/ψ cross section in the central rapidity region at a hadron collider. The differential cross section is shown in Fig. 5, and the integrated cross section is measured to be:

$$\sigma(p\bar{p} \to J/\psi X, |y(J/\psi)| \leq 0.6) \times \mathscr{B}(J/\psi \to \mu\mu)$$

$$= 240 \pm 1(\text{stat})^{+35}_{-28}(\text{syst}) \,\text{nb}.$$

Taking advantage of the large azimuthal coverage of

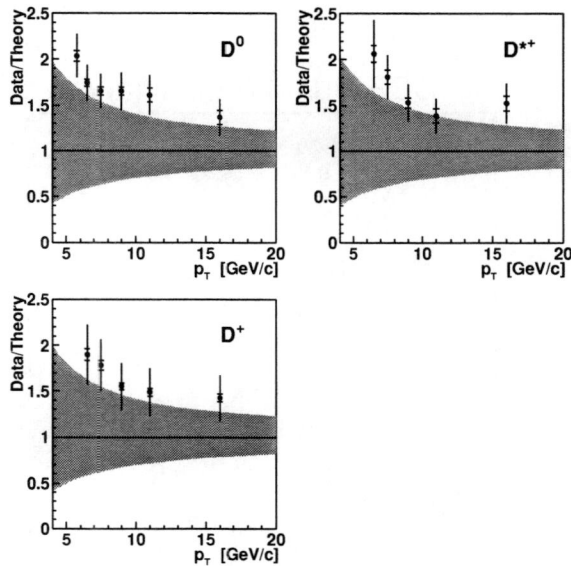

FIGURE 3. Ratio of the measured cross sections to the theoretical calculation from Cacciari and Nason. The inner error bar represents the statistical uncertainty, the outer error bar the quadratic sum of the statistical and systematic uncertainty. The hatched band represents the uncertainty from varying the renormalization and factorization scale.

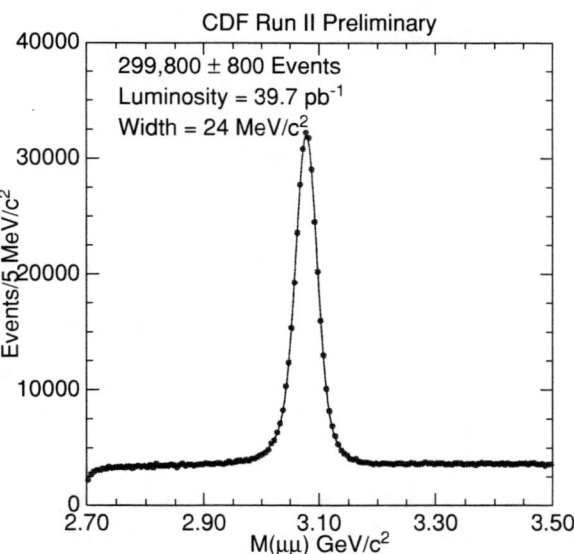

FIGURE 4. The invariant mass distribution of triggered J/ψ events reconstructed in 39.7 pb^{-1} of CDF II data.

its muon system, the DØ experiment measured the differential cross section of inclusive J/ψ production as a function of rapidity using 4.74 pb^{-1}. The distribution is shown in Fig. 6.

4. BOTTOM QUARK PRODUCTION CROSS SECTION

The Run I central b quark production cross sections measured by CDF and DØ have a minimum transverse momentum cut on the b hadrons due to the trigger requirements. Since those measurements only explore a small fraction ($\sim 10\%$) of the b hadron p_T spectrum, it is not clear from the data whether the excess over the theory is due to an overall discrepancy of the $b\bar{b}$ production rate, or it is caused by a shift in the spectrum toward higher p_T. An inclusive measurement of bottom quark production over all transverse momenta can certainly help resolve this ambiguity.

Notice that a large fraction of b hadrons, H_b, decays to J/ψ final states, where H_b denotes both hadron and anti-hadron. CDF performed an analysis to extract the inclusive b hadron cross section from the measured inclusive J/ψ production cross section. Due to the long lifetime of the b hadrons, the vertex of the J/ψ decay from a b hadron (secondary J/ψ) is usually hundreds of microns away from the primary vertex, while

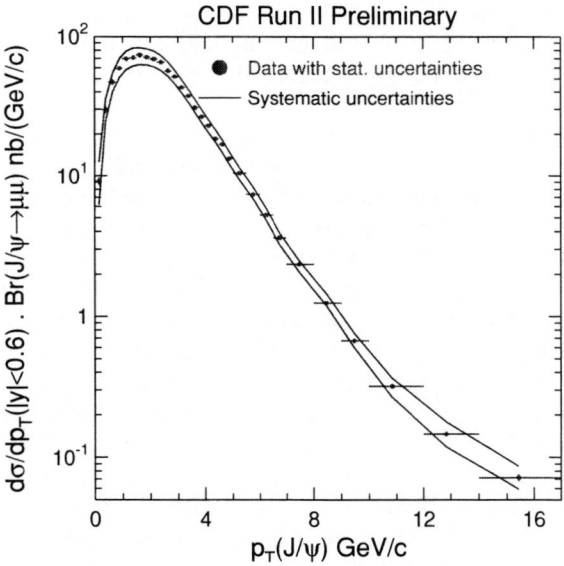

FIGURE 5. The differential cross section for inclusive J/ψ production as a function of p_T with $|y| \leq 0.6$.

for prompt J/ψ's, which are directly produced or decayed from higher charmonium states, their vertex is at the proton-antiproton interaction point. Therefore, we can statistically separate these two components by examining the projected J/ψ decay distance along its trans-

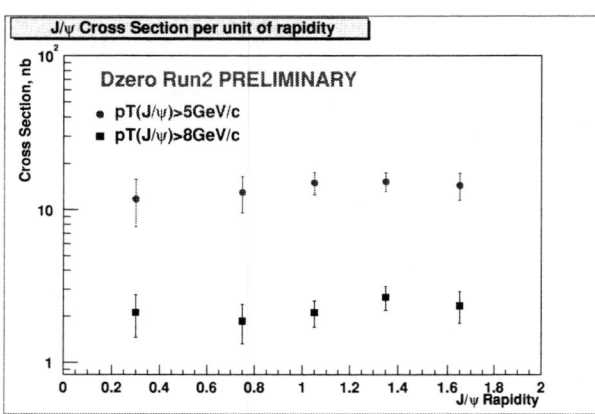

FIGURE 6. The differential J/ψ cross section as a function of $|y|$.

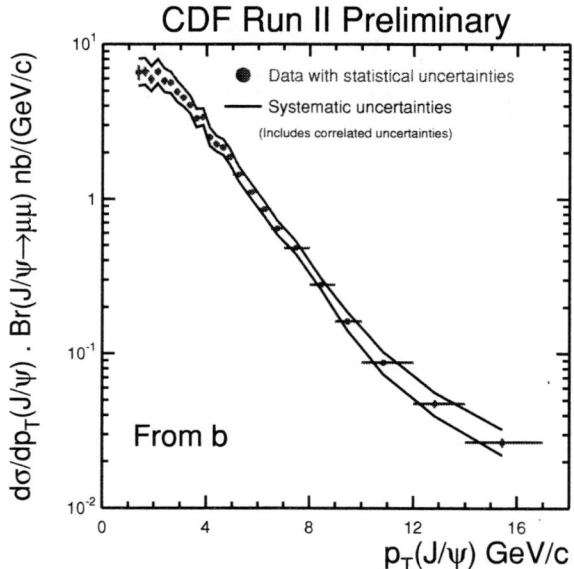

FIGURE 7. b hadron differential cross section as a function of $p_T(J/\psi)$.

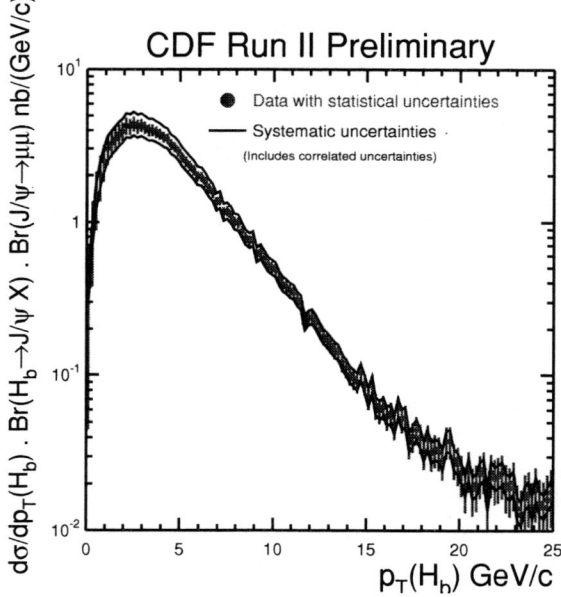

FIGURE 8. b hadron differential cross section as a function of $p_T(H_b)$ with $|y| \leq 0.6$.

verse momentum. The measured b fraction is then applied to the previously measured inclusive J/ψ cross section to obtain the differential production cross section of $H_b \to J/\psi X$ as a function of $p_T(J/\psi)$. However, the above algorithm to extract the b fraction does not work for J/ψ candidates with low transverse momentum, because a b hadron with low p_T does not travel far enough away from the primary vertex in the transverse plane. To obtain a reliable determination of the b fractions, a minimum $p_T(J/\psi) \geq 1.25$ GeV/c requirement is imposed for the b hadron cross section measurement as a function of $p_T(J/\psi)$. The measured differential cross section result is shown in Fig. 7.

Luckily, the b hadrons decaying at rest can still transfer a few (~ 1.7) GeV/c transverse momentum to its J/ψ daughter because of the large b hadron mass. The knowledge of secondary J/ψ production with transverse momenta less than 2.0 GeV/c allows us to probe b hadrons with low transverse momenta down to zero. Therefore using the measured b hadron differential cross section with $1.25 \leq p_T(J/\psi) \leq 17.0$ GeV/c, we are able to extract the b hadron differential cross section as a function of $p_T(H_b)$ down to $p_T(H_b) = 0$ GeV/c. To do so, we perform a convolution that is based on Monte Carlo templates of the b hadron transverse momentum distribution. The unfolding procedure is then repeated using the measured b hadron production spectrum as the input spectrum for the next iteration, until the χ^2 comparison between the input and output spectrum calculated after each iteration becomes stable. The measured b hadron cross section as a function of $p_T(H_b)$ is plotted in Fig. 8. The total b hadron cross section is found to be

$$\sigma(p\bar{p} \to H_b X, |y(H_b)| \leq 0.6) \times \mathscr{B}(H_b \to J/\psi X)$$

$$\times \mathscr{B}(J/\psi \to \mu\mu) = 24.5 \pm 0.5(\text{stat})^{+4.7}_{-3.9}(\text{syst})\,\text{nb},$$

The total single b quark cross section is obtained by

dividing this measurement by two, the corresponding branching fractions, and the rapidity correction factors

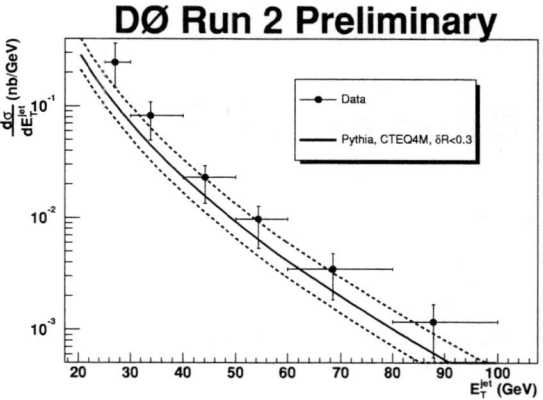

FIGURE 9. Measured Run II b-jet cross section compared to theoretical prediction.

obtained from MC. We find:

$$\sigma(p\bar{p} \to \bar{b}X, |y(b)| \le 0.6) =$$

$$= 18.0 \pm 0.4(\text{stat}) \pm 3.8(\text{syst}) \ \mu\text{b},$$

and

$$\sigma(p\bar{p} \to \bar{b}X, |y(b)| \le 1) =$$

$$= 29.4 \pm 0.6(\text{stat}) \pm 6.2(\text{syst}) \ \mu\text{b}.$$

5. b-JET PRODUCTION CROSS SECTION

Using 3.4 pb^{-1} of data, a b-jet production cross section analysis has been performed by DØ. Candidate events are selected by associating a muon track with a jet, which is defined by a $R = \sqrt{\Delta\eta^2 + \Delta\phi^2}$ cone algorithm. Notice the fact that muons from b quark decays have a higher transverse momentum with respect to the net momentum vector of combined muon plus jet. The corresponding distribution is then used to fit the b-jet fraction. The signal template is modeled from a $b \to \mu$ Monte Carlo simulation, and the background template is extracted from 1.5 million QCD events. After correcting the muon and jet reconstruction efficiency, and the calorimeter jet energy resolution, the b-jet cross section is obtained. The preliminary result is shown in Fig. 9 compared to the theoretical prediction. This measurement is consistent with the previous Run I measurement [14].

6. CONCLUSION

The understanding of heavy flavor production is currently one of the most important challenges faced by

QCD. In this paper, we present the most recent measurements from the CDF and DØ experiments. Given the large amounts of data that have already been collected, rapid improvement of Tevatron performance, and further understanding of the detector, we expect that much more precise measurements of heavy flavor production will be available in the near future. At the same time, several other analyses have also been carried out to explore other aspects of heavy flavor production mechanisms, such as studies of $b\bar{b}$ and $c\bar{c}$ correlations during production. In the next few years, a combination of better experimental data and improved theory should advance our knowledge of heavy quark production.

ACKNOWLEDGMENTS

We thank the CDF and DØ collaboration for their helps while preparing this paper. We also thank the conference organizers for a wonderful meeting.

REFERENCES

1. Abbott, B., et al., [DØ Collaboration], *Phys. Lett.* **B487**, 264–272 (2000).
2. Acosta, D., et al., [CDF Collaboration], *Phys. Rev.* **D65**, 052005 (2002).
3. Nason, P., Dawson, S., and Ellis, R. K., *Nucl. Phys.* **B327**, 49–92 (1989).
4. Albajar, C., et al., [UA1 Collaboration], *Phys. Lett.* **B186**, 237 (1987).
5. Berger, E. L., et al., *Phys. Rev. Lett.* **86**, 4231–4234 (2001).
6. Cacciari, M., and Nason, P., *Phys. Rev. Lett.* **89**, 122003 (2002).
7. Blair, R., et al., [CDF Collaboration], The CDF II Detector: Technical Design Report, Tech. Rep. FERMILAB-PUB-96-390-E (1996).
8. Abachi, S., et al., [DØ Collaboration], The DØ Upgrade: The Detector and its Physics, Tech. Rep. FERMILAB-PUB-96-357-E (1996).
9. Ashmanskas, W., et al., [CDF Collaboration], *Nucl. Instrum. Meth.* **A447**, 218–222 (2000).
10. Acosta, D., et al., [CDF Collaboration], *Phys. Rev. Lett.* **91**, 241804 (2003).
11. Cacciari, M., and Nason, P., *JHEP* **09**, 006 (2003).
12. Kniehl., B. (private communication), Their calculation employs the method described in B.A. Kniehl, G. Kramer, B. Potter, *Nucl. Phys.* **B597**, 337-369 (2001).
13. Binnewies, J., Kniehl, B. A., and Kramer, G., *Phys. Rev.* **D58**, 034016 (1998).
14. Abbott, B., et al., [DØ Collaboration], *Phys. Rev. Lett.* **85**, 5068–5073 (2000).

B Reconstruction and Spectroscopy at DØ

Eckhard von Toerne

(Representing the DØ Collaboration)

Kansas State University, Department of Physics, Manhattan, KS 66506, U.S.A.

Abstract. The recent upgrade of the DØ detector for the Tevatron Run II significantly improved the *B* reconstruction capabilities of DØ. Based on an integrated luminosity of about 115 pb^{-1}, we introduce the reader to DØ's *B* reconstruction techniques and present first results in *B* spectroscopy as well as an improved limit for $B_s \rightarrow \mu^- \mu^+$.

1. INTRODUCTION

The results presented here are based on a data set of about 115 pb^{-1}, collected between April 2002 and June 2003. These data allow us to study the reconstruction capabilities of the DØ detector which are demonstrated on benchmark *b* hadron decays. More information about the DØ *B* physics program can be found in the overview presentation by V. Jain [1].

1.1. *B* Physics at the Tevatron

The *B* factories BaBar and Belle together with many measurements from CLEO, have brought us ample knowledge about B^0 and B^+ meson decays.

More massive *b* quark states (excluding $b\bar{b}$-states) are at the moment only accessible at the Tevatron. One of the central topics of *B* physics at the Tevatron is the search for B_s^0-$\bar{B}_s^0$ mixing [2], which requires data based on a multitude of B_s^0 decay modes. Other important topics are searches for rare *B* decays, lifetime measurements, *B* spectroscopy and *b* production properties. Several higher *B* states, such as B_c, are yet to be discovered.

1.2. DØ Detector

The DØ detector has recently been upgraded for the Tevatron Run II [3]. The tracking system consists of a silicon microvertex detector [4] and a fiber tracker. Both devices are enclosed in a solenoidal magnetic field of 2 Tesla. Pre-shower counters in front of the calorimeter improve the electron and photon identification. The muon system consists of three layers; muons need a momentum larger than 3 GeV/c to be able to penetrate all layers. The muon system can identify muons with

$|\eta| < 2$. Requirements on the transverse momentum of reconstructed muons vary with η. In most of the results presented here, we require $p_T > 2$ GeV/c.

1.3. Data Set

The dataset corresponds to about 115 pb^{-1}. Initial event selection was based either on the single muon or the dimuon trigger. We required that muons have hits in all layers of the muon system. The single muon trigger was prescaled for instantaneous luminosities above $\mathscr{L} > 2.0 \times 10^{31}$ cm^{-2}s^{-1}.

2. RECONSTRUCTION OF *B* HADRONS

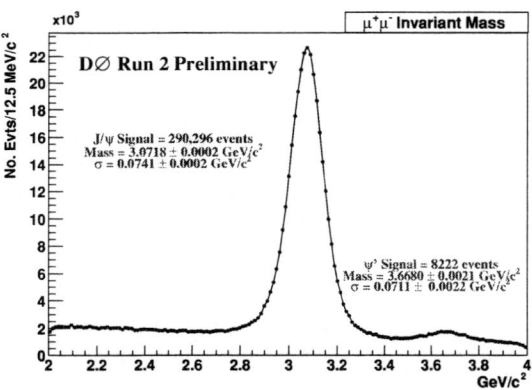

FIGURE 1. Invariant mass distribution for $J/\psi \rightarrow \mu^+\mu^-$. A small peak from $\psi' \rightarrow \mu^+\mu^-$ is also visible.

J/ψ **Reconstruction:** Fig. 1 shows the J/ψ peak in the $\mu^+\mu^-$ channel. A small peak from $\psi' \rightarrow \mu^+\mu^-$ is also visible. The experimental resolution for the

CP722, *B Physics at Hadron Machines*, edited by M. Paulini and S. Erhan

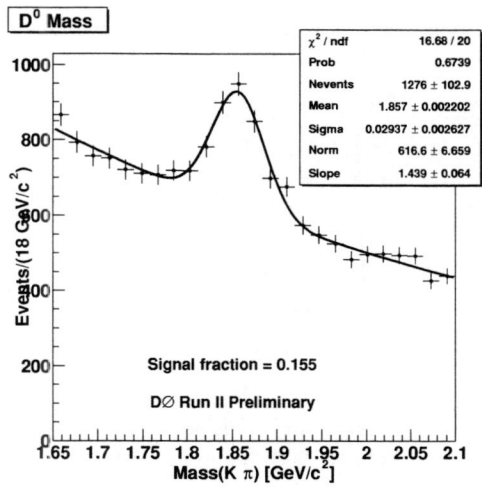

FIGURE 2. Invariant mass distribution for $D^0 \to K^- \pi^+$.

J/ψ is 74.1 ± 0.2 MeV/c^2. The good overall signal to background ratio demonstrates the lepton-id capabilities of the DØ detector.

D^0 Reconstruction: D^0 candidates are formed from unlike-sign track pairs. We obtain a mass resolution of about 30 MeV/c^2 (Fig. 2).

Building B Hadrons: To ensure optimal mass resolution when building B particles, we apply mass-constrained kinematic fits to all particles that were built from tracks ($J/\psi, D^0, \phi, \Lambda, K_S^0$).

Benchmark B Decays: Figures 3 and 4 demonstrate the B reconstruction capabilities of our detector. Typical B mass resolutions are around 35 MeV/c^2, depending upon the decay mode.

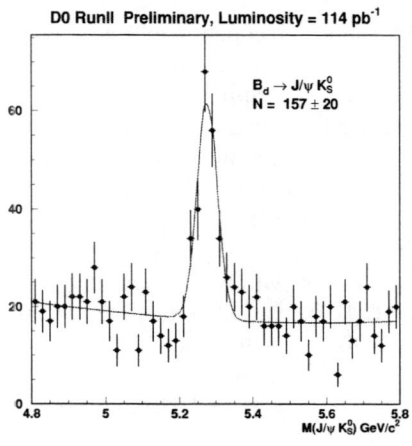

FIGURE 3. $B \to J/\psi K_s^0$.

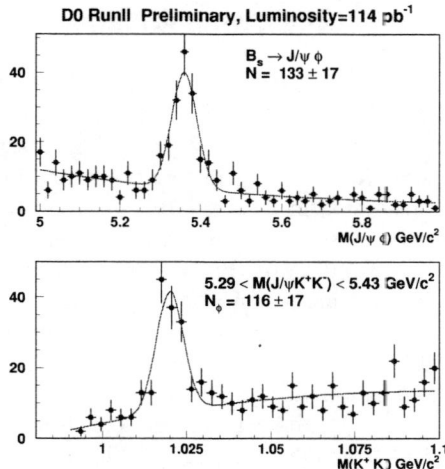

FIGURE 4. Top: $B_s^0 \to J/\psi\phi$. Bottom: $\phi \to K^+K^-$ from B_s^0 candidates in the B_s^0 signal mass window.

3. B SPECTROSCOPY

3.1. B^{**}

B^{**} is a collective name for four resonances expected to be close together and which have not been separated yet. Using 1193 ± 52 fully reconstructed B^+ and 463 ± 38 B^0 decays, we searched for the decays $B^{**0} \to B^+\pi^-$ and $B^{**+} \to B^0\pi^+$. In the B^{**0} mode, DØ observed a signal of 65 ± 17 events (Fig. 5). Most of the background is combinatorial and can be estimated using wrong-sign $B^-\pi$ combinations (red histogram in Fig. 5).

Besides its importance for B spectroscopy, the decay $B^{**+} \to B^0\pi^+$ can also be utilized as an important flavor tagging technique (Same Side Tagging).

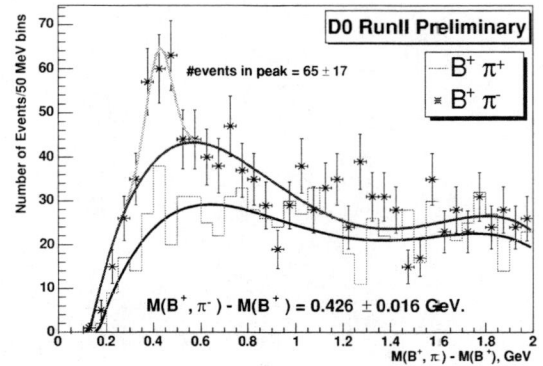

FIGURE 5. $B^{**+} \to B^0\pi^+$.

73

3.2. Single Muon Samples.

DØ's single muon trigger samples provide an abundant source of semi-leptonic B decays $B \to D\mu X$. Based on a small data sample of 6.2 pb^{-1}, Fig. 6 shows the reconstructed $K^- \pi^+$ mass in events where a muon has been reconstructed in the vicinity of the D^0. A cut on the invariant mass of the muon-D^0 system of $2.2 < M(\mu-D^0) < 5.5$ GeV/c^2 has reduced the background to a level where the signal is clearly dominant. As can be seen from the distribution of wrong sign μD^0 pairs (Fig. 6, right plot), most $(\mu^+ \bar{D}^0)$ combinations come indeed from b hadron decays.

B decays to D_s^+ and a muon are shown in Fig. 7. A satellite peak from $D^+ \to \phi\pi^+$ is visible to the left of the large peak from $D_s^+ \to \phi\pi^+$.

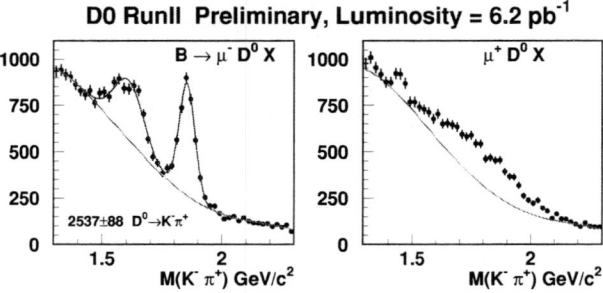

FIGURE 6. Invariant $K^- \pi^+$ mass distribution. Left plot: Right-sign pairs. Requiring a muon in the vicinity of the D^0 enhances $B \to D^0\mu X$. To the left of the main peak is a satellite peak from $D \to K^*\pi$. Right plot: Wrong-sign pairs. The combinatorial background in wrong-sign pairs has only a small $D^0 \to K^-\pi^+$ component.

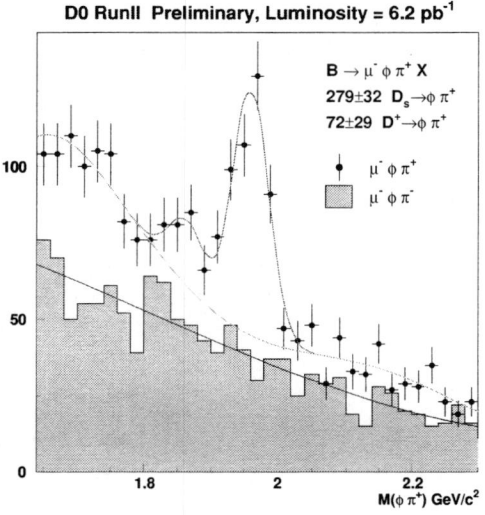

FIGURE 7. Invariant mass distribution for $D_s^+ \to \phi\pi^+$. Requiring a muon in the vicinity of the D^0 enhances $B \to D_s^+\mu X$. A satellite peak from $D^+ \to \phi\pi^+$ is visible to the left of the large peak from $D_s^+ \to \phi\pi^+$.

3.3. $B_s^0 \to \mu^+\mu^-$

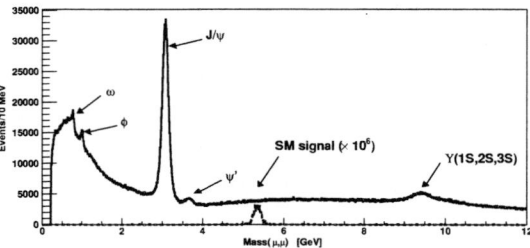

FIGURE 8. Invariant mass of $\mu^+\mu^-$ pairs. Mass peaks from $\omega, \phi, J/\psi$ and Υ are clearly visible. The expected Standard Model signal times 10^6 is shown as the dashed line (simulated signal).

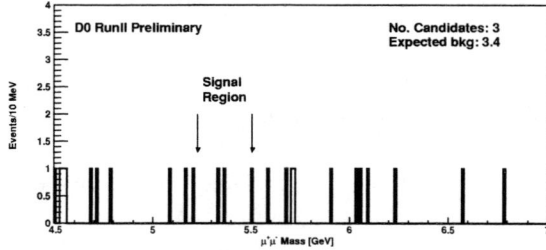

FIGURE 9. $B_s^0 \to \mu^+\mu^-$ candidates after all cuts. The signal region is indicated by arrows.

Flavor changing neutral currents are so small in the Standard Model that non-Standard Model amplitudes might contribute measurably. This explains the considerable theoretical interest in $B_s^0 \to \mu^+\mu^-$ [5, 6]. In some models, non-standard contributions are large enough such that the decay mode is possibly observable in Run II.

The spectrum of reconstructed muon pairs is shown in Fig. 8. The considerable background of direct muon pairs is reduced by requiring the two muon tracks to form a displaced vertex and selecting candidates with a minimum transverse momentum of $p_T > 4.0$ GeV/c, in addition to the requirement that each muon is isolated within a cone of $\Delta R < 1.0$.

After applying these cuts three candidates remain, which is consistent with the background expectation of 3.4 events (Fig. 9). Using the Feldman-Cousins method, we obtain a limit on the branching ratio of $\mathscr{B}(B_s^0 \to \mu^+\mu^-) < 1.6 \cdot 10^{-6}$ at the 90 % confidence limit. This limit is competitive with the CDF Run I limit.

4. OUTLOOK

The results presented here indicate the excellent B reconstruction capabilities of the DØ detector. A Silicon Track Trigger (STT) is currently being commissioned which will allow us to trigger better on secondary vertices from long-lived particles [7]. We are currently making impact parameter cuts as part of the Level 3 trigger. The STT will allow us to move trigger decisions to an earlier stage, allowing a significantly larger fraction of events to be accepted.

ACKNOWLEDGMENTS

I would like to thank my DØ colleagues, the members of the DØ B group and the conference organizers. Special thanks go to Vivek Jain and Guennadi Borissov.

REFERENCES

1. Jain, V., in these proceedings (2004).
2. Anikeev, K., et al., arXiv:hep-ph/0201071 (2001).
3. Abachi, S., et al., [DØ Collaboration], The DØ Upgrade: The Detector and its Physics, Tech. Rep. FERMILAB-PUB-96-357-E (1996).
4. Kajfasz, E., [DØ Collaboration], *Nucl. Instrum. Meth.* **A511**, 16–19 (2003).
5. Kane, G. L., Kolda, C., and Lennon, J. E., arXiv:hep-ph/0310042 (2003).
6. Dedes, A., Dreiner, H. K., and Nierste, U., *Phys. Rev. Lett.* **87**, 251804 (2001).
7. Steinbruck, G., [DØ Collaboration], *Nucl. Instrum. Meth.* **A511**, 145–149 (2003).

CHARM PHYSICS

Review of Recent Results in Charm Physics

Jürgen Engelfried

Instituto de Física, Universidad Autónoma de San Luis Potosí, San Luis Potosí 78000, México

Abstract. A (biased) review of recent results in charm physics is presented. New results on D^0-$\bar{D}^0$ mixing, rare decays of D^0 and $D^\pm$, scalar resonances in D^+ and D_s^+ decays, and new decay modes and mass measurements in Λ_c^+, $\Xi_c^{+,0}$, Ω_c^0, and Ξ_{cc}^+ are discussed.

1. INTRODUCTION

In contrary to the last five years or so, where mostly "traditional" charm experiments like E791, FOCUS, SE-LEX, WA89, WA92, CLEO, and H1/ZEUS published results about more "traditional" topics like production, lifetimes, rare decays, and limits on D^0-$\bar{D}^0$ mixing, accompanied by a small number of theory and phenomenology papers, in the last year, a shift in charm physics occurred. New players like BaBar, Belle and CDF entered the field, new charm states such as doubly charmed baryons, hidden double charm ($J/\psi\, c\bar{c}$), D_s^*, $X(3872)$, were discovered, and the first pentaquark was observed. All this triggered a large number of "theory" papers, pre- and post-dicting the spectroscopy and production of these new states. In most of these papers a (back-)shift to the diquark picture of charmed hadrons can be observed.

We will present here a (biased) selection of recent results in charm physics. In several other talks at this conference charm results were shown.

2. D^0-$\bar{D}^0$ MIXING

The usual observable for CP violation in the charm system is the lifetime difference between $D^0 \to K^-K^+$ and $D^0 \to K^-\pi^+$, defined as $y_{CP} = \tau(K^-\pi^+)/\tau(K^-K^+) - 1$, predicted in the Standard Model to $y_{CP} \sim 10^{-3}$. Another possible analysis is the "wrong-sign" Double Cabibbo Suppressed decay $D^0 \to K^+\pi^-$, with the observable y'. Recent results where published by Belle [1] and BaBar [2], and are compared with previous results in Table 1.

All measurements are compatible with zero, *e.g.*, no CP violation was observed yet in the charm system.

3. RARE DECAYS OF D^0 AND $D^\pm$ MESONS

FOCUS observed the rare decay $D^0 \to K^-K^-K^+\pi^+$ with a yield of 132 ± 19 events, and measured the relative branching ratio to be $\Gamma(D^0 \to K^-K^-K^+\pi^+)/\Gamma(D^0 \to K^-\pi^-\pi^+\pi^+) = 0.00257 \pm 0.00034 \pm 0.00024$ [8]. Resonant substructures with ϕ and $\bar{K}^{*0}$ are dominant.

Belle observed $D^0 \to \phi\pi^0$, $\phi\eta$, and $\phi\gamma$ [9].

CLEO performed a Dalitz plot analysis of $D^0 \to \pi^-\pi^+\pi^0$, and studied $D^0 \to K_S^0\eta\pi^0$ [10]. CLEO also observed the Cabibbo suppressed decays $D^+ \to \pi^+\pi^0$, $K^+\bar{K}^0$ and $K^+\pi^0$ [11]. The measured branching ratios are shown in Table 2.

FOCUS studied di-muon decays for D^+ and D_s^+ [12], and obtained new limits on these modes.

A new player in the field, CDF, set a limit for $D^0 \to \mu^+\mu^-$ at $< 2.5 \cdot 10^{-6}$ [13].

4. SCALAR RESONANCES IN D^+ AND D_s^+ DECAYS

For a few years now, E791 is studying the modes $D^+ \to K^-\pi^+\pi^+$, $D^+ \to \pi^-\pi^+\pi^+$, and $D_s^+ \to \pi^-\pi^+\pi^+$. To explain the resonant substructures in the decays, they need to include two scalar resonances, one for $K\pi$ (the κ) with mass $(797 \pm 19 \pm 43)$ MeV/c^2 and width $(410 \pm 43 \pm 87)$ MeV/c^2, and a second in $\pi\pi$ (the σ) with mass $(478^{+24}_{-23} \pm 17)$ MeV/c^2 and width $(324^{+42}_{-40} \pm 21)$ MeV/c^2 [14–16].

CP722, *B Physics at Hadron Machines,* edited by M. Paulini and S. Erhan

TABLE 1. Recent measurements of *CP* violation observables in the D^0 system.

Experiment	Measurement	Reference
Belle	$y_{CP} = (+1.15 \pm 0.69 \pm 0.38)\%$	[1]
BaBar	$y_{CP} = (-0.8 \pm 0.4 ^{+0.5}_{-0.4})\%$	[2]*
CLEO	$y_{CP} = (-1.2 \pm 2.5 \pm 1.4)\%$	[3]†
FOCUS	$y_{CP} = (3.42 \pm 1.39 \pm 0.74)\%$	[4]
E791	$y_{CP} = (0.8 \pm 2.9 \pm 1.0)\%$	[5]**
BaBar	$-0.056 < y' < 0.039$ (95% CL)	[6]
CLEO	$-0.058 < y' < 0.01$ (95% CL)	[7]

* also includes $D^0 \to \pi^+\pi^-$

† also includes $D^0 \to \pi^+\pi^-$

** Measured $\Delta\Gamma = (0.04 \pm 0.14 \pm 0.05)\,\mathrm{ps}^{-1}$

TABLE 2. Branching ratios for D^+ decays, measured by CLEO [11].

$\mathcal{B}(D^+ \to \pi^+\pi^0)$	$(1.31 \pm 0.17 \pm 0.09 \pm 0.09) \cdot 10^{-3}$
$\mathcal{B}(D^+ \to K^+\bar{K}^0)$	$(5.24 \pm 0.43 \pm 0.20 \pm 0.34) \cdot 10^{-3}$
$\mathcal{B}(D^+ \to K^+\pi^0)$	$< 4.2 \cdot 10^{-4}$ (90% CL)

TABLE 3. Masses M and Width Γ for Σ_c^{++} and Σ_c^0 as measured by CLEO [24].

$M(\Sigma_c^{++}) - M(\Lambda_c^+)$	$(167.4 \pm 0.1 \pm 0.2)\ \mathrm{MeV}/c^2$
$M(\Sigma_c^0) - M(\Lambda_c^+)$	$(167.2 \pm 0.1 \pm 0.2)\ \mathrm{MeV}/c^2$
$\Gamma(\Sigma_c^{++})$	$(2.3 \pm 0.2 \pm 0.3)\ \mathrm{MeV}/c^2$
$\Gamma(\Sigma_c^0)$	$(2.5 \pm 0.2 \pm 0.3)\ \mathrm{MeV}/c^2$

5. THE D_s SYSTEM

On April 12, 2003, BaBar announced the observation of a narrow resonance, decaying to $D_s\pi^0$, at 2.32 GeV/c^2 [17]. Shortly after, CLEO not only confirmed the observation, but observed an additional resonance, decaying to $D_s^*\pi^0$ [18, 19]. During the summer conferences, Belle confirmed both observations [20, 21]. The most likely nature of these states are excited D_s mesons. The search for similar states in the D^0 and $D^\pm$ system has already started. More details can be found in these proceedings [22].

6. CHARMED BARYONS: Λ_c^+ AND $\Sigma_c^{0,++}$

CLEO reports the observation of the decay $\Lambda_c^+ \to \Lambda\pi^+\pi^+\pi^-\pi^0$ [23], with $\mathcal{B} = (1.79 \pm 0.47 \pm 0.43)\%$, while most of the decay happens via $\Lambda_c^+ \to \Lambda\omega\pi^+$.

CLEO also measured the masses and widths of Σ_c^{++} and Σ_c^0 [24] (the results are shown in Table 3), updating previous results on the masses from E791 [25].

7. CHARMED BARYONS: Ξ_c^+ AND Ξ_c^0

FOCUS measured several new decay modes of the Ξ_c^+ and re-measured some previously observed ones. A summary is given in Table 4. FOCUS also includes upper limits for resonances in these decay modes.

CLEO obtained a new measurement of the Ξ_c^+ lifetime, $\tau(\Xi_c^+) = (503 \pm 47 \pm 18)$ fs [29].

CLEO also reports the first observation of the decay $\Xi_c^0 \to pK^-K^-\pi^+$ [30], with a relative branching ratio of $\mathcal{B}(\Xi_c^0 \to pK^-K^-\pi^+)/\mathcal{B}(\Xi_c^0 \to \Xi^-\pi^+) = 0.35 \pm 0.08 \pm 0.05$. In this decay, CLEO sees evidence for a resonant $\bar{K}^{*0}(892)$ substructure.

8. CHARMED BARYONS: THE Ω_c^0

Evidence for the Ω_c^0 in e^+e^- interactions was reported long time ago by ARGUS [31], and now first CLEO [32] and recently Belle [33, 34] confirm this observation. Both measure the mass of the Ω_c^0 (Belle: $(2693.9 \pm 1.1 \pm 1.4)$ MeV/c^2, CLEO: $(2694.6 \pm 2.6 \pm 1.9)$ MeV/c^2) significantly different from the PDG 2000: (2704 ± 4) MeV/c^2 [35]. Both observe the mode $\Omega_c^0 \to \Omega^-\pi^+$ and $\Omega_c^0 \to \Omega^-e^+\nu$, and Belle observes in addition the semileptonic muon mode.

9. DOUBLY CHARMED BARYONS: Ξ_{cc}^+

The SELEX experiment reported the first observation of a member of the doubly charmed baryon family, the Ξ_{cc}^+, in the decay mode $\Xi_{cc}^+ \to \Lambda_c^+K^-\pi^+$ [36]. Further work on various other decay modes is ongoing.

TABLE 4. Relative branching ratios for Ξ_c^+.

Decay Mode	FOCUS [26]	CLEO [27]	SELEX [28]
$\frac{\Gamma(\Xi_c^+ \to \Sigma^+ K^- \pi^+)}{\Gamma(\Xi_c^+ \to \Xi^- \pi^+ \pi^+)}$	$0.91 \pm 0.11 \pm 0.04$	$1.18 \pm 0.26 \pm 0.17$	$0.92 \pm 0.20 \pm 0.07$
$\frac{\Gamma(\Xi_c^+ \to \Sigma^+ K^+ K^-)}{\Gamma(\Xi_c^+ \to \Sigma^+ K^- \pi^+)}$	$0.16 \pm 0.06 \pm 0.01$		
$\frac{\Gamma(\Xi_c^+ \to \Lambda^0 K^- \pi^+ \pi^+)}{\Gamma(\Xi_c^+ \to \Xi^- \pi^+ \pi+)}$	$0.28 \pm 0.06 \pm 0.06$	$0.58 \pm 0.16 \pm 0.07$	
$\frac{\Gamma(\Xi_c^+ \to \Omega^- K^+ \pi^+)}{\Gamma(\Xi_c^+ \to \Xi^- \pi^+ \pi+)}$	$0.07 \pm 0.03 \pm 0.03$		
$\frac{\Gamma(\Xi_c^+ \to \Sigma^*(1385)^+ \bar{K}^0)}{\Gamma(\Xi_c^+ \to \Xi^- \pi^+ \pi+)}$	$1.00 \pm 0.49 \pm 0.24$		

ACKNOWLEDGMENTS

I would like to thank John Cumalat and Erik Gottschalk from FOCUS, Sheldon Stone and JC Wang from CLEO, Jeff Appel from E791, Masa Yamauchi and Karim Trabelsi from Belle, and Livio Lanceri from BaBar for providing me with information and figures from their respective experiments.

I would also like to thank the organizers of the conference for the opportunity to give this presentation.

REFERENCES

1. Abe, K., et al., [Belle Collaboration], arXiv:hep-ex/0308034 (2003).
2. Aubert, B., et al., [BaBar Collaboration], *Phys. Rev. Lett.* **91**, 121801 (2003).
3. Csorna, S. E., et al., [CLEO Collaboration], *Phys. Rev.* **D65**, 092001 (2002).
4. Link, J. M., et al., [FOCUS Collaboration], *Phys. Lett.* **B485**, 62–70 (2000).
5. Aitala, E. M., et al., [E791 Collaboration], *Phys. Rev. Lett.* **83**, 32 (1999).
6. Aubert, B., et al., [BABAR Collaboration], *Phys. Rev. Lett.* **91**, 171801 (2003).
7. Godang, R., et al., [CLEO Collaboration], *Phys. Rev. Lett.* **84**, 5038–5042 (2000).
8. Link, J. M., et al., [FOCUS Collaboration], *Phys. Lett.* **B575**, 190–197 (2003).
9. Abe, K., et al., [Belle Collaboration], *Phys. Rev. Lett.* **92**, 101803 (2004).
10. Dubrovin, M. S., [CLEO Collaboration], arXiv:hep-ex/0305006 (2003).
11. Arms, K., et al., [CLEO Collaboration], arXiv:hep-ex/0309065 (2003).
12. Link, J. M., et al., [FOCUS Collaboration], *Phys. Lett.* **B572**, 21–31 (2003).
13. Acosta, D., et al., [CDF Collaboration], *Phys. Rev.* **D68**, 091101 (2003).
14. Aitala, E. M., et al., [E791 Collaboration], *Phys. Rev. Lett.* **89**, 121801 (2002).
15. Aitala, E. M., et al., [E791 Collaboration], *Phys. Rev. Lett.* **86**, 770–774 (2001).
16. Bediaga, I., [Fermilab E791 Collaboration], *AIP Conf. Proc.* **688**, 252–265 (2004).
17. Aubert, B., et al., [BABAR Collaboration], *Phys. Rev. Lett.* **90**, 242001 (2003).
18. Besson, D., et al., [CLEO Collaboration], *Phys. Rev.* **D68**, 032002 (2003).
19. Stone, S., and Urheim, J., *AIP Conf. Proc.* **687**, 96–104 (2003).
20. Krokovny, P., et al., [Belle Collaboration], *Phys. Rev. Lett.* **91**, 262002 (2003).
21. Abe, K., et al., [Belle Collaboration], *Phys. Rev. Lett.* **92**, 012002 (2004).
22. Wang, J. C., in these proceedings (2004).
23. Cronin-Hennessy, D., et al., [CLEO Collaboration], *Phys. Rev.* **D67**, 012001 (2003).
24. Artuso, M., et al., [CLEO Collaboration], *Phys. Rev.* **D65**, 071101 (2002).
25. Aitala, E. M., et al., [E791 Collaboration], *Phys. Lett.* **B379**, 292–298 (1996).
26. Link, J. M., et al., [FOCUS Collaboration], *Phys. Lett.* **B571**, 139–147 (2003).
27. Bergfeld, T., et al., [CLEO Collaboration], *Phys. Lett.* **B365**, 431–436 (1996).
28. Jun, S. Y., et al., [SELEX Collaboration], *Phys. Rev. Lett.* **84**, 1857–1861 (2000).
29. Mahmood, A. H., et al., [CLEO Collaboration], *Phys. Rev.* **D65**, 031102 (2002).
30. Danko, I., et al., [CLEO Collaboration], arXiv:hep-ex/0309020 (2003).
31. Albrecht, H., et al., [ARGUS Collaboration], *Phys. Lett.* **B288**, 367–372 (1992).
32. Cronin-Hennessy, D., et al., [CLEO Collaboration], *Phys. Rev. Lett.* **86**, 3730–3734 (2001).
33. Ammar, R., et al., [CLEO Collaboration], *Phys. Rev. Lett.* **89**, 171803 (2002).
34. Abe, K., et al., [Belle Collaboration] (2003), Contributed paper to Lepton-Photon 2003, BELLE-CONF-0333.
35. Groom, D. E., et al., [Particle Data Group Collaboration], *Eur. Phys. J.* **C15**, 1–878 (2000).
36. Mattson, M., et al., [SELEX Collaboration], *Phys. Rev. Lett.* **89**, 112001 (2002).

The CLEO-c Research Program

David Asner

(Representing the CLEO Collaboration)

University of Pittsburgh, Department of Physics and Astronomy, Pittsburgh, PA 15260, U.S.A.

Abstract. The CLEO-c research program will include studies of leptonic, semileptonic and hadronic charm decays, searches for exotic and gluonic matter, and tests for physics beyond the Standard Model. In the summer of 2003, the experiment and the CESR accelerator were modified to operate at center-of-mass energies between 3 and 5 GeV. Data at the $\psi(3770)$ resonance were recorded with the CLEO-c detector in September 2003 beginning a new era in the exploration of the charm sector.

1. INTRODUCTION

The CLEO-c physics program [1] includes a variety of measurements that will improve the understanding of Standard Model processes as well as provide the opportunity to probe physics that lies beyond the Standard Model. The primary components of this program are measurements of absolute branching ratios for charm mesons with a precision of the order of 1-2%, determination of charm meson decay constants and of the CKM matrix elements $|V_{cs}|$ and $|V_{cd}|$ at the 1-2% level as well as investigation of processes in charm decays that are highly suppressed within the Standard Model. A $10 \mathrm{nb}^{-1}$ cross section for $e^+ e^- \rightarrow D\bar{D}$ is assumed throughout Ref. [1].

Beginning in 2003, the CESR accelerator will be operated at center-of-mass energies corresponding to $\sqrt{s} \sim$ 3770 MeV (ψ''), $\sqrt{s} \sim 4140$ MeV and $\sqrt{s} \sim 3100$ MeV (J/ψ). The luminosity over this energy region will range from $5 \times 10^{32} \mathrm{cm}^{-2} \mathrm{s}^{-1}$ down to about $1 \times 10^{32} \mathrm{cm}^{-2} \mathrm{s}^{-1}$ yielding 3 fb^{-1} each at the ψ'' and at $\sqrt{s} \sim 4140$ MeV above $D_s^+ D_s^-$ threshold and 1 fb^{-1} at the J/ψ. These integrated luminosities correspond to samples of 1.5 million $D_s^+ D_s^-$ pairs, 30 million $D\bar{D}$ pairs and one billion J/ψ decays [1]. These datasets will exceed those of the Mark III experiment by factors of 480, 310 and 170, respectively. Table 1 summarizes the run plan.

From fall 2001 to spring 2003, CLEO collected a total of 4 fb^{-1} of data on the $\Upsilon(1S)$, $\Upsilon(2S)$, $\Upsilon(3S)$ and $\Upsilon(5S)$ which is currently under analysis. These data sets will increase the available $b\bar{b}$ bound state data by more than an order of magnitude.

Only modest hardware modifications are required for low energy operation. The transverse cooling of the CESR beams will be enhanced by 16 meters of superconducting wiggler magnets. Half of the full complement

TABLE 1. The 3-year CLEO-c run plan [1].

Resonance	Anticipated Luminosity	Reconstructed Events
$\psi(3770)$	~ 3 fb^{-1}	30 M $D\bar{D}$
$\sqrt{s} \sim 4140$ MeV	~ 3 fb^{-1}	1.5 M $D_s^+ D_s^-$
$\psi(3100)$	~ 1 fb^{-1}	60 M radiative J/ψ

of 12 wigglers were installed in summer 2003 with the additional 6 wigglers scheduled for installation in 2004. The CLEO III silicon vertex detector was replaced by a small, low mass inner drift chamber. The solenoidal field was reduced from 1.5 T to 1.0 T. No other modifications are planned.

2. PHYSICS PROGRAM

The following sections will outline the CLEO-c physics program. The first section will focus on the Upsilon spectroscopy, the second section will describe the charm decay program, the third section will give an overview about the exotic and gluonic matter studies and the last section will describe the opportunities to probe physics beyond the Standard Model.

2.1. Upsilon Spectroscopy

The only established $b\bar{b}$ states below $B\bar{B}$ threshold are the three vector triplet Υ resonances ($^3S_1$) and the six χ_b and χ_b' (two triplets of ^3P_J) that are accessible from these parent vectors via E1 radiative transitions. CLEO will address a variety of outstanding physics issues with the data samples at the $\Upsilon(1S)$, $\Upsilon(2S)$ and $\Upsilon(3S)$ resonances.

CP722, *B Physics at Hadron Machines*, edited by M. Paulini and S. Erhan

2.1.1. Searches for the η_b and h_b

The η_b is the ground state of $b\bar{b}$. Most present theories [2] indicate the best approach would be the hindered M1 transition from the $\Upsilon(3S)$, with which CLEO might have a signal of 5σ significance in 1 fb^{-1} of data. In the case of the h_b, CLEO established an upper limit of $\mathscr{B}(\Upsilon(3S) \to \pi^+\pi^- h_b) < 0.18\%$ at 90% confidence level [3]. This result, based on ~ 110 pb^{-1}, already tests some theoretical predictions [4–6] for this transition which range from $< 0.01\% - 1.0\%$.

2.1.2. Observation of 1^3D_J States

The $b\bar{b}$ system is unique as it has states with $L = 2$ that lie below the open-flavor threshold. These states have been of considerable theoretical interest, as indicated by many predictions of the center-of-gravity of the triplet and by a recent review [7]. In an analysis of the $\Upsilon(3S)$ CLEO data sample, the $\Upsilon(1^3D_2)$ state could already be observed in the four-photon cascade $\Upsilon(3S) \to \gamma_1 \chi_b' \to \gamma_1 \gamma_2 \Upsilon(^3D_J) \to \gamma_1 \gamma_2 \gamma_3 \chi_b \to \gamma_1 \gamma_2 \gamma_3 \gamma_4 \ell^+ \ell^-$. The mass of the $\Upsilon(1^3D_2)$ state is determined to be $(10161.1 \pm 0.6 \pm 1.6)$ MeV/c^2 [8].

2.1.3. Glueball Candidates in Radiative $\Upsilon(1S)$ Decays

Signals for glueball candidates in radiative J/ψ decay - a glue-rich environment - might be observed in radiative $\Upsilon(1S)$ decays. Naively one would expect the exclusive radiative decay to be suppressed in Υ decay by a factor of roughly 40, which implies product branching fractions for Υ radiative decay of $\sim 10^{-6}$. With 1 fb^{-1} of data and efficiencies of around 30% one can expect ~ 10 events in each of the exclusive channels, which would be an important confirmation of the J/ψ studies.

2.2. Charm Decays

The observable properties of the charm mesons are determined by the strong and weak interactions. As a result, charm mesons can be used as a laboratory for the studies of these two fundamental forces. Threshold charm experiments permit a series of measurements that enable direct study of the weak interactions of the charm quark, as well as tests of our theoretical technology for handling the strong interactions.

2.2.1. Leptonic Charm Decays

Measurements of leptonic decays in CLEO-c will benefit from the fully tagged D^+ and D_s^+ decays available at the $\psi(3770)$ and at $\sqrt{s} \sim 4140$ MeV. The leptonic decays $D_s \to \mu\nu$ are detected in tagged events by observing a single charged track of the correct sign, missing energy, and a complete accounting of the residual energy in the calorimeter. The clear definition of the initial state, the cleanliness of the tag reconstruction, and the absence of additional fragmentation tracks make this measurement straightforward and essentially background-free. This will enable measurements of the poorly known leptonic decay rates for D^+ and D_s^+ to a precision of 3-4% and will allow the validation of theoretical calculations of the decay constants f_D and f_{D_s} at the 1-2% level. Table 2 summarizes the expected precision in the decay constant measurements.

TABLE 2. Expected decay constants errors for leptonic decay modes.

| | Decay Constant Error % | |
Decay Mode	PDG 2000	CLEO-c [1]
$D^+ \to \mu^+\nu$ (f_D)	Upper Limit	2.3
$D_s^+ \to \mu^+\nu$ (f_{D_s})	17	1.7
$D_s^+ \to \tau^+\nu$ (f_{D_s})	33	1.6

2.2.2. Semileptonic Charm Decays

The CLEO-c program will provide a large set of precision measurements in the charm sector against which the theoretical tools needed to extract CKM matrix information precisely from heavy quark decay measurements will be tested and calibrated.

CLEO-c will measure the branching ratios of many exclusive semileptonic modes, including $D^0 \to K^- e^+\nu$, $D^0 \to \pi^- e^+\nu$, $D^0 \to K^- e^+\nu$, $D^+ \to \bar{K}^0 e^+\nu$, $D^+ \to \pi^0 e^+\nu$, $D^+ \to \bar{K}^{0*} e^+\nu$, $D_s^+ \to \phi e^+\nu$ and $D_s^+ \to \bar{K}^{0*} e^+\nu$. The measurement in each case is based on the use of tagged events where the cleanliness of the environment provides nearly background-free signal samples, and will lead to the determination of the CKM matrix elements $|V_{cs}|$ and $|V_{cd}|$ with a precision level of 1.6% and 1.7%, respectively. Measurements of the vector and axial vector form factors $V(q^2)$, $A_1(q^2)$ and $A_2(q^2)$ will also be possible at the $\sim 5\%$ level. Table 3 summarizes the expected fractional errors on the branching ratios.

HQET provides a successful description of the lifetimes of charm hadrons and of the absolute semileptonic branching ratios of the D^0 and D_s^+ [9]. Isospin invariance in the strong forces implies $\Gamma_{SL}(D^0) \simeq \Gamma_{SL}(D^+)$ up to corrections of $\mathscr{O}(\tan^2\theta_C) \simeq 0.05$. Likewise, $SU(3)_{Fl}$ symmetry relates $\Gamma_{SL}(D^0)$ and $\Gamma_{SL}(D_s^+)$, but a priori

TABLE 3. Expected branching ratio fractional errors for selected semileptonic decay modes.

| Decay Mode | BR fractional error % | |
	PDG 2000	CLEO-c [1]
$D^0 \to K\ell\nu$	5	0.4
$D^0 \to \pi\ell\nu$	16	1.0
$D^+ \to \pi\ell\nu$	48	2.0
$D_s^+ \to \phi\ell\nu$	25	3.1

TABLE 5. Expected branching ratio fractional errors for hadronic decay modes [1].

| Decay Mode | BR fractional error % | |
	PDG 2000	CLEO-c [1]
$D^0 \to K\pi$	2.4	0.6
$D^+ \to K\pi\pi$	7.2	0.7
$D_s^+ \to \phi\pi$	25	1.9

would allow them to differ by as much as 30%. However, HQET suggests that they should agree to within a few percent. The charm threshold region is the best place to measure absolute inclusive semileptonic charm branching ratios, in particular $\mathscr{B}(D_s^+ \to X\ell\nu)$ and thus $\Gamma_{SL}(D_s^+)$.

The CLEO-c program of leptonic and semileptonic measurements has two components: one of calibrating and validating theoretical methods for calculating hadronic matrix elements, which can then be applied to all problems in CKM extraction in heavy quark physics; and one of extracting CKM elements directly from the CLEO-c data. The direct results of CLEO-c are the precise determination of $|V_{cd}|$, $|V_{cs}|$, f_D, f_{D_s}, and the semileptonic form factors. The precision knowledge of the decay constants f_D and f_{D_s}, together with the rigorous calibration of theoretical techniques for calculating heavy-to-light semileptonic form factors, are required for the direct extraction of CKM elements from CLEO-c. This also drives the indirect results, namely the precision extraction of CKM elements from experimental measurements of the B^0 mixing frequency, the B_s^0 mixing frequency, and the $B \to \pi\ell\nu$ decay rate measurements which will be performed by BaBar, Belle, CDF, DØ, BTeV, LHCb, ATLAS and CMS. In Table 4 the combined projections are presented [1]. In the determination of the CKM elements $|V_{cd}|$ and $|V_{cs}|$ from B^0 and B_s^0 mixing $|V_{tb}| = 1$ is used. The tabulation also includes improvement in the direct measurement of $|V_{tb}|$ expected from the Tevatron experiments [10].

TABLE 4. CKM elements at present and after CLEO-c [1].

Present Knowledge		
$\Delta V_{ud}/V_{ud} = 0.1\%$	$\Delta V_{us}/V_{us} = 1\%$	$\Delta V_{ub}/V_{ub} = 25\%$
$\Delta V_{cd}/V_{cd} = 7\%$	$\Delta V_{cs}/V_{cs} = 16\%$	$\Delta V_{cb}/V_{cb} = 5\%$
$\Delta V_{td}/V_{td} = 36\%$	$\Delta V_{ts}/V_{ts} = 39\%$	$\Delta V_{tb}/V_{tb} = 29\%$
After CLEO-c		
$\Delta V_{ud}/V_{ud} = 0.1\%$	$\Delta V_{us}/V_{us} = 1\%$	$\Delta V_{ub}/V_{ub} = 5\%$
$\Delta V_{cd}/V_{cd} = 1\%$	$\Delta V_{cs}/V_{cs} = 1\%$	$\Delta V_{cb}/V_{cb} = 3\%$
$\Delta V_{td}/V_{td} = 5\%$	$\Delta V_{ts}/V_{ts} = 5\%$	$\Delta V_{tb}/V_{tb} = 15\%$

2.2.3. Hadronic Charm Decays

The CLEO and ALEPH experiments by far provide the most precise measurements for the decay $D^0 \to K^-\pi^+$. They use the same technique by looking at $D^{*+} \to \pi^+D^0$ decays and taking the ratio of the D^0 decays into $K^-\pi^+$ to the number of decays with only the π^+ from the D^{*+} decay detected. The dominant systematic uncertainty is the background level in the latter sample. In both experiments, the systematic errors exceed the statistical errors. The D^+ absolute branching ratios are determined by using fully reconstructed D^{*+} decays, comparing $\pi^0 D^+$ with $\pi^+ D^0$ and using isospin symmetry. Hence, this rate cannot be determined any better than the absolute D^0 decay rate using this technique. The D_s^+ absolute branching ratios are determined by comparing fully reconstructed $B \to D^{(*)}D_s^{*+}$ to the partially reconstructed $B \to D^{(*)}D_s^{*+}$ requiring only the γ from the D_s^{*+} decay. Here the dominant systematic uncertainty is due to the background shape in the partially reconstructed sample. By using $D^0\bar{D}^0$, $D^+ D^-$ and $D_s^+ D_s^-$ decays, and tagging both D mesons, the background can be reduced to almost zero and the branching ratio fractional error can be improved significantly (see Table 5).

2.3. Exotic and Gluonic Matter

The approximately one billion J/ψ mesons produced at CLEO-c will be a glue factory to search for glueballs and other glue-rich states via $J/\psi \to gg \to \gamma X$ decays. The region of $1 < M_X < 3$ GeV/c^2 will be explored with partial wave analyses for evidence of scalar or tensor glueballs, glueball-$q\bar{q}$ mixtures, exotic quantum numbers, quark-glue hybrids and other new forms of matter predicted by QCD. This includes the establishment of masses, widths, spin-parity quantum numbers, decay modes and production mechanisms for any identified states, a detailed exploration of reported glueball candidates such as the scalar states $f_0(1370)$, $f_0(1500)$ and $f_0(1710)$, and the examination of the inclusive photon spectrum $J/\psi \to \gamma X$ with < 20 MeV photon resolution and identification of states with up to 100 MeV width and inclusive branching ratios above 1×10^{-4}.

In addition, spectroscopic searches for new states of

the $b\bar{b}$ system and for exotic hybrid states such as $cg\bar{c}$ will be made using the 4 fb^{-1} $\Upsilon(1S)$, $\Upsilon(2S)$, $\Upsilon(3S)$ and $\Upsilon(5S)$ data sets. Analysis of $\Upsilon(1S) \to \gamma X$ will play an important role in verifying any glueball candidates found in the J/ψ data.

2.4. Charm Beyond the Standard Model

CLEO-c will have the opportunity to probe for physics beyond the Standard Model. Three highlights - rare charm decays, $D^0\bar{D}^0$ mixing and CP violation - are discussed in the following sections.

2.4.1. Rare Charm Decays

Rare decays of charmed mesons and baryons provide "background-free" probes of new physics effects. In the framework of the Standard Model (SM), these processes occur only at one loop level. SM predicts vanishingly small branching ratios for processes such as $D \to \pi/K^{(*)}\ell^+\ell^-$ due to the almost perfect GIM cancellation between the contributions of strange and down quarks. This causes the SM predictions for these transitions to be very uncertain. In addition, in many cases annihilation topologies also give sizable contribution. Several model-dependent estimates exist indicating that the SM predictions for these processes are still far below current experimental sensitivities [11, 12].

2.4.2. $D^0\bar{D}^0$ Mixing

Neutral flavor oscillation in the D meson system is highly suppressed within the Standard Model. The time evolution of a particle produced as a D^0 or $\bar{D}^0$, in the limit of CP conservation, is governed by four parameters: $x = \Delta m/\Gamma$, $y = \Delta\Gamma/2\Gamma$ characterize the mixing matrix, δ the relative strong phase between Cabibbo favored (CF) and doubly-Cabibbo suppressed (DCS) amplitudes and R_D the DCS decay rate relative to the CF decay rate [13]. Standard Model based predictions for x and y, as well as a variety of non-Standard Model expectations, span several orders of magnitude [14, 15]. It is reasonable to assume that $x \approx y \approx 10^{-3}$ in the Standard Model. The mass and width differences x and y can be measured in a variety of ways. The most precise limits are obtained by exploiting the time-dependence of D decays [13]. Time-dependent analyses are not feasible at CLEO-c; however, the quantum-coherent $D^0\bar{D}^0$ state provides time-integrated sensitivity to x, y at $\mathcal{O}(1\%)$ level and $\cos(\delta) \sim 0.05$ [1, 16]. Although CLEO-c does not have sufficient sensitivity to observe Standard Model

charm mixing the projected results compare favorably with current experimental results (see Fig. 1 in Ref. [13]).

2.4.3. CP Violation

Standard Model CP violation is strongly suppressed in charm. While theoretical predictions have significant uncertainties, Standard Model predictions for the rate of CP violation in charm mesons are as large as 0.1% for D^0 decays and as large as 1% for certain D^+ and D_s^+ decays [17].

The production process $e^+e^- \to \psi(3770) \to D^0\bar{D}^0$ produces an eigenstate of CP_+, in the first step, since the $\psi(3770)$ has J^{PC} equal to 1^{--}. Now consider the case where both the D^0 and the $\bar{D}^0$ decay into CP eigenstates. Then the decays $\psi(3770) \to f_+^i f_+^j$ or $f_-^i f_-^j$ are forbidden, where f_+ denotes a CP_+ eigenstate and f_- denotes a CP_- eigenstate. This is because $CP(f_\pm^i f_\pm^j) = (-1)^\ell = -1$ for the $\ell = 1$ $\psi(3770)$. Hence, if a final state such as $(K^+K^-)(\pi^+\pi^-)$ is observed, one immediately has evidence of CP violation. Moreover, all CP_+ and CP_- eigenstates can be summed over for this measurement. The expected sensitivity to direct CP violation is $\sim 1\%$. This measurement can also be performed at higher energies where the final state $D^{*0}\bar{D}^{*0}$ is produced. When either D^* decays into a π^0 and a D^0, the situation is the same as above. When the decay is $D^{*0} \to \gamma D^0$ the CP parity is changed by a multiplicative factor of -1 and all decays $f_+^i f_-^j$ violate CP [18]. Additionally, CP asymmetries in CP even initial states depend linearly on x allowing sensitivity to CP violation in mixing of $\sim 3\%$ [1].

2.4.4. Dalitz Plot Analyses

A Dalitz plot analysis of multibody final states measures amplitudes and phases rather than the rates and so may provide greater sensitivity to CP violation. In the limit of CP conservation, charge conjugate decays will have the same Dalitz distribution. Although the D^+ and D_s^+ decays are self-tagging, there have been no reported Dalitz analyses that search for CP violation with charged D's. The decay $D^0 \to K_S^0\pi^+\pi^-$ proceeds through intermediate states that are CP_+ eigenstates, such as $K_S^0 f_0$, CP_- such as $K_S^0\rho$ and flavor eigenstates such as $K^{*-}\pi^+$ [19]. It is noteworthy that for uncorrelated D^0 mesons the interference between CP_+ and CP_- eigenstates integrates to zero across the Dalitz plot but for correlated D the interference between CP_+ and CP_- eigenstates is locally zero. The Dalitz plots for $\Psi(3770) \to D^0\bar{D}^0 \to f_+ K_S^0\pi^+\pi^-$ and $\Psi(3770) \to D^0\bar{D}^0 \to f_- K_S^0\pi^+\pi^-$ will be distinct and the Dalitz plot for the untagged sample $\Psi(3770) \to D^0\bar{D}^0 \to X K_S^0\pi^+\pi^-$

will be distinct from that observed with uncorrelated D's from continuum production at $\sim$ 10 GeV [19]. The sensitivity at CLEO-c to CP violation with Dalitz plot analyses has not yet been evaluated.

3. SUMMARY

The high-precision charm and quarkonium data will permit a broad suite of studies of weak and strong interaction physics as well as probes of new physics. In the threshold charm sector, measurements are uniquely clean and make possible the unambiguous determinations of physical quantities discussed above. The advances in strong interaction calculations enabled by CLEO-c will allow advances in weak interaction physics in all heavy quark endeavors and in future explorations for physics beyond the Standard Model.

REFERENCES

1. Briere, R. A., et al., [CLEO Collaboration] (2001), Tech. Rep. CLNS-01-1742.
2. Godfrey, S., and Rosner, J. L., *Phys. Rev.* **D64**, 074011 (2001).
3. Butler, F., et al., [CLEO Collaboration], *Phys. Rev.* **D49**, 40–57 (1994).
4. Kuang, Y.-P., and Yan, T.-M., *Phys. Rev.* **D24**, 2874 (1981).
5. Voloshin, M. B., *Sov. J. Nucl. Phys.* **43**, 1011 (1986).
6. Voloshin, M. B., and Zakharov, V. I., *Phys. Rev. Lett.* **45**, 688 (1980).
7. Godfrey, S., and Rosner, J. L., *Phys. Rev.* **D64**, 097501 (2001).
8. Skwarnicki, T., arXiv:hep-ph/0311243 (2003).
9. Bigi, I. I. Y., arXiv:hep-ph/0001003 (2000).
10. Swain, J., and Taylor, L., *Phys. Rev.* **D58**, 093006 (1998).
11. Fajfer, S., Prelovsek, S., Singer, P., and Wyler, D., *Phys. Lett.* **B487**, 81–86 (2000).
12. Burdman, G., Golowich, E., Hewett, J. L., and Pakvasa, S., *Phys. Rev.* **D52**, 6383–6399 (1995).
13. Asner, D., *Phys. Rev.* **D66**, 010001 (2002), $D^0 - \bar{D}^0$ Mixing minireview, in Particle Data Group, Review of Particle Physics.
14. Nelson, H. N., arXiv:hep-ex/9908021 (1999).
15. Petrov, A., arXiv:hep-ph/0311371 (2003).
16. Gronau, M., Grossman, Y., and Rosner, J. L., *Phys. Lett.* **B508**, 37–43 (2001).
17. Buccella, F., Lusignoli, M., and Pugliese, A., *Phys. Lett.* **B379**, 249–256 (1996).
18. Bigi, I. I. Y., and Sanda, A. I., *Cambridge Monogr. Part. Phys. Nucl. Phys. Cosmol.* **9**, 1–382 (2000).
19. Muramatsu, H., et al., [CLEO Collaboration], *Phys. Rev. Lett.* **89**, 251802 (2002).

Charmed-Strange Mesons Experimental Results

Jianchun Wang

Syracuse University, Department of Physics, Syracuse, NY 13244, U.S.A.

Abstract. Two new states in the charm strange sector, $D_{sJ}^*(2317)^+$ and $D_{sJ}(2460)^+$, have recently been discovered at e^+e^- collider experiments. The new states are first observed in the dominant $D_s^+\pi^0$ and $D_s^{*+}\pi^0$ modes respectively and are very narrow. They are consistent with 0^+ and 1^+ P-wave $c\bar{s}$ mesons. The $D_{sJ}(2460)^+$ meson is also observed in $D_s^+\gamma$ and $D_s^+\pi^+\pi^-$ modes. A review of the discoveries and possible explanations is given.

1. INTRODUCTION

In a simplified picture, the charmed-strange meson $c\bar{s}$ (generically denoted as D_{sJ} in this paper) is an atom of a massive charm quark and a light anti-strange quark. The mass splitting of different states is the result of interaction of the spin angular momenta of the two quarks, $\vec{s}_c$ and $\vec{s}_{\bar{s}}$, and the orbital angular momentum $\vec{L}$ between them. According to HQET [1, 2], in the limit that the charm quark is infinitively heavy, its spin is totally decoupled from the light degree of freedom. Then the spin of the charm quark $\vec{s}_c$ and $\vec{j} = \vec{L} + \vec{s}_{\bar{s}}$ are conserved separately by strong interactions. This is the so-called heavy quark symmetry (HQS).

The charm quark, however, is not infinitively heavy, but it is heavier than the QCD scale Λ_{QCD}. Thus taking $\vec{J} = \vec{L} + \vec{s}_{\bar{s}} + \vec{s}_c$ as a good quantum number, the two ground states ($L = 0$, $J^P = 0^-, 1^-$) can be considered as $j = 1/2$ doublets and the four first orbital excited states ($L = 1$) can be treated as $j = 1/2$ doublets ($J^P = 0^+, 1^+$) and $j = 3/2$ doublets ($J^P = 1^+, 2^+$) [2, 3].

Before this year, only four of these six states had been observed. All the observed ones are narrow. The 0^- state, D_s^+, is the lightest D_{sJ} meson and thus can decay only weakly [4]. The 1^- state, D_s^{*+}, was discovered in the electromagnetic radiative mode $D_s^{*+} \to D_s^+\gamma$ [5]. The kinematically allowed strong transition $D_s^{*+} \to D_s^+\pi^0$ is isospin suppressed, and has branching fraction of only $\sim$ 6% [6]. The two observed $L = 1$ states are $D_{s1}(2536)^+ \to D^*K$, and $D_{sJ}(2573)^+ \to DK$ [7, 8]. Being members of $j = 3/2$ doublets, they decay in D-wave not S-wave, explaining their relatively narrow widths.

The two missing $L = 1$ states (0^+ and 1^+) were predicted by most potential models [9–13] to be massive enough that they would decay to DK and D^*K in an S-wave, respectively. The widths were thus expected to be very broad, $\sim$200-300 MeV/c^2. There were, however, a

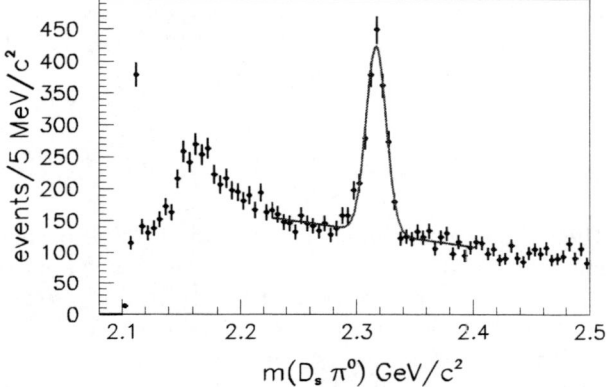

FIGURE 1. The $D_s^+\pi^0$ invariant mass distribution from BaBar.

few predictions that these states would have masses below $D^{(*)}K$ threshold that evidently were not paid much attention to [14–16]. Effectively "everyone" thought that $D^{(*)}K$ were the modes to look for these two states and they were difficult to find due to the large width. The recent discoveries reveal a different picture [17–22].

2. DISCOVERY OF $D_{sJ}^*(2317)^+$

The BaBar collaboration observed a $D_s^+\pi^0$ structure in their e^+e^- continuum event sample [17]. The center of peak is $(2317.3 \pm 0.4 \pm 0.8)$ MeV/c^2 as shown in Fig. 1. The width of the peak is 8.6 ± 0.4 MeV/c^2, consistent with their detector resolution. The structure is observed in different D_s^+ decay modes. It does not appear in their generic Monte Carlo simulated sample, and thus it is not a reflection of a previously known decay.

Since the decay products of this new state must contain a charm and an anti-strange quark, it is natural to think

CP722, *B Physics at Hadron Machines,* edited by M. Paulini and S. Erhan
© 2004 American Institute of Physics 0-7354-0203-5/04/$22.00

that this is one of the $L = 1$ D_{sJ} mesons that have not yet been observed. Thus, it is named $D^*_{sJ}(2317)^+$. Furthermore, the 1^+ meson is forbidden to decay into $0^- 0^-$, whereas the 0^+ meson is allowed in S-wave. The decay angular distribution is flat after reconstruction efficiency correction, which means either $D^*_{sJ}(2317)^+$ is generated unpolarized or it is a spin-0 state. So this new state is probably the 0^+ D_{sJ} meson, though higher spin is not ruled out.

The mass of the $D^*_{sJ}(2317)^+$, however, is much lighter than the 0^+ D_{sJ} meson predicted by most potential models. For example, the model in Ref. [13] worked quite well with known D and D_{sJ} mesons at the time it was created, and successfully predicted the mass of 0^+ and 1^+ D mesons that were later discovered. It predicted the mass of the 0^+ D_{sJ} meson to be 2487 MeV/c^2. The newly observed $D^*_{sJ}(2317)^+$ is 170 MeV/c^2 lower than the expectation, it is even ~ 40 MeV/c^2 below the DK threshold and the width is much narrower (< 10 MeV/c^2) than the prediction of ~ 200-300 MeV/c^2.

3. DISCOVERY OF $D_{sJ}(2460)^+$

The CLEO collaboration confirms the $D_s^+ \pi^0$ resonance observed by BaBar [18, 19]. They find that the measured width of the peak is $8.0^{+1.3}_{-1.2}$ MeV/c^2, somewhat broader than their detector resolution of 6.0 ± 0.3 MeV/c^2. More interestingly, they also observe another state, $D_{sJ}(2460)^+$, at 2463 MeV/c^2 that decays into $D_s^{*+} \pi^0$ (Figure 2).

Figure 2(a) shows the invariant mass difference, $\Delta M = M(D_s^+ \gamma \pi^0) - M(D_s^+ \gamma)$. Requiring $D_s^+ \gamma$ consistent with D_s^{*+}, they find 55 ± 10 events in the peak. The center of peak is measured to be 349.8 ± 1.3 MeV/c^2, similar to that of $D^*_{sJ}(2317)^+$ that CLEO finds at 349.4 ± 1.0 MeV/c^2 in the $\Delta M = M(D_s^+ \pi^0) - M(D_s^+)$ spectrum. The width of peak is 6.1 ± 1.0 MeV/c^2, close to the detector resolution of 6.6 ± 0.5 MeV/c^2. The BaBar data also shows excess in the $D_s^+ \gamma \pi^0$ invariant mass spectrum [17], although the conclusion reached in the publication was that further study is needed due to the complexity of the reflection from the $D^*_{sJ}(2317)^+$.

The ΔM values are very close for $D^*_{sJ}(2317)^+$ and $D_{sJ}(2460)^+$. When the D_s^+ from a $D^*_{sJ}(2317)^+$ decay picks up a random photon, the invariant mass of the two can fall in the selection window of D_s^{*+}. Because of the equality of the mass difference, when the π^0 of the same $D^*_{sJ}(2317)^+$ decay is added, the total invariant mass is consistent with $D_{sJ}(2460)^+$. Thus, the $D^*_{sJ}(2317)^+$ could reflect into the $D_{sJ}(2460)^+$ peak, but simulation shows that this peak has a width of ~ 15 MeV/c^2, much broader than the real $D_{sJ}(2460)^+$ signal peak. Checking the event sample from D_s^{*+} sidebands (Fig. 2b), CLEO finds that

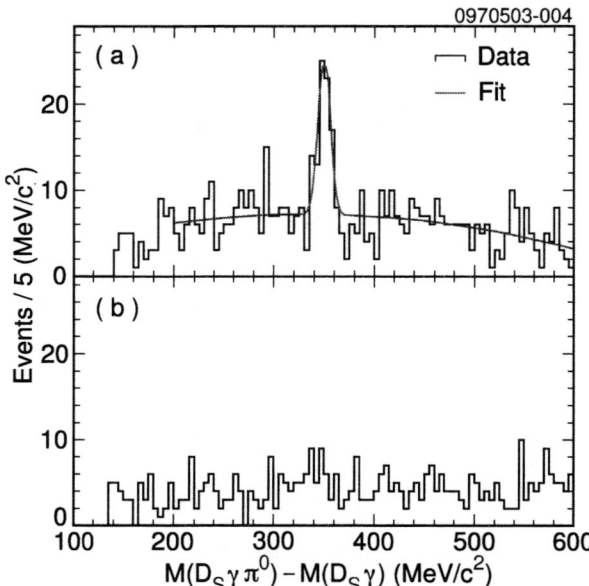

FIGURE 2. The $D_s^+ \gamma \pi^0$ mass distribution from CLEO for $D_s^+ \gamma$ candidates from (a) D_s^{*+} signal region, (b) D_s^{*+} sidebands.

the reflection of $D^*_{sJ}(2317)^+$ could only account for 1/5 to 1/4 of the events in the $D_{sJ}(2460)^+$ peak.

The reflection also exists in the opposite direction, when the single photon from the $D_{sJ}(2460)^+$ decay is "ignored" and a fake $D^*_{sJ}(2317)^+$ peak is created. With the MC simulation event sample, CLEO estimates the cross reflection efficiencies, and then extracts the true number of reconstructed $D^*_{sJ}(2317)^+$ and $D_{sJ}(2460)^+$ signal events. In both peaks, about 20% events are due to reflections. The number of $D_{sJ}(2460)^+$ signal is 41 ± 12, consistent with estimation using D_s^{*+} sidebands.

The Belle collaboration confirms both $D^*_{sJ}(2317)^+$ and $D_{sJ}(2460)^+$ states in continuum event sample as well as in B decays that will be discussed in next section [20, 21]. They also observed $D_{sJ}(2460)^+$ in $D_s^+ \gamma$ and $D_s^+ \pi^+ \pi^-$ modes. After careful study of the cross reflection, the BaBar collaboration also confirms the $D_{sJ}(2460)^+$ meson [22]. Thus, there is no doubt about the existence of the $D_{sJ}(2460)^+$ state. As $D_{sJ}(2460)^+$ decays to $1^- 0^-$, it is most probably the missing $J^P = 1^+$ state decaying into an S-wave. It can not be a 0^+ state, though other possibilities are not ruled out. Further investigation is needed.

4. OBSERVATION OF $D_{sJ}^*(2317)^+$ AND $D_{sJ}(2460)^+$ IN B DECAYS

Cross reflection of the two new D_{sJ} states in continuum data complicates the investigation. The cross reflec-

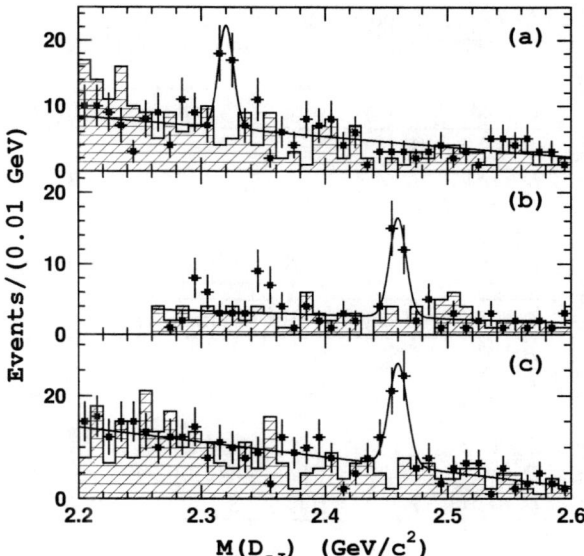

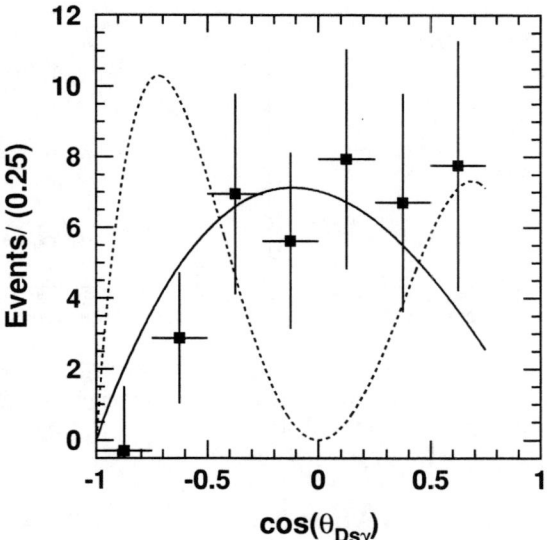

FIGURE 3. Invariant mass for D_{sJ} candidates produced in $B \to \bar{D}D_{sJ}^+$ decays for (a) $D_{sJ}^+ \to D_s^+\pi^0$, (b) $D_{sJ}^+ \to D_s^{*+}\pi^0$ and (c) $D_{sJ}^+ \to D_s^+\gamma$. The hatched regions are for ΔE sidebands.

FIGURE 4. Helicity angular distribution of $D_{sJ}(2460)^+$ in $B \to \bar{D}D_{sJ}(2460)^+$, $D_{sJ}(2460)^+ \to D_s^+\gamma$. The solid line is the expectation of a 1^+ state and the dotted line is for 2^+.

tion, however, is eliminated in B decays as extra constraints are applied. Belle searches for $B \to \bar{D}D_{sJ}^+$ decays of both charged and neutral B [21]. For events whose mass and beam energy constraints are consistent with the $B \to \bar{D}D_{sJ}^+$ decay, the invariant mass spectrum of $D_s^+\pi^0$, $D_s^{*+}\pi^0$ and $D_s^+\gamma$ are shown in Fig. 3. Belle observes both the $D_{sJ}^*(2317)^+$ and $D_{sJ}(2460)^+$ in B decays. The peak in Fig. 3(c) is the first observation of $D_{sJ}(2460)^+ \to D_s^+\gamma$ mode. The ratio of partial width of this mode to that of $D_{sJ}(2460)^+ \to D_s^{*+}\pi^0$ is measured to be $0.38 \pm 0.11 \pm 0.04$, consistent with $0.55 \pm 0.13 \pm 0.8$ measured in continuum data by Belle. The branching fractions are measured to be:

$$\mathscr{B}(B \to \bar{D}D_{sJ}^*(2317)^+) \times \mathscr{B}(D_{sJ}^*(2317)^+ \to D_s^+\pi^0)$$
$$= (8.5^{+2.6}_{-1.9} \pm 2.6) \times 10^{-4},$$

$$\mathscr{B}(B \to \bar{D}D_{sJ}(2460)^+) \times \mathscr{B}(D_{sJ}(2460)^+ \to D_s^{*+}\pi^0)$$
$$= (17.8^{+4.5}_{-3.9} \pm 5.3) \times 10^{-4},$$

$$\mathscr{B}(B \to \bar{D}D_{sJ}(2460)^+) \times \mathscr{B}(D_{sJ}(2460)^+ \to D_s^+\gamma)$$
$$= (6.7^{+1.3}_{-1.2} \pm 2.0) \times 10^{-4}.$$

The B decay provides a much better laboratory to study the spin parity of the new D_{sJ} states. In $B \to \bar{D}D_{sJ}^+$ decay, the D_{sJ} is totally longitudinally polarized as both B and D are spin-0 particles. Belle measures the helicity angular distribution of $D_{sJ}(2460)^+$ in $D_s^+\gamma$ mode shown in Fig. 4. The measurement strongly supports the 1^+ assignment.

5. POSSIBLE EXPLANATION AND SEARCH OF OTHER DECAY MODES

The world averaged mass difference are 349.1 ± 0.6 MeV/c^2 and 346.7 ± 0.8 MeV/c^2 for $D_{sJ}^*(2317)^+$ and $D_{sJ}(2460)^+$, respectively. Adding the PDG value of $M(D_s^+) = 1968.5 \pm 0.6$ MeV/c^2 and $M(D_s^{*+}) = 2112.4 \pm 0.7$ MeV/c^2, the masses are 2317.6 ± 0.8 MeV/c^2 and 2459.1 ± 1.0 MeV/c^2. The upper limits of width at 90% CL are 4.6 and 5.5 MeV/c^2, respectively, set by Belle [21].

Since the discovery of the $D_{sJ}^*(2317)^+$ state, several possible explanations appeared. Cahn and Jackson use non-relativistic vector and scalar exchange forces and recalculate within potential models to explain the mass [23]. Van Beveran and Rupp use a unitarized meson model to explain the low mass as a threshold effect [24]. Bardeen et al. explain that it is a normal $c\bar{s}$ state [14, 25]. Barnes et al. suggest that it is a DK molecule [26]. Several others propose different multi-quark models [27–31].

Due to the low mass and narrow width, the $D_{sJ}^*(2317)^+$ has difficulty to fit in the potential models, nor does $D_{sJ}(2460)^+$. They could be DK and D^*K molecules as they are about just 40 MeV/c^2 below the thresholds. The mass difference between D and D^* is ~ 140 MeV/c^2, explaining the mass difference between $D_{sJ}^*(2317)^+$ and $D_{sJ}(2460)^+$ of ~ 142 MeV/c^2. Inside the molecule, $D^{(*)}$ and K are pre-formed. As the direct decay mode $D^{(*)}K$ is closed, quark antiquark pairs of the

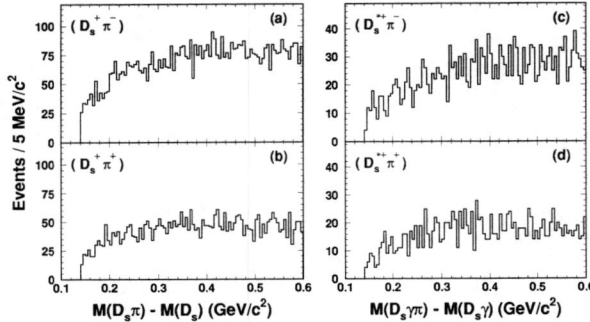

FIGURE 5. $D_s^{(*)+}\pi^{\pm}$ mass distribution from CLEO for (a) opposite-signed $D_s\pi$, (b) same-signed $D_s\pi$, (c) opposite-signed $D_s^*\pi$ and (d) same-signed $D_s^*\pi$.

TABLE 1. Ratio of branching fractions of different $D_{sJ}^*(2317)^+$ and $D_{sJ}(2460)^+$ modes. Limits are with 90% CL.

$D_{sJ}^*(2317)^+$ decay	BEH	Belle	CLEO
$D_s^+\pi^0$	$\equiv 1$	$\equiv 1$	$\equiv 1$
$D_s^+\pi^+\pi^-$	0	$< 4\times10^{-3}$	< 0.019
$D_s^+\gamma$	0	< 0.05	< 0.052
$D_s^{*+}\pi^0$	0		< 0.01
$D_s^{*+}\gamma$	0.08	< 0.18	< 0.059

$D_{sJ}(2460)^+$ decay	BEH	Belle	CLEO
$D_s^{*+}\pi^0$	$\equiv 1$	$\equiv 1$	$\equiv 1$
$D_s^{*+}\gamma$	0.22	< 0.31	< 0.16
$D_s^+\pi^0$	0	< 0.21	
$D_s^+\pi^+\pi^-$	0.20	0.14 ± 0.04	< 0.08
$D_s^+\gamma$	0.24	0.44 ± 0.09	< 0.49
$D_{sJ}^*(2317)^+\gamma$	0.13		< 0.58

two have to be broken to form a D_s^+ and a π^0, thus the decay is weak.

The molecule picture suggests the existence of $D_s^{(*)+}\pi^{\pm}$ resonances. Observation of these resonances would strongly support the molecule hypothesis as they are not conventional $q\bar{q}$ mesons due to their quark content. The CDF collaboration studies $D_s^+\pi^{\pm}$ modes and finds no narrow structure. The CLEO collaboration has searched for $D_s^{(*)+}\pi^{\pm}$ structures as shown in Fig. 5. No narrow structure is found. The production of narrow $D_s^{(*)+}\pi^{\pm}$ states are at least a factor of ten lower than the $D_s^{(*)+}\pi^0$ modes. This proves that $D_{sJ}^*(2317)^+$ and $D_{sJ}(2460)^+$ are iso-scalers. It, however, does not totally rule out the molecule scenario as an iso-vector molecule is expected to be broad, although there is no indication of the existence of such structure in B decay samples.

The new D_{sJ} states fit well into the quark model as normal 0^+ and 1^+ $c\bar{s}$ mesons except for maybe the low masses. Bardeen et al. couple chiral perturbation theory with a quark model representing HQET, and in fact predicted the masses of the 0^+ and 1^+ $c\bar{s}$ mesons below the $D^{(*)}K$ thresholds. The narrow widths are due to isospin violation in the decays. They infer that $D_{sJ}^*(2317)^+$ is indeed the 0^+ $c\bar{s}$ meson. It has an 1^+ partner with mass splitting identical to that between 0^- and 1^- $c\bar{s}$ mesons, which is backed up by the measurements. They also calculate the partial width of other decay modes as shown in Table 1. The measured ratios and limits at 90% CL from CLEO and Belle are also listed. The predictions are consistent with the measurements, and thus this explanation is favored.

Factorization implies that the branching fractions of $B \to \bar{D}D_{sJ}^+$ for the new D_{sJ} states be similar to that of D_s^+ and D_s^{*+}, which are $\sim 1\%$. The measurements are about a factor of ten lower. This casts a shadow on the favored conventional $c\bar{s}$ explanation. Four-quark or molecule states, however, would have branching fractions consistent with the measurements [28, 31, 32].

Browder et al. propose that these states are mixtures of $c\bar{s}$ and four-quark states [33]. More experimental measurements and theoretical ideas are needed to reveal the true identity of these two new states.

ACKNOWLEDGMENTS

The author would like to thank all those, especially Prof. Sheldon Stone, for helping prepare this note.

REFERENCES

1. De Rujula, A., Georgi, H., and Glashow, S. L., *Phys. Rev. Lett.* **37**, 785 (1976).
2. Isgur, N., and Wise, M. B., *Phys. Rev. Lett.* **66**, 1130 (1991).
3. Godfrey, S., and Kokoski, R., *Phys. Rev.* **D43**, 1679 (1991).
4. Chen, A., et al., [CLEO Collaboration], *Phys. Rev. Lett.* **51**, 634 (1983).
5. Albrecht, H., et al., [ARGUS Collaboration], *Phys. Lett.* **B146**, 111 (1984).
6. Gronberg, J., et al., [CLEO Collaboration], *Phys. Rev. Lett.* **75**, 3232 (1995).
7. Albrecht, H., et al., [ARGUS Collaboration], *Phys. Lett.* **B230**, 162 (1989).
8. Kubota, Y., et al., [CLEO Collaboration], *Phys. Rev. Lett.* **72**, 1972 (1994).
9. Godfrey, S., and Isgur, N., *Phys. Rev.* **D32**, 189 (1985).
10. Zeng, J., Van Orden, J. W., and Roberts, W., *Phys. Rev.* **D52**, 5229 (1995).
11. Gupta, S. N., and Johnson, J. M., *Phys. Rev.* **D51**, 168 (1995).
12. Ebert, D., Faustov, R. N., and Galkin, V. O., arXiv:hep-ph/9809285 (1998).
13. Di Pierro, M., and Eichten, E., *Phys. Rev.* **D64**, 114004 (2001).

14. Bardeen, W. A., and Hill, C. T., *Phys. Rev.* **D49**, 409 (1994).
15. Fayyazuddin, and Riazuddin, *Phys. Rev.* **D48**, 2224 (1993).
16. Deandrea, A., Gatto, R., Nardulli, G., Polosa, A. D., and Tornqvist, N. A., *Phys. Lett.* **B502**, 79 (2001).
17. Aubert, B., et al., [BaBar Collaboration], *Phys. Rev. Lett.* **90**, 242001 (2003).
18. Stone, S., and Urheim, J., *AIP Conf. Proc.* **687**, 96–104 (2003).
19. Besson, D., et al., [CLEO Collaboration], *Phys. Rev.* **D68**, 032002 (2003).
20. Abe, K., et al., *Phys. Rev. Lett.* **92**, 012002 (2004).
21. Krokovny, P., et al., [Belle Collaboration], *Phys. Rev. Lett.* **91**, 262002 (2003).
22. Aubert, B., et al., [BABAR Collaboration], *Phys. Rev.* **D69**, 031101 (2004).
23. Cahn, R. N., and Jackson, J. D., *Phys. Rev.* **D68**, 037502 (2003).
24. van Beveren, E., and Rupp, G., *Phys. Rev. Lett.* **91**, 012003 (2003).
25. Bardeen, W. A., Eichten, E. J., and Hill, C. T., *Phys. Rev.* **D68**, 054024 (2003).
26. Barnes, T., Close, F. E., and Lipkin, H. J., *Phys. Rev.* **D68**, 054006 (2003).
27. Szczepaniak, A. P., *Phys. Lett.* **B567**, 23 (2003).
28. Cheng, H.-Y., and Hou, W.-S., *Phys. Lett.* **B566**, 193 (2003).
29. Terasaki, K., *Phys. Rev.* **D68**, 011501 (2003).
30. Nussinov, S., arXiv:hep-ph/0306187 (2003).
31. Datta, A., and O'Donnell, P. J., *Phys. Lett.* **B567**, 273 (2003).
32. Chen, C.-H., and Li, H.-n., *Phys. Rev.* **D69**, 054002 (2004).
33. Browder, T. E., Pakvasa, S., and Petrov, A. A., *Phys. Lett.* **B578**, 365–368 (2004).

Charm Physics at Belle

Karim Trabelsi

(Representing the Belle Collaboration)

Department of Physics and Astronomy, University of Hawaii at Manoa, Honolulu, HI 96822, U.S.A.

Abstract. We present recent charm results from the Belle Collaboration: an updated study of double $c\bar{c}$ production in the e^+e^- continuum, the production of D meson excited states in B decays, measurements of D_{SJ} resonance properties and its production in B decays and the observation of the radiative decay $D^0 \to \phi\gamma$.

1. DOUBLE CHARM PRODUCTION

Charmonium production in the e^+e^- annihilation provides a test of both perturbative and non-relativistic QCD (NRQCD). According to NRQCD predictions, the $e^+e^- \to J/\psi gg$ process should be the leading mechanism, with a cross-section as high as 1 pb. The color-octet $e^+e^- \to J/\psi g$ contribution should also be significant. In contrast, neither of these processes have been observed so far, while the process $e^+e^- \to J/\psi c\bar{c}$ has been measured by Belle with an unexpectedly large cross-section [1].

Recently, Belle updated its study of charmonium production with an additional $c\bar{c}$ pair using a data sample of 101.8 fb^{-1}. In this analysis, production of J/ψ together with another charmonium state, $e^+e^- \to J/\psi(c\bar{c})_{res}$, is studied by calculating the mass of the system recoiling against the J/ψ, $M_{recoil} = [(E_{CMS} - E_{J/\psi})^2 - p_{J/\psi}^2]^{1/2}$. The spectrum is plotted in Fig. 1(a). In an attempt to explain, at least partially, the discrepancy between the observed $e^+e^- \to J/\psi\eta_c$ rate and the theoretical prediction, it was suggested [2] that processes proceeding via two virtual photons may be important and the signal observed might include double J/ψ events. Using $e^+e^- \to \psi(2S)\gamma$, $\psi(2S) \to J/\psi\pi^+\pi^-$ events as a control sample, it was checked that no momentum scale bias confused the interpretation of the peaks in the M_{recoil} spectrum; the shift in the M_{recoil} distribution is estimated to be less than 3 MeV/c^2. Clear peaks are observed around the η_c, χ_{c0} and $\eta_c(2S)$ masses. A fit of this spectrum is performed including all the known narrow charmonium states: the η_c, χ_{c0} and $\eta_c(2S)$ masses are free parameters, those of J/ψ, χ_{c1}, χ_{c2} and $\psi(2S)$ are fixed to their nominal values. The fit results (Table 1) give negative yields for the J/ψ, χ_{c1}, χ_{c2} and $\psi(2S)$ and confirm that the cross-section for $e^+e^- \to J/\psi\eta_c$ is an order of

magnitude larger that the NRQCD prediction. In a similar way, the $\psi(2S)$ recoil mass spectrum is obtained in Fig. 1(b): the $\eta_c(2S)$ peak is significant, but only hints for η_c and χ_{c0} signals are seen (Table 1).

In the previous study [1], the $e^+e^- \to J/\psi c\bar{c}$ cross-section was inferred from the measured significant excess of D^0 and D^{*+} mesons produced in prompt J/ψ events, relying on the Lund fragmentation model. This time, Belle reconstructed as many ground state charm hadrons as possible, to reduce the model dependence of the result. The $D^0 \to K^-\pi^+$, $D^0 \to K^-\pi^+\pi^+\pi^-$, $D_s^+ \to K^+K^-\pi^+$ and $\Lambda_c \to pK^-\pi^+$ decay modes are used. To extract the number of charmed hadrons produced jointly with J/ψ, the charmed hadron signals are fitted in bins of the dilepton mass. The number of $e^+e^- \to J/\psi c\bar{c}$ events is calculated as a sum over the D^0, D^+, D_s^+, Λ_c and $(c\bar{c})_{res}$ yields corrected for efficiency. Belle then updated the cross-section for $e^+e^- \to J/\psi c\bar{c}$ with better accuracy and with reduced model dependence :

$$\frac{\sigma(e^+e^- \to J/\psi c\bar{c})}{\sigma(e^+e^- \to J/\psi X)} = 0.82 \pm 0.15 \pm 0.14.$$

2. D^{**} MESONS

Belle has studied the production of D meson excited states, collectively referred as D^{**}'s [3], which are P-wave excitations of quark-antiquark systems containing one charmed and one light (u, d) quark. In the heavy quark limit, the heavy quark spin $\vec{s}_c$ decouples from the other degrees of freedom and the total angular momentum of the light quark $\vec{j}_q = \vec{L} + \vec{s}_c$ is a good quantum number. There are four P-wave states, two $j_q = 3/2$ states, labeled as $D_1(1^+)$ and $D_2^*(2^+)$ which are narrow with a width of about 20 MeV/c^2. They have already been observed. In addition, there are two $j_q = 1/2$ states, $D_0^*(0^+)$

CP722, *B Physics at Hadron Machines*, edited by M. Paulini and S. Erhan

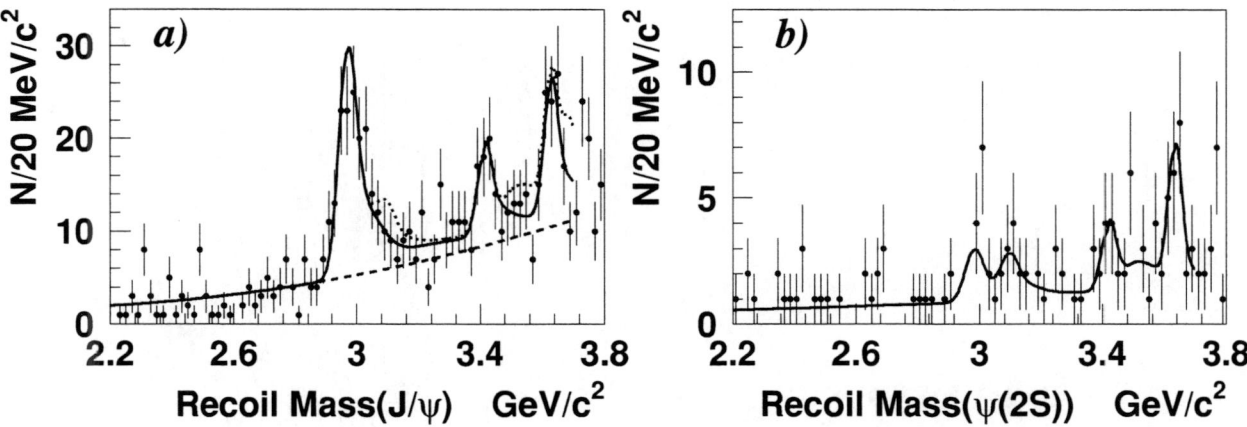

FIGURE 1. The mass of the system recoiling against the reconstructed (a) J/ψ and (b) $\psi(2S)$. In (a), the solid line is the result of a fit with the contributions of J/ψ, χ_{c1}, χ_{c2} and $\psi(2S)$ fixed at zero, the dotted line corresponds to the case where these contributions are set at their 90% confidence level upper limit values.

TABLE 1. Summary of the signal yields, masses and significances for $e^+e^- \to J/\psi(\psi(2S))(c\bar{c})_{res}$.

$(c\bar{c})_{res}$	N (evts)	Mass [GeV/c^2]	σ	N (evts)	σ
η_c	175 ± 23	2.972 ± 0.007	9.9	15 ± 7	2.6
J/ψ	-9 ± 17	fixed	–	12 ± 7	–
χ_{c0}	61 ± 21	3.409 ± 0.010	2.9	18 ± 9	2.4
$\chi_{c1} + \chi_{c2}$	-15 ± 19	fixed	–	7 ± 9	–
$\eta_c(2S)$	107 ± 24	3.630 ± 0.008	4.4	31 ± 10	3.7
$\psi(2S)$	-38 ± 21	fixed	–	-4 ± 7	–

and $D_1'(1^+)$, decaying via S-wave which are expected to be quite broad. These two last states have not yet been directly observed.

In a sample of 65 million $B\bar{B}$ mesons, Belle searched for the decays $B^- \to D^{(*)+}\pi^-\pi^-$ obtaining a yield of 1101 ± 46 (D^+) and 578 ± 30 (D^{*+}) events. The branching ratios derived from these yields are $(1.02 \pm 0.04 \pm 0.15) \times 10^{-3}$ and $(1.25 \pm 0.08 \pm 0.22) \times 10^{-3}$, respectively. In each case, an unbinned fit to the Dalitz plot has been performed to disentangle the contributions from various intermediate resonances. For the $D^+\pi^-\pi^-$ sample, apart from the narrow D_2^{*0} resonance, a significant contribution from a broad resonance is necessary in order to describe the data. It is identified as the D_0^{*0} resonance. The mass and width obtained for these two resonances are summarized in Table 2 and the product of the branching fractions are found to be

$$\mathcal{B}(B^- \to D_2^{*0}\pi^-) \times \mathcal{B}(D_2^{*0} \to D^+\pi^-) =$$
$$(3.4 \pm 0.3 \pm 0.6 \pm 0.4) \times 10^{-4},$$
$$\mathcal{B}(B^- \to D_0^{*0}\pi^-) \times \mathcal{B}(D_0^{*0} \to D^+\pi^-) =$$
$$(6.1 \pm 0.6 \pm 0.9 \pm 1.6) \times 10^{-4}.$$

The $D^{*+}\pi^-\pi^-$ is well described by the production of $D_2^*\pi$, $D_1'\pi$ and $D_1\pi$ with $D^{**} \to D^*\pi$. From a coherent

TABLE 2. Summary of the four P-wave states of the D meson. The first error is statistical, the second is systematic and the third is the model dependent error.

	Mass [MeV/c^2]	Width [MeV]
D_2^{*0}	$2461.6 \pm 2.1 \pm 0.5 \pm 3.3$	$45.6 \pm 4.4 \pm 6.5 \pm 1.6$
D_0^{*0}	$2308 \pm 17 \pm 15 \pm 28$	$276 \pm 21 \pm 18 \pm 60$
D_1^0	$2421.4 \pm 1.5 \pm 0.4 \pm 0.8$	$23.7 \pm 2.7 \pm 0.2 \pm 4.0$
$D_1'^0$	$2427 \pm 26 \pm 20 \pm 15$	$384^{+107}_{-75} \pm 24 \pm 70$

amplitude analysis, Belle obtained the masses and widths of these resonances (Table 2) plus the product branching fractions:

$$\mathcal{B}(B^- \to D_1\pi^-) \times \mathcal{B}(D_1 \to D^{*+}\pi^-) =$$
$$(6.8 \pm 0.7 \pm 1.3 \pm 0.3) \times 10^{-4},$$
$$\mathcal{B}(B^- \to D_2^{*0}\pi^-) \times \mathcal{B}(D_2^{*0} \to D^{*+}\pi^-) =$$
$$(1.8 \pm 0.3 \pm 0.3 \pm 0.2) \times 10^{-4},$$
$$\mathcal{B}(B^- \to D_1'^0\pi^-) \times \mathcal{B}(D_1'^0 \to D^{*+}\pi^-) =$$
$$(5.0 \pm 0.4 \pm 1.0 \pm 0.4) \times 10^{-4}.$$

TABLE 3. Product branching fractions for $B \to \bar{D}D_{sJ}$.

B mode	D_{sJ} mode	$\mathcal{B}(10^{-4})$	Sig (σ)
$\bar{D}D_{sJ}(2317)$	$D_s\pi^0$	$8.5^{+2.1}_{-1.9} \pm 2.6$	6.1
	$D_s^*\gamma$	< 7.5	1.8
$\bar{D}D_{sJ}(2457)$	$D_s^*\pi^0$	$17.8^{+4.5}_{-3.9} \pm 5.3$	6.4
	$D_s\gamma$	$6.7^{+1.3}_{-1.2} \pm 2.0$	7.4
	$D_s^*\gamma$	< 7.3	2.1
	$D_s\pi^0$	< 1.8	–
	$D_s\pi^+\pi^-$	< 1.6	–

3. D_{sJ} MESONS

Recently, BaBar discovered [4] a new narrow resonance $D_{sJ}(2317)$ decaying to $D_s\pi^0$. A natural interpretation is that this is a P-wave $c\bar{s}$ quark state that is below the DK threshold, which accounts for the small width. This interpretation is supported by CLEO's observation [5] of a second resonance $D_{sJ}(2457)$ decaying to $D_s^*\pi^0$. The mass difference between the two observed states is consistent with the expected hyperfine splitting of the P-wave D_s meson doublet with total light-quark angular momentum $j_q = 1/2$. However, the masses of these states are considerably below potential model expectations and are nearly the same as those of the corresponding $c\bar{u}$ states mentioned in the previous section. The Belle Collaboration, in addition to confirming these results, observed new decay modes of these resonances [6] and made the first observation of both states in B meson decays [7]. In order to determine quantum numbers and branching ratios, the additional constraints in decays of B mesons are important. In a sample 123.8×10^6 $B\bar{B}$ events, Belle searched for decays of the type $B \to \bar{D}D_{sJ}$ which are expected to be the dominant exclusive D_{sJ} production mechanism in B decays. The results are summarized in the Table 3. Signals for the $B \to \bar{D}D_{sJ}(2317)[D_s\pi^0]$ and $B \to \bar{D}D_{sJ}(2457)[D_s^*\pi^0, D_s\gamma]$ channels have greater than 5σ statistical significance.

The new observation $D_{sJ}(2457) \to D_s\gamma$ eliminates the $J=0$ hypothesis for $D_{sJ}(2457)$. The angular analysis of this decay supports the hypothesis that the $D_{sJ}(2457)$ is a 1^+ state. Searching for these particles in e^+e^- continuum annihilation, Belle confirmed the existence of the radiative decay mode $D_{sJ}(2457) \to D_s\gamma$ (Fig. 2) and observed for the first time the dipion decay mode $D_{sJ}(2457) \to D_s\pi^+\pi^-$ (Fig. 3). The ratio of branching fractions is determined to be

$$\frac{\mathcal{B}(D_{sJ}^+(2457) \to D_s^+\pi^+\pi^-)}{\mathcal{B}(D_{sJ}^+(2457) \to D_s^{*+}\pi^0)} = 0.14 \pm 0.04 \pm 0.02.$$

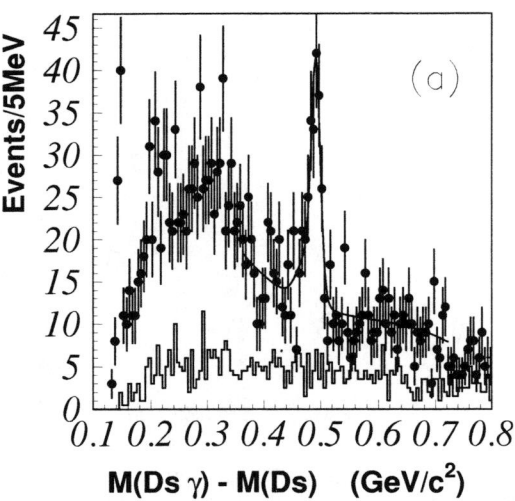

FIGURE 2. The $\Delta M(D_s^+\gamma)$ distribution fitted by a double Gaussian for the signal and a third-order polynomial for the background.

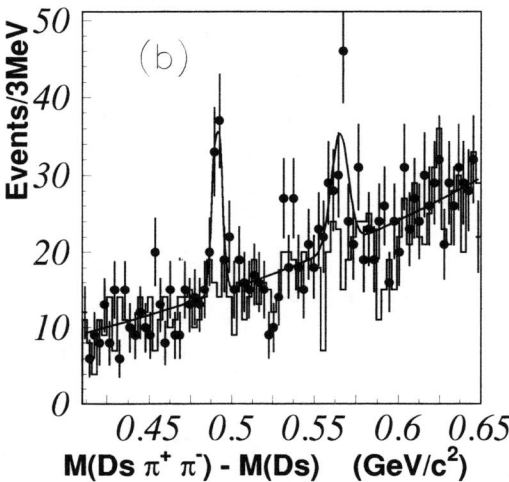

FIGURE 3. The $\Delta M(D_s^+\pi^+\pi^-)$ distribution fitted by Gaussian for the signals and a third-order polynomial for the background.

4. RADIATIVE DECAY $D^0 \to \phi\gamma$

Flavor-changing radiative decays of the charmed meson system, $D \to V\gamma$, where V is a vector meson, have not previously been observed. The Belle Collaboration has conducted a search for $D^0 \to \phi\gamma$ [8]. The branching fraction of this mode, dominated by $D \to VV' \to V\gamma$ (where $V^{(')}$ are vector mesons), is estimated to lie in the range $(0.04 - 3.4) \times 10^{-5}$, well below the previous limit of 1.9×10^{-4} but partially overlapping the sensitivity of the B factories.

To reduce the combinatorial background, D^*-tagging

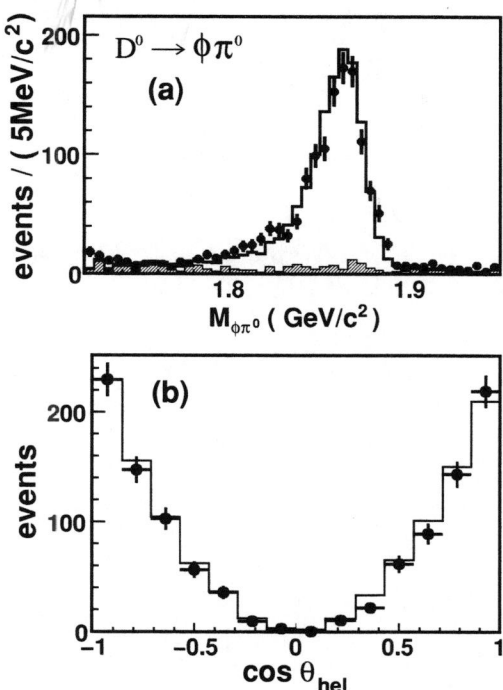

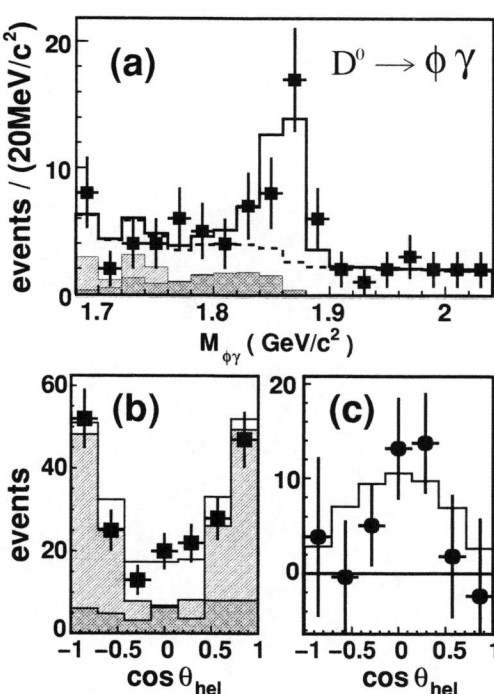

FIGURE 4. (a) $\phi\pi^0$ mass distribution for data (points) and Monte Carlo (histogram); the ϕ mass sideband is also shown (shaded). (b) Helicity distribution.

FIGURE 5. (a) $\phi\gamma$ mass distribution for data (points), the ML fit (open histogram), the background component of the fit (dashed), and the sum of $\phi\pi^0$, $\phi\eta$ and $\phi\pi^+\pi^0$ backgrounds (shaded). (b) $\cos\theta_{\text{hel}}$ distribution: data (points), and MC predictions for the total (open histogram), total background (light), and non-$\phi\pi^0$ background (dark). (c) Background-subtracted distribution for the data (points) and MC (histogram).

TABLE 4. Branching fraction of the $\phi\gamma$, $\phi\pi^0$ and $\phi\eta$ modes.

Mode	$\mathscr{B}$
$\phi\gamma$	$(2.60^{+0.70}_{-0.61}{}^{+0.15}_{-0.17}) \times 10^{-5}$
$\phi\pi^0$	$(8.01 \pm 0.26 \pm 0.47) \times 10^{-4}$
$\phi\eta$	$(1.48 \pm 0.47 \pm 0.09) \times 10^{-4}$

ACKNOWLEDGMENTS

I would like to thank the Beauty 2003 organizers for the invitation to present this review and my colleagues of the Belle collaboration, in particular the charm studies group.

and cuts on the D^* momentum and γ energy are used. The dominant background is then due to the Cabibbo- and color-suppressed decays $D^0 \to \phi\pi^0$ and $\phi\eta$, which have not previously been observed. Using analogous cuts to select $D^0 \to \phi\pi^0$, the reconstructed $M_{\phi\pi^0}$ distribution shows a clear enhancement at the D^0 mass (Fig. 4). A smaller but still clear signal of 31.1 ± 9.8 events is seen for $D^0 \to \phi\eta$. The contribution of these decays to the $\phi\gamma$ spectrum is suppressed by a helicity angle cut $|\cos\theta_{\text{hel}}| < 0.4$, favoring transversely polarized ϕ of $D^0 \to \phi\gamma$ over the longitudinally polarized ϕ of the $\phi\pi^0$ and $\phi\eta$ modes.

A clear peak is then observed at the D^0 mass in the $M_{\phi\gamma}$ invariant mass distribution (Fig. 5.a) with a yield of $27.6^{+7.4}_{-6.4}(\text{stat})^{+0.5}_{-1.0}(\text{syst})$. Table 4 summarizes the branching fractions measured for the three decays.

The $\phi\gamma$ mode is the first radiative decay and the first FCNC decay observed in the D meson system.

REFERENCES

1. Abe, K., et al., [Belle Collaboration], *Phys. Rev. Lett.* **89**, 142001 (2002).
2. Bodwin, G. T., Lee, J., and Braaten, E., *Phys. Rev. Lett.* **90**, 162001 (2003).
3. Abe, K., et al., [Belle Collaboration], arXiv:hep-ex/0307021 (2003).
4. Aubert, B., et al., [BaBar Collaboration], *Phys. Rev. Lett.* **90**, 242001 (2003).
5. Besson, D., et al., [CLEO Collaboration], *AIP Conf. Proc.* **698**, 497–502 (2004).
6. Abe, K., et al., *Phys. Rev. Lett.* **92**, 012002 (2004).
7. Krokovny, P., et al., [Belle Collaboration], *Phys. Rev. Lett.* **91**, 262002 (2003).
8. Abe, K., et al., [Belle Collaboration], *Phys. Rev. Lett.* **92**, 101803 (2004).

DETECTORS, HARDWARE, AND COMPUTING

The Status of the ATLAS Inner Detector

Hans-Günther Moser

(Representing the ATLAS Collaboration)

Max-Planck-Institut für Physik, Föhringer Ring 6, D-80805 Munich, Germany

Abstract. The ATLAS Inner Detector uses three subdetectors for tracking of charged particles from $r = 5$ cm to $r = 107$ cm inside a solenoid magnet of 2 T. The innermost detector is a high resolution silicon pixel detector. It provides precise 3D tracking information close to the interaction point allowing secondary vertex reconstruction and hence b identification. It is followed by the SCT, a large area tracking device based on silicon strip detectors. The TRT, based on straw tubes, provides continuous tracking and improves electron identification due to its ability to detect transition radiation. These detectors are presently under construction. This report presents a brief report on the design, construction status and expected performance of the Inner Detector.

1. INTRODUCTION

The ATLAS Inner Detector is the central tracking device of the ATLAS detector, one of the two major general purpose detectors for the Large Hadron Collider (LHC) at CERN. The detector has passed the design and prototyping phase and most components are presently in production to match the LHC schedule, which foresees data taking in 2007. This report gives a brief review of the design, construction status and expected performance of the ATLAS Inner Detector.

2. TRACKING IN ATLAS

Tracking detectors at LHC have to face two major challenges: high occupancy and radiation damage. A typical high luminosity interaction creates about 700 charged particles within $|\eta| < 2.5$. Depending on the radius, up to 10^{14} neutrons/cm^2/year (1 MeV equivalent) cause substantial radiation damage. Furthermore, the short bunch crossing interval of 25 ns demands fast detectors with fast readout electronics. Driven by physics performance, survival and costs, the following requirements were identified in the Technical Design Report [1] for the Inner Detector:

- Rapidity coverage $|\eta| < 2.5$.
- Momentum resolution for isolated tracks:

$$\sigma\left(\frac{1}{p_T}\right) = \left(0.4 \oplus \frac{13}{p_T \sin\theta}\right) \times 10^{-3} \quad (1)$$

(p_T in GeV/c).

- Impact parameter resolution:

$$\sigma_{r-\phi}\left(11 \oplus \frac{60}{p_T\sqrt{sin\theta}}\right)\mu\text{m} \quad (2)$$

$$\sigma_z = \left(70 \oplus \frac{100}{p_T\sqrt{sin\theta}}\right)\mu\text{m} \quad (3)$$

(p_T in GeV/c).

- Track reconstruction efficiency $> 95(90)\%$ for isolated tracks (tracks in jets) at high p_T.
- Less than 1% ghost tracks (for isolated tracks).
- Lowest possible material budget in order to minimize multiple scattering and not to compromise calorimeter performance.
- A lifetime of at least 10 years under LHC conditions.

These requirements lead to a choice of three detector technologies which will be described in the following sections.

3. THE ATLAS INNER DETECTOR

Occupancy and radiation damage depend strongly on the radius. In order to allow for an optimized technical solution for each radial region, the ATLAS Inner Detector (ID) is segmented in three subdetectors each using a different detector technology:

- In the innermost region, from $r = 5$ cm to 12.5 cm, a high granularity silicon pixel detector is used with excellent 3D space resolution and low occupancy.

CP722, *B Physics at Hadron Machines*, edited by M. Paulini and S. Erhan
© 2004 American Institute of Physics 0-7354-0203-5/04/$22.00

The price to pay is a large number of readout channels. The high resolution of this detector is essential for secondary vertex reconstruction and *b* identification.

- In the intermediate region, from $r = 30$ cm to 52 cm, a silicon strip detector is used. The strips are oriented to measure with high precision the r-ϕ coordinates of a track. A small stereo angle between two faces of a detector module allows to measure the orthogonal coordinate with less precision. The detector will measure up to four space points (i.e. eight single hits) of a track. The total area of silicon is about 70 m^2.

- Finally, from 56 cm to 107 cm, the transition radiation tracker uses straw tubes. The large number of straws allows continuous tracking with many hits, which, despite the moderate space resolution of single hits, results in a high precision measurement at large radius. In addition, the ability to detect transition radiation improves electron/pion separation. The challenge of this detector is to cope with the high occupancy.

The layout of the ID is shown in Figure 1 and a summary of the subdetector parameters is given in Table 1. The complete Inner Detector is a cylinder of 2 m diameter and 7 m length. It is operated in a 2 T axial field. Detailed descriptions and specifications of the ATLAS inner detector and its components can be found in Refs. [1, 2].

3.1. Pixel Detector

For the pixel detectors, high granularity and extreme radiation hardness are the main challenges. Each silicon sensor has 46080 pixels of 50 μm $\times$ 400 μm. The sensor size is 16.4 mm $\times$ 60.8 mm and it is made in n-on-n[1] technology. The small pixel size results in low capacitance and an excellent signal-to-noise ratio even if the silicon bulk is only partially depleted.

The sensors are presently in production. A first series of 600 have been produced by CIS in Germany and 400 more are in production. A second series will be produced by Tesla in the Czech Republic. A pre-series of 50 prototype sensors has been successfully produced and series production is about to start.

The readout electronics is made in Deep Submicron (DSM) Technology. First attempts to produce chips in DMILL technology were abandoned because of low yield. In DSM, yields of $> 90\%$ were reached. The present prototype is working and can be used for test

modules. However, another iteration is necessary to fix some minor problems, then the main series production can be submitted. Each sensor is read out by 16 frontend chips, which are bump-bonded onto the sensor.

Each module is controlled by one MCCI2 chip. A new version has been designed with triple logic in order to improve the tolerance with respect to single event upset (flip of a register status due to an ionizing particle). The chip is currently in production.

The pixel detector modules are arranged on three cylindrical barrel layers in the central region and 2 $\times$ 3 endcap disks. Due to schedule constraints, the detector can be inserted at the end of the ATLAS assembly. Only a support tube has to be installed at the beginning inside the silicon tracker. The pixel detector with services and the beampipe will then be inserted later. This also allows removal and therefore possible upgrades of the device. The mechanical support structures are in production.

In testbeam measurements, it could be demonstrated that the modules survive the expected radiation dose (60 Mrad) and still show the required performance. For example, the single hit efficiency, which is 99.7% for an unirradiated detector, drops to 97.7%, which is still more than sufficient.

3.2. Silicon Tracker

The basic building block of the SCT, the module, is made from four single-sided silicon strip detectors arranged in two back-to-back planes. The strip directions on the two planes are rotated by a stereo angle of 40 mrad with respect to each other. The detectors are mounted on a mechanical carrier made out of TPG (thermal pyrolythic graphite). TPG has an excellent thermal conductivity (1500-1700 W/m/K) needed for efficient cooling of the silicon [3]. The strips are read out by 12 ASIC chips supported by a hybrid made out of a Kapton foil laminated on a carbon substrate. Whilst the hybrid of the endcap module is truly double sided and mounted at one end of the detectors, the barrel module has two single sided hybrids mounted centrally like a bridge over the silicon detectors.

The silicon detectors are rather conventional single sided AC-coupled detectors made in p-on-n[2] technology. The barrel detectors measure about 6 cm $\times$ 6 cm and the strip pitch is 80 μm. The detectors for the endcap modules are wedge shaped of similar size with a mean pitch of 80 μm. Depending on the radial position of the module, five different types of endcap wedge detectors are needed. The detectors closest to the interaction region

---

[1] n^+ implant on n-type silicon.

[2] p^+ implant on n-type silicon.

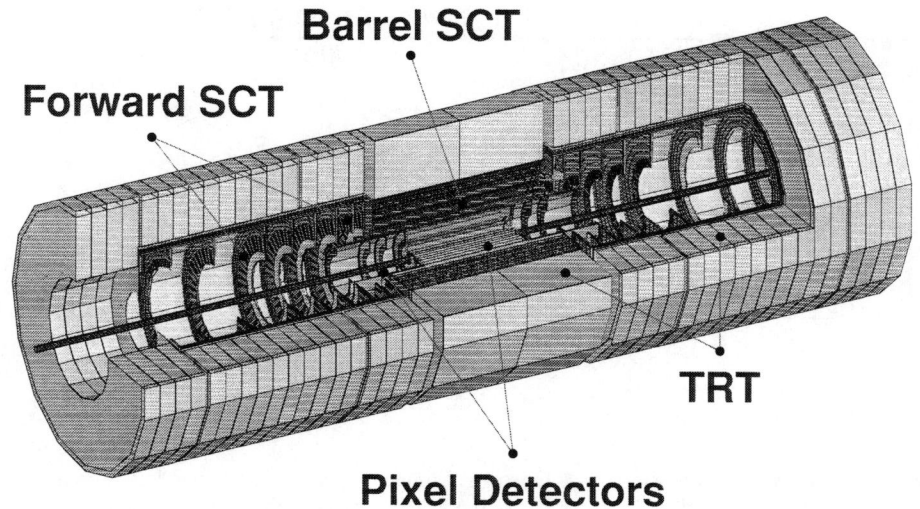

FIGURE 1. The ATLAS Inner Detector.

TABLE 1. Properties of the ATLAS Inner Detector subdetectors.

subdetector	radius [cm]	element size	resolution	hits/track	channels
Pixel (Silicon)	5-12.5	$50~\mu m \times 400~\mu m$	$12~\mu m \times 60~\mu m$	3	93×10^6
SCT (Silicon strip)	30-52	$80~\mu m \times 12~cm$	$16~\mu m \times 580~\mu m$	4	6×10^6
TRT (Straw tubes)	56-107	$4~mm \times 74~cm$	$170~\mu m$	36	0.4×10^6

are made using oxygenated silicon which significantly improves the (reverse-) annealing properties of the devices: oxygenation slows down processes that lead to an increase of the depletion voltage after irradiation if the detectors are left at room temperature, see Fig. (2). In order to reduce this reverse annealing and reduce leakage currents, the silicon detectors (like for the pixels) have to be operated at $-10°$ C.

In the SCT, the detector modules are arranged in four cylindrical barrel layers and two times nine disks in both endcaps. About 4000 modules are used. The 20,000 silicon detectors are already produced. The readout electronics has been produced in DMILL technology. Each ABCD chip has 128 charge sensitive amplifiers, a comparator and a digital pipeline.

Series production of barrel modules has already started and is about to start for endcap modules.

The mechanical support structures for the modules are in production (and partially finished). These support structures, made out of honeycomb material and carbon fiber, must be light but nevertheless stiff and extremely precise. For example, the disk used to hold the SCT endcap modules is flat within $60~\mu m$ over a diameter of 1 m.

The SCT modules have been tested in various testbeams. It has been demonstrated that the specifications of $> 99\%$ efficiency with $< 5 \times 10^{-4}$ noise occupancy can be met for the barrel modules. The endcap modules

have a slightly higher noise level and this specification can only be met in a very narrow threshold range [4].

3.3. Transition Radiation Tracker

The TRT uses two different module types:

- The barrel module use axial straw tubes and a foil radiator.
- In the endcap region, radial straws embedded in a foam radiator are used.

The straw tubes are up to 1.44 m long. In case of the barrel TRT the wires are separated in the middle and read out at both ends. Each track has up to 36 hits in the TRT. This is equivalent to a single hit of $50~\mu m$ precision. In addition, by detection transition radiation, pions can by suppressed by a factor of 100 with respect to electrons. Transition radiation hits can be distinguished from higher mass particle hits using logic with two thresholds.

The production of the barrel modules is completed, and the modules are being tested now. The production of the endcap modules is delayed due to problems with the front-end boards and should be completed in May 2005.

Recently, a problem with the TRT gas mixture has been discovered. The original gas mixture of $Xe(70\%), CF_4(20\%), CO_2(10\%)$ destroyed the glass wire joints due to CF_4 radicals created by radiation. A

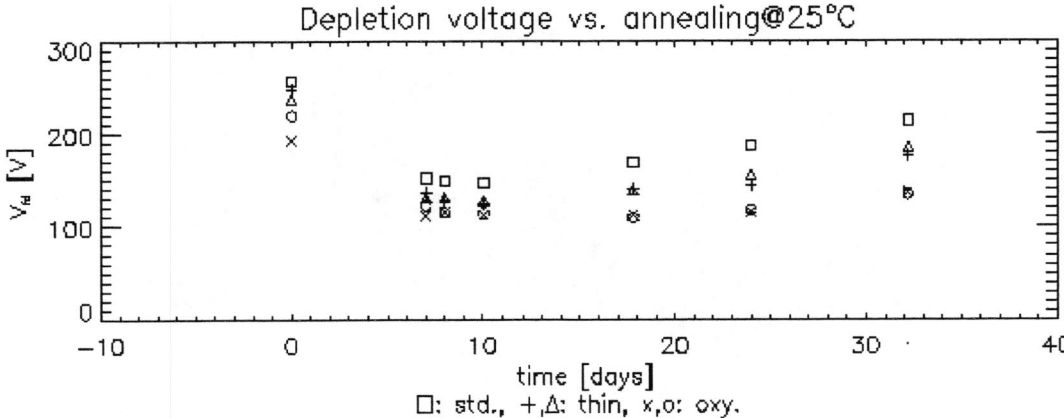

FIGURE 2. Depletion voltage of silicon detectors as function of time after irradiation. When left at room temperature annealing processes first decrease the depletion voltage. After several days (10), however, the depletion voltage increases again (reverse annealing). Shown are data obtained from detectors made on standard substrate ($\square$, $+$, $\triangle$) and on oxygenated substrate ($\times$, $\circ$) which show a substantially reduced reverse annealing.

possible solution would have been to use epoxy wire joints, however, since more than 20% of the straws were already produced, this option was discarded. After a large R&D program, it could be shown that a gas mixture of $Xe(70\%), CO_2(27\%), O_2(3\%)$ yields acceptable performance (except for being slower) without damaging the wire joints. However, this requires periodical cleaning of the wires (to remove deposits) using a Ar, CO_2, CF_4 mixture. This can be done during normal operation, but, lacking Xe, transition radiation cannot be detected during these periods.

4. RECENT CHANGES

Compared to the original layout given in the Technical Design Report [1], some important changes have to be reported:

- The material of the inner detector increased, partially due to an increased thickness of the pixel sensors, partially due to more realistic engineering of the services.
- The radius of the inner pixel layer moved from 4.3 cm to 5.0 cm, to allow for a wider beampipe.

Clearly this has an impact on the expected p_T and impact parameter resolution. Especially for low momentum tracks, the increased multiple scattering becomes important. For example, the impact parameter resolution is now parameterized as:

$$\sigma_{r-\phi} = \left(17.7 \oplus \frac{108}{p_T \sqrt{sin\theta}} \right) \mu\text{m}. \qquad (4)$$

In addition, due to funding and schedule constraints, some items need to be staged, hence the Initial Detector will not be equipped with:

- The middle layer of the pixel detector, at $r = 9$ cm.
- The middle disks of the endcap pixel detector at $z = \pm 58$ cm.
- The endcap TRT "C" wheels, at $|\eta| > 1.7$.

This will also have some impact on the p_T resolution at large rapidity, impact parameter resolution and pattern recognition. The consequences for B physics are discussed in Refs. [5, 6].

5. ASSEMBLY AND INTEGRATION

The assembly of the detector components to a larger detector system is a major task which requires substantial engineering effort. A special surface building at CERN ('SR-building'), close to the underground cavern, is presently being prepared for the assembly and test of the SCT barrel, the TRT and the pixel detector. The SCT endcaps will be assembled in Liverpool and NIKHEF and shipped to CERN in order to be integrated in the TRT endcaps. This integration will also happen in the SR-building. Further engineering effort is needed to design and construct the support structures, services, gas system and the cooling system. As an example, the cooling system will be discussed in more detail: The total power of the on-detector electronics is 96.5 kW (12.5 kW Pixel, 39 kW SCT and 46 kW TRT), the cables inside the detector volume dissipate 7.6 kW of power and the thermal enclosures needed to insulate the detectors from each other need further 7.3 kW. Furthermore, the system is complicated by the different operating temperatures of the subdetectors. Pixel and SCT are operated at low temperatures (Coolant at $-30°$ C), while the TRT operates at room temperature. For the TRT and most of the cables, cooling with a monophase liquid is sufficient, while for the SCT and Pixel detectors an evaporative cooling system using C_3F_8 is needed.

6. SCHEDULE

Since ATLAS should see first physics events in 2007, the ID must be ready for installation in ATLAS in March 2006. The assembly of the major subdetectors in the SR-building is scheduled to start in April 2004. The SCT Barrel should be ready in January 2005 together with the TRT barrel. The SCT barrel will be inserted in the TRT barrel and the ID barrel will be ready for installation in ATLAS in July 2005. The first SCT endcap will be assembled in April 2005, the TRT endcap in November 2005, and the first ID endcap will be ready in October 2005. For the second endcap, the respective dates are August 2005 (SCT), September 2005 (TRT) and March 2006 (ID). The staged items will be completed later in 2006 and will not be ready for installation in the initial ATLAS detector.

7. CONCLUSIONS

The ATLAS Inner Detector project has completed the design and prototype phase. Practically all technical problems have been solved and most of the detector components are now in production. The infrastructure for the detector integration is being prepared. The main concern is the extremely tight schedule which foresees the initial Inner Detector to be ready for installation in ATLAS at the latest in March 2006. However, the 3rd pixel layer and the TRT "C" wheels will be staged and will not be available in the first data taking period. The implications for B physics are under study.

ACKNOWLEDGMENTS

I would like to thank the organizers of the conference for providing a pleasant stay in Pittsburgh. I would also like to thank my ATLAS colleagues for providing the information needed to prepare this presentation.

REFERENCES

1. ATLAS Collaboration, Inner Detector Technical Design Report, Tech. Rep. CERN/LHCC/97-16,97-17 (1997).
2. ATLAS Collaboration, ATLAS Detector and Physics Performance TDR, Tech. Rep. CERN/LHCC/99-14 and 99-15 (1999).
3. Heusch, C. A., Moser, H. G., and Kholodenko, A., *Nucl. Instrum. Meth.* **A480**, 463–469 (2002).
4. Mangin-Brinet, M., et al., [ATLAS Collaboration], Electrical test results from ATLAS-SCT end-cap modules, Tech. Rep. ATL-INDET-2003-004 (2003).
5. Eerola, P., in these proceedings (2004).
6. Parodi, F., in these proceedings (2004).

Status of the Compact Muon Solenoid Detector at the Large Hadron Collider

Homer Neal

(Representing the CMS Collaboration)

Yale University, Department of Physics, New Haven, CT 06520, U.S.A.

Abstract. The Compact Muon Solenoid (CMS) detector is being constructed at the Large Hadron Collider (LHC) at CERN for the purpose of exploring new physics in proton-proton collisions at 14 TeV. The detector is to be designed, constructed and commissioned in time for the start of the LHC in April 2007. This article summarizes the current status of the detector and foreseen program for completion.

1. INTRODUCTION

The Compact Muon Solenoid (CMS) detector is being constructed at the Large Hadron Collider (LHC) at CERN for the purpose of exploring new physics in the highest energy collisions ever achieved at an accelerator. The particle decay products from the 14 TeV proton-proton collision will be studied for evidence of Higgs boson production, supersymmetric particles, black holes and quark gluon plasmas (from heavy ion collisions) as well as numerous interesting Standard Model studies such as the physics of the top quark and m_W. The intense background radiation and very high rate of events resulting from bunch crossings every 25 nanoseconds and overlapping events impose many demanding constraints on the detector. To fulfill the physics and tolerance requirements, the detector employs a variety of technological solutions within its 12,500 tons. The target completion date for the detector is set by the foreseen April 2007 turn-on date for the LHC. This article summarizes the progress to date that the CMS collaboration (which consists of nearly 2000 scientists from over 35 nations) has made towards achieving this goal.

1.1. Infrastructure

The CMS detector will be located at access point 5 on the 27 km LHC ring near Cessy, France. At a depth of about 100 m below the surface hall (SX5) is the 26.5 m high and 53 m wide experimental cavern (UXC5) where the detector will ultimately be located on the beam line. So far, the floor and over half of the walls have been concreted. The caverns will be completed by the middle

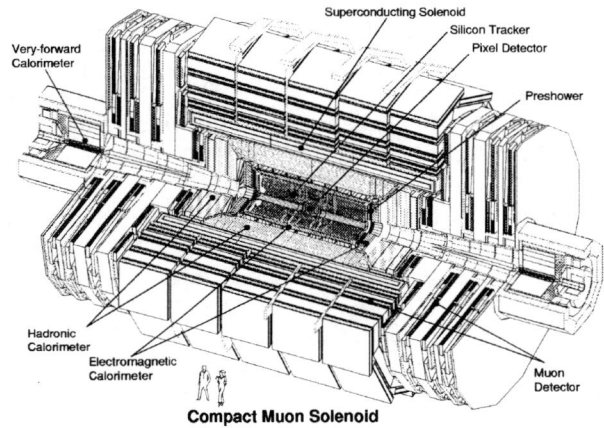

FIGURE 1. The layout of the CMS detector.

of 2004. The lowering of the major pieces of CMS into UXC5 will occur between the end of May and the end of September 2005.

1.2. Detector Overview

Starting from the innermost systems and going out, the CMS detector (Fig. 1) consists of pixel and strip trackers followed by a lead tungstate electromagnetic calorimeter, a brass hadron calorimeter, a solenoid and finally a resistive plate, drift tube and cathode strip muon system. These systems are now briefly described and their current status and foreseen completion schedule given.

CP722, *B Physics at Hadron Machines,* edited by M. Paulini and S. Erhan

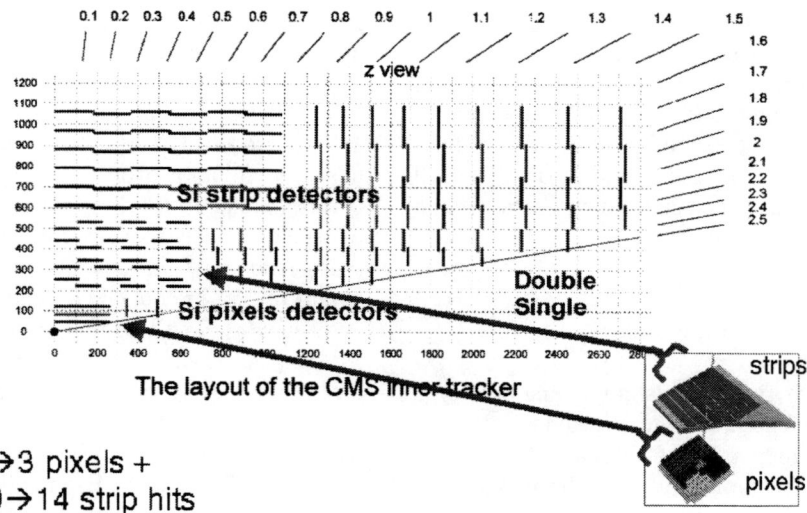

FIGURE 2. The layout of the CMS tracker showing the arrangement of the silicon pixel and silicon strip detectors.

1.3. Coil

The tracking and calorimeter barrel systems are in a uniform 4 Tesla field provided by the 5000 ton solenoid coil [1]. The coil is 13 m long with a free inner diameter of 5.9 m. The magnetic flux is returned by a 1.5 m thick saturated iron yoke which has a mass of 7000 tons. The iron in the yoke also forms part of the muon detector which is discussed in a later section.

1.3.1. Coil Status

As of the writing of this article, all 21 reinforced coil conductor lengths have been successfully produced by Techmeta in Annecy. The winding is 50% complete and the final module delivery is expected in June 2004 The first test of the magnet in the surface hall should occur during the first half of 2005.

1.4. The Tracker

The need for precise position and momentum measurements in a harsh radiation environment leads to a unique multi-technological solution for the tracker involving silicon pixels and silicon strips. The pixel tracker [2, 3] occupies the innermost volume given by radius(R) < 20 cm. This is followed by the silicon strip tracker (SST) [4, 5] for 20 cm < R < 110 cm. The layout of the tracker is shown in Fig. 2.

1.4.1. Pixel Tracker

The unambiguous three-dimensional space points provided by the pixel tracker are essential for correct pattern recognition at small radii. The pixel system has layers at 4.3 cm, 7.2 cm and between 10 to 11 cm from the beam axis. The 100×150 μm pixels yield spatial resolutions of about 15 microns in the azimuthal (ϕ) and longitudinal (z) directions. The improvement over the typical 35 μm expected for the pixel size is accomplished by purposely not compensating for the large Lorentz angle thus leading to significant charge sharing. The pixel barrel detector has an area of 0.8 m^2 and provide three high resolution hits out to a pseudorapidity of $|\eta| = 2.2$. In addition, there are two endcap disks at $|z| = 32.5$ cm and 46.5 cm covering an additional 0.28 m^2. The total pixel count for the detector is around 50 million pixels. The detector is designed to endure fluencies up to 6×10^{14} neutrons/cm^2.

1.4.2. Silicon Strip Tracker

The SST will occupy the radial range 20 to 110 cm and the longitudinal range $|z| < 280$ cm, thus covering pseudorapidities out to $|\eta| < 2.5$. It consists of 10 concentric layers of silicon strip detectors (four in an Inner Barrel (TIB) and six in an Outer Barrel (TOB)). The strips are oriented along the beam axis for measurement of the azimuthal coordinates for the "single-sided" modules. The two innermost TIB and TOB layers are equipped with "double-sided" detector modules with one side having strips at a stereo angle of 100 mrad with respect to the longitudinal strips for providing measurement of the longitudinal coordinate as well. The TIB is complemented by three inner disks (TID) on either side. Nine End-Cap (TEC) disks are located between $|z|$=120-280 cm. All endcap layers have strips oriented radially to allow measurement of the ϕ coordinate. In addition, "double-sided" layers at the inner and outer radii of the end-cap disks permit measurements of the radial coordinate. The full SST has roughly 10 million channels distributed over an active surface of approximately 200 m^2.

1.4.3. Tracker Status

The pixel system is currently testing its first prototype modules. For the SST, all final contracts for module parts are active. Module production and validation has started. A prototype of the front-end driver has been designed. Much of the mechanical structures have been procured. The tracker inner barrel is foreseen to arrive and be tested at CERN by April 2005. The tracker outer barrel is expected to be completed by this time. The tracker endcap is expected to be tested at CERN in October 2005. These dates for the SST include a contingency of two months.

1.5. Electromagnetic Calorimeter

The electromagnetic calorimetry is located just outside the tracker. It includes a crystal calorimeter and a preshower detector.

The electromagnetic calorimeter (ECAL) is composed of 75000 lead tungstate (PbWO$_4$) crystals (60000 in the barrel, 15000 in the endcaps combined) which have a scintillation spectrum peaking at 440 nm [6]. These transparent crystals have a density (8.28 g/cm^3) exceeding that of iron. The light from the scintillation produced by relativistic particles passing through the crystals is collected and converted to an analog signal by avalanche photodiodes (APD) in the barrel and vacuum phototriodes (VPT) in the endcaps. A primary reason for the selection of APD's for the readout is that they can provide gain in the presence of the high magnetic field. The radiation environment of up to 30 kGy and 3×10^{14} N/cm^2 in the endcaps leads to the choice of VPT's for the endcaps. After digitization, the data is transferred to the counting room by fiber-optic links. One fiber is used for 25 channels (a trigger tower). This represents a change to the original design where one fiber per channel was envisioned. Each trigger tower consists of a motherboard for the distribution of the low voltage to the very front-end (VFE) cards, five VFE cards each containing the amplifiers and ADC's for five channels, front-end cards (for distributing the clock and control, collecting and shipping the data and calculating and shipping the trigger

primitives), and a low voltage regulator board. The overall low voltage current that is used by the front-end electronics is estimated to be approximately 50000 Amps.

The barrel covers $|\eta|$<1.48 and the endcaps cover 1.48<$|\eta|$<3.00. The ECAL barrel uses trapezoidal crystals. The front face of these crystals is 22×22 mm^2 corresponding to the Molière radius. In ϕ, the angular coverage of each crystal corresponds to 0.0175 radians (=1°), and in η the coverage is 0.0175. They have a length of 23 cm equivalent to a thickness of 25.8 radiation lengths. The barrel contains 36 super-modules of 1700 crystals each. In the endcaps, the dimensions of the face (24.7×24.7 mm^2) remain the same as the pseudorapidity and azimuthal angle vary but the angular coverage increases to $\Delta\eta \times \Delta\phi = 0.05 \times 0.05$ as $|\eta|$ increases. The crystal are 22 cm long. Each endcap contains 2 dees of 3662 crystals each. The crystals are supported by 0.4 mm thick alveolar structures made from carbon-fiber (in the endcaps) and glass fiber (in the barrel). Further details concerning the CMS electromagnetic calorimeter can be found in the design reports [7, 8].

The preshower detector [9] is located in front of the ECAL endcaps in the pseudorapidity range $1.65 < \eta < 2.61$ and consists of two lead/silicon detector layers for π/γ separation. Its design was endorsed in March 2003. It will be installed in two steps to allow scheduling flexibility. The support structure will be installed before the beam pipe and the lead planes with electronics will be installed after.

1.5.1. Electromagnetic Calorimeter Status

For the summer 2003 test beam, two super-modules were tested. All aspects of the system were verified including the crystals plus APD's, new front-end boards, low voltage regulation and optical control and readout. The first super-module (SM0 - Fig. 3), tested in September, contained 100 channels of FPPA's (the original version of the floating point pre-amplifiers) but with the new architecture for the rest of the system. The second supermodule (SM1) used the complete new architecture including 50 channels of 0.25 μm Multi-Gain Pre-Amplifiers (MGPA's). The successful test of the supermodules using beam demonstrated the functionality of the electronics and cooling and noise levels within acceptable limits. Some minor design changes are currently being implemented and a new set of full system tests will occur in summer 2004 including supermodules with all 1700 channels active.

About 32% of the crystals have been delivered. A production rate of 3800 crystals per quarter is expected in 2004. The alveolus should be completed by the end of 2003. The APD production and screening should reach completion by April 2004.

FIGURE 3. The first of two CMS ECAL supermodules beam tested during fall 2003.

FIGURE 4. A photo of the completed CMS hadron barrel structure.

1.6. The Hadron Calorimeter

To identify and measure the energy of particles (primarily hadrons and muons) that are not stopped in or before the ECAL, a brass calorimeter is used. This hadron calorimeter (HCAL) [10] consists of a barrel section covering $|\eta| < 1.3$ and r=1.81 to 2.95 m and endcaps which cover $1.3 < |\eta| < 3.0$. The barrel is 79 cm deep, which at η=0 corresponds to 5.15 absorption lengths in thickness. It has two half barrels of 18 calorimeter "wedges". Each is 4.3 meters long in z and weighs 25.7 metric tonnes. The brass plates are interleaved with plastic scintillator embedded with wavelength shifting optical fibers.

The endcap has 10 interaction lengths (19 active layers). The brass absorber sampling thickness is 8 cm and the front and back plates are made of stainless steel to increase strength. The absorber plates are bolted together to form a single monolithic structure, with gaps for scintillator insertion.

In the region $|\eta| < 3.0$ the first muon absorber layer is instrumented with scintillator tiles to form an Outer Hadron Calorimeter (HO).

To improve the hermeticity, the region $3.0 < |\eta| < 5.0$ is instrumented with a quartz fiber calorimeter.

1.6.1. Hadron Calorimeter Status

The HB (Fig. 4) and HE structures are complete. The mechanics and optical links for all sectors has been completed. The installation and burn-in of the readout boxes for the HB will occur during 2004.

2. MUON DETECTOR

The muon system consists of the iron flux return yoke of the magnet instrumented with detectors for triggering and position measurements. In the barrel, four layers of Resistive Plate Chambers (RPC) and Drift Tubes (DT) are used in the gaps between the iron layers. The RPC's provide good timing for the trigger and drift tubes provide accurate position measurements. The endcap has four layers of RPC's combined with four stations of Cathode Strip Chambers (CSC's) each containing six layers for the position measurements and main trigger. The full muon system covers $|\eta| < 2.4$, and provides three to four track segments along a muon track. The depth will be at least 16 radiation lengths down to $|\eta|$=2.4 [11]. The expected global momentum resolution is 1% to 4% depending on p_T. A schematic of the muon detector showing the locations of the DT, CSC and RPC chambers is shown in Fig. 5. Details concerning the muon system can be found in the design report [12].

2.1. Muon Detector Status

For the RPC's, as of Sep. 2003, a total of 114 chambers have been assembled and 74 installed. As of December 2003, of the 186 drift tube chambers needed for the barrel, 128 have been built, and 99 have been tested. For the CSC's, as of September 2003, 439 of the 482 chambers have been assembled. Of these, 223 have been assembled with electronics and tested. There were 125 chambers at CERN and 105 of them ready for installation. At that time, 90 chambers had already been installed.

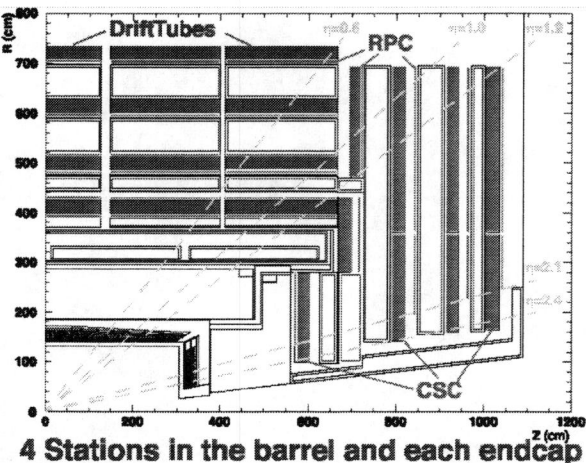

4 Stations in the barrel and each endcap

FIGURE 5. The layout of one quarter the muon detector showing the position of the Drift Tubes, Resistive Plate Chambers (RPC's), and Cathode Strip Chambers (CSC's).

2.2. Level-1 Trigger

Given the short period between beam crossings and the high backgrounds, the task of the Level-1 Trigger [13] becomes very difficult. The trigger must reduce the event rate from 40 MHz to a maximum of 100 kHz while retaining a very high efficiency for potentially interesting physics events.

To date, prototypes of the Level-1 trigger have been manufactured and final validation tests have been completed. Test are ongoing on the integration of the trigger with the detector and DAQ.

In 2004, further tests involving integration and using a structured LHC-like beam will be performed. All systems will enter the production phase.

3. SUMMARY

The CMS detector design and construction is on target for completion before the commissioning of the LHC. However, the schedule is very tight for several systems. The yokes are finished as well as the assembly of the HCAL. The foreseen completion dates for the other subsystems are as follows. The muon barrel should be completed by summer 2004 followed by the tracker at the end of 2005 and finally the ECAL barrel in spring 2006 and the ECAL endcaps at the end of 2006. To insure completion by the required date and to respond to budgetary pressures, new techniques for detector preparation (such as the ECAL crystal cutting) and new technological solutions such as the 0.25 μm ASIC's are being implemented. Successful prototype test runs indicate that production mode will be starting in the near term for those systems that have not already started. CMS will be ready for circulating beams by 1 April 2007.

ACKNOWLEDGMENTS

Many thanks to all the CMS collaborators who are working hard to insure the success of this major effort and those who are preparing the LHC. This work was supported by the Department of Energy contract DE-FG02-92ER40704 (Yale).

REFERENCES

1. Acquistapace, G., et al., [CMS Collaboration], The Magnet Project: Technical Design Report, Tech. Rep. CERN-LHCC-97-10 (1997).
2. Rohe, T., et al., arXiv:physics/0312009 (2003).
3. Cremaldi, L., [CMS Collaboration], *Nucl. Instrum. Methods* **A511**, 64–67 (2003).
4. Angarano, M. M., [CMS Tracker Collaboration], *Nucl. Instrum. Methods* **A501**, 93–99 (2003).
5. Abbaneo, D., et al., [CMS Collaboration], Layout and Performance of the CMS Silicon Strip Tracker, Tech. Rep. CMS CR 2003-032 (2003).
6. Gascon-Shotkin, S., [CMS ECAL Collaboration], Recent Developments in Crystal Calorimeters (Featuring the CMS PbWO$_4$ Electromagnetic Calorimeter), Prepared for ICHEP 2002 Conference, Amsterdam, 2002, Tech. Rep. CMS CR 2003-001 (2002).
7. CMS Collaboration, The Electromagnetic Calorimeter Technical Design Report, Tech. Rep. CERN-LHCC-97-33 (1997).
8. CMS Collaboration, Addendum to the CMS ECAL Technical Design Report: Changes to the CMS ECAL electronics, Tech. Rep. CERN-LHCC-2002-027 (2002).
9. Tournefier, E., [CMS Collaboration], *Nucl. Instrum. Methods* **A461**, 355–360 (2001).
10. CMS Collaboration, The Hadron Calorimeter Technical Design Report, Tech. Rep. CERN-LHCC-97-31 (1997).
11. Giacomelli, P., *Nucl. Instrum. Methods* **A478**, 147–152 (2002).
12. CMS Collaboration, CMS Muon Technical Design Report, Tech. Rep. CERN-LHCC-97-32 (1997).
13. Dasu, S., et at., [CMS Collaboration], The TriDAS Project. Technical Design Report, Vol. 1: The Trigger Systems, Tech. Rep. CERN-LHCC-2000-038 (2000).

BTeV Detector Update

David Cinabro

(Representing the BTeV Collaboration)

Wayne State University, Department of Physics and Astronomy, Detroit, MI 48202, U.S.A.

Abstract. I update the status of the BTeV detector project.

1. INTRODUCTION: P5 ENDORSEMENT

The BTeV project has been coalescing for more than a decade. Just before this conference, it finally received its first official endorsement beyond Fermilab. The Particle Physics Project Prioritization Panel (P5) released its first report (http://doe-hep.hep.net/p5/index.html) on 29 September 2003. Quoting from that report:

> P5 supports the construction of BTeV as an important project in the world-wide quark flavor physics area. Subject to constraints within the HEP budget, we strongly recommend an earlier BTeV construction profile, and enhanced C0 optics.

This speaks for itself and implies that BTeV should enter the construction phase in the latter half of 2004. Thus this contribution will describe the present design for BTeV and is based on a Technical Design Report which was prepared for a Department of Energy review which took place just after the conference.

In this contribution, I will discuss the drivers of the BTeV design, briefly cover the components of the detector, and give a sketch of the planned activity for the collaboration over the next year. I will not cover one of BTeV's most important components, the trigger, as it is covered in a separate contribution by Michael Wang [1]. I also will not cover the physics capabilities of BTeV as these are covered in a separate contribution by Brad Cox [2].

2. DESIGN DRIVERS

The primary production mechanism for b quarks in high energy proton-antiproton collisions is gluon-gluon fu-
sion. This implies that b quark production is flat out to large rapidities, b's at large rapidity have high momentum, and a large fraction of $B\text{-}\bar{B}$ correlated production is collinear. These imply that a detector with a single arm in the forward direction has good acceptance for fast, co-produced $B\text{-}\bar{B}$. On the other hand the b cross section is 1/500 of the total inelastic cross section and thus such a detector must have a high performance trigger to avoid being overwhelmed by events not of interest.

Table 1 is a list of B decays that focuses on achieving a complete understanding of CP violation in the B system. To study these decays, a set of detector capabilities are needed which are cross referenced for each mode in the table. All these capabilities would be valuable to flavor tag the other B to enable CP studies in events with co-produced B's. These capabilities are very ambitious and imply a general purpose detector with broad reach well beyond the study of CP violation in B decays.

BTeV will be located at the TeVatron. Its characteristics and our assumptions about cross sections, etc., define an environment that has implications for the detector. These are summarized in Table 2. For the things not based on measurements, we have made conservative assumptions. The detector and its readout system are designed to handle up to double this luminosity with 132 ns bunch spacing. The Tevatron now runs with a bunch spacing of 396 ns, and this is not likely to change. In this case the number of interactions per crossing would go to 6.0 at the start of a run and average 3.5. When we asses the physics capabilities of the BTeV design, we use this more pessimistic running scenario rather than what appears in Table 2.

BTeV will be located in the C0 collision hall. Its size is a physical limitation which has some impact on the muon system. That is, we would build a larger muon system if we could fit it into the already constructed hall.

CP722, *B Physics at Hadron Machines,* edited by M. Paulini and S. Erhan
© 2004 American Institute of Physics 0-7354-0203-5/04/$22.00

TABLE 1. A list of B decays useful for studying CP violation cross referenced with detector capabilities needed to study them.

Decay Mode	Detector Property				
	Vertex Trigger	K/π Separation	γ Detection	Superb τ Resolution	Lepton ID
$B^0 \rightarrow \rho\pi \rightarrow \pi^+\pi^-\pi^0$	$\checkmark$	$\checkmark$	$\checkmark$		
$B_s^0 \rightarrow D_s^+ K^-$	$\checkmark$	$\checkmark$		$\checkmark$	
$B^+ \rightarrow D^0 K^+$	$\checkmark$	$\checkmark$			
$B^0 \rightarrow K\pi$	$\checkmark$	$\checkmark$	$\checkmark$		
$B \rightarrow \pi^+\pi^-, B_s^0 \rightarrow K^+K^-$	$\checkmark$	$\checkmark$		$\checkmark$	
$B_s^0 \rightarrow J/\psi\eta', J/\psi\eta$	$\checkmark$	$\checkmark$	$\checkmark$	$\checkmark$	$\checkmark$
$B^0 \rightarrow J/\psi K_s$					$\checkmark$
$B_s^0 \rightarrow \phi K_s, \eta' K_s, J/\psi\phi$	$\checkmark$	$\checkmark$	$\checkmark$		$\checkmark$
$B^0 \rightarrow J/\psi K^*, B_s^0 \rightarrow J/\psi\phi$					$\checkmark$
$B_s^0 \rightarrow D_s^+\pi^-$	$\checkmark$	$\checkmark$		$\checkmark$	
$B_s^0 \rightarrow J/\psi\eta', K^+K^-, D_s^+\pi^-$	$\checkmark$	$\checkmark$	$\checkmark$		$\checkmark$

TABLE 2. Characteristics of the Tevatron and our assumed cross sections for the heavy flavors.

Luminosity	$2 \times 10^{32} \text{cm}^{-2}\text{sec}^{-1}$
$b\bar{b}$ Cross Section	$100\ \mu$b
b's in 10^7 sec	4×10^{11}
$\sigma(b\bar{b})/\sigma(\text{Total})$	$\sim 0.15\%$
$\sigma(c\bar{c})$	$> 500\ \mu$b
Bunch Spacing	132 ns
Length of Luminous Region	10-20 cm
Transverse Width of Luminous Region	$\sigma_x \approx \sigma_y \approx 50\ \mu$m
Interactions/crossing	$< 2.0 >$

BTeV Detector Layout

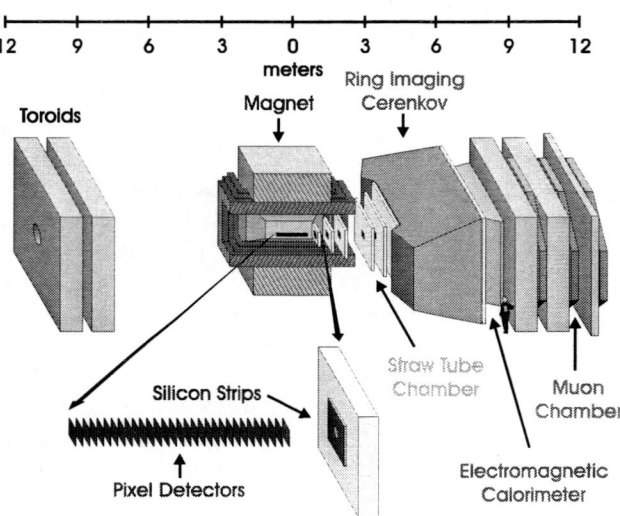

FIGURE 1. Layout of the BTeV detector.

3. DETECTOR COMPONENTS

Figure 1 gives a diagram of the detector which satisfies the design drivers discussed above. In this section, I will briefly describe each of the detector components and their status with an emphasis on what R&D remains to be done.

The collision region sits in a 1.6 T dipole field provided by the existing SM3 magnet, which has been used in two previous fixed target experiments. The field integrates to 5.2 T-m and gives a 1.5 GeV kick. New pole inserts are designed to keep the field uniform to 1% in the tracking volume of the pixel detector.

The key sub-detector is a pixel tracker that sits in the poles of the dipole. Its output feeds the vertex trigger and is used to do precision vertexing. It lies within the vacuum of the beam with the nearest pixels within 0.5 cm of the beam to achieve the best possible resolution with no degradation by scattering in the beampipe. It consists of 30 planar stations and has 23 million individual pixels.

A great deal of R&D has been completed on this detector giving us confidence that it can be realized. Sensors are bump-bonded to readout chips which are grouped into Multi-Chip Modules (MCM) which are formed into L-shaped half stations. Each half station has two sets of modules rotated 90° from each other to minimize dead regions. These are attached to a substrate made of Thermal Pyrolytic Graphite (TPG), which has a thermal conductivity of ~ 2000 W/K-m. Each station has a radiation length of $1.3\% = \chi/\chi_0$, which is kept as low as possible to maintain resolution. Signals are brought to vacuum feed through boards via flex circuits. The detector runs at $-5°$ C or lower to minimize radiation damage. Modules are connected by flexible Pyrolytic Graphite Sheet (PGS) which also has high thermal conductivity to liquid nitrogen cooling lines that run along the exterior of the detector's vacuum box. Each half of the detector is mounted on a common support which moves away from the beam during injection.

Figure 2 shows one side of the detector with two half stations installed.

The detector achieves a hit resolution of better than 10 μm, verified with prototypes in test beam, and a momentum resolution ($\Delta p/p$) of 4-5% for 3-40 GeV/c tracks in a complete simulation. These are both vital inputs to our simulation of how well the trigger will perform.

Ongoing pixel R&D includes test beam runs of MCM's, tests of running the detector at cryogenic temperatures, the effect of the detector and its support structure on the Tevatron's vacuum, shielding from beam RF and other EM effects, system issues in a 10% prototype, as well as assembly and installation.

The tracking system is extended by seven planes, three within the dipole, of silicon strips, close to the beampipe,

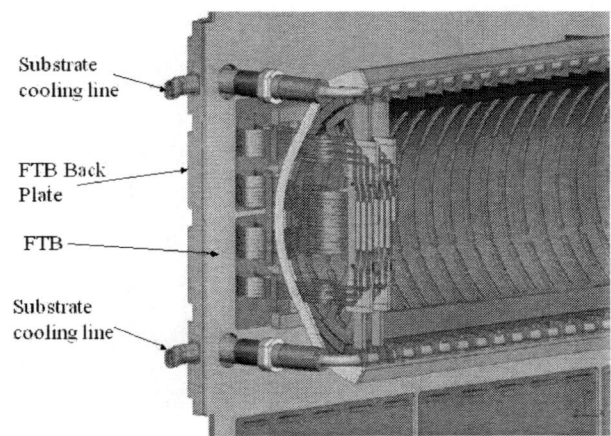

FIGURE 2. Cut away view of the BTeV pixel detector.

and straws. These improve the momentum resolution and connect pixel tracks to the downstream detectors. These detectors are standard technology but are challenged by high occupancy and radiation levels. Ongoing R&D includes the development of radiation hard silicon readout chips, the performance of a silicon prototype, studies of copper (rather than aluminum) coated straws to reduce gas permeability, the choice of gas and straw material to avoid degradation with age, and manufacturing glass capillaries which divide the wires within the straw.

Identification of hadrons is provided by a RICH. It is a two part system with a liquid (C_5F_{12}, n = 1.24) radiator for low momentum, and gas (C_4F_{10}, n = 1.00138) for high momentum. The ID of hadrons at low momentum is very beneficial for flavor tagging. Cerenkov photons from the liquid go directly to the side and are detected with PMT's. Those from the gas bounce off mirrors and are detected with an array of HPD's or MAPMT's.

The detector has been extensively simulated based on R&D results on mirrors, photon detectors, engineering plans, and backgrounds. The $K\pi/\pi\pi$ separation for the gas radiator is shown in Figure 3. A large number of interactions/crossings noticeably degrades the performance of the RICH due to stray hits and occupancy. For example, fake rates go up by 25% at 8.0 interactions per crossing (recall, we expect an average of 3.5 interaction per crossing). This is taken into account when we asses physics reach.

Ongoing R&D for the RICH includes beam tests of photodetectors and the liquid radiator, development of carbon fiber mirrors, studies on how to align the detector, extensive plans for component quality assurance, and making the choice between HPD's and MAPMT's for the gas photo detector.

A very capable EM calorimeter is a unique feature of BTeV. It will consist of 10,000, 2.8x2.8x22cm tapering $PbWO_4$ (PWO) crystals with PMT readout, as it is

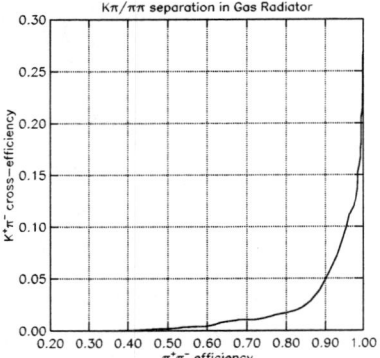

FIGURE 3. $K\pi/\pi\pi$ separation for the BTeV RICH gas radiator.

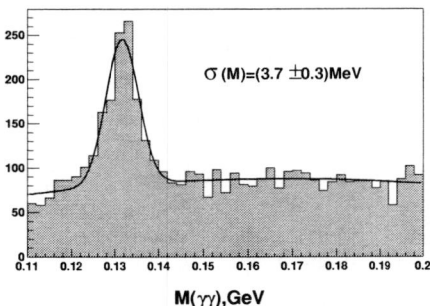

FIGURE 4. Predicted π^0 peak in $B \to \rho^+\pi^-$ in the BTeV EM calorimeter.

beyond the field of the vertex magnet. The crystals are supported from a frame by a lattice of 0.3 mm thick Al strips. They point off vertex to minimize dead regions. Extensive test beam studies have been done to evaluate the radiation hardness of the crystals from many PWO vendors. This leads to strict requirements and a rigorous testing regime for crystals to be used in construction. An extensive calibration and light pulsar monitoring system is also needed to maintain performance.

The detector's response to $B \to \rho^+\pi^-$ has been simulated based on the observed crystal energy resolution and variation plus expected backgrounds. Figure 4 shows the expected π^0 peak. The expected performance is excellent allowing analysis of $B \to \pi\pi\pi$ Dalitz plots.

Ongoing R&D for the calorimeter includes continued test beam studies of the radiation hardness of PWO crystals, development of methods of crystal quality assurance, and the design of an LED calibration and monitoring system.

The muon system consists of two toroidal magnet and three Ar-CO_2 proportional tube tracking stations inter-leaved with iron absorbers. The basic element is a two layer plank that is formed into a four layer octant at each station. There are a total of 37,000 channels and the system measures muon momentum with a resolution of $(19 \oplus 0.6)\%$ as input to the trigger. This resolution is based on beam test measurements. There are compensating dipoles within the muon system. The uninstrumented arm also has matching toroids and compensating dipoles.

Ongoing R&D for the muon system includes test beam studies with production planks, high radiation dose lifetime tests, tests of production front end electronics, mechanical tests of the octants, development of mass production techniques, construction of a prototype gas system, and background measurements in the C0 hall.

The BTeV design is mature for a detector that has not yet entered production. As can be seen, the R&D effort is focused on issues of performance and construction rather than testing and development of concepts.

4. SCHEDULE AND PLANS

BTeV is in the midst of the review process. The project has been thoroughly reviewed by the Fermilab Program Advisory Committee and P5. At the moment funding agencies are conducting a series of detailed reviews. Two of these have already been completed and BTeV did very well. A major milestone will be a Critical Decision 1, "Lehman", Review recently scheduled for the end of April 2004. We expect, based on our past performance and the state of our design and construction plans, that this review will lead to swift approval and the beginning of construction in the latter half of 2004.

Our present TDR assumes a five year construction period. The P5 report could lead to an acceleration of this plan, but this would require a front loaded funding profile, which I think is unlikely to happen.

In any case, it is safe to assume that BTeV will start taking data with a mostly complete detector in mid-2009, or earlier if fortune is kind.

5. SUMMARY

In summary, after ten years of activity BTeV took a major step towards realization with the very favorable P5 report released at the end of September. This should lead to a very capable detector taking data within six years. The projected capabilities of BTeV are summarized in Table 3. Complete details on BTeV for which this contribution is far too condensed can be found at the project web site, URL http://www-btev.fnal.gov/.

TABLE 3. Summary of the projected performance of BTeV.

Angular Acceptance	10-300 mrad
Charged Particle Momentum Acceptance	> 3 GeV/c
Mass Resolution (all charged B decay)	< 50 MeV/c^2
Tracking Efficiency	$> 98\%$
Primary Vertex Resolution	100 μm
Proper Time Resolution	< 50 fs
Trigger Efficiency (vertexable B decay)	$> 50\%$
Trigger Efficiency (single prong B decay)	$> 20\%$
Rejection of Light Quark Events	$> 99.8\%$
Data Rate	< 200 Mbytes/sec
$K - \pi$ Separation (3-70 GeV/c)	$> 4\sigma$
$p - K$ Separation (3-70 GeV/c)	$> 4\sigma$
EM Resolution	$< 2\%/\sqrt{E}$
EM Energy Range	> 1 GeV
EM Acceptance	10-200 mrad
Muon ID	5-100 GeV/c
Muon mis-ID	$< 0.1\%$
Muon Momentum Resolution	$< 20 \oplus 0.6p\%$
Time Response	100 ns
Radiation Hardness	> 10 years

ACKNOWLEDGMENTS

Thanks to all my BTeV collaborators. Many thanks to Sheldon Stone and Joel Butler and other BTeV colleagues who gave useful comments on the draft of the presentation. Thanks to Manfred Paulini for being patient, and to all of the organizers for a very good conference.

REFERENCES

1. Wang, M., in these proceedings (2004).
2. Cox, B., in these proceedings (2004).

LHCb: Reoptimized Detector and Tracking Performance

Gerhard Raven

(Representing the LHCb Collaboration)

NIKHEF and VU, Amsterdam, The Netherlands

Abstract. LHCb is an experiment dedicated to the study of *CP* violation and other rare processes in the *B* mesons system at the LHC. In order to achieve these goals, the detector must not only provide a trigger system which is capable of triggering on both leptonic and hadronic final states, but also have a high track reconstruction efficiency, $K - \pi$ separation capability over a wide momentum range, and an excellent proper time resolution. Recently, the detector has been reoptimized to reduce the material budget and to improve the trigger performance and robustness. The construction of the various detector components is well advancing and the experiment is expected to be ready for data taking at the start of the LHC operation.

1. INTRODUCTION

The aim of the LHCb experiment is to study *CP* violation and rare *B* meson decays with high precision. To do this, it takes advantage of the huge $b\bar{b}$ production rate at the LHC collider. This rate is the product of the large cross-section, about 500 μb, for $b\bar{b}$ production in pp collisions with a center-of-mass energy of 14 TeV and the luminosity of 2×10^{32} cm^{-2} s^{-1}. This results in the production of 10^{12} $b\bar{b}$ pairs per year. However, this should be compared to the total inelastic cross section, which is about 160 times larger. As a result, triggering will be crucial.

The detector is a single-arm spectrometer with a forward angular coverage from approximately 15 mrad to 300 (250) mrad in the bending (non-bending) plane. This choice is motivated by the fact that at LHC energies both *b* and $\bar{b}$ hadrons are predominantly produced collinear in the same forward cone, a feature exploited in the flavour tag.

2. EVOLUTION SINCE THE TECHNICAL PROPOSAL

At the time of the Technical Proposal [1], the material budget up to the second Ring Imaging Cherenkov detector (RICH2) was 40% of X_0 (10% of λ_I), where X_0 (λ_I) is the radiation (nuclear interaction) length. Due to various technological constraints, this increased to 60% (20%) by the time the Outer Tracker Technical Design Report [2] was submitted in September 2001. The additional material not only deteriorates the detection capabilities of electrons and photons, but also increases the multiple scattering of charged particles and increases the occupancy of the tracking system. The increased nuclear interaction length implies that more kaons and pions interact before completing their traversal through the tracking system. This in turn decreases the number of reconstructible *B* mesons, even if the track finding efficiency for tracks which do traverse the entire tracking system remains high. For these reasons an effort has been made to reduce the material budget back to the level of the time of the Technical Proposal. The detector layout of this reoptimized detector is shown in Fig. 1.

In addition, the trigger strategy has been optimized. It was realized that both the robustness and efficiency of the second trigger level (Level-1) could be improved by not only using information on the presence of displaced vertices, as described in the Technical Proposal, but also by adding p_T information to tracks with a large impact parameter with respect to the primary vertex. This requirement has been taken into account during the reoptimization process, and has had an impact on the layout of the tracking system of the reoptimized detector.

The design and performance of the reoptimized detector are described in detail in Ref. [3], whereas the trigger performance is described in Ref. [4].

3. TRACKING SYSTEM

The tracking system of LHCb can globally be divided into three sub-systems. First, around the interaction point, a silicon Vertex Locator (Velo) is placed. Second, downstream of the interaction point, but still in front of the magnet, the so-called Trigger Tracker (TT), instrumented with silicon strip sensors, is located. Finally, be-

CP722, *B Physics at Hadron Machines,* edited by M. Paulini and S. Erhan
© 2004 American Institute of Physics 0-7354-0203-5/04/$22.00

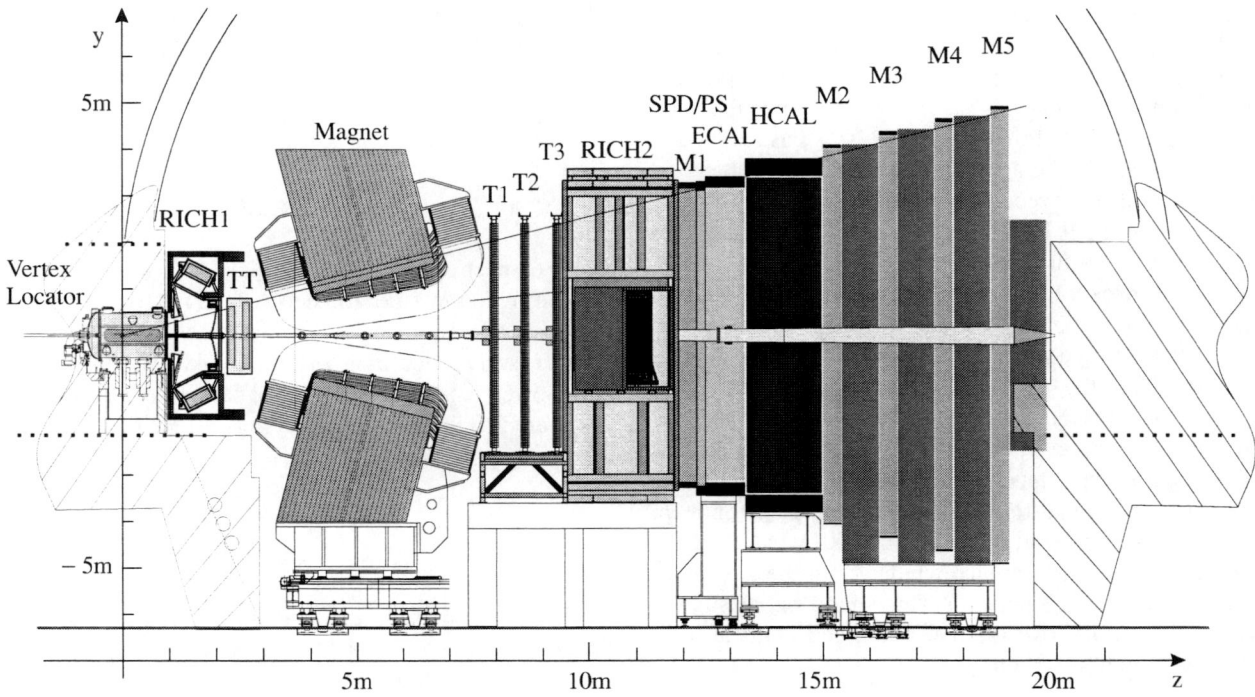

FIGURE 1. Reoptimized LHCb detector layout, showing the Vertex Locator (Velo), the dipole magnet, the two RICH detectors, the tracking stations TT and T1–T3, the Scintillating Pad Detector (SPD), Preshower (PS), Electromagnetic (ECAL) and Hadronic (HCAL) calorimeters, and the muon stations M1–M5. Also shown is the direction of the y- and z-coordinate axis; the x-axis completes the right-handed coordinate system.

hind the magnet, three tracking stations, T1, T2 and T3, are found. These three stations are constructed utilizing two different detector technologies. The interior part, close to the beamline, consists of silicon strip detectors whereas the exterior part is made from straw tube drift cells.

3.1. Velo

The Velo has to provide precise measurements of the track coordinates close to the interaction region. For this, the Velo features a series of disk-shaped silicon stations placed perpendicular along the beam direction. The inner radius of the sensors is at a radial distance of 8 mm from the beam, which is smaller than the aperture required by the LHC during injection. As a result the detectors must be retractable, which is achieved by mounting them in a setup similar to Roman pots (see Fig. 2). To minimize the material between the interaction region and the detectors, the silicon sensors are inside a thin aluminum box with a pressure of less than 10^{-4} mbar. The side of the box facing the beam also serves to guide the wakefield of the beams, and as a shield against RF pickup. This side, the so-called RF-foil, has a corrugated shape, which minimizes the amount of material before the first measure-

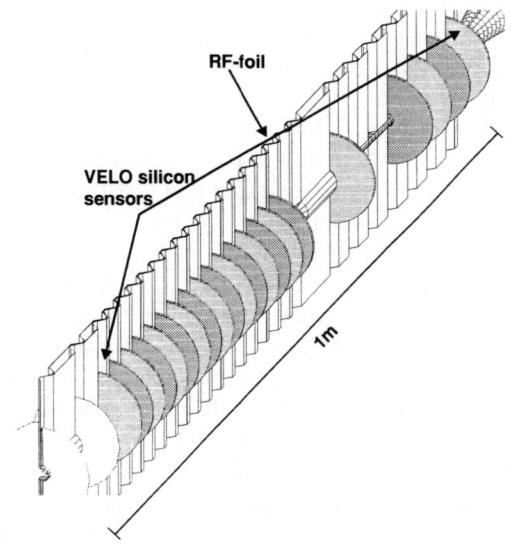

FIGURE 2. Arrangements of the Velo detectors along the beam axis.

ment and allows for overlapping sensors. The foil has a thickness of 300 μm, and is made from an aluminum alloy with 3% of magnesium.

During the studies for the overall LHCb optimization,

the role of the Velo has changed. In addition to its original task of providing precise coordinate measurements to allow the determination of the location of primary and secondary vertices, it has now also become the main tracking detector before the magnet. This has resulted in a few changes in the sensor design to give a better performance for track reconstruction and for the L1 trigger with only a small loss in vertex and impact parameter resolution. Each of the 21 stations in the Velo consists of back-to-back r- and ϕ-measuring layers. The pitch of the strips ranges from 37 μm to 103 μm, depending on the distance to the beam. The detectors are single-sided and, thanks to the 45° segmentation (which reduces the strip capacitance at larger radii) and the performance of the current version of the Beetle front-end chip, it has been possible to reduce the thickness to 220 μm, while maintaining a signal-to-noise ratio in excess of fourteen. As part of the material reduction, the exit window of the Velo also serves as entrance window into the gas radiator of the first RICH detector. This window consists of a 2 mm thick foil, covering 390 mrad, and is made from an aluminum alloy forging.

3.2. Trigger Tracker

The Trigger Tracker (TT) is located downstream of the first RICH detector, and in front of the entrance to the magnet. It also serves a dual purpose. First, it will be used in the L1 trigger to assign transverse momentum information to tracks with a large impact parameter. Second, it is crucial in the off-line reconstruction to determine the trajectories of both long-lived particles which decay outside the Velo volume, and low momentum particles which are swept out of the experiment before reaching the downstream tracking station T1–T3.

To allow this station to fulfill its main function of providing momentum information to the trigger, it is crucial that there is a magnetic field of sufficient bending power between the Velo and TT. This is accomplished by eliminating the shielding plate in front of the magnet. The resulting bending power available between Velo and TT is 0.15 Tm, sufficient for trigger purposes.

However, by construction, this will create a non-zero magnetic field strength in the location of RICH1, situated between the Velo and TT. As a result, the RICH1 detector has undergone major changes to insure that the photodetectors are sufficiently shielded, while maximizing the field inside the tracking acceptance [3, 5].

The sensitive area of this tracking station is instrumented with silicon strip detectors with a thickness of 500 μm, a pitch of 198 μm, and strip length up to 33 cm. The detectors are arranged in four layers; the two exterior layers are perpendicular to the bending plane, whereas

the two interior layers have stereo angles of +5° and −5°, respectively. The total active surface area amounts to approximately 8.3 m^2 of silicon, and 180,000 readout channels.

3.3. Magnet

The spectrometer dipole is placed close to the interaction region in order to keep its size relatively small. It consists of a warm Al coil, which dissipates 4.2 MW of power, surrounded by a 1450 ton yoke. The total bending power generated is 4 Tm. All parts have already been delivered to the experimental area, where the magnet is currently under construction.

3.4. T Stations

The T stations, located behind the magnet, cover a large area of about 6×5 m^2. Over this area, there is a large variation in particle flux. To deal with this, the detector is split into two different technologies: the smaller, inner part is instrumented with silicon strip detectors, whereas the region further from the beam is instrumented with straw-tube chambers. The area covered by the inner tracker is only 1.3% of the T stations, but it corresponds to 20% of the particle fluence.

3.4.1. Inner Tracker

To take into account the effect of the bending plane of the magnet, the inner tracker has a swiss cross shaped outline. The individual stations each contain four layers of silicon strip detectors, of which two measure the coordinate in the bending plane and two have stereo angles. The detectors are 320 μm thick, and have a readout pitch of 198 μm. In total, this leads to 130,000 readout channels.

3.4.2. Outer Tracker

The Outer Tracker consists of straw tube detectors with a diameter of 5 mm. Each of the three stations consists of four double layers of staggered straws, for a total of 101,000 readout channels. The wires in the 4.7 m long modules are split in the middle, and readout at both the top and bottom. To guarantee a signal collection within 2 LHC beam crossings, $i.e.$ 50 ns, a fast drift gas Ar(75)/CF$_4$(15)/CO$_2$(10) is chosen. The spatial resolution obtained during test beam operation with this gas is 200 μm.

4. EVENT GENERATION AND SIMULATION

To evaluate the performance of the reoptimized detector, large samples of minimum bias and generic $b\bar{b}$ events have been simulated and reconstructed. The simulation takes into account that several inelastic proton-proton collisions may occur in the same bunch crossing. This "pile-up" phenomenon is simulated assuming that the number of inelastic interactions in a single bunch crossing can be obtained from a Poisson distribution with a mean given by $\mathscr{L}\sigma_{\mathrm{inel}}/\nu$, where $\mathscr{L}$ is the instantaneous luminosity, σ_{inel} is the cross section, taken to be 80 mb, and $\nu = 29.49$ MHz is the average non-empty bunch crossing frequency. The luminosity $\mathscr{L}$ is assumed to fall exponentially with a 10-hour lifetime during the course of 7-hour fills, with an average value per fill of $2 \times 10^{32}\,\mathrm{cm}^{-2}\,\mathrm{s}^{-1}$. For events which contain B hadrons, this leads to an average probability of 64.9% that there is no pile-up, 27.8% for a single pile-up, and 6.2% for two pile-up interactions. The average number of interactions in events in which B hadrons are produced is 1.42.

After the generation, the particles are tracked through the detector material and the surrounding environment using the Geant3 package. The geometry and material of the detectors are described in detail, using the results of the various sub-detector TDR's. The detector response is simulated using the results obtained from beam tests of prototypes, *e.g.* detection efficiencies and resolution. Electronic noise and cross-talk effects are also included.

The detector response is simulated as a function of the arrival time of each particle, and, depending on the detection technology and electronics, may span several consecutive bunch crossings. This effect is referred to as "spill-over". To take this into account, not only particles produced in the specific bunch-crossing are considered, but also those in the two previous crossings, and in the following crossing. The probability that one of these crossings produces particles is also determined using the instantaneous luminosity.

5. TRIGGER STRATEGY

The trigger is one of the biggest challenges of the LHCb experiment. It is designed to distinguish minimum-bias events from events containing B mesons through the presence of particles with a large transverse momentum and the existence of secondary vertices. The trigger system consists of several levels. The first trigger, Level-0, requires at least one lepton or hadron with a p_T exceeding 1 to 3 GeV/c and reduces the rate to 1 MHz. The Level-0 trigger is implemented as dedicated hardware, and runs synchronous to the LHC beam crossings. The

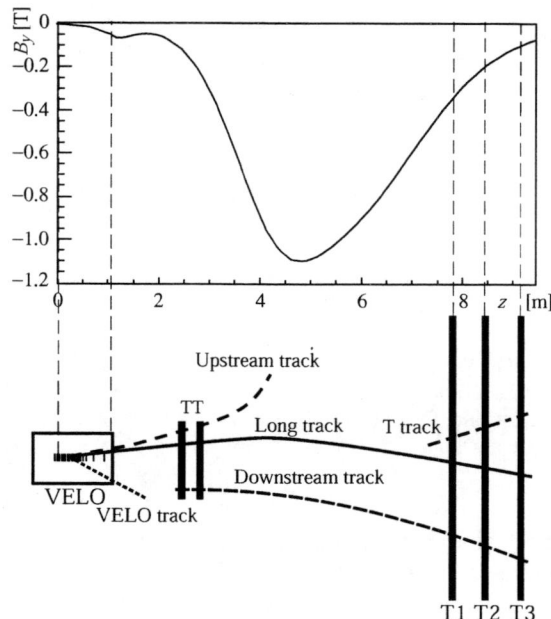

FIGURE 3. A schematic illustration of the various track types: long, upstream, downstream, Velo and T tracks. For reference, the main B-field component (B_y) is plotted above as a function of the z coordinate

next level of trigger, the Level-1 trigger, is implemented in software. With respect to the Technical Proposal, the robustness of the Level-1 trigger has been improved by not only requiring the presence of tracks with a large impact parameter, but by also considering the transverse momentum of these tracks. This can be done in two complementary ways. First, by associating the tracks found by the Level-1 trigger to the high-p_T calorimeter clusters and muons obtained at Level-0. Second, which is more efficient for hadrons, is to extrapolate those tracks which have a large impact parameter to the TT station. As mentioned earlier, this requires sufficient bending power between the Velo and the TT station, and has lead to the removal of the shielding plate in front of the magnet. More details on the implementation of the trigger and its performance can be found in Refs. [4, 6].

6. TRACKING STRATEGY AND PERFORMANCE

Track finding starts by reconstructing tracks in the Velo. These tracks are extrapolated through the detector, and, using an optical algorithm, are matched to hits in the T stations. Next, the unused hits in the T stations are used to reconstruct so-called T seeds. These T seeds are then matched to tracks in the Velo which did not get T hits added in the first step. The tracks found this way, con-

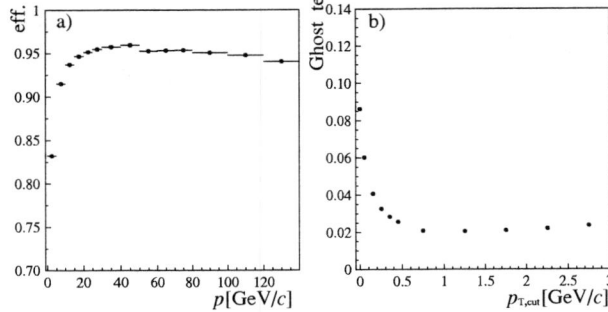

FIGURE 4. Performance of the long track finding: (a) efficiency as a function of the momentum of the generated particle; (b) ghost rate, for tracks with reconstructed transverse momentum greater than p_T^{cut}.

taining hits over the full length of the tracking system, are referred to as 'long tracks'. These tracks have both a good momentum resolution as the deflection angle in the magnet is well determined, and a good impact parameter measurement. Tracks from B meson decays typically fall in this category. To find tracks originating from the decay of long-lived particles (*i.e.* those which decay beyond the Velo volume) are found by extrapolating the T seeds, and looking for hits in the TT station. Tracks found this way are referred to as downstream tracks. Finally, there is a class of tracks which are swept out of the acceptance of the T stations by the magnet. These are typically low momentum tracks, and a search for these is performed by extrapolating the Velo tracks into the TT station. The various track types are summarized in Fig. 3.

On average, an event contains 72 reconstructed tracks, which can be divided into 26 long, 11 upstream, 4 downstream, 5 T and 26 Velo tracks. Shown in Fig. 4 are the efficiency of finding long tracks as a function of momentum, and the ghost rate as a function of transverse momentum. Tracks originating from B decays typically have momenta in excess of 10 GeV/c and a p_T larger then 0.5 GeV/c.

The high track-finding efficiency allows one to take full advantage of the large B hadron sample produced at the LHC by exclusively reconstructing B decays. In the case of fully reconstructed $B_s^0 \to D_s(\to KK\pi)\pi$ decays, the excellent track momentum resolution (shown in Fig. 5) leads to a resolution on the B_s^0 mass of 14 MeV/c^2. Combining the momentum resolution with the impact parameter resolution of the reconstructed decay products (also indicated in Fig. 5) results in a proper time resolution of better than 40 fs. This resolution, in combination with the large signal yield and well-controlled backgrounds, allows for a 5σ measurement of the B_s^0 oscillation frequency up to 68 ps^{-1} using data collected during a nominal running year. Further performance results can be found in Refs. [3, 7].

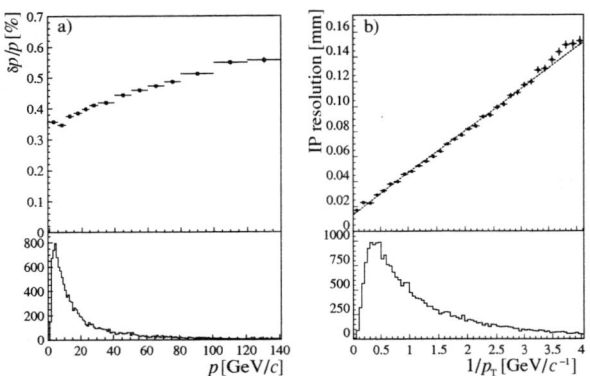

FIGURE 5. Resolution on the reconstructed track parameters at their production vertex: (a) momentum resolution as a function of momentum, (b) impact parameter resolution as a function of $1/p_T$. The momentum and transverse momentum spectra of B decay daughters are shown in the lower parts of the plots.

7. SUMMARY

Detailed simulation studies show that the reoptimized LHCb detector fulfills the requirements set by the physics goals. Charged particles are reconstructed with high efficiency, and a sufficiently low ghost rate which does not introduce significant combinatorial background in the signal samples. Utilizing the excellent momentum and vertex resolutions, a proper time resolution of 40 fs can be reached. This resolution is sufficient to observe Δm_s up to 68 ps^{-1}, well beyond the Standard Model expectations, meeting the requirement for resolving time-dependent CP asymmetries in the B_s^0 system.

The updated construction schedule shows that the full physics program can start when the LHC will become operational in 2007.

REFERENCES

1. Amato, S., et al., [LHCb Collaboration], Technical Proposal, Tech. Rep. CERN/LHCC 98-4 (1998).
2. Barbosa Marinho, P. R., et al., [LHCb Collaboration], Outer Tracker Technical Design Report, Tech. Rep. CERN/LHCC 2001-24 (2001).
3. Antunes Nobrega, R., et al., [LHCb Collaboration], Reoptimized Detector Design and Performance Technical Design Report, Tech. Rep. CERN/LHCC 2003-30 (2003).
4. Antunes Nobrega, R., et al., [LHCb Collaboration], Trigger Technical Design Report, Tech. Rep. CERN/LHCC 2003-31 (2003).
5. Adinolfi, M., in these proceedings (2004).
6. Callot, O., in these proceedings (2004).
7. Uwer, U., in these proceedings (2004).

Status of the LHCb RICH and Hadron Particle Identification

(Representing the LHCb Collaboration)

University of Oxford, Sub-department of Particle Physics, Oxford, OX1 3RH, United Kingdom

Abstract. The LHCb experiment will use a Ring Imaging Cherenkov detector (RICH) in order to identify hadrons with momentum between 1 and 100 GeV/c. In this paper, the development of the RICH detector and the particle identification algorithms will be presented. It will also be shown that with the current design, it is possible to discriminate pions from kaons in the selected momentum range

1. INTRODUCTION

Particle identification is one of the major challenges the LHCb experiment [1], shown in Figure 1, has to face. In order to measure with high precision *CP* violation in *B* meson decay, the ability to distinguish pions from kaons in the final states is particularly important. For example, in the analysis of the $B^0 \to \pi^+\pi^-$ decay, the background from other 2-body decays such as $B^0 \to K^+\pi^-$ and $B_s^0 \to K^+K^-$ has to be rejected. It is found that without having efficient particle identification, the contamination from the different background sources is larger than the statistics available from the $B^0 \to \pi^+\pi^-$ decay itself. Furthermore, since the charge of the kaons produced in the $b \to c \to s$ decay chain of the accompanying *b* hadron depends on the *b* quark flavour, flavour tagging also benefits from an efficient particle identification. In order to achieve the required particle identification, LHCb has chosen to use a system based on Ring Imaging Cherenkov detectors (RICH) [2].

2. THE LHCB RICH DETECTORS

A major step in the design of the RICH detectors is the selection of the radiator, which needs to have a refractive index covering the whole momentum range of interest. The upper momentum range to 100 GeV/c is set by the hadrons produced in 2-body *B* meson decays. The lower range is defined by the tagging kaons which have momenta as low as 1 GeV/c. In order to cover the whole momentum spectrum, LHCb has chosen to use three different radiators: CF_4 is used to cover the highest momentum tracks, the intermediate range is covered using C_4F_{10} and silica aerogel is used for the lowest momentum tracks (Table 1).

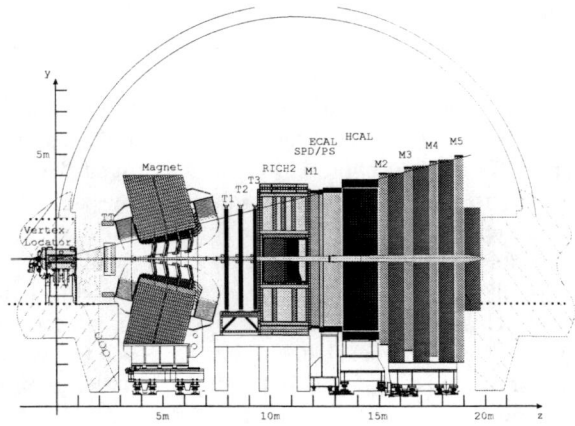

FIGURE 1. The LHCb detector has been designed to study *B* meson decays. Close to the interaction region is the VErtex LOcator which allows direct observation of the decay vertex of short-lived particles. This is followed by the first RICH detector and the first tracking chamber. The magnet and the downstream tracking chambers allow measurement of particle momenta, and are followed by the second RICH detector. The apparatus is completed by the electromagnetic and hadronic calorimeters and the muon chambers.

C_4F_{10} and CF_4 are used in their gaseous state at STP. Aerogel is a foam whose useful wavelength range of the Cherenkov radiation is limited by Rayleigh scattering. The transmission T of photons of wavelength λ through a length L of aerogel is described by the equation

$$T = Ae^{-CL/\lambda^4} \qquad (1)$$

where C is the clarity coefficient. Since aerogel is a solid, it is important to verify its optical properties do not vary appreciably when exposed to high doses of radiation as found in the LHCb environment. This has been studied by exposing aerogel tiles to radiation. Figure 2

CP722, B Physics at Hadron Machines, edited by M. Paulini and S. Erhan
© 2004 American Institute of Physics 0-7354-0203-5/04/$22.00

119

TABLE 1. The LHCb RICH employs 3 different radiators to cover the momentum range of interest. Silicon aerogel, C_4F_{10} and CF_4 are used to cover different regions. Their Cherenkov emission threshold for pions and kaons, refractive index, Cherenkov angle and radiation length in the LHCb RICH are given.

	Aerogel	C_4F_{10}	CF_4
Length (mm)	50	850	1670
n	1.03	1.0014	1.0005
ϑ_c (mrad)	242	53	32
π [GeV/c]	0.6	2.6	4.4
K [GeV/c]	2.0	9.3	15.6

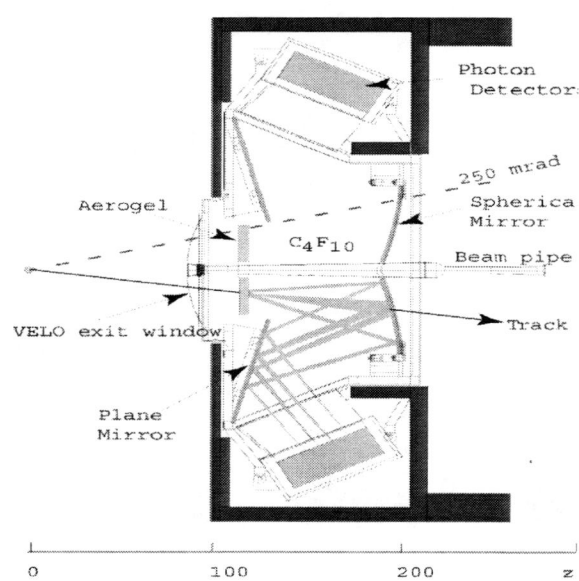

FIGURE 3. RICH 1 is used to identify particles in the low and intermediate range. Cherenkov photons produced when particles cross the aerogel and C_4F_{10} radiators are focused outside the acceptance of the detectors using spherical mirrors, where a second set of flat mirrors directs them onto the photo-detector planes.

shows that after exposing the aerogel to a total fluency of 52×10^{12} particles/cm^2 of either protons or neutrons, corresponding to several years of LHCb running, the clarity of the tile has not changed significantly. Aerogel is an hygroscopic medium, thus it can be expected that its optical properties vary with water absorption. This has been verified exposing a tile to humid air flow and measuring a considerable increase of the clarity coefficient as function of water absorption. However, it is found that it is possible to restore the initial conditions by baking the tile at a temperature of 500 C for four hours.

Tracks with momenta less than 5 GeV/c will be swept out of the spectrometer acceptance by magnetic field, thus a RICH detector, RICH 1, Figure 3, is placed upstream of the magnet. RICH 1 has to cover the whole angular acceptance of LHCb, therefore, in order to minimize the photo-detector area, it is placed as close as possible to the interaction region. High momentum tracks are produced almost entirely at small angles, thus making the identification of their Cherenkov rings difficult in the RICH 1. It uses aerogel and C_4F_{10} as radiators. Therefore, a second RICH detector, RICH 2, downstream of the magnet, is used to identify high momentum tracks using CF_4. The angular coverage of RICH 2 is limited to

the region 120 (hor.) $\times$ 100 (ver.) mrad^2 again minimizing the photo-detector area.

Tracks generated in the interaction region will enter RICH 1 and cross a 5 cm thick layer of aerogel and a 85 cm thick gas vessel filled with C_4F_{10}. In order to reduce the background caused by Rayleigh scattered light in the aerogel, a glass filter, 100 μm thick, is placed between the two radiators. The Cherenkov photons are focused using spherical mirrors to a second system of flat mirrors placed outside the geometrical acceptance of the spectrometer. These plane mirrors direct the photons onto the photo-detector planes placed above and below the gas vessel. In this way, it is possible to minimize the amount of material within the detector geometrical acceptance. The spherical mirrors are required to be as light as possible: LHCb will use carbon-fiber/epoxy composite spherical mirrors similar to the ones used by the HERMES collaboration [3]. A quarter-size prototype of the mirrors has been tested and the most critical characteristics, such as spot size and reflectivity, have been measured to be well within LHCb specifications. While HERMES has used this particular kind of mirrors in a C_4F_{10} environment with no reported problem, the LHCb RICH will have a higher resolution, leaving the possibility that effects not seen in HERMES may degrade the performance at LHCb. In order to exclude this possibility, the prototype is currently undergoing long term tests in C_4F_{10}.

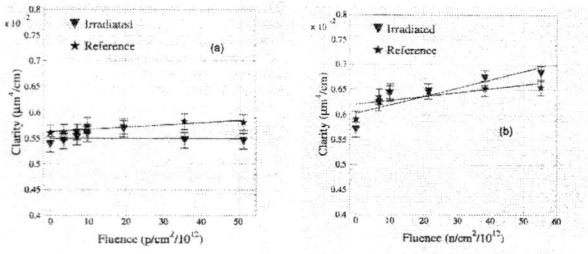

FIGURE 2. The effects of radiation of the optical properties of aerogel can be studied monitoring the clarity coefficient of a tile of aerogel exposed to protons (a) and neutrons (b). In both cases the clarity does not change significantly for fluencies up to 52×10^{12} p/cm^2, corresponding to several LHCb years.

The flat mirrors are not required to be particularly light, and it is therefore possible to minimize the cost using 6 mm thick glass mirrors coated with a 900 nm of Al and MgF_2. The total radiation length of the RICH-1 within the detector acceptance is 7.3% X_0, where 5.9% X_0 is accounted for by the radiators.

In a similar way to RICH 1, Cherenkov photons generated by particles crossing the 167 cm of CF_4 in RICH 2 are focused on the photo-detector planes using spherical and flat mirrors. However the spherical mirrors can be built of the cheaper 6 mm thick glass, since RICH 2 is behind the tracking chambers and the requirements on total material are not as strict as for the case of RICH 1.

LHCb will use Pixel Hybrid Photon detectors (HPD) [4] to detect the Cherenkov photons. Each HPD has a silicon diode sensor with 1024 0.5×0.5 mm^2 pixels, with a demagnification of 5, corresponding to 2.5×2.5 mm^2 on the photocathode. The array of HPD gives a 70% active area. The read-out of the pixels is achieved through a 0.25 μm deep sub-micron CMOS chip, directly bump-bonded onto the silicon sensor.

3. RICH PERFORMANCE

3.1. Cherenkov Angle Precision

The performance of the RICH detector has been studied within the framework of the LHCb GEANT-3 based simulation program. The Cherenkov process is described with an in-house code which has been verified against test-beam data. The transmissions, reflectivities and HPD quantum efficiency have been set to the nominal values, the aerogel clarity coefficient has been set to 0.008 μm^4/cm, to be compared with the measured clarity of the aerogel tiles used in the radiation tests which was 0.0064 μm^4/cm. All known background sources have been simulated. They include:

- photons from secondary particles;
- Rayleigh scattered photons;
- photons from charged particles striking the HPD window;
- late arriving photo-electrons from previous beam crossings.

In the Cherenkov angle reconstruction [5], only the tracks reconstructed in the tracking detectors [6] are taken into account.

The performance of the LHCb RICH detector can be quantified by studying the number of photo-electrons N_{pe} detected per track with $\beta \simeq 1$, and the corresponding Cherenkov angle resolution for single photo-electrons, $\Delta\vartheta_C$ (Table 2). Several effects, Table 3, contribute to the resolution on the Cherenkov angle. These include: the

TABLE 2. The number of photo-electrons detected per track with $\beta \simeq 1$, and the corresponding Cherenkov angle resolutions for single photo-electron for the three radiators used in the LHCb RICH.

Radiator	N_{pe}	$\Delta\vartheta_c$ (mrad)
Aerogel	6.8	1.89
C_4F_{10}	30.3	1.27
CF_4	23.2	0.59

error on the emission point (currently assumed to be at the mid-point of the track through the radiator), the chromatic dispersion of the radiator, the error due to the finite size of the pixels and, the resolution on the track direction. It is clear from the results that the dominant contribution in the case of aerogel is the chromatic dispersion of the medium. The two gaseous radiators, and in particular for the CF_4, the different effects contribute roughly in the same amount.

3.2. Hadron Identification

Particle identification is performed by comparing the pattern of the hit pixels in the RICH HPD's with the the pattern that would be expected under a given set of mass hypotheses for the reconstructed tracks crossing the RICH. A likelihood is calculated from the comparison and the mass hypothesis is modified so as to maximize the likelihood [5].

In order to evaluate the efficiency of the particle identification, tracks passing through the whole spectrometer, so called "long-tracks", have been studied in a sample of $B_s^0 \rightarrow D_s^- K^+$ events. The ratio of the likelihood between the pion and the kaons hypotheses is calculated:

$$\Delta\mathscr{L}_{K\pi} = \ln\mathscr{L}(K) - \ln\mathscr{L}(\pi). \qquad (2)$$

It is found that $\Delta\mathscr{L}_{K\pi}$ has positive values for kaons and negative values for pions. The significance of the π-K

TABLE 3. The contribution for the error on the emission point, the chromatic error, the finite pixel size and the resolution on the track direction are combined to give the total resolution for the 3 radiators.

Contribution	Aerogel	C_4F_{10}	CF_4
Emission	0.29	0.69	0.28
Chromatic	1.61	0.81	0.36
Pixel	0.62	0.62	0.17
Tracking	0.52	0.40	0.53
Total	1.89	1.27	0.59

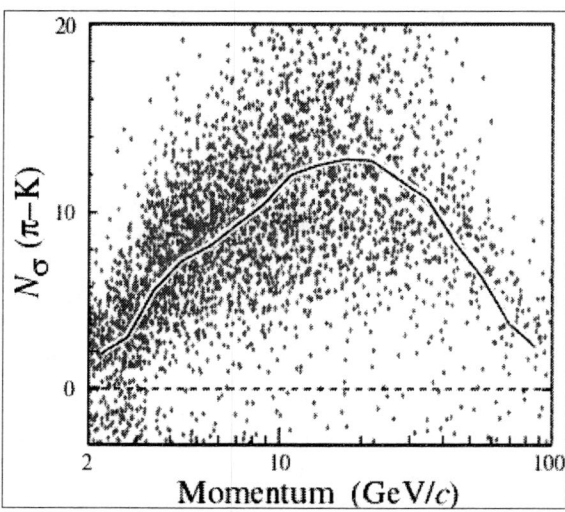

FIGURE 4. The π-K separation defined by equation 3 is plotted for a sample of pions produced in the $B^0 \to \pi^+\pi^-$ decay. It is clear that an average separation (solid line) better than 2σ is achieved over the full momentum range.

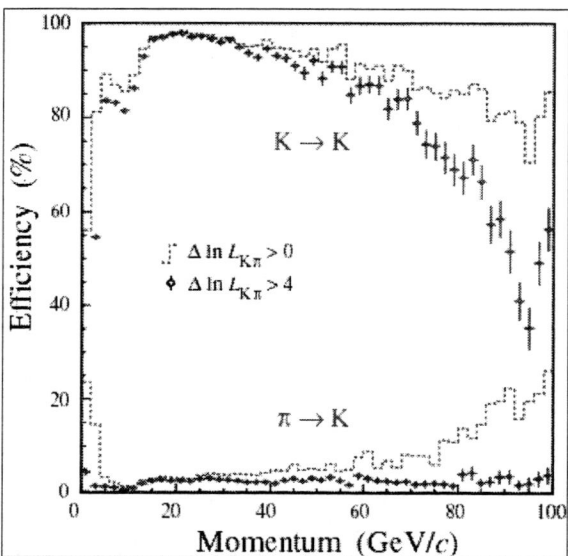

FIGURE 5. The kaon identification efficiency and the pion misidentification fraction as a function of momentum for $\Delta \mathcal{L}_{K\pi} > 0$, histogram, and $\Delta \mathcal{L}_{K\pi} > 4$, points.

separation can then be assigned:

$$N_\sigma^2 = \sqrt{2\Delta \mathcal{L}_{K\pi}}, \qquad (3)$$

signed according to $\Delta \mathcal{L}_{K\pi}$. By studying the pions reconstructed in a sample of $B^0 \to \pi^+\pi^-$ decays, it is possible to show that over the desired momentum range of 1-100 GeV/c, the separation is better than 2σ. Furthermore, it is better than 3σ over the 3-80 GeV/c range, shown in Figure 4. However, it is clear from the figure that the behavior of the separation is not Gaussian. It is therefore interesting to compare the efficiency for reconstructing kaons with the pion misidentification rate. Using again a sample of long-tracks produced in $B_s^0 \to D_s^- K^+$ events, and assuming that a particle is a kaon if $\Delta \mathcal{L}_{K\pi} > 0$, otherwise it is a pion, one finds that the average efficiency for identifying a kaon is 91%, while the average pion misidentification rate is 5.6%, shown in Figure 5. The figure also shows that the efficiency is lower for low momenta, where the only radiator available is aerogel. By changing the cut on $\Delta \mathcal{L}_{K\pi}$, the purity of the kaon sample can be increased at the expense of a reduction in the kaon identification efficiency. Figure 5 shows the effect of a cut at $\Delta \mathcal{L}_{K\pi} > 4$, which reduces the average pion misidentification to 1.7%, to be compared with an average kaon efficiency of 85%.

4. CONCLUSIONS

The LHCb collaboration will use two RICH detectors to efficiently identify hadrons produced in the decays of B mesons. These employ three different radiators in order

to cover the momentum range between 1 and 100 GeV/c. The design has been optimized in order to minimize material in the experiment geometrical acceptance. The results of the Monte-Carlo simulation including all possible background effects, show that it is possible to achieve an identification efficiency for kaons of the order of 90% with a pion misidentification of about 5%.

REFERENCES

1. LHCb Collaboration, Technical Proposal, Tech. Rep. CERN-LHCC/98-4 (1998).
2. LHCb Collaboration, LHCb RICH, Tech. Rep. CERN-LHCC/2000-7 (2000).
3. Akopov, N., et al., *Nucl. Instrum. Meth.* **A479**, 511–530 (2002).
4. T. Gys et al., [LHCb Collaboration], The Use of Pixel Hybrid Photon Detectors in the RICH Counters of LHCb, Tech. Rep. LHCb/2000-63 (2000).
5. Forty, R., and Schneider, O., [LHCb Collaboration], RICH Pattern Recogintion, Tech. Rep. LHCb/1998-40 (1998).
6. Raven, G., in these proceedings (2004).

LHCb Calorimeters and Muon System
Lepton Identification

Frédéric Machefert

(Representing the LHCb Collaboration)

Laboratoire de l'Accélérateur Linéaire, Université Paris-Sud, Orsay, France

Abstract. The calorimeter and muon systems of the LHCb detector are described. The methods used to identify electrons and muons are explained and the expected performances are given.

1. CALORIMETER SYSTEM

The LHCb calorimeter [1] is built around three sub-detectors, following the particle stream, the preshower (PRS) combined with the scintillating pad detector (SPD), the electromagnetic (ECAL) and finally the hadronic calorimeter (HCAL). On the one hand, the goal of the calorimeter system is to provide a fast response to the first trigger (L0) on the nature of a meson decay. Hence, the SPD/PRS provides a good γ/charged particle and electron/π separation and the electromagnetic and hadronic calorimeters give a fast E_T measurement. On the other hand, it provides precision measurements in order to perform particle identification and, for example, to reconstruct π^0 mesons.

This imposes stringent constraints on the calorimeter system :

- the shaping of the signal should be performed at the LHC frequency (40 MHz),
- the granularity should follow the expected occupancy on the face of the calorimeter,
- the detector should be able to cope with high radiation doses, especially in the beam pipe vicinity.

The overall calorimeter system is segmented into three zones (only two for the HCAL) with different cell sizes according to the occupancy varying from 40.4 to 60.6 and 121.2 mm in the outer region. The acceptance and segmentation of the sub-detectors are designed pseudo-projective to facilitate the shower reconstruction at trigger level and offline. Table 1 shows some of the characteristics required for the LHCb calorimeter system – characteristics which are fulfilled by a sampling detector based on photomultipliers with wave-length-shifter (WLS) fiber readout.

1.1. SPD / PRS

The detector consists of a 2.5 radiation length (X^0) thick lead wall sandwiched between two scintillator tile planes, the SPD and the PRS tiles located respectively before and after the lead absorber. The light produced in the scintillator is collected by helicoidal WLS fibers embedded into the body of the tiles up to clear fibers that transport the signal to multi-anode photomultipliers installed at the bottom and at the top of the calorimeters. Charged particles induce a signal in the SPD while neutrals do not. The leading edge triggers the electron showers which then produce a large signal in the PRS detector. The SPD/PRS is a compact system (less than 20 cm thick), whose performances allow to reject online 95% of π's and retain 95% of electrons after applying a threshold at 0.7 MIP on the SPD, to keep all the electrons while rejecting 98.5% of photons.

1.2. ECAL

The ECAL is a shashlik detector made of 67 layers (4 mm thick) of scintillator and 66 layers of 2 mm thick lead absorber. The light produced in the scintillator is re-emitted and transported by WLS fiber up to the back of the ECAL and is finally converted by a photomultiplier. A chemical matting of the edge of the cells provides an improved light yield and a better global uniformity ($< 3\%$) compared to the aluminization technique. The fiber density is chosen to give a good local uniformity without degrading the global uniformity. The ECAL resolution has been measured on series modules and in test beam to be $\sigma(E)/E = 9.4\%/\sqrt{E} \oplus 0.8\%$.

CP722, *B Physics at Hadron Machines*, edited by M. Paulini and S. Erhan
© 2004 American Institute of Physics 0-7354-0203-5/04/$22.00

TABLE 1. Requirements on the calorimeter sub-detectors.

Sub-detector	SPD/PRS	ECAL	HCAL
number of channels	2×5952	5952	1468
lateral size	$6.2 \times 7.6 \, \text{m}^2$	$6.3 \times 7.8 \, \text{m}^2$	$6.8 \times 8.4 \, \text{m}^2$
longitudinal depth	180 mm	835 mm	1655 mm
	$2 \, X^0, \, 0.1 \, \lambda_I$	$25 \, X^0, \, 1.1 \, \lambda_I$	$5.6 \, \lambda_I$
basic requirement	20/30 photo-electrons per MIP	$10\%/\sqrt{(E)} \oplus 1.5\%$	$80\%/\sqrt{(E)} \oplus 10\%$
dynamic range	0-100 MIP's	0-10 GeV E_T	0-10 GeV E_T
	1 bit (SPD), 10 bits (PRS)	12 bits	12 bits

1.3. HCAL

Unlike the SPD/PRS and the ECAL, the HCAL is segmented into two zones with cell sizes of 131.1 and 262.26 mm. The sampling is performed by 6 mm thick iron master plates and 4 mm thick iron spacers interleaved with 3 mm thick scintillator tiles parallel to the beam. WLS fibers run along the edges of the tiles up to the PMT located at the back of the cells. The resolution of the series module measured in test beam is $75\%/\sqrt{E} \oplus 10\%$.

The ECAL and HCAL front-end electronics are identical and installed above the calorimeter system. It uses the narrow pulse shape of the detectors and a clipping technique to measure every bunch crossing with hardly no pile-up effect. The fast and low noise reset of the integrated signal is performed by the integration of the opposite signal after 25 ns. Finally, the radiation tolerance of the electronics has been intensively estimated in test beams. Triple voting and anti-fuse technology are used to protect the digital components from single event effects.

2. MUON SYSTEM

Muon reconstruction and identification are crucial for LHCb both at the first level trigger and offline. The muon system [2, 3] uses the penetration power of the particle. The main requirements for the trigger are a good time resolution to identify each bunch crossing and a p_T resolution of at least 20%. The offline analysis muon identification requires an efficiency better than 90% down to $p = 3 \, \text{GeV}/c$ while retaining less than 1.5% of pions.

The detector consists of five tracking stations interspersed with an absorber to attenuate hadrons, electrons and photons. The first station (M1) is located in front of the SPD, the calorimeter system serving as a shield while the others are placed behind the HCAL, three 80 cm thick iron planes being used as filters.

2.1. Technology

The technology choice is determined by the physics goals and the background conditions of the detector and, more precisely, by the aging and rate capabilities as well as the time and spacial resolutions. Multi Wire Proportional Chambers (MWPC) have been adopted on the full detector surface. A time resolution of 4 ns is obtained thanks to a fast gas (mixture of $Ar/CO_2/CF_4$, 40:50:10) and a 2 mm wire spacing (see Figure 1). An operating voltage of 2.8 kV leads to an efficiency of 99% in 20 ns for a 2 gap chamber, the electronics providing both a wire and cathode readout. M1 is made of such chambers (2 gaps) with a special honeycomb in order to reduce the amount of matter in front of the calorimeter to $15\% \, X^0$. The four stations located behind the HCAL will be 4 gap chambers.

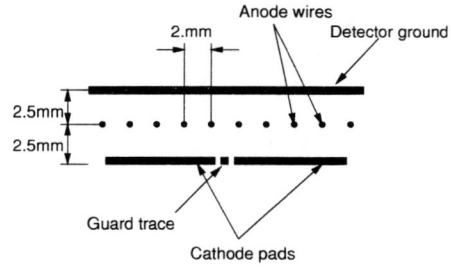

FIGURE 1. Schematic view of one sensitive gap MWPC.

2.2. Layout

The muon detector layout (see Figure 2) is projective. Each station is divided in four regions with different logical pad dimensions, region and pad sizes doubling from one region to the next. The granularity has been driven by the p_T requirement and the background rejection :

- the x-dimension is chosen to perform a good p_T measurement at the trigger level,
- the y-dimension is determined to reject background which does not originate from the interaction point.

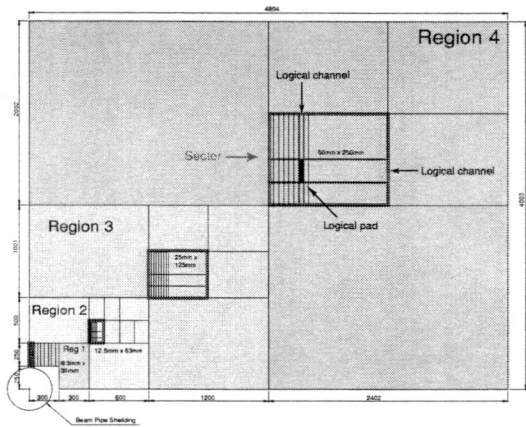

FIGURE 2. View of a quadrant of Station 2. Each region is shown with a sector. The intersection of logical channels makes the logical pads. The region and channel dimensions are doubled from one region to the next one.

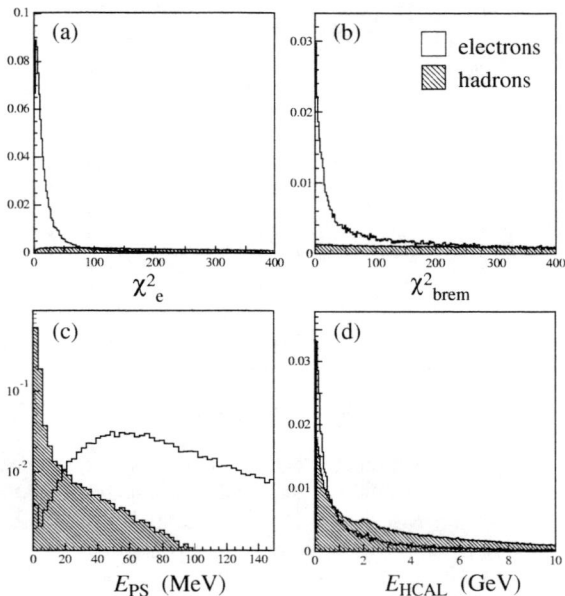

FIGURE 3. The four estimators constructed from the calorimeter system and the tracker used to identify electrons.

The total number of logical pads is 55296. They are obtained from the crossing of horizontal and vertical strips made from the front-end channels or directly from the physical pad channels of the chambers. The logical AND of the crossing is performed by the trigger processor. This technique reduces the number of channels to be handled by the following electronics and gives a precise information on the hit positions on the stations.

3. LEPTON IDENTIFICATION

Lepton identification at LHCb [4] is primarily performed for electrons by the calorimeters and for muons by the muon system. However, the RICH improves the lepton identification and a combined likelihood is built to extract the most efficient information from the LHCb detector.

3.1. Electrons

In LHCb, electron candidates are looked at among the tracks whose extrapolation corresponds to a calorimeter cluster. The cluster-track position estimator χ_γ^2 takes into account the track's covariance matrix and the second-order moment of the cluster to quantify the matching. A simple cut on χ_γ^2 removes most of the photon background and doesn't affect the electron signal. Then, electron identification is performed on the remaining clusters by combining four variables :

- The estimator χ_e^2 (Fig. 3a) is built from the matching of the tracking system and the calorimeter. It includes the balance between the ECAL energy measurement and the track momentum reconstructed as well as the matching between the corrected cluster position and the track extrapolation up to the calorimeter front face.

- There is little material in the magnet area. Hence, the Bremsstrahlung photon impacts on the calorimeter face are well defined. The estimator χ_{Brem}^2 (Fig. 3b) is evaluated from the track extrapolations and the reconstructed neutral clusters.

- The energy deposition in the preshower E_{PS} (Fig. 3c) and the hadronic calorimeter E_{HCAL} (Fig. 3d) along the charged particle trajectory.

Figure 3 shows the distributions of the four estimators for electrons and hadrons extracted from a selection of B decay channels. Normalized two-dimensional likelihood histograms are calculated for each variable versus the particle momentum both for the true electron and wrong track hypothesis. For a given particle, the log-likelihood differences between the two hypotheses are summed on the four estimators. The calorimeter information obtained is finally combined with the RICH likelihood based on the mass hypothesis.

The performance of the overall identification is illustrated in Figure 4. The electron efficiency and the pion misidentification rate are shown for a particular log-likelihood cut ($\Delta \ln \mathcal{L}_{e\pi} > 0$) with respect to the particle momentum. The average efficiency and mistag rate are 95% and 0.7%, respectively.

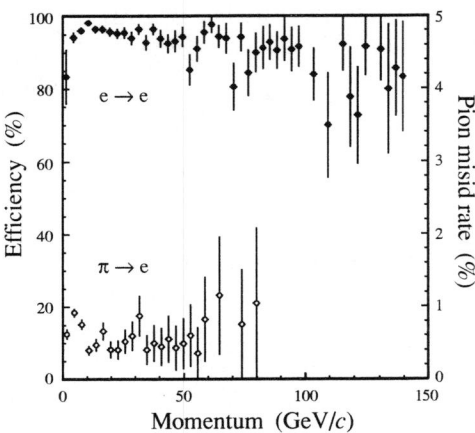

FIGURE 4. Electron efficiency (solid points) and misidentification rates (open points, right scale) as a function of momentum.

TABLE 2. Stations for which hits are required in the corresponding FOI of the track as a function of the particle momentum.

Momentum [GeV/c]	Muon Stations
$3 < p < 6$	M2+M3
$6 < p < 10$	M2+M3+(M4 or M5)
$p > 10$	M2+M3+M4+M5

3.2. Muons

Muons are searched among reconstructed tracks whose momentum is greater than 3 GeV/c. Fields of Interest (FOI) are then defined around the extrapolation of tracks in each muon station. Muon candidates are required to have at least a hit in the FOI of a minimum number of stations according to their momentum (see Table 2). The FOI's are parameterized in order to increase the muon identification efficiency while leading to a reasonable pion misidentification. The efficiency and pion mistag of the method (measured on a sample of $B^0 \to J/\psi K^0_S$) are $(94.3 \pm 0.3)\%$ and $(2.9 \pm 0.1)\%$, respectively. Figure 5 shows the performance as a function of the track momentum.

The muon contamination is then reduced by using two additional discriminating variables :

- the slopes measured in the tracker and with the muon station hits are compared,
- the average track-hit distance for the hits in the FOI's.

For each muon candidate, the log-likelihood difference in the two hypotheses is evaluated and summed with the calorimeter and RICH likelihood (if available). A simple cut on the final $\Delta \ln \mathcal{L}_{\mu\pi}$ parameter reduces the pion misidentification to 1% and maintains the muon

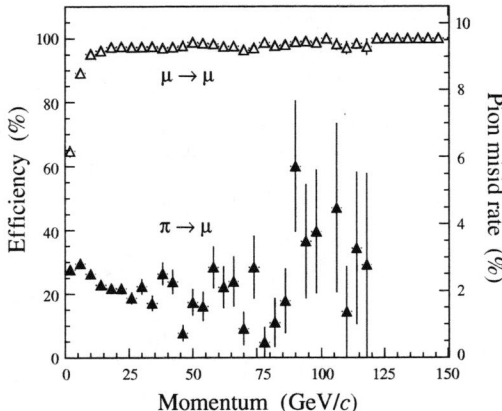

FIGURE 5. Muon efficiency (open triangle) and pion misidentification (solid triangle, right scale) as a function of the momentum of track.

efficiency at 93%.

4. CONCLUSION

The sensitivity of the lepton identification with respect to the detector performances has been thoroughly studied. Hence, among other simulated degradations, the levels of incoherent and coherent noise has been varied, dead channels have been simulated, wrong gains and mis-calibrations have been applied to the calorimeters, the muon background and the track multiplicity has been artificially increased, etc. Those studies [4] demonstrated the robustness both of the algorithms and of the calorimeter and muon systems.

ACKNOWLEDGMENTS

I would like to thank the organizers of *Beauty 2003* for their hospitality during the conference and the members of LHCb for their help.

REFERENCES

1. LHCb Collaboration, LHCb Calorimeters Technical Design Report, Tech. Rep. CERN/LHCC 2000-036 (2000).
2. LHCb Collaboration, LHCb Muon System Technical Design Report, Tech. Rep. CERN/LHCC 2001-010 (2001).
3. LHCb Collaboration, Addendum to the LHCb Muon System Technical Design Report, Tech. Rep. CERN/LHCC 2003-002 (2003).
4. LHCb Collaboration, LHCb Reoptimized Detector Design and Performance, Tech. Rep. CERN/LHCC 2003-0030 (2003).

ATLAS Software for *B* Physics

Fabrizio Parodi

(Representing the ATLAS Collaboration)

Università di Genova, Dipartimento di Fisica, 16146 Genova, Italy

Abstract. An overview of the software developed for *B* physics studies with the ATLAS detector is given, focusing on the architecture and on the main reconstruction algorithms. The developers environment is also briefly described.

1. INTRODUCTION

The ATLAS detector, which will be operating at the LHC collider (data taking start due in 2007), has been primarily designed to search for new particles in proton-proton collisions at 14 TeV energy in the center-of-mass. Nevertheless, *B* physics requirements have been accommodated in the design and precise measurements are foreseen in several clean decay channels: *e.g.* $\sin(2\beta)$ ($B^0 \rightarrow J/\psi K_S^0$), Δm_s ($B_s^0 \rightarrow D_s \pi(a_1)$), $\Delta\Gamma_s$ ($B_s^0 \rightarrow J/\psi \phi$) and rare muonic decay branching ratios ($B^0(B_s^0) \rightarrow (X)\mu\mu$).

The tools needed for the ATLAS *B* physics program are: soft lepton identification (electrons and muons), charged track reconstruction and vertex reconstruction. In this contribution, I will describe how the offline and online ATLAS software has been designed to accomplish these tasks, focusing on the architecture and the main reconstruction algorithms.

2. ATLAS DETECTOR

The ATLAS detector (described in details elsewhere [1]) is a multipurpose experiment for the LHC collider (see Fig. 1). It has overall cylindrical barrel and end-caps geometry. The Inner Detector immersed in a 2 Tesla solenoid magnetic field provides tracking up to $|\eta| = 2.5$. Beyond the solenoid coil, the liquid argon calorimeter provides electromagnetic calorimetry in the barrel region ($\eta \leq 2.5$) and both electromagnetic and hadronic calorimetry in the end-cap region ($\eta \leq 5$). Hadronic calorimetry in the barrel is provided by the iron tile scintillator calorimeter. These elements are surrounded by the air toroids of the muon spectrometer. At the design luminosity, $\mathscr{L} = 10^{34} \, \mathrm{cm}^2 \, \mathrm{s}^{-1}$, each bunch crossing, every 25 ns, yields in average 23 minimum bias pp collisions. The data taking rate is expected to be 100 Hz,

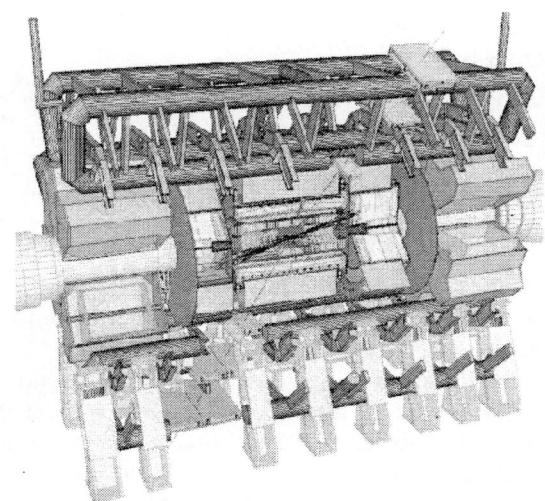

FIGURE 1. Cross-section of the ATLAS detector.

totaling 1 PByte of data per year.

3. SOFTWARE FRAMEWORK AND WORK MODEL

The ATLAS reconstruction is now being developed in the Athena framework, based on Gaudi, initially developed by the LHCb Collaboration. Athena relies on the separation between data (handled in the Transient Event Store, TES) and algorithms (sketched in Fig. 2).

Algorithms read data (*e.g.* silicon hits), process them and write new data (*e.g.* silicon clusters). The sequencing and configuration of algorithms are specified at run time through an ASCII file. This makes the configuration of a job very flexible, so that one can easily switch between a very detailed simulation of the data flow and a coarser

CP722, *B Physics at Hadron Machines,* edited by M. Paulini and S. Erhan
© 2004 American Institute of Physics 0-7354-0203-5/04/$22.00

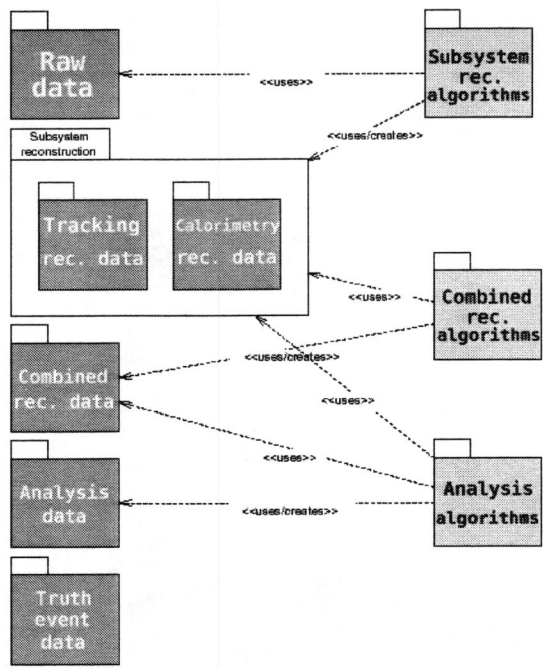

FIGURE 2. Sketch of the separation between data and algorithms, from raw data to physics analysis.

one, or switch between two different clustering or track reconstruction algorithms.

All the reconstruction algorithms, initially developed in Fortran, have now being ported to C++. Besides the high level algorithms, special attention has also been paid to implement a realistic and detailed simulation of the raw data flow from the front-end board byte-stream to pre-reconstructed quantities.

All the other ATLAS software activities besides reconstruction (detailed trigger simulation, event simulation, complete and fast detector simulation) can be performed in the same framework, allowing correlations to be studied.

This common framework allows an independent development of several packages by hundreds of people world-wide. The development follows a six month cycle for major validated releases to be used by physicists and in large statistics data productions. There are also developer releases every three weeks and a nightly build which facilitates vastly the integration effort.

4. *B* PHYSICS TRIGGER SELECTION

At the LHC, about one collision in every hundred will produce a $b\bar{b}$ quark pair. Therefore, in addition to rejecting non-$b\bar{b}$ events, the *B* trigger (described in more details in Ref. [2]) must have the ability to identify and se-

lect those events containing *B* decay channels of specific interest.

A flexible trigger strategy has been developed based on a di-muon trigger at the start of higher luminosity $(2 \times 10^{33}\,\mathrm{cm^{-2}\,s^{-1}})$ LHC fills and introducing further triggers later in the beam coast as well as for low luminosity fills. The additional triggers will require a jet or an electromagnetic cluster to be identified in addition to a single muon by the Level-1 trigger. All Level-1 objects will guide reconstruction at Level-2 and the results of the Level-2 trigger will seed reconstruction in the Event Filter. Performing the reconstruction only in the limited region of the Inner Detector containing the *B* candidate strongly reduces the processing power requirements.

5. TRACKING

Tracking in ATLAS is performed up to $|\eta| = 2.5$ with three layers of pixels and four layers of stereo silicon strip detectors (SCT), followed by a straw tracker (TRT) providing typically 30 drift time measurements per track, in addition to electron transition radiation detection capability (details on the ATLAS Inner Detector can be found in Ref. [3]).

5.1. Pattern Recognition and Track Reconstruction

Two different algorithms of pattern recognition for charged tracks have been used at ATLAS:

- **IPatRec** searches for tracks using the three dimensional space points (SP's) in pixels and SCT; candidates are extrapolated to the TRT and drift time hits added using a χ^2 fit;
- **XKalman** searches for tracks in TRT using a fast histogramming algorithm based on the straw hits. Candidates are extrapolated to the SCT and pixel detectors. They are fit using a Kalman filter. The obtained tracks are then extrapolated back to the TRT and drift-time hits are added.

The two approaches have different advantages: XKalman is faster because it benefits from the fast histogramming of TRT hits; IPatRec is less sensitive to interactions or bremsstrahlung. The overall performance of the two algorithms turns out to be very similar. The modular structure of the new framework will help merging the good features of the two algorithms.

The track efficiency is about 0.95 for single tracks and 0.90 for tracks in jets (the fake track fraction is about 0.003). The resolutions for the transverse impact parameter (d_0) and for the transverse momentum (p_T)

can be parametrized as

$$\sigma(d_0) = 12 \otimes \left(\frac{107}{p_T}\right) \ \mu m, \qquad (1)$$

$$\sigma\left(\frac{1}{p_T}\right) = 0.6 \otimes \left(\frac{18}{p_T}\right) \ \mathrm{TeV}^{-1}. \qquad (2)$$

These resolutions depend strongly on the radius of the innermost pixel layer and on the radius and thickness of the beampipe. The impact of a reduced configuration at the startup (second pixel layer missing) is almost negligible and concerns only the resolution on the d_0 parameter.

5.2. Primary Vertex Reconstruction

At design luminosity, the expected 23 interactions per beam crossing are spread out along the beam line by $\sigma(z) = 5.6 \, \mathrm{cm}$. For any physics analysis it is thus very important to reconstruct the z-coordinate of the primary vertex in order to eliminate tracks not coming from the main interaction vertex.

The primary vertex is reconstructed by histogramming the z-impact parameter (z_0) of the reconstructed tracks. Clusters of tracks are used as indicators of vertex candidates. The z_{beam}-coordinate is then computed by fitting together tracks in the same cluster. The average resolution on the z_{beam}-coordinate is 48 μm.

5.3. Secondary Vertex Reconstruction

The reconstruction of a secondary vertex, detached from the primary interaction vertex, is very important for B mixing and lifetime studies.

In particular, the reconstruction of the decay chain $B_s^0 \rightarrow D_s(\phi\pi)\pi \rightarrow KK\pi\pi$ relies on the identification of the intermediate resonances whose resolution is determined by the track quality. The final sensitivity to the oscillation frequency Δm_s crucially depends on the precision of the B proper time reconstruction:

$$\sigma_t = \sqrt{\sigma_L^2 + (\sigma_p/p)^2 t^2}, \qquad (3)$$

where σ_L is the uncertainty due to the flight length resolution while σ_p/p is the relative uncertainty on the B momentum.

The proper time resolution is expected to be 0.071 ps, dominated by the flight length resolution ($\sigma_p/p = 0.6\%$), allowing to extend the sensitivity up to $\Delta m_s = 36 \ \mathrm{ps}^{-1}$ in the first year of data taking (10 fb^{-1}).

Another important ingredient for B physics is the reconstruction of K_S^0 decay vertexes. This task is quite challenging because of the long K_S^0 lifetime which leads to

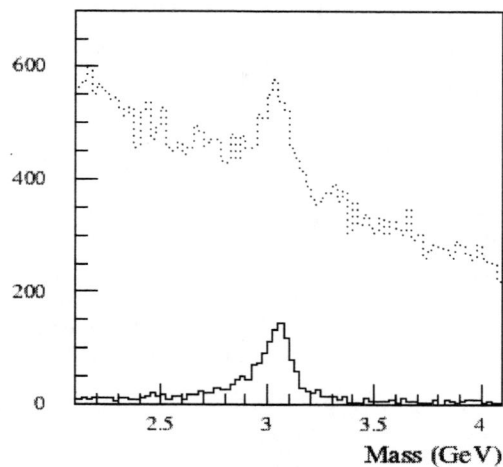

FIGURE 3. Rejection of combinatorial background using electron identification in TRT. Invariant mass of all track combinations in events containing $J/\psi \rightarrow e^+e^-$ before (dashed) and after (solid) identification cuts.

tracks with a limited number of hits in the silicon detector. The average efficiency of the K_S^0 reconstruction (in the channel $B^0 \rightarrow J/\psi K_S^0$) is expected to be 41%.

5.4. Electron Identification

Electromagnetic clusters are searched for with a sliding window algorithm. Rectangular clusters of granularity $\Delta\eta \times \Delta\phi = 0.025 \times 0.025$ are preferred for their robustness against pile-up, underlying event and material effects.

The electromagnetic clusters are then matched with a charged track reconstructed in the Inner Detector (fitted allowing for bremsstrahlung). Transition Radiation (TR) hits in the straw tracker bring additional rejection. The impact of the TR hits alone on the selection of J/ψ mesons decaying to e^+e^- is given in Fig. 3.

For low p_T electron reconstruction (below 7 GeV/c), tracks satisfying TR hit identification are extrapolated to the electromagnetic calorimeter, where the energy deposition around the extrapolated track is tested against the electron shower hypothesis.

Using the combined identification from the Inner Detector and the calorimeter, inclusive $b\bar{b} \rightarrow \mu eX$ events (sample of candidates for B decays containing J/ψ decays in e^+e^-) can be selected with 70% efficiency with a rejection toward $b\bar{b} \rightarrow \mu X$ of about 600.

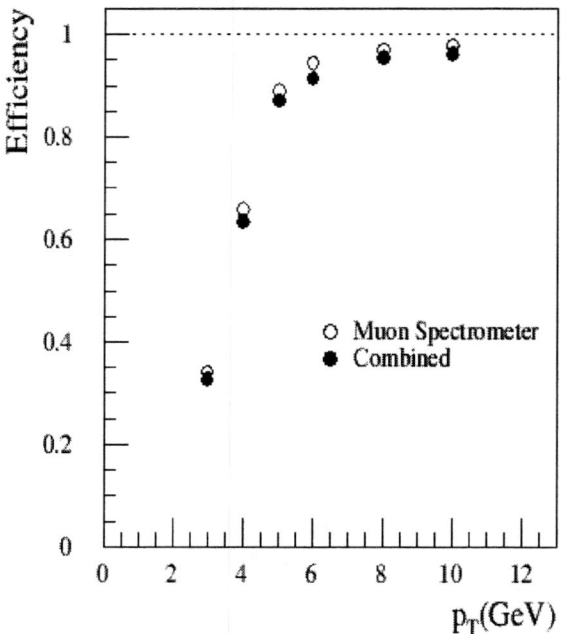

FIGURE 4. Efficiency of muon reconstruction using the muon chambers and the inner detector as a function of p_T.

5.5. Muon Identification

The muon spectrometer provides standalone muon identification and measurement from typically three stations in the toroids (fitted with tracking detectors using four different technologies), each capable of reconstructing a 3D segment of the muon trajectory. The efficiency is typically 95%, due to holes for detector support and services.

The reconstructed muon is backtracked to the interaction point through the calorimeter, corrected for its estimated energy loss, and combined with its Inner Detector track in order to improve the momentum resolution for p_T up to 20 GeV/c. The combined muon reconstruction efficiency as a function of p_T for Level-1 muons (threshold at 6 GeV/c) is shown in Fig. 4. The efficiency of muon matching with the inner detector track is remarkable.

Typical mass resolution for $J/\psi \to \mu^+\mu^-$ produced in B decays is 40 MeV/c^2.

6. CONCLUSIONS

A wide variety of algorithms are being developed in order to exploit the ATLAS B physics potential.

Studies are so far based on a very detailed Monte-Carlo simulation, with the performances of each detector carefully tuned on test beam data.

In 2004, a barrel wedge with all the ATLAS detectors will be put in a test beam giving the opportunity to test the algorithms in a real environment.

A repackaging is on-going so that all algorithms performing the same tasks with different strategies share the same interface. This will ease algorithm improvements and new algorithm development, as well as their optimization for different physics channels and running conditions.

REFERENCES

1. ATLAS Collaboration, ATLAS Detector and Physics Performance: Technical Design Report Vol 1, Tech. Rep. CERN/LHCC/99-14 (1999).
2. George, S., in these proceedings (2004).
3. Moser, H.-G., in these proceedings (2004).

Grid Computing in High Energy Physics

Paul Avery

University of Florida, Department of Physics, Gainesville, FL 32611-8440, U.S.A.

Abstract. Over the next two decades, major high energy physics (HEP) experiments, particularly at the Large Hadron Collider, will face unprecedented challenges to achieving their scientific potential. These challenges arise primarily from the rapidly increasing size and complexity of HEP datasets that will be collected and the enormous computational, storage and networking resources that will be deployed by global collaborations in order to process, distribute and analyze them.

Coupling such vast information technology resources to globally distributed collaborations of several thousand physicists requires extremely capable computing infrastructures supporting several key areas: (1) *computing* (providing sufficient computational and storage resources for all processing, simulation and analysis tasks undertaken by the collaborations); (2) *networking* (deploying high speed networks to transport data quickly between institutions around the world); (3) *software* (supporting simple and transparent access to data and software resources, regardless of location); (4) *collaboration* (providing tools that allow members full and fair access to all collaboration resources and enable distributed teams to work effectively, irrespective of location); and (5) *education, training and outreach* (providing resources and mechanisms for training students and for communicating important information to the public).

It is believed that computing infrastructures based on Data Grids and optical networks can meet these challenges and can offer data intensive enterprises in high energy physics and elsewhere a comprehensive, scalable framework for collaboration and resource sharing. A number of Data Grid projects have been underway since 1999. Interestingly, the most exciting and far ranging of these projects are led by collaborations of high energy physicists, computer scientists and scientists from other disciplines in support of experiments with massive, near-term data needs. I review progress in this important area, outlining new research directions and practical experiences in deploying national and international scale Data Grids.

1. HIGH ENERGY PHYSICS COMPUTING CHALLENGES

Over the next two decades, major high energy physics (HEP) experiments will face unprecedented challenges to achieving their scientific potential. These challenges arise primarily from the rapidly increasing size and complexity of HEP datasets that will be collected and the enormous computational, storage and networking resources that must be deployed by their international collaborations in order to process, distribute and analyze them. While data volumes (raw, simulated, processed and analyzed) for major experiments at SLAC, KEK, Fermilab and Brookhaven are approaching or (in some cases) have surpassed the petabyte (10^{15} bytes) threshold, the collective data from experiments at the Large Hadron Collider (LHC) at CERN is expected to reach 100 petabytes by the end of the decade and to cross the exabyte (10^{18} bytes) barrier a few years later. These enormous dataset volumes will be accompanied by massive computational resources and ultra-high speed networks.

Coupling such vast information technology resources to globally distributed collaborations of several thousand physicists requires extremely capable computing infrastructures supporting several key areas: (1) *comput-* *ing* (providing sufficient computational and storage resources for all processing, simulation and analysis tasks undertaken by the collaborations); (2) *networking* (deploying high speed networks to transport data quickly between institutions around the world); (3) *software* (supporting simple and transparent access to data and software resources, regardless of location); (4) *collaboration* (providing tools that allow members full and fair access to all collaboration resources and enable distributed teams to work effectively, irrespective of location); and (5) *education, training and outreach* (providing resources and mechanisms for training students and for communicating important information to the public).

While the challenges confronting LHC collaborations are especially daunting, an increasing number of enterprises and projects are discovering similar barriers to their efficient operation as they grow in size and rely more heavily on shared data. These include scientific collaborations (digital astronomy, gravitational wave searches, genomics and proteomics research), engineering projects (space exploration, earthquake simulations), medical teams (distributed medical imaging), governmental organizations (state extension programs, federal agencies) and corporations (particularly multinational companies). Thus tools and mechanisms that provide ef-

CP722, *B Physics at Hadron Machines,* edited by M. Paulini and S. Erhan

fective resource sharing and collaboration across significant geographic and organizational boundaries will have wide applicability.

2. MEETING HEP COMPUTING CHALLENGES WITH DATA GRIDS

2.1. The Grid Concept

Computing infrastructures based on "Grids" [1] provide a general framework for meeting the challenges presented in the previous section of resource sharing by large distributed collaborations. An excellent introduction to Grids [2] includes the following description:

"The real and specific problem that underlies the Grid concept is coordinated resource sharing and problem solving in dynamic, multi-institutional virtual organizations. The sharing that we are concerned with is not primarily file exchange but rather direct access to computers, software, data, and other resources, as is required by a range of collaborative problem-solving and resource brokering strategies emerging in industry, science, and engineering. This sharing is, necessarily, highly controlled, with resource providers and consumers defining clearly and carefully just what is shared, who is allowed to share, and the conditions under which sharing occurs. A set of individuals and/or institutions defined by such sharing rules form what we call a virtual organization (VO)."

This explanation shows clearly that Grid computing encompasses much more than distributed computing, which is concerned chiefly with the technical tools needed to run applications across multiple computers. Grid computing, in contrast, is oriented fundamentally around providing frameworks and tools for preserving *ownership* of resources while *sharing* them with other entities under conditions that they can control.

In a high energy physics experiment like CMS, for example, the VO might be the entire CMS collaboration and the resources to be shared could include, among others, the control room, the subdetector calibration and testing systems, the slow control system, the online control and data collection system, the information technology resources at the different university and laboratory sites, the interfaces and switches to the wide area networks that connect the sites, the software and tools for reconstruction, simulation, visualization and analysis, and various catalogs and directories that permit data and resources to be located. Any Grid framework providing access to these resources must obviously be built with great care to respect and enforce the multiple priorities and policies of the CMS collaboration concerning these resources. In fact, multiple, overlapping sub-VO's

would probably be used, each with its resources, authorized people and policies.

2.2. Globus Based Grids

One cannot build Grid-based computing infrastructures without a basic implementation of Grid tools and services. The Globus Project [3] has for several years defined and maintained the Globus Toolkit, the basis of almost all Grid projects in the world today and the one used by the major high energy and nuclear physics (HENP) collaborations. Leading members of Globus collaborate closely on the projects described here.

2.3. Data Grids

The existence of large distributed data collections adds significant new dimensions to enterprise-wide resource sharing, including scheduling and staging the movement of large data sets, caching frequently used data, providing catalog services to find datasets, cleaning up scratch space, accommodating both files and databases, etc. These challenges have spurred substantial research and development efforts on "Data Grid" infrastructures, capable of supporting this more complex collaborative environment. This work has taken on special urgency for the LHC collaborations, which are reaching global proportions and are taking part in increasingly difficult "data challenges" as part of their general buildup to data-taking in 2007. Several groundbreaking Data Grid projects are underway, with HEP physicists collaborating with computer scientists, network specialists, members of other scientific and engineering fields and industry to conduct research and deploy Grid testbeds and production-scale environments. These projects are discussed later.

3. DATA GRIDS IN HIGH ENERGY PHYSICS

3.1. The LHC Data Grid Hierarchy

In 1998, the MONARC group [4] studied the problem of optimizing LHC computing resources and determined that the most effective organization of resources for globally distributed LHC collaborations was a hierarchical structure [5]. This layout was adopted by the LHC experiments and has since been mapped to a Data Grid hierarchy of five "Tiers" of globally distributed computing and storage resources connected by high speed regional, national and international networks [6].

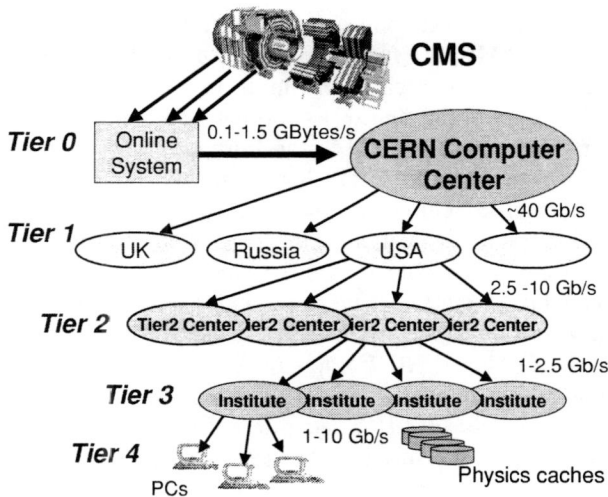

FIGURE 1. Data Grid Hierarchy for an LHC Experiment.

In this model, data at the experiment are stored at the rate of 100−1500 Mbytes/sec throughout the year, resulting in petabytes per year of stored and processed binary data that are accessed and processed repeatedly by worldwide collaborators. The global resources then fulfill the following roles (see Figure 1):

Tier 0 The central facility at CERN where the experimental data is taken, and where all the raw data is stored and initially processed;

Tier 1 A major national center supporting the full range of computing, data handling and support services required for effective analysis by a community of several hundred physicists. Much of the analysis-specific reconstruction as well as the analysis of the data will be carried out at these centers, which for the most part will be located at national laboratories;

Tier 2 A smaller system with significant computing and RAID storage facilities (the use of tape resources will depend on the site) supporting analysis, simulation and reconstruction on demand, sited at a medium to large university or research laboratory;

Tier 3 A work-group cluster specific to a university department or a single high energy physics group, of the sort traditionally used to support local needs for data analysis;

Tier 4 An access device such as individual user's desktop, laptop or even mobile device.

The Data Grid hierarchy shown in Figure 1 should be viewed as a political organization of resources rather than as a strict model for data flows between institutions. The collaboration must, in fact, implement and enforce policies that maximize resource use in order to achieve its scientific goals. Highly organized teams will have prior-

ity use of the relatively few Tier 0 and Tier 1 facilities for large-scale tasks such as systematic data processing, archiving and distribution. At the smaller and more numerous Tier 2 and Tier 3 facilities, individuals and small groups will have greater control over how these resources are allocated to small and medium-sized tasks of special interest to them. Data flow among the Tiers will therefore include dynamic and opportunistic data transfers, as thousands of physicists vie for shares of local and remote facilities of different sizes. These tasks will differ widely in priority, turnaround times (from seconds to hours), computational requirements (from processor-seconds to processor-decades) and data volumes.

3.2. LHC Network Requirements

It is now realized that data flows between sites for currently running experiments as well as LHC collaborations will require network bandwidths along principal links of 10 Gb/s within the next two to three years, followed by a need for scheduled and dynamic use of multiple 10-Gb/s wavelengths by the time the LHC begins operation in 2007.

The driving forces behind LHC network bandwidth requirements are (1) that typical requests for data samples will often exceed a Terabyte in the early years of LHC operation, and could easily reach 10−100 Terabytes in the years following, and (2) the number of requests from the global HEP community is expected to reach hundreds per day. These transactions must complete in a reasonably short time (minutes to hours) to avoid high failure rates, leading to estimated network link speeds of several 10s to 100s of Gb/s, and even Tb/s by the next decade. These and other estimates of HEP networking needs, progress reports, measurements and other documents are maintained by the Standing Committee on Inter-regional Connectivity [7].

4. HEP DATA GRID PROJECTS

I describe in this section some of the principal projects that have been undertaken to develop Data Grids. These efforts have been driven by the overwhelming and near-term needs of current and planned scientific experiments. High energy physics has been at the forefront of these projects, a fact that can be attributed to its historical standing as both a highly data intensive and highly collaborative discipline. However, collaborations in gravitational research, digital sky surveys, virtual astronomical observatories and biomedical research, activities which also face challenges associated with large distributed data collections and a dispersed user community, are also adopting Data Grid computing infrastructures. Scientists from these disciplines are exploiting new initiatives and partnering with high energy physicists, computer scientists and each other to develop new Grid technologies and deploy production-scale Data Grids.

PPDG [8] (Particle Physics Data Grid), funded by the U.S. Department of Energy, is a collaboration of physicists and computer scientists from HENP experiments at SLAC, Fermilab, Brookhaven, Jefferson Lab and LHC. Unlike other Grid projects which develop new tools, PPDG works on hardening existing Grid tools and technologies and integrating them into end-to-end systems capable of serving current and future HENP experiments. Examples of their work include Grid security, monitoring, storage management, data transport and systems for Grid-enabled analysis. PPDG's practical deployments provide important feedback to Grid developers.

GriPhyN [9], supported by the U.S. National Science Foundation, is a collaboration of computer scientists and physicists from four forefront experiments, including ATLAS, CMS, LIGO [10] (Laser Interferometer Gravitational-wave Observatory) and SDSS [11] (Sloan Digital Sky Survey). GriPhyN has adopted "virtual data" as a unifying theme for its investigations of Data Grid concepts and technologies, and is conducting research into virtual data cataloging, execution planning, execution management and performance analysis. The results of this work, and other relevant technologies, are being integrated into its Virtual Data Toolkit (VDT [12]), which is being disseminated to other disciplines. Twelve VDT releases have been made as of January 2004.

IVDGL [13] (International Virtual Data Grid Laboratory), supported by the U.S. National Science Foundation, also supports the ATLAS, CMS, LIGO and SDSS experiments as well as the emerging U.S. National Virtual Observatory (NVO [14]). Its primary activities include: (1) deploying a Grid laboratory consisting of university and laboratory sites around the world plus a Grid Operations Center; (2) providing computing and personnel resources for the U.S. LHC Tier 2 universities; (3) integrating Data Grid tools in the software infrastructures of its constituent experiments; and (4) providing research, education and outreach opportunities to under-represented groups and remote universities.

DataGrid [15] is a European Union collaboration of computer scientists and researchers from the LHC experiments, molecular genomics and earth observations whose goal is to research and deploy Data Grid technologies in support of European science. DataGrid, funded by the European Commission, is led by CERN together with several main partners and a number of smaller associated partners that together make up one of the largest coherent Grid efforts in the world. Its effort is comprised of twelve work packages, of which the first nine involve

technologies, networks and testbeds relating to LHC.

DataTAG [16], funded by the European Commission, created a large-scale intercontinental Grid testbed that focuses upon advanced networking issues and interoperability between the U.S. and European Grid domains. The project also studies high performance inter-Grid networking, including sustained and reliable high performance data replication, end-to-end advanced network services, and novel monitoring techniques. DataTAG and partners at Caltech have set "land speed records" for transferring data between the U.S. and Europe.

LCG [17] (LHC Computing Grid) is preparing the computing infrastructure for the simulation, processing and analysis of LHC data for all four LHC experiments. This includes both the common infrastructure of libraries, tools and frameworks required to support the physics application software, and the development and deployment of the computing services needed to store and process the data, providing batch and interactive facilities for the worldwide community of physicists involved in LHC. LCG depends mainly on leveraged resources from other projects, particularly the European Grid projects, and discussions are still ongoing about participation in testbeds, milestones, etc.

EGEE [18] (Enabling Grids for E-science in Europe), which begins in April 2004, extends the work started by DataGrid but will focus on developing production-quality Grids in Europe. Funded by the EC, EGEE will integrate a number of national and regional Grid programs and attempt to re-engineer middleware tools to better support production Grids. For the first two years, it will provide effort for two significant pilot projects, the LHC Computing Grid, described above, and biomedical Grids. These projects will be undertaken with collaboration with U.S. and Asian grid groups. Later milestones and funding will depend on the assessment of these projects at the end of the two year period.

5. RECENT HEP GRID ACTIVITIES

I summarize in this section recent highlights and interesting directions being taken by HEP Grid projects. While the main developments are covered, my review is necessarily incomplete given the large number of projects and participants and the widening of research interests.

5.1. Grid Project Coordination

Coordinating the efforts of the growing number of HEP Grid projects has been recognized for some time as necessary to avoid divergence of Grid software and tools. While all projects use the same Globus Toolkit, there are important differences in packaging and in the implementation of higher level Grid services such as scheduling, monitoring, resource discovery, etc. These differences are most pronounced between the U.S. and European Grid projects.

Trillium: Since early 2002, the U.S. Grid projects Gri-PhyN, iVDGL and PPDG have worked jointly as the "Trillium" consortium in researching, developing and deploying Grid technologies and infrastructure for the multidisciplinary experiments that they serve. The close integration of the Trillium projects has put Grid testbed users in close contact with Grid developers and significantly tightened the feedback loop between them, allowing large testbeds to be built (using GriPhyN's VDT as the software distribution method) that can support some degree of production work. Most importantly, the close partnership of physicists and computer scientists from the three projects is leading to a strategic view for creating Grid infrastructure that is gaining wide influence with other disciplines and with U.S. funding agencies.

HICB: [19] The HEP Inter-Grid Coordination Board, formed in 2001, was the earliest HEP coordination body. It includes Grid project leaders and the computing heads of major HEP experiments from the U.S., Europe and Asia. HICB has met several times during gatherings of the Global Grid Forum [20]. HICB's Joint Technical Board (JTB) provides a forum for technical discussions between the U.S. and European based projects. The JTB established the GLUE [21] (Grid Laboratory Uniform Environment) project, which developed the joint database schema [22] that were later incorporated into the Globus monitoring framework.

WorldGrid: [23] The WorldGrid project [24, 25] successfully achieved interoperability between the U.S. and EU Grids in late 2002 and demonstrations were shown at the SuperComputing2002 and IST2002 conference venues. ATLAS and CMS simulation jobs were submitted by different virtual organizations and run at 18 sites split across both continents. The GLUE schema provided a common base of information exchange.

PNPA-RG: [26] GGF recently approved the Particle and Nuclear Physics Applications Research Group (PNPA-RG), which provides a forum for discussion of issues related to HENP applications and production grids. The research group has three specific goals: (1) to explain and inform the wider grid community of the specific needs and issues of HENP; (2) to strengthen participation of the HENP community in grid standardization efforts, particularly as they relate to HENP Grids; and (3) to provide early feedback to GGF technical working groups on the success or failure of various grid software components as used in high performance production activities by HENP experiments. PNPA meetings commence at the March 2004 GGF meeting.

LCG: [17] Collaborative activities of HEP Grid efforts are becoming increasingly centered around the LCG project, which provides a persistent structure well suited to the decades-long LHC research program. LCG Phase I (2002–2005) is concerned with the application support environment and common application elements, the computing services and a series of computing data challenges of increasing size and complexity that will demonstrate the effectiveness of the software and computing models. Phase I will conclude with a Computing System Technical Design Report, providing a blueprint for the computing services that will be required when the LHC turns on. LCG Phase II (2006–2008) will oversee the construction, commissioning and first years of operation of the initial LHC computing system.

5.2. Grid-Enabled Analysis

SAM [27] (*Sequential data Access via Metadata*), developed by DØ, is an operational prototype of many of the concepts being developed for Grid computing. SAM, a PPDG subproject, provides a transparent mechanism for users to make requests for relevant data collections (stored worldwide) and to submit simulation, reconstruction or analysis jobs on available computing resources (also located worldwide). Sam programs transform data by consuming data files and producing new data files in a different "data tier". Some DØ-specific components are being replaced by standard components emanating from the various Grid research projects. Others will be modified to conform to Grid protocols and APIs.

The LHC collaborations have for some time been carrying out their production tasks using Grid tools, but Grid-enabled analysis efforts have proceeded more unevenly. Many of these endeavors have recently been organized by LCG under ARDA [28] (see Figure 2), or Architectural Roadmap for Distributed Analysis in order to devise a common strategy for Grid-enabled analysis.

Individual experiment activities include CMS' CAIGEE system [30], ATLAS' DIAL project [31], ALICE's AliEn framework [32] and LHCb's DIRAC system [33]. Other projects are extending the ROOT analysis system [34] to work in Grid environments and to take advantage of recently developed virtual data technologies from GriPhyN [35, 36]. A comprehensive set of user requirements [37] has been collected and are being used to determine common Grid requirements.

5.3. Grid2003, Open Science Grid and LCG

Over the past two years, the U.S. and Europe Grid projects have deployed a series of Grid testbeds. The ma-

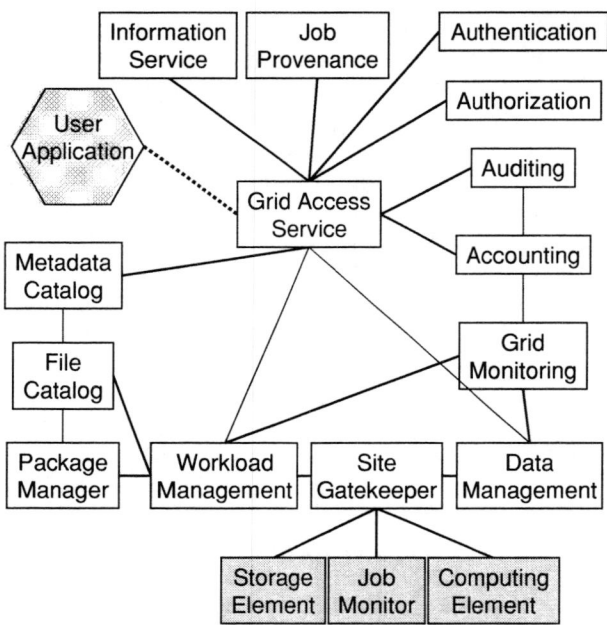

FIGURE 2. Interaction between a set of proposed core ARDA Services [29] and a user client.

jority of these have been experiment specific and allowed their participants to gain experience with Grid technologies and develop tools for their experiment software infrastructures.

Since 2003, however, Grid deployments have become more sophisticated and able to support a wider variety of applications as technologies have matured and become adopted by multiple experiments (even outside HEP). DataGrid, for example, recently deployed Testbed 3 [38] featuring their EDG-2 software release across more than 15 sites in Europe. Similarly, the LCG has deployed a Grid testbed [17] across 25 different sites and in 13 countries in the U.S., Europe and Asia (primarily at the large Tier 1 centers), providing a computational resource capacity of several hundred processors. The LCG testbed is expected to grow rapidly in 2004 to several thousand processors.

The Grid2003 Project [39] has deployed one of the largest and most capable Data Grids to date. Operated jointly by the Trillium Grid projects as well as US-CMS and US-ATLAS, Grid2003 had at its peak about 2700 processors at 26 sites in the U.S. and one site in Korea. Although it was originally intended to be a demonstration project for Supercomputing 2003, upgrades to middleware tools and improvements in operations have allowed Grid2003 to run continuously since October 2003.

Grid2003's central goal was the operation of a shared, multi-Virtual Organization, multi-application Grid laboratory whose performance could be monitored and compared against targeted goals. Project personnel included

application experts, site system managers (at Fermilab, Brookhaven, US-CMS and US-ATLAS Tier 2 Centers, physics departments and large computing centers) and groups responsible for delivery and support of Grid system services and operations. The overall approach of the project was "end-to-end" in terms of giving equal attention to the application, organization, and site infrastructure and system services needed to achieve science applications running on a shared Grid.

Grid2003 highlights include (1) a total of 27 sites at universities and national laboratories, including 2100–2700 CPUs; (2) continuous operation since October 2003 with 400–1100 jobs running simultaneously; (3) ability to run 10 separate applications, including high energy physics (CMS, ATLAS, BTeV), a gravitational wave search, galaxy cluster finding, molecular genomics, bio-molecular analysis, monitoring and data movement; and (4) ability to incorporate and use large resources not dedicated to HENP experiments, including large-scale CMS production simulations that exploit significant non-CMS resources [40] (see Figure 3)

Like earlier Grid testbeds built by experiment groups that exposed shortcomings in basic Grid tools, Grid2003 has demonstrated the tremendous value of operating a significant Grid infrastructure for identifying where current Grid technologies fall short and what challenges need to be overcome in scaling up current efforts to create production-quality Grids. A frank assessment Grid2003's performance and of the advances needed to significantly increase its scale have been collected in a "lessons learned" document [41].

Grid2003 stakeholders have proposed the creation of Open Science Grid (OSG [42]), a large-scale production Grid that would provide the critical computing resources needed by the U.S. LHC research program but will also serve the broad U.S. science community and educate a new generation of students. This new infrastructure would be constructed in stages of increased functionality and scale over the next several years, while participating in LHC data challenges and maintaining compatibility with LCG efforts. A program is being developed to federate grid resources at labs and universities, and potentially at other large centers [43], in this new national Grid resource. The program would extend, develop and deploy common interfaces and adapters, based on current and future Grid technologies, to the services provided by the center facilities.

New Grid deployments are expected to benefit from the powerful packaging framework [44] used in Grid2003 that is being extended to simplify further the installation and configuration of Grid tools and applications. Advanced packaging will allow larger Grids to be built, facilitate software uniformity across sites and enable small institutes to participate more easily.

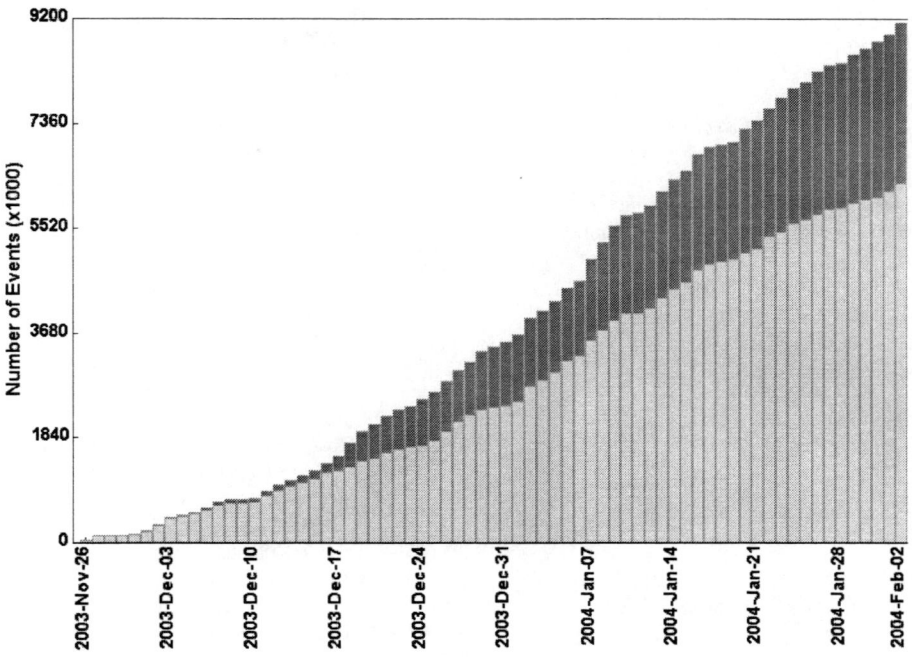

FIGURE 3. Daily cumulative summary of US-CMS production simulations [40] on Grid2003 using experiment specific resources (light color) and non-dedicated resources (dark color).

6. THE FUTURE OF HEP GRIDS

Grid technologies are developing rapidly as they are increasingly deployed in science, engineering, medicine, government, defense, industry and commerce. In high energy and nuclear physics, Data Grid computing architectures offer the only realistic mechanism for global collaborations of thousands of physicists to efficiently utilize their massive distributed computing resources and reach their scientific potential.

However, in applying Grid tools to HENP experiments many daunting technical and political challenges persist. Complicating matters in the Grid arena is the need for HENP Grid efforts to be aligned with new Grid developments such as OGSA [45] compliant architectures, which are being widely adopted by many disciplines and by industry. At the same time major advances are needed in knowledge management, computing and storage technologies, advanced monitoring tools, collaborative technologies, network protocols and optical network deployments. New global information systems [46] integrating computing, storage and optical network resources will need to be developed to fully support the LHC scientific mission. The reality of the "Digital Divide" [47, 48] must also be addressed so that scientists and students from poor or remote areas have access to all experiment resources and can participate fully in the scientific mission of these experiments.

In their attempts to meet these diverse challenges, high energy physicists have arrived at the front ranks of new Grid developments, but they have done so only by forging new alliances with computer and computational scientists, network specialists, educators and teachers, scientists and engineers from other disciplines and industry partners. New laboratory and university partnerships have formed, with scientists, laboratory staff and students building significant Grids that are already being exploited for large-scale data challenges. These collaborative successes are influencing funding agency agendas while demonstrating the broad relevance of high energy physics research to societal concerns.

ACKNOWLEDGMENTS

I am grateful for the outstanding support I have received from the U.S. National Science Foundation, which has funded my work on Data Grids, including the GriPhyN project (ITR-0086044) and iVDGL (PHY-0122557).

REFERENCES

1. Foster, I. and Kesselman, C. (eds.) "The Grid: Blueprint for a New Computing Infrastructure", Morgan Kaufmann, 1999.
2. Foster, I. Kesselman, C. Tuecke C., "The Anatomy of the Grid: Enabling Virtual Scalable Organizations, International Journal of High Performance

Computing Applications", 15(3), 200-222 (2001); http://www.globus.org/anatomy.pdf.

3. Globus Project, http://www.globus.org/.
4. MONARC, http://www.cern.ch/MONARC
5. MONARC Report CERN/LCB-2000-001, http://www.cern.ch/MONARC/docs/phase2report/Phase2Report.pdf
6. Bunn J. and Newman H., "Data-intensive Grids for high energy physics", in *Grid Computing: Making the Global Infrastructure a Reality*, F. Berman, G.E. Fox, A.J.C. Hey, eds., pp. 859–906, Wiley, 2002.
7. See, for example, Newman, H., "Networking for High Energy and Nuclear Physics", Jan. 26, 2004, http://icfa-scic.web.cern.ch/
8. PPDG, http://www.ppdg.net/
9. GriPhyN project, http://www.griphyn.org/
10. LIGO exeriment, http://www.ligo.caltech.edu/
11. Sloan Digital Sky Survey, http://www.sdss.org/
12. Virtual Data Toolkit, http://www.griphyn.org/vdt
13. iVDGL project, http://www.ivdgl.org/
14. U.S. National Virtual Observatory, http://www.us-vo.org/
15. DataGrid, http://www.eu-edg.org/
16. DataTag project, http://www.datatag.org/
17. LCG project, http://www.cern.ch/lcg/
18. EGEE project, http://egee-ei.web.cern.ch/egee-ei
19. HICB, http://www.hicb.org/
20. Global Grid Forum, http://www.gridforum.org/
21. GLUE, http://www.hicb.org/glue/glue.htm
22. GLUE schema descriptions, http://www.hicb.org/glue/glue-schema/schema.htm
23. WorldGrid project, http://www.ivdgl.org/worldgrid/
24. Gardner R. et al., "The WorldGrid Design", GriPhyN/iVDGL document 2003-22.
25. Donno, F. et al, "The WorldGrid transatlantic testbed: a successful example of Grid interoperability across EU and U.S. domains", in Proceedings of the Conference on Computing in High Energy and Nuclear Physics, La Jolla, CA, March 24–28, 2003, GriPhyN Document 2003-25.
26. PNPA working group at GGF, http://force.gridforum.org/projects/pnpa-rg/
27. SAM Grid system, http://d0db.fnal.gov/sam/
28. ARDA, http://project-arda.web.cern.ch/
29. "Architectural Roadmap Towards Distributed Analysis", CERN-LCG-2003-033, Oct. 31, 2003, http://lcg.web.cern.ch/LCG/documents/ARDA-report-final.pdf
30. CMS CAIGEE project, http://hpcbunn.cithep.caltech.edu/GAE/CAIGEE/default.htm
31. ATLAS DIAL project, http://www.usatlas.bnl.gov/~dladams/dial//
32. ALICE AliEn project, http://alien.cern.ch/
33. LHCb DIRAC framework, http://lhcb-comp.web.cern.ch/lhcb-comp/Production/
34. ROOT analysis system, http://root.cern.ch/

35. CAVES project, http://ufgrid02.phys.ufl.edu/~bourilkov/
36. Chimera Virtual Data System, http://www.griphyn.org/chimera/
37. HEPCAL documents, HEPCAL, LHC-SC2-20-2002, http://lcg.web.cern.ch/LCG/sc2/RTAG4/finalreport.doc; HEPCAL II, LCG-SC2-2003-032 http://lcg.web.cern.ch/LCG/SC2/GAG/HEPCAL-II.doc
38. DataGrid testbed 3, http://mapcenter.in2p3.fr/datagrid-rgma/
39. Grid2003 project, http://www.ivdgl.org/grid2003/
40. US-CMS production simulations, http://www.phys.ufl.edu/~prescott/uscms/oscar_prod.html
41. "Grid2003 Lessons Learned", Grid2003 document 2004-1, http://www.ivdgl.org/grid2003/
42. Open Science Grid, http://www.opensciencegrid.org/
43. TeraGrid project, http://www.teragrid.org/
44. Pacman packaging system, http://physics.bu.edu/~youssef/pacman/
45. Open Grid Services Architecture (OGSA), http://http://www-fp.globus.org/ogsa/
46. UltraLight information system, http://ultralight.caltech.edu/
47. Digital Divide subcommittee of SCIC, http://icfa-scic.web.cern.ch/ICFA-SCIC/Committees/digdivide.htm
48. Conference on Grids and the Digital Divide, http://www.uerj.br/lishep2004

$|V_{ucb}|$, $|V_{ub}|$ AND FACTORIZATION

Factorization and the Soft-Collinear Effective Theory: Color-Suppressed Decays

Sonny Mantry*, Dan Pirjol* and Iain W. Stewart[1] *

*Center for Theoretical Physics, Laboratory for Nuclear Science and Department of Physics,
Massachusetts Institute for Technology, Cambridge, MA 02139

Abstract. In this talk, we discuss the soft-collinear effective theory and the role of kinematic expansions in B decays. In particular we focus on reviewing the recent results for $\bar{B}^0 \to D^{(*)0}\pi^0$, as well as other color-suppressed decays that are not yet tested by the data.

1. INTRODUCTION

The soft-collinear effective theory [1–4] (SCET) provides a formalism for systematically investigating processes with both energetic and soft hadrons based solely on the underlying structure of QCD. Essentially all known methods for simplifying QCD predictions, without introducing model dependent assumptions, depend on exploiting hierarchies of mass scales. For predictions based on SU(3) symmetry, we exploit the fact that $m_{u,d,s}/\Lambda \ll 1$, and expect corrections at the $\sim 30\%$ level. In lattice QCD simulations, we choose our lattice spacing $a \ll 1/\Lambda$ and volume $V \gg 1/\Lambda^3$ so that we can focus on non-perturbative effects at scales $\sim \Lambda$. In SCET, we expand in $\Lambda/Q \ll 1$, with the large momentum of an energetic hadron or jet being $\sim Q$. For B decays, corrections will be at the $\sim 20\text{-}30\%$ level depending on the energy scale Q.

Most effective theories that we are familiar with are designed to separate the physics for hard $p_h^2 \simeq Q^2$ and soft $p_s^2 \ll Q^2$ momenta. Examples include the electroweak Hamiltonian, chiral perturbation theory, heavy quark effective theory, and non-relativistic QCD. In SCET, we incorporate an additional possibility, namely energetic hadrons where the constituents have momenta p_c^μ nearly collinear to a light-like direction n^μ. Both the energetic hadron and its collinear constituents have $\bar{n} \cdot p_c \sim Q$, where we have made use of light-cone coordinates $(p_c^+, p_c^-, p_c^\perp) = (n \cdot p_c, \bar{n} \cdot p_c, p_c^\perp)$. The collinear constituents still have small offshellness $p_c^2 \sim p_s^2$. The process of disentangling the interactions of hard-collinear-soft particles is known as factorization, and is simplified by the SCET framework.

Much like any effective theory, the basic ingredients of SCET are its field content, power counting and symmetries. The Lagrangian and operators, are organized in a series where only $\mathscr{L}^{(0)}$ and $\mathscr{O}^{(0)}$ are relevant at LO, an additional $\mathscr{L}^{(1)}$ or $\mathscr{O}^{(1)}$ is needed at NLO, etc. The expansion parameter will be $\lambda = \sqrt{\Lambda_{\rm QCD}/Q}$ or $\eta = \Lambda_{\rm QCD}/Q$ depending on whether the collinear fields describe an energetic jet of hadrons or an individual energetic hadron. The effective theory with an expansion in λ is called SCET$_{\rm I}$, while the one with an expansion in η is called SCET$_{\rm II}$. In processes such as color-suppressed decays, the separation of scales is $Q^2 \gg Q\Lambda \gg \Lambda^2$ and the chain QCD–SCET$_{\rm I}$–SCET$_{\rm II}$ proves to be useful. The intermediate theory SCET$_{\rm I}$ provides the dynamics to re-arrange soft and collinear quark lines so that they can end up in soft and energetic hadrons. The final theory SCET$_{\rm II}$ describes the universal low energy hadronic matrix elements. In the case of color-suppressed decays $B \to D^{(*)}M$, these are light-cone distribution functions $\phi_M(x)$ where $M = \pi, \rho, K$, or K^* and two generalized parton distribution functions $S^{(0,8)}(k_1^+, k_2^+)$ for the $B \to D^{(*)}$ transition.

2. COLOR-SUPPRESSED DECAYS AND SCET

Color-suppressed decays were investigated in Ref. [5] using SCET. For $B \to D\pi$ decays the four quark operators which contribute are [$\Gamma = V - A$]

$$H_W = \frac{G_F V_{cb} V_{ud}^*}{\sqrt{2}} \left[C_1 (\bar{c}b)_\Gamma (\bar{d}u)_\Gamma + C_2 (\bar{c}_i b_j)_\Gamma (\bar{d}_j u_i)_\Gamma \right], \quad (1)$$

---

[1] Plenary talk given by I.S. at the 9th International Conference on B Physics at Hadron Machines (Beauty 2003).

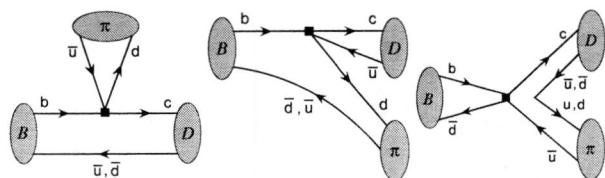

FIGURE 1. $B \rightarrow D\pi$ flavor topologies, T, C, and E, respectively. For the color-suppressed decays only C and E contribute.

with flavor contractions shown by the Fig. 1 diagrams. For the amplitudes, we use $A_{+-} = A(\bar{B}^0 \rightarrow D^+\pi^-)$, $A_{0-} = A(B^- \rightarrow D^0\pi^-)$, and $A_{00} = A(\bar{B}^0 \rightarrow D^0\pi^0)$. Written in terms of isospin amplitudes

$$
\begin{aligned}
A_{+-} &= T + E = \frac{1}{\sqrt{3}}A_{3/2} + \sqrt{\frac{2}{3}}A_{1/2}, \\
A_{0-} &= T + C = \sqrt{3}A_{3/2}, \\
A_{00} &= \frac{C - E}{\sqrt{2}} = \sqrt{\frac{2}{3}}A_{3/2} - \frac{1}{\sqrt{3}}A_{1/2}.
\end{aligned} \tag{2}
$$

The amplitudes for decays to $B \rightarrow D^{(*)}\rho$ are defined in a similar fashion.

In the large N_c limit, $C/T \sim E/T \sim 1/N_c$ (where we take $C_1 \sim 1$ and $C_2 \sim 1/N_c$). The color-allowed amplitudes A_{+-} and A_{0-} are described by a factorization theorem [6–8], proven with SCET [9]

$$
A^{(*)} = N^{(*)} \xi(w_{max}) \int_0^1 dx\, T^{(*)}(x, m_c/m_b)\, \phi_\pi(x) + \dots, \tag{3}
$$

where $\xi(w_{max})$ is the Isgur-Wise function at maximum recoil, $\phi_\pi(x)$ is the light-cone distribution function for the pion, $T = 1 + O(\alpha_s)$ is the hard scattering kernel, and $N^{(*)} = \frac{G_F}{\sqrt{2}}V_{cb}V_{ud}^* E_\pi f_\pi \sqrt{m_{D^{(*)}} m_B}(1 + m_B/m_{D^{(*)}})$. The ellipses in Eq. (3) denote terms suppressed by Λ/Q where $Q = \{m_b, m_c, E_\pi\}$. In the heavy quark limit, Eq. (3) predicts $A = A^*$, so $\mathscr{B}(\bar{B}^0 \rightarrow D^+\pi^-) = \mathscr{B}(\bar{B}^0 \rightarrow D^{*+}\pi^-)$ and $\mathscr{B}(B^- \rightarrow D^0\pi^-) = \mathscr{B}(B^- \rightarrow D^{*0}\pi^-)$. This agrees well with the experimental results [10, 11], which yield

$$
\begin{aligned}
\frac{\mathscr{B}(\bar{B}^0 \rightarrow D^{*+}\pi^-)}{\mathscr{B}(\bar{B}^0 \rightarrow D^+\pi^-)} &= 1.03 \pm 0.14, \\
\frac{\mathscr{B}(B^- \rightarrow D^{*0}\pi^-)}{\mathscr{B}(B^- \rightarrow D^0\pi^-)} &= 0.93 \pm 0.11.
\end{aligned} \tag{4}
$$

Eq. (3) also predicts $A_{+-} = A_{0-}$, however experimentally $|A_{0-}/A_{+-}| = 0.77 \pm 0.05$ for $D\pi$ and 0.81 ± 0.05 for $D^*\pi$. We will see that the reduction of these numbers from 1 are explained by a SCET power correction.

The color-suppressed amplitude A_{00} has contributions from C and E, but not T. With large N_c, very little can be said about the C and E contributions, besides the fact

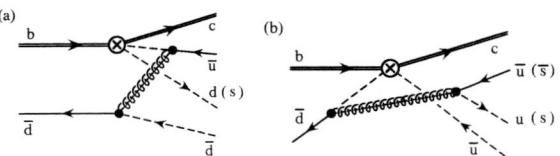

FIGURE 2. Diagrams in SCET$_I$ for tree level matching. $\otimes$ denotes the operator $Q_j^{(0,8)}$ and the dots are insertions of $\mathscr{L}_{\xi q}^{(1)}$. The solid lines and double solid lines carry momenta $p^\mu \sim \Lambda$ and form the B and D. The dashed lines are energetic collinear quarks that form the light meson M.

that we expect $A_{00} < A_{+-} \sim A_{0-}$. In SCET, the amplitudes C and E are suppressed by Λ/E_π relative to T. Despite this power suppression, predictive power is retained since only a single type of SCET$_I$ time ordered product contributes to give the proper quark rearrangement, $T(Q_j^{(0,8)}(0), i\mathscr{L}_{\xi q}^{(1)}(x), i\mathscr{L}_{\xi q}^{(1)}(y))$ [5]. This combination contributes to both C and E as shown in Fig. 2.

When matched onto SCET$_{II}$, the time-ordered product gives a product of soft $O_s^{(0,8)}$ and collinear O_c operators. Thus $\langle D^{(*)}\pi|O_s^{(0,8)}O_c|B\rangle = \langle D^{(*)}|O_s^{(0,8)}|B\rangle\langle\pi|O_c|0\rangle$. The soft operator $[P_L = (1 - \gamma_5)/2]$

$$
O_s^{(0)} = (\bar{h}_{v'}^{(c)}S)\slashed{n}P_L(S^\dagger h_v^{(b)})(\bar{d}S)_{k_1^+}\slashed{n}P_L(S^\dagger u)_{k_2^+}, \tag{5}
$$

while $O_s^{(8)}$ is identical but with color structure $T^A \otimes T^A$. In addition, there are operators encoding "long" distance contributions in SCET$_{II}$ that are the same order in Λ/Q. These come from the region of momentum space for Fig. 2 where the gluon still has $p^2 \sim Q\Lambda$, but the quark propagator has $p^2 \sim \Lambda^2$.

Using heavy quark symmetry one can prove

$$
\langle D^{(*)}|O_s^{(0,8)}|B\rangle = S_L^{(0,8)}(k_1^+, k_2^+), \tag{6}
$$

so that the matrix elements for $\bar{B}^0 \rightarrow D^0\pi^0$ and $\bar{B}^0 \rightarrow D^{*0}\pi^0$ are the same [5]. Furthermore, $S_L^{(0,8)}$ are complex from their dependence on n^μ, the *direction* of the light meson, and encode a non-perturbative strong phase shift. This leads to the predictions

$$
\begin{aligned}
\delta(D^*\pi) &= \delta(D\pi), \\
\mathscr{B}(\bar{B}^0 \rightarrow D^{*0}\pi^0) &= \mathscr{B}(\bar{B}^0 \rightarrow D^0\pi^0),
\end{aligned} \tag{7}
$$

where $\delta = \arg(A_{1/2}A_{3/2}^*)$ is the strong phase shift between isospin amplitudes. The predictions in Eq. (7) have corrections at $\mathscr{O}(\alpha_s(Q))$ and $\mathscr{O}(\Lambda/Q)$. For $M = \pi^0, \rho^0$ the long distance amplitude is suppressed by $\alpha_s(Q)$. The current experimental data [12–14] gives the following world averages

$$
\begin{aligned}
\mathscr{B}(D^{*0}\pi^0) &= 0.29 \pm 0.03, \quad \delta(D^*\pi) = 31.0 \pm 5.0°, \\
\mathscr{B}(D^0\pi^0) &= 0.26 \pm 0.05, \quad \delta(D\pi) = 30.4 \pm 4.8°,
\end{aligned} \tag{8}
$$

142

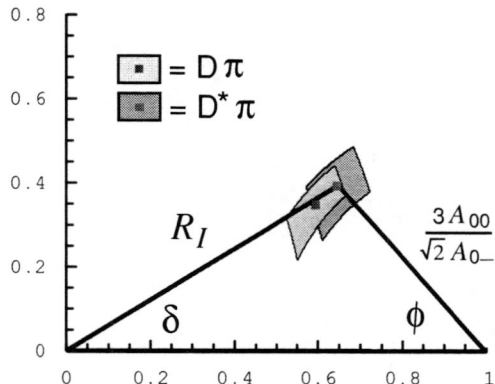

FIGURE 3. Experimental data for the $D\pi$ and $D^*\pi$ isospin triangles. This figure updates the one in Ref. [5] to include the recent BaBar data.

showing good agreement with Eq. (7). This agreement can also be shown graphically. The isospin relation between amplitudes implies that

$$1 = R_I + \frac{3}{\sqrt{2}}\frac{A_{00}}{A_{0-}}, \qquad (9)$$

where $R_I = A_{1/2}/(\sqrt{2}A_{3/2}) = (A_{+-} - A_{00}/\sqrt{2})/A_{0-}$. Eq. (9) can be represented by a triangle in the complex plane. The current world averages for $D\pi$ and $D^*\pi$ are shown in Fig. 3, where the overlap of the 1σ regions indicates the agreement.

It is useful to note that Eq. (7) provides a sensitive test of SCET-factorization, and not just heavy quark symmetry. The basic reason is that a priori "soft" gluon exchange between the b or c and the light quarks in the pion spoils the prediction. To see this more clearly, we can consider using just HQET with full QCD for the light quarks. In this case, the amplitude would be

$$\langle D^{(*)0}\pi^0|(\bar{h}_{v'}^{(c)}\gamma^\mu P_L h_v^{(b)})(\bar{d}\gamma_\mu P_L u)|B^0\rangle/\sqrt{m_B m_D}$$
$$= \text{Tr}[\bar{H}_{v'}^{(c)}\gamma^\mu P_L H_v^{(b)} X_\mu] \qquad (10)$$
$$+ \frac{1}{m_c}\text{Tr}[\bar{H}_{v'}^{(c)} i\sigma^{\alpha\beta}\frac{1+\slashed{v}'}{2}\gamma^\mu P_L H_v^{(b)} R_{\mu\alpha\beta}] + \cdots$$

where H_v and $H_{v'}$ are HQET superfields and X_μ and $R_{\mu\alpha\beta}$ are the most general tensor functions compatible with the symmetries of QCD. Here, the $R_{\mu\alpha\beta}$ term has a chromomagnetic operator insertion, $\bar{h}_{v'}\sigma_{\alpha\beta}G^{\alpha\beta}h_{v'}$ on the charm quark. Usually in HQET the $R_{\mu\alpha\beta}$ term would be suppressed relative to the X_μ term. However, in Eq. (10) the pion momentum $p_\pi^\mu = E_\pi n^\mu$ is an allowed four-vector in $R_{\mu\alpha\beta}$. Since $E_\pi/m_c \simeq 1.5$ the two terms are the same size (and this will also be the case for all other terms in the $1/m_c$ heavy quark expansion, ie., the expansion does not converge). Since $E_\pi \simeq 2.3$ GeV,

the "soft" gluons are carrying hard momenta. Terms like $R_{\mu\alpha\beta}$ break the heavy quark spin symmetry and give $A(\bar{B}^0 \to D^{*0}\pi^0) \neq A(\bar{B}^0 \to D^0\pi^0)$. In contrast, with SCET we can expand in Λ/E_π and factorize away the energetic pion. Thus the matrix element in Eq. (6) has no E_π dependence and is part of a convergent expansion.

The SCET analysis also gives predictions for several channels where the data is not yet available. For instance, the analysis above also applies for the ρ, predicting

$$\delta(D^*\rho) = \delta(D\rho), \qquad (11)$$
$$\mathscr{B}(\bar{B}^0 \to D^{*0}\rho^0) = \mathscr{B}(\bar{B}^0 \to D^0\rho^0).$$

A similar prediction can be made for decays to $D_s^{(*)}K^{(*)}$ except in this case the long distance contributions to the amplitudes are not suppressed. This means that both longitudinal and perpendicular polarizations occur at the same order. The analog of Eq. (11) is therefore:

$$\mathscr{B}(\bar{B}^0 \to D_s^* K^-) = \mathscr{B}(\bar{B}^0 \to D_s K^-), \quad (12)$$
$$\mathscr{B}(\bar{B}^0 \to D_s^* \bar{K}_\parallel^{*-}) = \mathscr{B}(\bar{B}^0 \to D_s \bar{K}_\parallel^{*-}),$$

where these color-suppressed decays are not part of an isospin triangle. Cabibbo suppressed decays to kaons are more analogous to $D\pi$ and $D\rho$, except that they also have long distance contributions which are not suppressed. In this case the analog of Eq. (11) is

$$\delta(D^* \bar{K}^0) = \delta(D\bar{K}^0), \qquad (13)$$
$$\mathscr{B}(\bar{B}^0 \to D^{*0}\bar{K}^0) = \mathscr{B}(\bar{B}^0 \to D^0\bar{K}^0),$$
$$\delta(D^* \bar{K}_\parallel^{*0}) = \delta(D\bar{K}_\parallel^{*0}),$$
$$\mathscr{B}(\bar{B}^0 \to D^{*0}\bar{K}_\parallel^{*0}) = \mathscr{B}(\bar{B}^0 \to D^0\bar{K}_\parallel^{*0}).$$

The predictions in Eqs. (11), (12), and (13) will be tested once data on $\bar{B}^0 \to D^{*0}\rho^0$, $\bar{B}^0 \to D_s^* K^{(*)-}$, and $\bar{B}^0 \to D^{*0}\bar{K}^{(*)0}$ become available. The significance of the long distance terms will be tested by comparing $\mathscr{B}(\bar{B}^0 \to D_{s\parallel}^* \bar{K}_\parallel^{*-})$ to $\mathscr{B}(\bar{B}^0 \to D_{s\perp}^* \bar{K}_\perp^{*-})$ or $\mathscr{B}(\bar{B}^0 \to D_\parallel^{*0}\bar{K}_\parallel^{*0})$ to $\mathscr{B}(\bar{B}^0 \to D_\perp^{*0}\bar{K}_\perp^{*0})$.

The full factorization theorem for color suppressed decays takes the form [5]

$$A_{00}^{D^{(*)}M} = N_0^M \int dx\, dz\, dk_1^+ dk_2^+ T_{L\mp R}^{(i)}(z) J^{(i)}(z, x, k_1^+, k_2^+)$$
$$\times S^{(i)}(k_1^+, k_2^+)\, \phi_M(x) + A_{long}^{D^{(*)}M}. \qquad (14)$$

where $T_{L\mp R}^{(i)}$ are hard scattering kernels and $N_0^M = G_F V_{cb} V_{ud}^* f_M \sqrt{m_B m_{D^{(*)}}}/2$. The non-perturbative dynamics is contained in ϕ_M, the light-cone distribution function for meson M, and $S^{(i)}$, $i = 0, 8$, a generalized parton distribution function for the $B \to D^{(*)}$ transition with k_1^+ and k_2^+ being momentum fractions of the light spectator quarks. Finally, the jet function $J^{(i)}$ is sensitive to physics at the $\mu^2 \sim E_M \Lambda$ scale and is responsible for the quark rearrangement.

The predictions discussed above are all valid independent of the form of $J^{(i)}$, meaning to all orders in $\alpha_s(\mu_0)$ at the intermediate scale, $\mu_0^2 \sim E_M \Lambda$. If we expand J in powers of $\alpha_s(\mu_0)$, then this introduces additional uncertainty, but gives further predictions. At lowest order, we have

$$A_{00}^{D^{(*)}M} = N_0^M C_L^{(0)} \frac{16\pi\alpha(\mu_0)}{9} s_{\text{eff}}(\mu_0)\langle x^{-1}\rangle_M, \quad (15)$$

where $C_L^{(0)} = C_1 + C_2/3$, $\langle x^{-1}\rangle_M = \int dx/x\, \phi_M(x)$, and $s_{\text{eff}} = -s^{(0)} + C_2/(4C_1 + 4/3C_2)s^{(8)}$ with $s^{(0,8)} = \int dk_1^+ dk_2^+/(k_1^+ k_2^+)\, S^{(0,8)}(k_1^+,k_2^+)$. Corrections to Eq. (15) are $O(\alpha_s(\mu_0))$ and $O(\Lambda/Q)$. At this order, the strong phase ϕ in Fig. 3 is generated by s_{eff} and so is independent of whether $M = \pi$ or $M = \rho$. Therefore, we predict that ϕ is universal for $D^{(*)}\pi$ and $D^{(*)}\rho$.

For the ratio of charged amplitudes, Eq. (15) can be used to predict the leading power correction,

$$\begin{aligned} R_c^{D^{(*)}M} &= \frac{A(\bar{B}^0 \to D^{(*)+}M^-)}{(B^- \to D^{(*)0}M^-)} \quad (16)\\ &= 1 - \frac{16\pi\alpha_s m_{D^{(*)}}}{9(m_B + m_{D^{(*)}})} \frac{\langle x^{-1}\rangle_M}{\xi(\omega_0)} \frac{s_{\text{eff}}}{E_M}. \end{aligned}$$

A value $s_{\text{eff}} \simeq (430\ \text{MeV})e^{i44°}$ gives $|R_c^{D\pi}| \simeq 0.8$, fitting the experimental values given below Eq. (4) with parameters of natural size. In naive factorization, the correction term in R_c would depend on the decay constant f_M, however for the true factorization theorem that gives Eq. (16) this turns out not to be the case. The observed similarity between R_c for $D\pi$ and $D\rho$ can be explained by having $\langle x^{-1}\rangle_\pi \simeq \langle x^{-1}\rangle_\rho$ and is not spoiled by the fact that $f_\rho/f_\pi \simeq 1.6$. Experimentally [11]

$$\frac{|R_c^{D\pi}|}{|R_c^{D\rho}|} = 0.96 \pm 0.13. \quad (17)$$

With this approximate equality and the $\phi^\pi = \phi^\rho$ prediction, we would expect that the strong phase $\delta^{D\rho} \simeq \delta^{D\pi}$, or in other words that the $D\pi$ and $D\rho$ triangles (as in Fig. 3) will be similar. If this turns out not to be the case, then it would indicate that there are substantial $\alpha_s^2(\mu_0)$ corrections to $J^{(0,8)}$. This would mean that the subset of predictions that follow from Eq. (15), which depend on a perturbative expansion for $J^{(0,8)}$, should not be trusted.

3. BARYON DECAYS

Recently, the authors in Ref. [15] have used SCET to make model-independent factorization predictions for

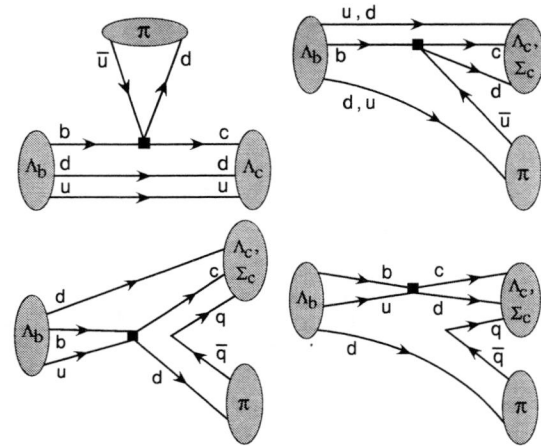

FIGURE 4. Classes of diagrams for Λ_b decays, giving amplitudes T (tree) and C (color-commensurate) for the top two diagrams, and E (exchange), and B (bow-tie) for the bottom two. Bow-tie diagrams are unique to baryon decays [15].

baryon decays. The main results for $\Lambda_b \to \Lambda_c\pi$, $\Lambda_c\rho$, $\Sigma_c^{(*)}\pi$, $\Sigma_c^{(*)}\rho$, and $\Xi_c^{(\prime,*)}K$ are briefly summarized here. The notation is $\Sigma_c = \Sigma_c(2455)$ and $\Sigma_c^* = \Sigma_c(2520)$.

The electroweak Hamiltonian for these baryon decays is in Eq. (1). The diagrams for the flavor contractions differ from the meson decays and are shown in Fig. 4. The decays to Λ_c get contributions from T, C, E, and B, decays to $\Sigma_c^{(*)}$ have contributions from C, E, and B, and decays to Ξ_c only from the E and B amplitudes.

In the large N_c limit $C/T \sim E/T \sim N_c^0$, while $B/T \sim N_c$. Thus, even $\Lambda_b \to \Lambda_c\pi$ decays do not factorize in the large N_c limit. The extra factors of N_c arise from the choice of which of the N_c quarks in the Λ_b and/or Λ_c participate in the weak interaction.

Expanding in Λ/Q where $Q = \{m_b, m_c, E_\pi\}$ using SCET, one finds $C/T \sim E/T \sim \Lambda/Q$ and $B/T \lesssim \Lambda^2/Q^2$. For $\Lambda_b \to \Lambda_c\pi$ the leading order result is from T and gives

$$\begin{aligned} A^{\Lambda_c\pi} &= N_L\, \zeta(w_{max}) \int_0^1 dx\, T_L(x, m_c/m_b)\, \phi_\pi(x) \quad (18)\\ &\quad + N_R\, \zeta(w_{max}) \int_0^1 dx\, T_R(x, m_c/m_b)\, \phi_\pi(x) + \dots, \end{aligned}$$

where $\zeta(w_{max})$ is the $\Lambda_b \to \Lambda_c$ Isgur-Wise function at maximum recoil, $T_{L,R}$ are hard scattering kernels, and $N_{L,R} = \sqrt{2}G_F V_{cb}V_{ud}^* E_\pi f_\pi \sqrt{m_{\Lambda_c} m_{\Lambda_b}}\, \bar{u}(v')\slashed{n} P_{L,R} u(v)$ with the normalization $\bar{u}(v)u(v) = 2$. The other factors are the analogs of those in Eq. (3). The value of $\zeta(w_{max})$ will be determined by q^2 spectrum measurements of $\Lambda_b \to \Lambda_c \ell\bar{\nu}_\ell$ which are not yet available. If $\zeta(w_{max})$ is similar to $\xi(w_{max})$, then Eq. (18) predicts that $\mathcal{B}(\Lambda_b \to \Lambda_c\pi) > \mathcal{B}(\bar{B}^0 \to D^+\pi^-)$ in agreement with the measurements from CDF [16].

For baryons, the analog of the color-suppressed decays $\bar{B}^0 \to D^{(*)0}\pi^0$ are $\Lambda_b \to \Sigma_c^{(*)0}\pi^0$ and its isospin partner $\Lambda_b \to \Sigma_c^{(*)+}\pi^-$. The leading amplitudes are $C \sim E$, while B is suppressed by an additional Λ/Q. Using heavy quark symmetry on $\Lambda_b \to \Sigma_c^{(*)}$ matrix elements of the SCET operators $O_s^{(0,8)}$ gives [15]

$$\frac{\mathscr{B}(\Lambda_b \to \Sigma_c^* \pi)}{\mathscr{B}(\Lambda_b \to \Sigma_c \pi)} = 2, \tag{19}$$

up to corrections suppressed by Λ/Q or $\alpha_s(Q)$. Here $\Sigma_c^{(*)}\pi = \Sigma_c^{(*)0}\pi^0$ or $\Sigma_c^{(*)+}\pi^-$. A similar prediction is also made for decays to a ρ,

$$\frac{\mathscr{B}(\Lambda_b \to \Sigma_c^* \rho)}{\mathscr{B}(\Lambda_b \to \Sigma_c \rho)} = 2. \tag{20}$$

Using $\Sigma_c^{(*)0}\rho^0 \to \Lambda_c \pi^- \pi^+ \pi^-$, Eq. (20) may be easier to test experimentally than Eq. (19). For decays involving cascades there can be sizable long distance contributions, but we still expect

$$\frac{\mathscr{B}(\Lambda_b \to \Xi_c^* K)}{\mathscr{B}(\Lambda_b \to \Xi_c' K)} = 2,$$

$$\frac{\mathscr{B}(\Lambda_b \to \Xi_c^* K_\parallel^*)}{\mathscr{B}(\Lambda_b \to \Xi_c' K_\parallel^*)} = 2. \tag{21}$$

The branching ratio $\mathscr{B}(\Lambda_b \to \Xi_c K)$ is also expected to be of the same order of magnitude since it occurs at this order in the power counting.

4. CONCLUSION

In this talk, we have reviewed the SCET predictions for non-leptonic decays with charmed hadrons in the final state from Refs. [5, 15]. This included the decays $\bar{B}^0 \to D^{(*)+}\pi^-$, $B^- \to D^{(*)0}\pi^-$, and $\Lambda_b \to \Lambda_c \pi$ which occur at leading order, as well as decays which are power suppressed, $\bar{B}^0 \to D^{(*)0}\pi^0$ and $\Lambda_b \to \Sigma_c^{(*)}\pi$. Analogous decays where the π is replaced by a ρ or kaon were also discussed. For $\bar{B}^0 \to D^{(*)0}\pi^0$, we updated the experimental comparison in Fig. 3 and Eq. (8) to take into account the new BaBar results [14].

ACKNOWLEDGMENTS

I.S. would like to thank A. Leibovich, Z. Ligeti, and M. Wise for collaboration on the baryon results discussed here. This work was supported in part by the Department of Energy under the cooperative research agreement DF-FC02-94ER40818. I.S. was also supported by a DOE Outstanding Junior Investigator award.

REFERENCES

1. Bauer, C. W., Fleming, S., and Luke, M. E., *Phys. Rev.* **D63**, 014006 (2001).
2. Bauer, C. W., Fleming, S., Pirjol, D., and Stewart, I. W., *Phys. Rev.* **D63**, 114020 (2001).
3. Bauer, C. W., and Stewart, I. W., *Phys. Lett.* **B516**, 134–142 (2001).
4. Bauer, C. W., Pirjol, D., and Stewart, I. W., *Phys. Rev.* **D65**, 054022 (2002).
5. Mantry, S., Pirjol, D., and Stewart, I. W., *Phys. Rev.* **D68**, 114009 (2003).
6. Dugan, M. J., and Grinstein, B., *Phys. Lett.* **B255**, 583–588 (1991).
7. Politzer, H. D., and Wise, M. B., *Phys. Lett.* **B257**, 399–402 (1991).
8. Beneke, M., Buchalla, G., Neubert, M., and Sachrajda, C. T., *Nucl. Phys.* **B591**, 313–418 (2000).
9. Bauer, C. W., Pirjol, D., and Stewart, I. W., *Phys. Rev. Lett.* **87**, 201806 (2001).
10. Ahmed, S., et al., [CLEO Collaboration], *Phys. Rev.* **D66**, 031101 (2002).
11. Hagiwara, K., et al., [Particle Data Group Collaboration], *Phys. Rev.* **D66**, 010001 (2002).
12. Coan, T. E., et al., [CLEO Collaboration], *Phys. Rev. Lett.* **88**, 062001 (2002).
13. Abe, K., et al., [Belle Collaboration], *Phys. Rev. Lett.* **88**, 052002 (2002).
14. Aubert, B., et al., [BaBar Collaboration], *Phys. Rev.* **D69**, 032004 (2004).
15. Leibovich, A. K., Ligeti, Z., Stewart, I. W., and Wise, M. B., arXiv:hep-ph/0312319 (2003).
16. CDF Collaboration, Internal note 6396 (2003), URL: http://www-cdf.fnal.gov/physics/new/bottom/030702.blessed-lblcpi-ratio.

Semileptonic Branching Ratios and Moments

Giuseppe Della Ricca

University of Trieste and INFN, I-34127, Trieste (TS), Italy

Abstract. Recent B physics results from the BaBar, Belle, CLEO, and LEP collaborations are reviewed. In particular, results on semileptonic B meson branching ratios and spectral moments are presented.

1. INTRODUCTION

Studying the decays of B hadrons enables important tests of the Standard Model and Heavy Quark Effective Theory to be made. A few recent experimental results on semileptonic B meson branching ratios and spectral moments are presented in this paper.

Measuring the semileptonic decay of B mesons to narrow orbitally-excited charm mesons is of particular interest, as the sum of the measured exclusive semileptonic B decay branching ratios is less than the measured inclusive branching ratio. The uncertainty in $\mathcal{B}(B \to D^{**}\ell\bar{\nu}X)$ is also a large source of uncertainty in measurements of $|V_{cb}|$. Thus, it is desirable to measure this decay as precisely as possible.

Moments of the hadronic mass spectrum and of the lepton energy spectrum are sensitive to the masses of the b and c quarks as well as to the non-perturbative parameters of the Heavy Quark Expansion, which are needed to extract the fundamental parameters $|V_{ub}|$ and $|V_{cb}|$ from the measured rates.

2. SEMILEPTONIC B DECAYS TO D^{**0} WITH THE OPAL DETECTOR

The OPAL collaboration studied the two neutral narrow D^{**0} states: D_1^0 ($J^P = 1^+$) and D_2^{*0} ($J^P = 2^+$). Events from the decay chain $b \to \bar{B} \to D^{**0}\ell\bar{\nu}X, D^{**0} \to D^{*+}\pi^-, D^{*+} \to D^0\pi_{slow}^+, D^0 \to (K\pi \text{ or } K3\pi)$ are reconstructed requiring a high momentum lepton (e, μ) candidate. An unbinned maximum likelihood fit of the Δm^{**} distribution is performed to determine the semileptonic B to D^{**0} branching ratios (Fig. 1). The largest source of systematic uncertainty are the parameterizations of the signal and background functions. The fit gives the product branching ratio [1]: $\mathcal{B}(b \to \bar{B}) \times \mathcal{B}(\bar{B} \to D_1^0\ell\bar{\nu}X) \times \mathcal{B}(D_1^0 \to D^{*+}\pi^-) = (2.64 \pm 0.79(\text{stat}) \pm 0.39(\text{syst})) \times$

10^{-3}. No evidence is found for decays into the $J^P = 2^+$ state, and a limit is set on the product branching ratio: $\mathcal{B}(b \to \bar{B}) \times \mathcal{B}(\bar{B} \to D_2^{*0}\ell\bar{\nu}X) \times \mathcal{B}(D_2^{*0} \to D^{*+}\pi^-) < 1.4 \times 10^{-3}$ at the 95% confidence level. The result is consistent with similar results from ARGUS [2], CLEO [3], and other LEP experiments [4, 5].

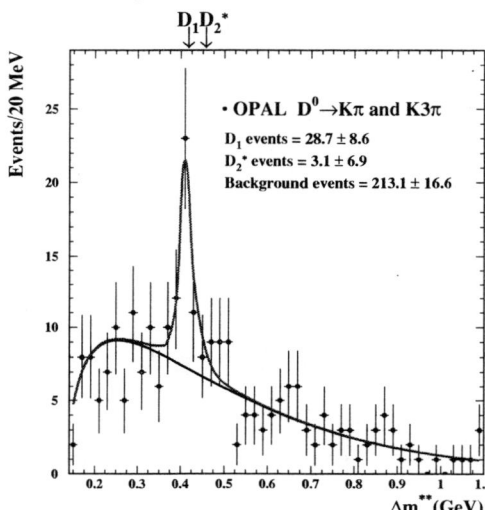

FIGURE 1. Mass difference Δm^{**} between D^{**0} and D^{*+} candidates. The expected values for the D_2^{*0} and D_1^0 states are shown. The curves show the signal plus background and separate background shapes.

3. MEASUREMENT OF $|V_{cb}|$ FROM $\bar{B}^0 \to D^{*+}\ell^-\bar{\nu}_\ell$

3.1. Using the *BABAR* Detector

A preliminary measurement of $|V_{cb}|$ and the branching fraction $\mathcal{B}(\bar{B}^0 \to D^{*+}\ell^-\bar{\nu}_\ell)$ has been performed by the BaBar collaboration, based on a sample of about 55,700 $\bar{B}^0 \to D^{*+}\ell^-\bar{\nu}_\ell$ decays. The decays are iden-

CP722, *B Physics at Hadron Machines*, edited by M. Paulini and S. Erhan
© 2004 American Institute of Physics 0-7354-0203-5/04/$22.00

tified in the $D^{*+} \to D^0 \pi^+$ final state, with the D^0 reconstructed in three different decay modes ($K\pi$, $K3\pi$, $K\pi\pi^0$). The differential decay rate is measured as a function of the relativistic boost of the D^{*+} in the $\bar{B}^0$ rest frame (Fig. 2). The value of the differential decay rate at 'zero recoil', namely the point at which the D^{*+} is at rest in the $\bar{B}^0$ rest frame, is predicted in Heavy Quark Effective Theory as a kinematic factor times $\mathscr{F}(1)|V_{cb}|$, where $\mathscr{F}$ is the unique form factor governing the decay. The measured differential decay rate is extrapolated to the zero recoil point, obtaining $\mathscr{F}(1)|V_{cb}| = (34.03 \pm 0.24 \pm 1.31) \times 10^{-3}$, $\rho_{A_1}^2 = 1.23 \pm 0.02 \pm 0.28$ [6]. Using $\mathscr{F}(1) = 0.913^{+0.030}_{-0.035}$, the value $|V_{cb}| = (37.27 \pm 0.26(\text{stat}) \pm 1.43(\text{syst})^{+1.5}_{-1.2}(\text{theo})) \times 10^{-3}$ was obtained. From the integrated decay rate, the branching fraction $\mathscr{B}(\bar{B}^0 \to D^{*+}\ell^-\bar{\nu}_\ell) = (4.68 \pm 0.03 \pm 0.29)\%$ has been derived.

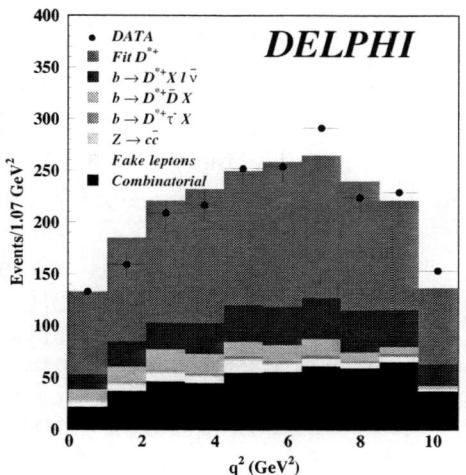

FIGURE 3. The experimental q^2 (the square of the four momentum transfer from the $\bar{B}^0$ to the D^{*+}) distribution compared to the fit result.

3.2. Using the DELPHI Detector

Measurements of $\mathscr{F}(1)|V_{cb}|$, $\rho_{A_1}^2$ and $\mathscr{B}(\bar{B}^0 \to D^{*+}\ell^-\bar{\nu}_\ell)$ have been obtained using exclusively reconstructed D^{*+} decays by the DELPHI collaboration. Variables have been defined which allow different decay mechanisms producing D^{*+} mesons in the final states to be separated. The following values have been obtained [7]: $\mathscr{F}(1)|V_{cb}| = (39.2 \pm 1.8 \pm 2.2) \times 10^{-3}$, $\rho_{A_1}^2 = 1.32 \pm 0.15 \pm 0.33$ (Fig. 3), which correspond to a branching fraction: $\mathscr{B}(\bar{B}^0 \to D^{*+}\ell^-\bar{\nu}_\ell) = (5.90 \pm 0.22 \pm 0.48)\%$. The b quark semileptonic branching fraction into a D^{*+} emitted from higher mass charmed excited states has been measured to be: $\mathscr{B}(b \to D^{*+}X\ell^-\bar{\nu}_\ell) = (0.67 \pm 0.08 \pm 0.10)\%$. Combining this measurement with the previous DELPHI results yields: $\mathscr{F}(1)|V_{cb}| = (37.7 \pm 1.1 \pm 1.9) \times 10^{-3}$, $\rho_{A_1}^2 = 1.39 \pm 0.10 \pm 0.33$, $\mathscr{B}(\bar{B}^0 \to D^{*+}\ell^-\bar{\nu}_\ell) = (5.39 \pm 0.11 \pm 0.33)\%$. Using $\mathscr{F}(1) = 0.91 \pm 0.04$, this gives: $|V_{cb}| = (41.4 \pm 1.2 \pm 2.1 \pm 1.8) \times 10^{-3}$, where the last uncertainty corresponds to the systematic error from theory.

4. SPECTRAL MOMENTS IN SEMILEPTONIC *B* DECAYS WITH THE DELPHI DETECTOR

The production characteristics of D^{**} mesons in B hadron semileptonic decays have been studied using exclusively reconstructed decay channels by the DELPHI collaboration. The corresponding branching fraction has been measured to

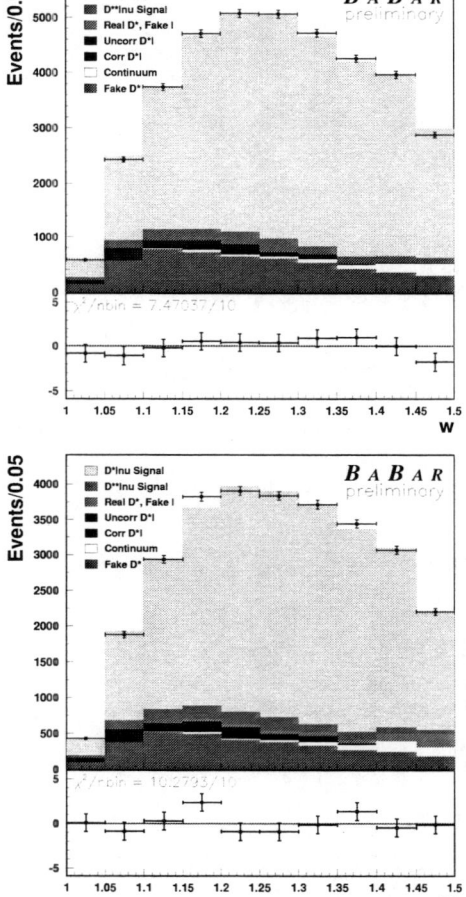

FIGURE 2. The experimental distributions of w (the product of the $\bar{B}^0$ and D^{*+} four velocities) compared to fit results, separately for electrons (top) and muons (bottom). The points are the data and the histograms are the results of the fit to the sum of the signal and the different sources of background. The fit residuals are also shown below.

be [8]: $\mathscr{B}(\bar{B}^0 \to D^{**+}\ell^-\bar{\nu}_\ell) = (2.7 \pm 0.7 \pm 0.2)\%$. The dominant contributing channel is the broad D_1^* state, whose mass and total width have been obtained as: $m_{D_1^*} = (2445 \pm 34 \pm 10)$ MeV/c^2, $\Gamma_{D_1^*} = (234 \pm 74 \pm 25)$ MeV/c^2. Moments of the total hadronic mass distribution have been measured to be: $< m_H^2 - m_{\bar{D}}^2 > = (0.647 \pm 0.046 \pm 0.093)$ GeV/c^2, $< (m_H^2 - m_{\bar{D}}^2)^2 > = (1.98 \pm 0.23 \pm 0.29)(\text{GeV}/c^2)^2$, $< (m_H^2 - < m_H^2 >)^2 > = (1.56 \pm 0.18 \pm 0.17)(\text{GeV}/c^2)^2$, $< (m_H^2 - < m_H^2 >)^3 > = (4.05 \pm 0.74 \pm 0.31)(\text{GeV}/c^2)^3$. Moments of the lepton energy spectrum in semileptonic B decays have also been measured to be: $< E_\ell^* > = (1.383 \pm 0.012 \pm 0.009)$ GeV, $< (E_\ell^* - < E_\ell^* >)^2 > = (0.19 \pm 0.005 \pm 0.008)$ GeV^2, $< (E_\ell^* - < E_\ell^* >)^3 > = (-0.029 \pm 0.005 \pm 0.006)$ GeV^3. These results have been interpreted in terms of constraints on the values of heavy quark masses, of the b quark kinetic energy and of the values of the parameters contributing at order $1/m_b^3$.

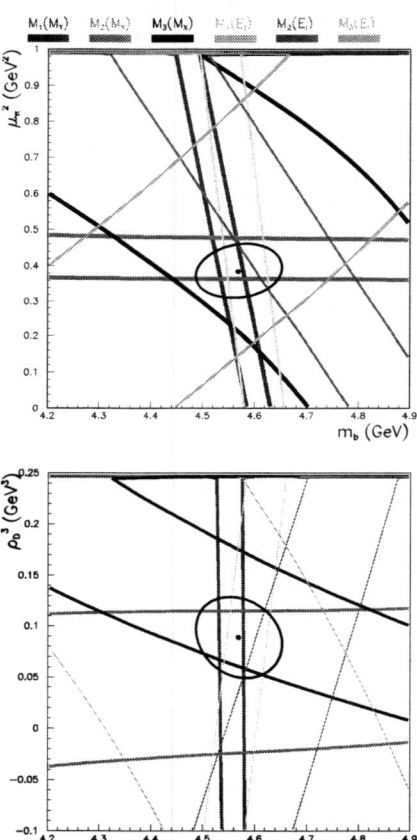

FIGURE 4. Projections of the constraints of the six measured moments on the $m_b - \mu_\pi^2$ (top) and $m_b - \mu_D^3$ plane (bottom). The ellipses represent the $1\,\sigma$ contours.

The comparison of these results with different measurements of the same parameters provides a test of the consistency of the theoretical predictions for inclusive semileptonic B decays and of the underlying assumptions (Fig. 4, 5). Using inclusive measurements of the B hadron lifetime and semileptonic branching fraction obtained at LEP, an accurate determination of the value of the $|V_{cb}|$ matrix element has been obtained: $|V_{cb}| = (42.4 \pm 0.6_{meas} \pm 0.8_{fit} \pm 0.4_{pert}) \times 10^{-3}$, where the first uncertainty reflects the accuracy on the semileptonic width determination.

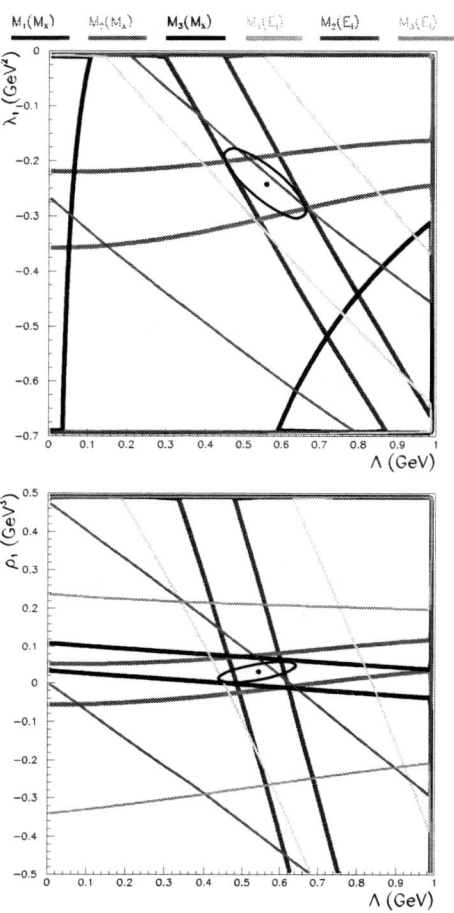

FIGURE 5. Projections of the constraints of the six measured moments on the $\bar{\Lambda}$-λ_1 (top) and $\bar{\Lambda}$-ρ_1 plane (bottom). The ellipses represent the $1\,\sigma$ contours.

5. HADRONIC MASS MOMENTS IN SEMILEPTONIC B DECAYS WITH THE BABAR DETECTOR

The BaBar collaboration performed a measurement of the first and second moment $< M_X >$ and $< M_X^2 >$ of the hadronic mass distribution in semileptonic B decays. These measurements have been carried out as a function of the threshold lepton momentum, p_{min}^*, extending from 0.9 GeV/c to 1.6 GeV/c, obtaining [9]:

$$< M_X >_{p^*_{min}=0.9 \text{ GeV}/c} = (2.072 \pm 0.013 \pm 0.013) \text{ GeV}/c^2,$$

$$< M_X^2 >_{p^*_{min}=0.9 \text{ GeV}/c} = (4.366 \pm 0.049 \pm 0.057) \text{ GeV},$$

$$< M_X >_{p^*_{min}=1.6 \text{ GeV}/c} = (2.026 \pm 0.013 \pm 0.011) \text{ GeV}/c^2,$$

$$< M_X^2 >_{p^*_{min}=1.6 \text{ GeV}/c} = (4.146 \pm 0.042 \pm 0.034) \text{ GeV},$$

where the uncertainties are from statistics and systematic effects, respectively.

From a fit to the measured hadronic moments $< M_X^2 >$, using the predictions of the HQE expansion [10] and the measured semileptonic decay width $\Gamma_{SL} = (4.37 \pm 0.18) \times 10^{-11}$ MeV, the parameters $\bar{\Lambda}^{\overline{MS}}$ and λ_1 (in the $\overline{MS}$ scheme), and $|V_{cb}|$, m_b^{1S}, and λ_1 (in the $1S$ scheme) were extracted, obtaining: $\bar{\Lambda}^{\overline{MS}} = 0.53 \pm 0.09$ GeV, $\lambda_1 = -0.36 \pm 0.09$ GeV^2, and $|V_{cb}| = (42.10 \pm 1.04 \pm 0.52 \pm 0.50) \times 10^{-3}$, $m_b^{1S} = 4.638 \pm 0.094 \pm 0.062 \pm 0.065$ GeV/c^2, $\lambda_1 = -0.26 \pm 0.06 \pm 0.04 \pm 0.04$ GeV^2, where the uncertainties are from experimental, theory, and $1/m_B^3$ effects, respectively (Figs. 6, 7 and 8).

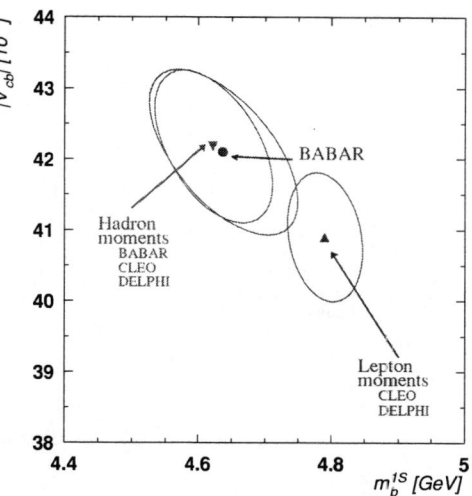

FIGURE 7. Constraints on the HQE parameters $|V_{cb}|$ and m_b^{1S} from hadronic moments as measured by BaBar [9], all hadronic moments combined (BaBar [9], CLEO [11] and DELPHI [12]), and the combined lepton moments (CLEO [11] and DELPHI [12]).

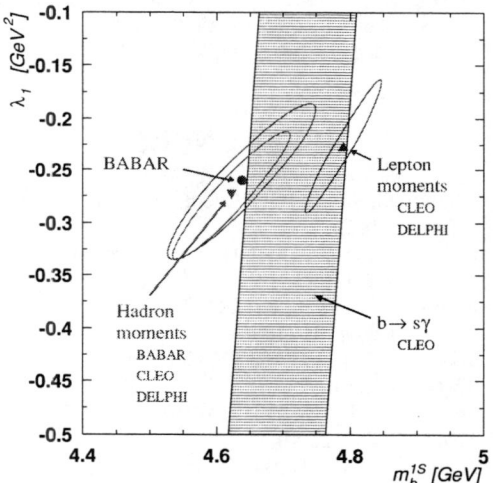

FIGURE 6. Constraints on the HQE parameters $\bar{\Lambda}^{\overline{MS}}$ and λ_1 from: the fit to the BaBar $< M_X^2 >$ measurements ($\Delta\chi^2 = 1$ ellipse) [9], the CLEO [11] and DELPHI [12] $< M_X^2 >$ measurements (1 σ bands), and the CLEO [11] $b \to s\gamma$ energy moment (1 σ band).

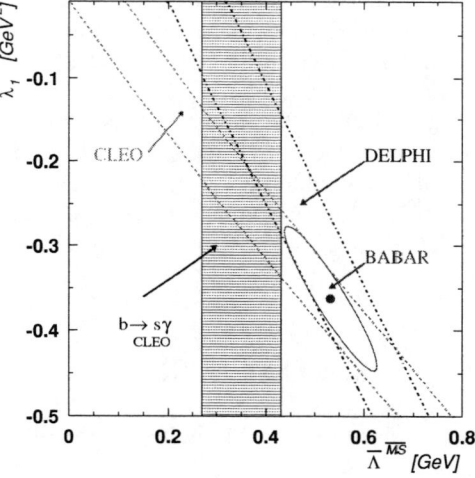

FIGURE 8. Constraints on the HQE parameters λ_1 and m_b^{1S} from hadronic moments as measured by BaBar [9], all hadronic moments combined (BaBar [9], CLEO [11] and DELPHI [12]), and the combined lepton moments (CLEO [11] and DELPHI [12]). Also indicated is the constraint from the $b \to s\gamma$ photon energy measured by CLEO [13].

6. HADRONIC RECOIL MASS MOMENTS IN SEMILEPTONIC *B* DECAYS WITH THE CLEO DETECTOR

The CLEO collaboration measured the hadronic recoil mass moments in semileptonic B decays obtaining the fractional contributions from the $B \to X_c \ell \nu$ processes with $X_c = D, D^*, D^{**}$, and non resonant X_c, and the pro-

cess $B \to X_u l \nu$. The measurement of the moment $< M_X^2 - \bar{M}_D^2 >$ was performed as a function of the lepton energy cut, with a minimum cut of 1.0 GeV, obtaining [14]: $< M_X^2 - \bar{M}_D^2 >_{E_\ell > 1.0 \text{ GeV}} = (0.456 \pm 0.014 \pm 0.045 \pm 0.109)$ GeV$^2/c^4$, and $< M_X^2 - \bar{M}_D^2 >_{E_\ell > 1.5 \text{ GeV}} = (0.293 \pm 0.012 \pm 0.033 \pm 0.048)$ GeV$^2/c^4$, where the uncertainties are from statistics, detector systematic effects, and model dependence, respectively.

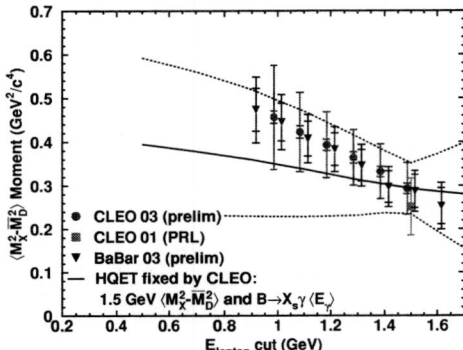

FIGURE 9. The CLEO results [14] compared to previous CLEO measurements [9, 11] and the HQET prediction [10].

7. ELECTRON MOMENTUM SPECTRUM ENDPOINT IN SEMILEPTONIC *B* DECAYS WITH THE BELLE DETECTOR

The Belle collaboration measured the inclusive charmless semileptonic *B* decay rate at the endpoint of the electron momentum spectrum in the $\Upsilon(4S)$ rest frame, in the momentum region $p^e_{CM} = 2.3 - 2.6$ GeV/c, which contains roughly 10% of the full phase space for $B \to X_u e v$ (Fig. 10). The value $\Delta\mathscr{B}(B \to X_u e v) = (1.19 \pm 0.11 \pm 0.10) \times 10^{-3}$ was obtained [15], where the uncertainties are statistical and systematic, respectively. The CKM matrix element $|V_{ub}| = (3.99 \pm 0.64) \times 10^{-3}$ is extracted from the result.

8. CONCLUSIONS

The topics discussed in this writeup represent only a small fraction of the recent results in heavy flavour physics. As this paper is being written, many other measurements are still being finalized. These measurements, performed using different approaches and experimental techniques, will allow us to gain deeper insight into experimental and theoretical details of how to best determine two of the fundamental parameters of the Standard Model, $|V_{ub}|$ and $|V_{cb}|$.

ACKNOWLEDGMENTS

I would like to thank the BaBar, Belle, CLEO, DELPHI and OPAL colleagues for providing the results reviewed here. I also wish to thank the organizers of the "Beauty 2003" conference for a very exciting and interesting event.

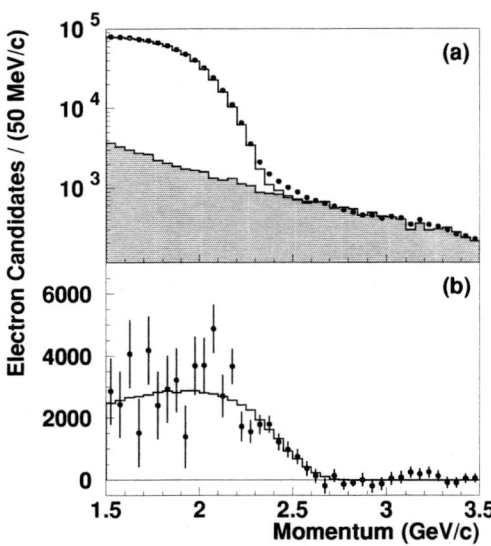

FIGURE 10. The electron momentum spectrum in the $\Upsilon(4S)$ rest frame: (a) on-resonance data (points), scaled off-resonance data (shaded histogram) and estimated $B\bar{B}$ backgrounds (unshaded histogram); (b) on-resonance data after subtraction of backgrounds and efficiency correction (points) and model spectrum $B \to X_u e v$ decays (histogram).

REFERENCES

1. Abbiendi, G., et al., [OPAL Collaboration], CERN-EP-2002-094 (2002).
2. Albrecht, H., et al., [ARGUS Collaboration], *Z. Phys.* **C57**, 533–540 (1993).
3. Anastassov, A., et al., [CLEO Collaboration], *Phys. Rev. Lett.* **80**, 4127–4131 (1998).
4. Buskulic, D., et al., [ALEPH Collaboration], *Z. Phys.* **C73**, 601–612 (1997).
5. Abreu, P., et al., [DELPHI Collaboration], *Phys. Lett.* **B475**, 407–428 (2000).
6. Aubert, B., et al., [BaBar Collaboration], arXiv:hep-ex/0308027 (2003).
7. Oyanguren, A., et al., [DELPHI Collaboration], DELPHI 2003-011 CONF 631 (2003).
8. Battaglia, M., et al., [DELPHI Collaboration], DELPHI 2003-028 CONF 648 (2003).
9. Aubert, B., et al., [BaBar Collaboration], arXiv:hep-ex/0307046 (2003).
10. Bauer, C. W., Ligeti, Z., Luke, M., and Manohar, A. V., *Phys. Rev.* **D67**, 054012 (2003).
11. Cronin-Hennessy, D., et al., [CLEO Collaboration], *Phys. Rev. Lett.* **87**, 251808 (2001).
12. Calvi, M., [DELPHI Collaboration], arXiv:hep-ex/0210046 (2002).
13. Chen, S., et al., [CLEO Collaboration], *Phys. Rev. Lett.* **87**, 251807 (2001).
14. Huang, G. S., et al., [CLEO Collaboration], arXiv:hep-ex/0307081 (2003).
15. Abe, K., et al., [Belle Collaboration], BELLE-CONF-0325 (2003).

Present Status of Our Knowledge of $|\mathbf{V_{cb}}|$

Marina Artuso

Syracuse University, Department of Physics, Syracuse, NY 13244, U.S.A.

Abstract. The Cabibbo-Kobayashi-Maskawa parameter $|V_{cb}|$ plays an important role in our effort to challenge the Yukawa sector of the Standard Model. The present status of our knowledge will be summarized with particular emphasis to the interplay between theoretical and experimental advances needed to improve upon present uncertainties.

1. INTRODUCTION

In the framework of the Standard Model, the quark sector is characterized by a rich pattern of flavor-changing transitions, described by the Cabibbo-Kobayashi-Maskawa (CKM) matrix

$$V_{CKM} = \begin{pmatrix} V_{ud} & V_{us} & V_{ub} \\ V_{cd} & V_{cs} & V_{cb} \\ V_{td} & V_{ts} & V_{tb} \end{pmatrix}. \qquad (1)$$

Since the CKM matrix must be unitary, it is determined by only four independent parameters. Wolfenstein proposed an approximate parameterization [1] that reflects the hierarchy between the magnitude of matrix elements belonging to different generations. Very frequently it is quoted in the approximation valid only to λ^3. We need to carry out this expansion further, in order to incorporate CP violation in neutral K decays. This expression, accurate to λ^3 for the real part and λ^5 for the imaginary part, is given by:

$$\begin{pmatrix} 1-\lambda^2/2 & \lambda & A\lambda^3(\rho - i\eta(1-\lambda^2/2)) \\ -\lambda & 1-\lambda^2/2 - i\eta A^2\lambda^4 & A\lambda^2(1+i\eta\lambda^2) \\ A\lambda^3(1-\rho-i\eta) & -A\lambda^2 & 1 \end{pmatrix}.$$

The Wolfenstein parameter λ is well measured as 0.2196 ± 0.0023 [2] and constraints exist on the parameters ρ and η from measurements of $|V_{ub}|$ and $B^0\bar{B}^0$ mixing. This paper focuses on the magnitude of the CKM element $|V_{cb}|$, related to the Wolfenstein parameter A [1].

2. $|\mathbf{V_{cb}}|$ FROM THE EXCLUSIVE DECAY $B \to D^* \ell \bar{\nu}$

HQET predicts that the differential partial decay width for this process, $d\Gamma/dw$, is related to $|V_{cb}|$ through:

$$\frac{d\Gamma}{dw}(B \to D^*\ell\nu) = \frac{G_F^2 |V_{cb}|^2}{48\pi^3} \mathcal{K}(w)\mathcal{F}(w)^2,$$

where w is the inner product of the B and D^* meson 4-velocities, $\mathcal{K}(w)$ is a known phase space factor and the form factor $\mathcal{F}(w)$ is generally expressed as the product of a normalization factor $\mathcal{F}(1)$ and $g(w)$, the Isgur-Wise function, whose shape is constrained by dispersion relations [3]. The analytical expression of $g(w)$ is not known a-priori. This introduces an additional uncertainty in the determination of $\mathcal{F}(1)|V_{cb}|$. First measurements of $|V_{cb}|$ were performed assuming a linear approximation for $\mathcal{F}(w)$. It has been shown [4] that this assumption is not justified, and that linear fits systematically underestimate the extrapolation at zero recoil ($w = 1$) by about 3%. Most of this effect is related to the curvature of the form factor, and does not depend strongly upon the details of the non-linear shape chosen [4]. All recent results use a non-linear shape for $g(w)$, approximated with an expansion near $w = 1$ [5], and parameterized in terms of the variable ρ^2, which is the slope of the form factor at zero recoil given in [5].

Considerable theoretical work has been devoted to the parameter $\mathcal{F}(1)$. Ultimately a precise value for it may be determined by lattice gauge calculations. Presently only a quenched lattice evaluation is available and gives $0.913^{+0.024}_{-0.017} \pm 0.016^{+0.003}_{-0.014} {}^{+0.000}_{-0.016} {}^{+0.006}_{-0.014}$. The errors reflect the statistical accuracy, the matching error, the finite lattice size, the uncertainty in the quark masses and an estimate of the error induced by the quenched approximation, respectively. The central value obtained with OPE sum rules is similar, with an error of ± 0.04 [6]. Consequently, I will use $\mathcal{F}(1) = 0.91 \pm 0.04$ [6].

The main contributions to the $\mathcal{F}(1)|V_{cb}|$ systematic error are from the uncertainty on the $B \to D^{**}\ell\nu$ shape and $\mathcal{B}(b \to B^0)$, (0.57×10^{-3}), fully correlated among the LEP experiments, the branching fraction of D and D^* decays, (0.4×10^{-3}), fully correlated among all the experiments, and the slow pion reconstruction from Belle, CLEO, and BaBar which are uncorrelated. The main contribution to the ρ^2 systematic error is from the uncer-

CP722, *B Physics at Hadron Machines*, edited by M. Paulini and S. Erhan
© 2004 American Institute of Physics 0-7354-0203-5/04/$22.00

TABLE 1. Experimental results for $\mathscr{F}(1)|V_{cb}|$ and ρ^2 rescaled to common inputs [7].

Experiment	$\mathscr{F}(1)|V_{cb}|(\times 10^3)$	ρ^2	Corr$_{\text{stat}}$
ALEPH update	33.6± 2.1± 1.6	0.75± 0.25± 0.37	94%
OPAL(partial reconstruction)	38.4± 1.2± 2.4	1.25± 0.14± 0.39	77%
OPAL (excl)	39.1± 1.6 ± 1.8	1.49± 0.21± 0.26	95%
DELPHI (partial reco.)	36.8± 1.4± 2.5	1.52± 0.14± 0.37	94%
DELPHI (excl. prelim.)	38.5± 1.8± 2.1	1.32± 0.15± 0.34	89%
Belle	36.7± 1.9± 1.9	1.45± 0.16± 0.20	91%
CLEO	43.6± 1.3± 1.8	1.61± 0.09± 0.21	87%
BaBar	34.1± 0.2± 1.3	1.23± 0.02± 0.28	92%
World average	36.7 ± 0.8	1.44 ± 0.14	91%

tainties in the measured values of R_1 and R_2 (0.12), fully correlated among experiments. Because of the large contribution of this uncertainty to the non-diagonal terms of the covariance matrix, the averaged ρ^2 is higher than one would naively expect.

Using $\mathscr{F}(1) = 0.91 \pm 0.04$ [6], this method gives $|V_{cb}| = (40.2 \pm 0.9_{exp} \pm 1.8_{theo}) \times 10^{-3}$. The dominant error is theoretical, but there are good prospects that lattice gauge calculations will significantly improve their accuracy.

3. $|\mathbf{V_{cb}}|$ FROM THE INCLUSIVE DECAY $B \to D\ell\bar{\nu}$

The study of the decay $B \to D\ell\nu$ poses new challenges both from the theoretical and experimental point of view.

The differential decay rate for $B \to D\ell\nu$ can be expressed as:

$$\frac{d\Gamma_D}{dw}(B \to D\ell\nu) = \frac{G_F^2 |V_{cb}|^2}{48\pi^3} \mathscr{K}_{\mathscr{G}}(w)\mathscr{G}(w)^2, \quad (2)$$

where w is the inner product of the B and D meson 4-velocities, $\mathscr{K}_{\mathscr{G}}(w)$ is the phase space and the form factor $\mathscr{G}(w)$ is generally expressed as the product of a normalization factor $\mathscr{G}(1)$ and a function, $g_D(w)$, constrained by dispersion relations [3].

The strategy to extract $\mathscr{G}(1)|V_{cb}|$ is identical to that used for the $B \to D^*\ell\nu$ decay. However both theory and experiments have additional difficulties in dealing with this channel. From the theoretical standpoint, the non-perturbative expansion includes $1/m_b$ and $1/m_c$ terms, as there is no suppression mechanism. Moreover, this is a decay that is experimentally challenging as it is difficult to isolate from the larger $D^*\ell\nu$ final state.

Belle [8] and ALEPH [9] studied the $\bar{B}^0 \to D^+\ell^-\bar{\nu}$ channel, while CLEO [10] studied both $B^+ \to D^0\ell^+\bar{\nu}$ and $\bar{B}^0 \to D^+\ell^-\bar{\nu}$ decays. The results scaled to common inputs are shown in Table 2. Averaging the data in Table 2, using the procedure of Ref. [11], we obtain $\mathscr{G}(1)|V_{cb}| = (41.3 \pm 4.0) \times 10^{-3}$ and $\rho_D^2 = 1.19 \pm 0.19$,

TABLE 2. Experimental results corrected to common inputs and world average [7]. ρ_D^2 is the slope of the form-factor given in Ref. [5] at zero recoil.

Experiment	$\mathscr{G}(1)|V_{cb}|(\times 10^3)$	ρ_D^2
ALEPH	40.0± 10.0± 6.4	1.02 ± 0.98± 0.37
Belle	41.8± 4.4± 5.2	1.12± 0.22± 0.14
CLEO	44.9± 5.8± 3.5	1.27± 0.25± 0.14
World average	42.1 ± 3.7	1.15± 0.16

where ρ_D^2 is the slope of the form-factor at zero recoil given in [5].

The theoretical predictions for $\mathscr{G}(1)$ are consistent: a quark model evaluation gives 1.03 ± 0.07 [12], and a recent heavy quark sum rule calculation [13] gives $1.04 \pm 0.02 \pm \delta_{exp}$, where δ_{exp} represents the error in $\mu_\pi^2(1 \text{ GeV})$. A quenched lattice calculation gives $\mathscr{G}(1) = 1.058_{-0.017}^{+0.021}$ [14], where the errors do not include the uncertainties induced by the quenching approximation and lattice spacing. Using $\mathscr{G}(1) = 1.04 \pm 0.07$, we get $|V_{cb}| = (40.5 \pm 3.6_{exp} \pm 2.4_{theo}) \times 10^{-3}$, consistent with the value extracted from $B \to D^*\ell\nu$ decays, but with a larger uncertainty.

4. $|\mathbf{V_{cb}}|$ FROM THE INCLUSIVE DECAY $B \to X_c\ell\bar{\nu}$

The decay $B \to X_c\ell\bar{\nu}$ is an alternative experimental approach to extract $|V_{cb}|$. In this case, the Operator Product Expansion (OPE) is the theoretical tool used. It yields the heavy quark inclusive decay rates as an asymptotic series in inverse powers of the heavy quark mass. More precisely, several mass scales are relevant: the b quark mass m_b, the c quark mass m_c and the energy release $E_r \equiv m_b - m_c$ [15]. The uncertainties in the predicted $\Gamma_{sl}/|V_{cb}|$ have been discussed in numerous theoretical papers [16]. However, the theory needs to provide predictions on independent observables that can be used to validate its accuracy. Experimental input includes the semileptonic width, as well as the determination of the theoretical parameters governing the hadronic matrix element, discussed below.

The key parameter in the theoretical expression for the semileptonic width is m_b. As the bare quark mass is affected by perturbative and non-perturbative contributions, considerable attention has been devoted to its proper definition [17, 18]. Similarly, m_c is a parameter in the hadronic matrix element and, recently, it has been argued [16] that extracting it from the relationship between $(m_b - m_c)$ and the spin averaged meson mass difference $(\bar{M}_B - \bar{M}_D)$ [19] may be inadequate.

The leading non-perturbative corrections arise only to order $1/m_b^2$ and are parameterized by the quantities μ_π^2 (or $-\lambda_1$) [19, 20] related to the expectation value of the kinetic energy of the b quark inside the b hadron, and μ_G^2 (or λ_2) [19, 20] related to the expectation value of the chromomagnetic operator. Quark-hadron duality is an important *ab initio* assumption in these calculations. While several authors [21] argue that this ansatz does not introduce appreciable errors as they expect that duality violations affect the semileptonic width only in high powers of the non-perturbative expansion, other authors recognize that an unknown correction may be associated with this assumption [22]. Arguments supporting a possible sizable source of errors related to the assumption of quark-hadron duality have been proposed [23].

I will start the discussion with the experimental studies of the moments of inclusive distributions. Most of the experimental studies have focused on the lepton energy and the invariant mass M_X of the hadronic system recoiling against the lepton-$\bar{\nu}$ pair. CLEO published the first measurement of the moments of the M_X^2 distributions. This analysis includes a 1.5 GeV/c lepton momentum cut, that allows them to single out the desired $b \to c\ell^-\bar{\nu}$ signal from the "cascade" $b \to c \to s\ell^+\nu$ background process. The hermeticity of the CLEO detector is exploited to reconstruct the ν 4-momentum vector. Moreover, the $B\bar{B}$ pair is produced nearly at rest and thus it allows a determination of M_X from the ν and ℓ momenta. They obtain $\langle M_X^2 - \bar{M}_D^2 \rangle = 0.251 \pm 0.066$ GeV^2 and $\langle (M_X^2 - \bar{M}_D^2)^2 \rangle = 0.576 \pm 0.170$ GeV^4, where $\bar{M}_D$ is the spin-averaged mass of the D and D^* mesons. The lepton momentum cut may reduce the accuracy of the OPE predictions, because restricting the kinematic domain may increase quark-hadron duality violations. The shape of the lepton spectrum provides further constraints on OPE. Moments of the lepton momentum with a cut $p_\ell^{CM} \geq 1.5$ GeV/c have been measured by the CLEO collaboration [24]. The two approaches give consistent results, although the technique used to extract the OPE parameters has still relatively large uncertainties associated with the $1/m_b^3$ form factors. The sensitivity to $1/m_b^3$ corrections depends upon which moments are con-

sidered. Bauer and Trott [25] have performed an extensive study of the sensitivity of lepton energy moments to non-perturbative effects. In particular, they have proposed "duality moments", very insensitive to neglected higher order terms. The comparison between the CLEO measurement of these moments [24] and the predicted values shows a very impressive agreement:

$$D_3 \equiv \frac{\int_{1.6 \text{ GeV}} E_\ell^{0.7} \frac{d\Gamma}{dE_\ell} dE_\ell}{\int_{1.5 \text{ GeV}} E_\ell^{1.5} \frac{d\Gamma}{dE_\ell} dE_\ell} = \begin{cases} 0.5190 \pm 0.0007 & \text{(T)} \\ 0.5193 \pm 0.0008 & \text{(E)} \end{cases}$$

$$D_4 \equiv \frac{\int_{1.6 \text{ GeV}} E_\ell^{2.3} \frac{d\Gamma}{dE_\ell} dE_\ell}{\int_{1.5 \text{ GeV}} E_\ell^{2.9} \frac{d\Gamma}{dE_\ell} dE_\ell} = \begin{cases} 0.6034 \pm 0.0008 & \text{(T)} \\ 0.6036 \pm 0.0006 & \text{(E)} \end{cases}$$

where "T" and "E" denote theory and experiment, respectively.

More recently, both CLEO and BaBar explored the moments of the hadronic mass M_X^2 with lower lepton momentum cuts. In order to identify the desired semileptonic decay from background processes including cascade decays, continuum leptons and fake leptons, CLEO performs a fit for the contributions of signal and backgrounds to the full three-dimensional differential decay rate distribution as a function of the reconstructed quantities q^2, M_X^2, $\cos\theta_{W\ell}$. The signal includes the components $B \to D\ell\bar{\nu}$, $B \to D^*\ell\bar{\nu}$, $B \to D^{**}\ell\bar{\nu}$, $B \to X_c\ell\bar{\nu}$ non-resonant and $B \to X_u\ell\bar{\nu}$. The backgrounds considered are: secondary leptons, continuum leptons and fake leptons. BaBar uses a sample where the hadronic decay of one B is fully reconstructed and the charged lepton from the other B is identified. In this case, the main sources of systematic errors are the uncertainties related to the detector modeling and reconstruction.

Fig. 1 shows the extracted $\langle M_X^2 - \bar{M}_D^2 \rangle$ moments as a function of the minimum lepton momentum cut from these two measurements, as well as the original measurement with $p_\ell \geq 1.5$ GeV/c. The results are compared with theory bands that reflect experimental errors, $1/m_b^3$ correction uncertainties and uncertainties in the higher order QCD radiative corrections [26]. The CLEO and BaBar results are consistent and show an improved agreement with theoretical predictions with respect to earlier preliminary results [27]. Moments of the M_X distribution without an explicit lepton momentum cut have been extracted from preliminary DELPHI data [28] and give consistent results.

The second element needed to extracted $|V_{cb}|$ with this method is the semileptonic width. Experiments operating at the $\Upsilon(4S)$ center-of-mass energy use a dilepton sample to separate the decay process $b \to c\ell^-\bar{\nu}$ (primary leptons) from the $b \to c \to s\ell^+\nu$ (cascade leptons). This technique allows a direct determination of the primary lepton spectrum over almost all the kinematically allowed range. Thus, the semileptonic branching fraction

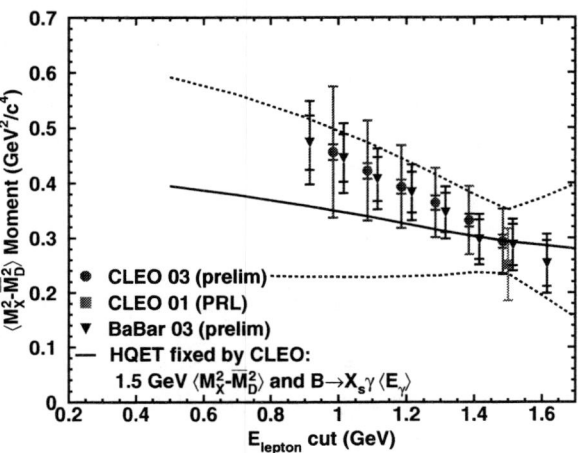

FIGURE 1. The results of the recent CLEO analysis [29] compared to previous measurements [30, 31] and the HQET prediction. The theory bands shown in the figure reflect the variation of the experimental errors on the two constraints, the variation of the third-order HQET parameters by the scale $(0.5 \text{ GeV})^3$, and variation of the size of the higher order QCD radiative corrections [26].

extracted from this measurement has almost no model dependence. Fig. 2 shows a summary of the $\Upsilon(4S)$ measurements of inclusive semileptonic branching fractions. The overall experimental error is of the order of 2%. Different extractions of the HQE cannot be combined in a straightforward manner and do not fully agree [31]. I will choose a representative set of parameters [29] and obtain:

$$|V_{cb}| = (41.5 \pm 0.4|_{\Gamma_{sl}} \pm 0.4|_{\lambda_1, \bar{\Lambda}} \pm 0.9|_{th}) \times 10^{-3}, \quad (3)$$

where the first uncertainty is from the experimental value of the semileptonic width, the second uncertainty is from the HQE parameters (λ_1 and $\bar{\Lambda}$) and the third uncertainty is the theoretical uncertainty obtained as described above. No quantitative account is given for possible quark hadron duality violation, but comparison between the two $|V_{cb}|$ determinations can possibly be used as an estimate of the uncertainties introduced by this assumption. The present difference between the two values of $|V_{cb}|$ obtained from $B \rightarrow D^* \ell \bar{\nu}$ and from inclusive semileptonic branching fraction measurements can be used to make a very rough estimate of the non quantified errors in the inclusive determination of $|V_{cb}|$ and these data imply about 6% uncertainty for non-quantified assumptions in the inclusive determination.

5. CONCLUSIONS

The values of $|V_{cb}|$ obtained both from the inclusive and exclusive method agree within errors. The value of $|V_{cb}|$

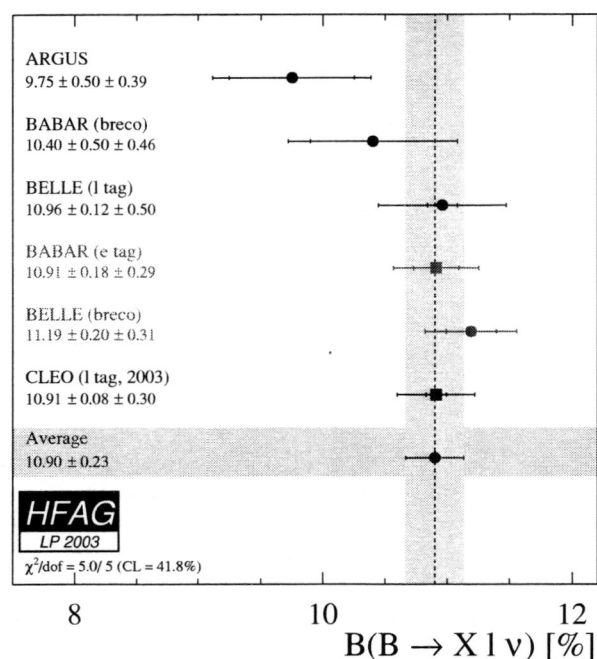

FIGURE 2. Summary of model independent semileptonic branching fraction measurements performed at the $\Upsilon(4S)$ center-of-mass energy.

obtained from the analysis of the $B \rightarrow D^* \ell \nu$ decay is:

$$|V_{cb}|_{exclusive} = (40.2 \pm 0.9_{exp} \pm 1.8_{theo}) \times 10^{-3}, \quad (4)$$

where the first error is experimental and the second error is from the $1/m_b^2$ corrections to $\mathscr{F}(1)$. The value of $|V_{cb}|$, obtained from inclusive semileptonic branching fractions is:

$$|V_{cb}|_{incl} = (41.4 \pm 0.5_{exp} \pm 0.4_{\lambda_1, \bar{\Lambda}} \pm 0.9_{theo}) \times 10^{-3},$$
$$(5)$$

where the first error is experimental, the second error is from the measured values of λ_1, and $\bar{\Lambda}$, assumed to be universal up to higher orders, and the last from $1/m_b^3$ corrections and α_s. Non-quantified uncertainties are associated with a possible quark-hadron duality violation. An estimate through a comparison between these two results implies an additional uncertainty of the order of 3%. For this reason, I choose not to average these two numbers, but quote a conservative estimate based on $|V_{cb}|$ exclusive.

High precision tests of HQET and more precise assessment of quark-hadron duality in inclusive semileptonic decays are needed to achieve the ultimate accuracy in this measurement.

ACKNOWLEDGMENTS

I would like to thank the organizers for a very interesting and lively conference. I would also like to thank C. Boulahouache and S. Stone for interesting discussions. This work was supported from the US National Science Foundation.

REFERENCES

1. Wolfenstein, L., *Phys. Rev. Lett.* **51**, 1945 (1983).
2. Gilman, F. J., Kleinknecht, K., and Renk, B., *Phys. Rev.* **D66**, 010001 (2002).
3. Boyd, C. G., Grinstein, B., and Lebed, R. F., *Phys. Rev.* **D56**, 6895–6911 (1997).
4. Stone, S., *B Decays*, 2nd Edition, World Scientific, 1994, p. 283.
5. Caprini, I., Lellouch, L., and Neubert, M., *Nucl. Phys.* **B530**, 153–181 (1998).
6. Working Group 1 Summary, CKM Workshop, CERN, CH (2002), URL: http://ckm-workshop.web.cern.ch/ckm-workshop/.
7. Heavy Flavour Averaging Group (HFAG) (2003), URL: http://www.slac.stanford.edu/xorg/hfag/semi/index.html.
8. Abe, K., et al., [Belle Collaboration], *Phys. Lett.* **B526**, 247–257 (2002).
9. Buskulic, D., et al., [ALEPH Collaboration], *Phys. Lett.* **B395**, 373–387 (1997).
10. Adam, N. E., et al., [CLEO Collaboration], *Phys. Rev.* **D67**, 032001 (2003).
11. LEP V_{cb} Working Group, Internal Note (2003), URL: http://lepvcb.web.cern.ch/LEPVCB/.
12. Scora, D., and Isgur, N., *Phys. Rev.* **D52**, 2783–2812 (1995).
13. Uraltsev, N., *Phys. Lett.* **B585**, 253–262 (2004).
14. Hashimoto, S., et al., *Phys. Rev.* **D61**, 014502 (2000).
15. Uraltsev, N., *Mod. Phys. Lett.* **A17**, 2317–2326 (2002).
16. Benson, D., Bigi, I. I., Mannel, T., and Uraltsev, N., *Nucl. Phys.* **B665**, 367–401 (2003).
17. El-Khadra, A. X., and Luke, M., *Ann. Rev. Nucl. Part. Sci.* **52**, 201–251 (2002).
18. Hoang, A. H., *Phys. Rev.* **D59**, 014039 (1999).
19. Ligeti, Z., arXiv:hep-ph/9904460 (1999).
20. Uraltsev, N., *J. Phys.* **G27**, 1081–1100 (2001).
21. Bigi, I. I. Y., and Uraltsev, N., *Int. J. Mod. Phys.* **A16**, 5201–5248 (2001).
22. Buchalla, G., arXiv:hep-ph/0202092 (2002).
23. Isgur, N., *Phys. Lett.* **B448**, 111–118 (1999).
24. Mahmood, A. H., et al., *Phys. Rev.* **D67**, 072001 (2003).
25. Bauer, C. W., and Trott, M., *Phys. Rev.* **D67**, 014021 (2003).
26. Bauer, C. W., Ligeti, Z., Luke, M., and Manohar, A. V., *Phys. Rev.* **D67**, 054012 (2003).
27. Luth, V. G., *Nucl. Phys. Proc. Suppl.* **117**, 546–550 (2003).
28. Battaglia, M., et al., *ECONF* **C0304052**, WG102 (2003).
29. Huang, G. S., et al., [CLEO Collaboration], arXiv:hep-ex/0307081 (2003).
30. Cronin-Hennessy, D., et al., [CLEO Collaboration], *Phys. Rev. Lett.* **87**, 251808 (2001).
31. Aubert, B., et al., [BaBar Collaboration], arXiv:hep-ex/0307046 (2003).

The Status of $|V_{ub}|$

Lawrence Gibbons

(Representing the CLEO Collaboration)

Cornell University, Department of Physics, Ithaca, NY 14850, U.S.A.

Abstract. I survey the theoretical and experimental information available for the determination of $|V_{ub}|$ with inclusive and exclusive techniques. Based on recent improvements, both experimental and theoretical, I propose an inclusive averaging procedure that combines measurements to limit the theoretical uncertainties of unknown magnitude. This procedure accounts more fully for uncertainties in the inclusive determination, and allows for a more meaningful average with the exclusive determinations.

1. INTRODUCTION

The magnitude of the Cabibbo-Kobayashi-Maskawa (CKM) [1, 2] matrix element V_{ub} remains a crucial input into tests of the unitarity of the 3-generation CKM matrix, yet a $|V_{ub}|$ averaging procedure that results in a defensible uncertainty remains elusive. Recent experimental and theoretical progress provides us with an opportunity to improve this situation. Here, I review the theoretical and experimental concerns in extraction of $|V_{ub}|$, and outline a potential inclusive "averaging" procedure in which information from all regions of phase space can be combined to bound experimentally the missing theoretical contributions that plague our averaging attempts.

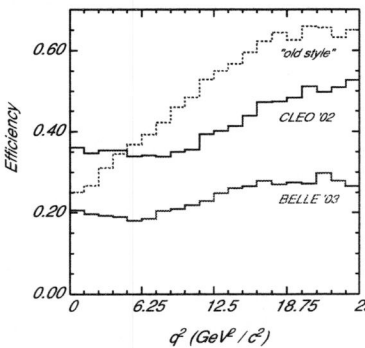

FIGURE 1. Efficiency as a function q^2 for the "traditional" lepton endpoint analysis (dotted curve) and more recent CLEO and Belle analyses (solid curves).

Inclusive theory has progressed significantly in categorization of the important corrections in different regions of phase space, and in determination of some of the hitherto unknown corrections. Experimentally, measurements have significantly reduced the model depen-

dence in the extraction of rates. For example, the reduced dependence of efficiency on dilepton mass $q^2 = M_{W*}^2$ in new rate measurements near the E_ℓ endpoint [3–5] (Figure 1) has a reduced model dependence by a factor of three. Recent measurements have also minimized the reliance on detailed $d\Gamma(B \to X_u \ell v)/dE_\ell dM_{X_u} dq^2$ modeling to separate the $b \to u \ell v$ signal from $b \to c \ell v$. We thus have rate measurements with well-defined sensitivities within well-defined regions of phase space for which we can categorize the important theoretical uncertainties. Both the theoretical and experimental improvements are key to a more robust evaluation of $|V_{ub}|$.

2. INCLUSIVE $b \to u \ell v$

Theoretically, issues regarding the calculation of the total semileptonic partial width $\Gamma(B \to X_u \ell v)$ via the operator product expansion (OPE) are well-controlled [6–11]. This non-perturbative power series in $1/m_b$ and perturbative expansion in α_s has, at order $1/m_b^2$, two non-perturbative parameters: λ_1 or μ_π^2, which is related to the Fermi momentum of the b quark in the meson, and λ_2 or μ_G^2, which parameterizes the hyperfine interaction between the heavy quark and the light degrees of freedom. The λ and μ parameters differ in their infrared behavior. In terms of these parameters, the OPE at $1/m_B^2$ yields

$$\Gamma(B \to X_u e \bar{v}) = \frac{G_F^2 |V_{ub}|^2}{192\pi^3} m_b^5 \qquad (1)$$
$$\times \left[1 - \frac{9\lambda_2 - \lambda_1}{2m_b^2} + \ldots - \mathcal{O}\left(\frac{\alpha_s}{\pi}\right) \right].$$

The perturbative corrections are known to order α_s^2 [12]. The error induced by uncertainties in the

CP722, *B Physics at Hadron Machines,* edited by M. Paulini and S. Erhan
© 2004 American Institute of Physics 0-7354-0203-5/04/$22.00

non-perturbative parameters $\lambda_{1,2}$ is relative small, and an evaluation by the LEP Heavy Flavors [13] working group yielded

$$|V_{ub}| = 0.00445 \left(\frac{\mathrm{B}(b \to u\ell\bar{\nu})}{0.002} \frac{1.55\mathrm{ps}}{\tau_b} \right)^{1/2} (2)$$
$$\times \left(1 \pm 0.020_\lambda \pm 0.052_{m_b} \right)$$

The mass $m_b^{1S}(1\,GeV) = 4.58 \pm 0.09$ GeV, which agrees with a recent survey [14], dominates the uncertainty. Because the OPE is a quark-level calculation, the estimate rests on the assumption of global quark-hadron duality, which is well-motivated [15] for $b \to u\ell\nu$ (particularly with its broad range of hadronic final states). We can bound $\sigma_{\mathrm{duality}}$ by comparison of the more precise inclusive and exclusive evaluations of $|V_{cb}|$. The difference [16] of $(0.8 \pm 1.6) \times 10^{-3}$ implies $\sigma_{\mathrm{duality}} = 4\%$, for a total uncertainty of 6.8% on the total rate.

To overcome the 100 times larger $b \to c\ell\nu$ background, inclusive $b \to u\ell\nu$ measurements utilize restricted regions of phase space in which the background is suppressed. Extraction of $|V_{ub}|$ then requires the fraction of the total $b \to u\ell\nu$ rate that lies within the given region of phase space. This complicates the theoretical issues and uncertainty considerably. We first consider measurements restricted either to the region $p_\ell \gtrsim 2.2$ GeV/c or to the region of low hadronic mass ($M_X \lesssim M_D$). Within both, the parton–level OPE fails because its expansion parameter $E_X \Lambda_{QCD}/m_x^2 \sim 1$. We will return to this issue and its impact on rate estimations.

Table 1 summarizes the rate measurements by CLEO [3], BaBar [4] and Belle [5] near the p_ℓ endpoint. These analyses must suppress background from continuum processes, and can reach down to about 2.2 GeV/c before control of $b \to c\ell\nu$ becomes problematic. The continuum suppression induces the efficiency variation with q^2 discussed above. The remainder of the model dependence could be eliminated by coarsely binning in q^2, or, preferably, through use of a tagged B sample.

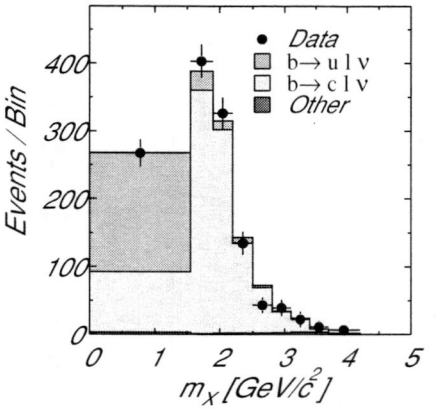

Source: Aubert et al. [17]

FIGURE 2. Reconstructed M_X distribution from the BaBar low M_X analysis.

BaBar [17] and Belle [18] have new analyses of the low M_X region (combined with a moderate $p_\ell > 1.0$ GeV/c requirement), which was first studied by DELPHI [19]. Finite M_X resolution smears the $b \to c\ell\nu$ background below its theoretical limit of M_D, forcing more stringent M_X requirements that approach a pole in the parton level spectrum. The preliminary Belle analysis utilizes a $D^{(*)}\ell\nu$ tag, with M_X calculated directly from all particles after removal of tag and lepton. This tag still results in a significantly larger background than signal level, and the systematic estimates (very preliminary) appear quite aggressive. The BaBar analysis uses fully reconstructed hadronic B tags, again with direct M_X calculation. This analysis achieves a beautiful $b \to u\ell\nu$ signal in the region $M_X < 1.55$ GeV/c^2 with signal to background ratio (see Figure 2) approaching 2:1. This ratio shows the anticipated power of the hadronic B tags, which afford unsurpassed resolution on M_X. The efficiency versus M_X, while not featureless, appears promisingly uniform. BaBar has also extracted the yields with a fit to the integrated $M_X < 1.55$ GeV/c^2 interval. Both features minimize dependence of the extracted rate on detailed modeling of the $b \to u\ell\nu$, which allows for cleaner theoretical interpretation and simplifies improved determination of $|V_{ub}|$ as theory advances.

Determination of the fraction of the $b \to u\ell\nu$ rate in the p_ℓ endpoint or the low M_X region requires re-summation of the OPE to all orders in $E_X \Lambda_{QCD}/m_x^2$ [6, 20–23]. The re-summation results, at leading-twist order, in a non-perturbative shape function $f(k_+)$. The argument $k_+ = k^0 + k_\parallel$, where $k^\mu = p_b^\mu - m_b v^\mu$ is the b quark momentum with the "mechanical" portion subtracted. The spatial components $k_\parallel$ and $k_\perp$ are defined relative to the $m_b v^\mu - q^\mu$ (roughly the recoiling u quark) direction. At leading twist, effects like the "jiggling" of $k_\perp$ are

TABLE 1. Partial branching fractions for $b \to u\ell\nu$ near the p_l endpoint. The rates are integrated up to $p_\ell^{\max} = 2.6$ GeV/c ($\Upsilon(4S)$ frame). The estimated rate fraction f_E is given for each range. The dagger (†) indicates where the QED radiative correction was applied.

$p_\ell^{\min}$ (GeV/c)	$\Delta\mathscr{B}_u(p)$ (10^{-4})	f_E	
2.0	4.22 ± 1.81	†0.266 ± 0.048	CLEO (e, μ)
2.1	3.28 ± 0.77	†0.198 ± 0.040	CLEO (e, μ)
2.2	2.30 ± 0.38	†0.130 ± 0.028	CLEO (e, μ)
2.3	1.43 ± 0.16	†0.074 ± 0.017	CLEO (e, μ)
	†1.52 ± 0.20	0.078 ± 0.017	BaBar (e)
	1.19 ± 0.15	†0.072 ± 0.016	Belle (e)
2.4	0.64 ± 0.09	†0.037 ± 0.008	CLEO (e, μ)

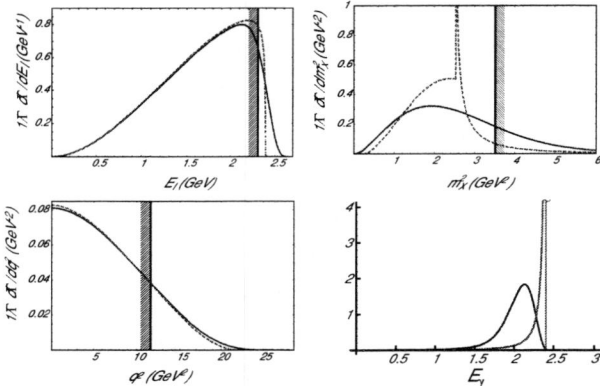

FIGURE 3. Parton level (blue) and model convoluted (red) spectra for $B \to X_u \ell \nu$ (top, bottom left; from Ref. [24]) and for $B \to X_s \gamma$ (bottom right).

ignored, and the differential partial width is given by the convolution of the shape function with the parton level differential distribution (see Figure 3):

$$d\Gamma = \int dk_+ \, f(k_+) \, d\Gamma^{(parton)}_{m_b \to m_b + k_+}. \qquad (3)$$

Because the shape function depends only on parameters of the B meson, this description holds for any B decay to a light quark, such as $B \to s\gamma$.

It has been known for some time [6, 21] that measurement of the E_γ spectrum in $b \to s\gamma$ can yield (at leading twist) $f(k_+)$. Ideally, $|V_{ub}|$ would be determined directly from integrated spectra [21, 25–29] without extraction of an intermediate shape function. For the lepton spectrum, for example, one takes

$$\left|\frac{V_{ub}}{V_{tb}V_{ts}^*}\right|^2 = \frac{3\alpha}{\pi} K_{\text{pert}} \frac{\widehat{\Gamma}_u(E_0)}{\widehat{\Gamma}_s(E_0)} + O(\Lambda_{\text{QCD}}/M_B), \qquad (4)$$

where K_{pert} is a calculable perturbative kernel, and $\widehat{\Gamma}_u(E_0)$ and $\widehat{\Gamma}_s(E_0)$ are appropriately weighted integrals over, respectively, the E_ℓ and E_γ spectra above the cutoff E_0. A similar expression exists for the M_X spectrum [26]. Practical application by the endpoint analyses is currently difficult because of integration of the rate in the $\Upsilon(4S)$ frame, which introduces significant smearing. With tagged B samples, endpoint measurements can be made in the B frame (with relaxed continuum suppression), which would allow for direct application of the weighted integral approach. In principle, current M_X analyses can already use this approach, though experimental efficiency and p_ℓ cutoffs must be incorporated into the integrals. All current analyses have, instead, relied on modeling an intermediate shape function.

The role of the shape function can be mitigated by restricting measurement to regions of large q^2, where

validity of the OPE expansion is restored [30, 31]. Restriction to the region kinematically forbidden to $b \to c\ell\nu$, $q^2 > (M_B - M_D)^2$, however, introduces a low mass scale [32, 33] into the OPE, introducing uncertainties of order $(\Lambda_{\text{QCD}}/m_c)^3$. However, the combination of an M_X with a looser q^2 requirement can suppress $b \to c\ell\nu$ background yet still reduce shape function contributions. Furthermore, the q^2 restriction moves the parton level pole away from typical M_X cuts. For a practical choice of region, $f(k_+)$ contributions are suppressed but not negligible. The elimination of high energy hadronic final states by the q^2 restriction may exacerbate duality concerns.

A recent Belle analysis [34] in this region employs a $p_\ell > 1.2 \text{ GeV}/c$ requirement followed by an "annealing" procedure to sort reconstructed particles into the "signal" and "other" B. The analysis integrates over the rate in the region $M_X < 1.7 \text{ GeV}/c^2$ and $q^2 > 8 \text{ GeV}^2$ to extract $|V_{ub}|$, which again has the desired effect of minimizing dependence of the analysis on detailed modeling of the $b \to u\ell\nu$ process. The signal to background ratio of the analysis, about 1:6, does not approach that of the B tag technique. Belle finds a rate $\Delta\mathcal{B}$ in that q^2–M_X region of

$$\frac{\Delta\mathcal{B}}{10^{-4}} = 7.37 \pm 0.89_{\text{stat}} \pm 1.12_{\text{sys}} \pm 0.55_{c\ell\nu} \pm 0.24_{u\ell\nu}. \qquad (5)$$

I look forward to analysis of this region with the significantly cleaner B tag technique.

Extraction of $|V_{ub}|$ from these measurements has required modeling of $f(k_+)$ for estimation of the fraction (f) of the rate in each region of phase space. The endpoint analyses use fractions estimated by CLEO [3] based on an $f(k_+)$ derived from the CLEO $b \to s\gamma \, E_\gamma$ spectrum. CLEO employed several two–parameter functional forms $f[\Lambda^{SF}, \lambda_1^{SF}](k_+)$ [35, 36] that were convolved with the parton-level E_γ calculation in fits to the measured spectrum between 1.5 and 2.8 GeV. These parameters are related to the HQET parameters of similar name, and play a similar role in evaluation of the rates. At this time, however, we do not know the relationship between the shape function parameters (or the moments of the shape function) and the HQET non-perturbative parameters $\overline{\Lambda} = m_B - m_b$ and λ_1 [37, 38]. The fact that Λ^{SF} and λ_1^{SF} depend on the functional ansatz adopted, while the HQET parameters depend on the renormalization scheme, underscores the current ambiguity.

Figure 4 shows the best fit parameters [39] and the one standard deviation contour for the exponential form [36]. The strong correlation between the two parameters results from the interplay between the b quark's effective mass (controlled by Λ^{SF}) and kinetic energy (controlled by λ_1^{SF}) in determining the *mean* of the E_γ spectrum (and the mean energy available for the final state in $b \to u\ell\nu$). The best fit corresponds to $(\Lambda^{SF}, \lambda_1^{SF}) = (0.545, -0.342)$ and the rate fractions $f_E = 0.14$, $f_M =$

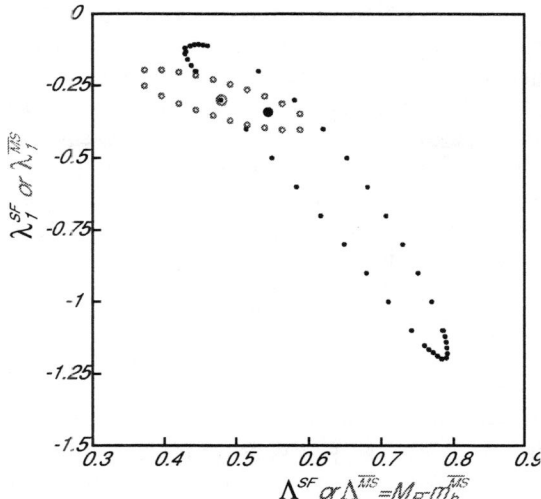

FIGURE 4. Shape function parameters from best fit to $b \rightarrow s\gamma$ photon energy spectrum and one standard deviation "ellipse" (solid blue circles). Also shown is the $(\overline{\Lambda}^{\overline{MS}}, \lambda_1^{\overline{MS}})$ ellipse corresponding to the BaBar shape function parameter choice (open red circles).

TABLE 2. Summary of inclusive $|V_{ub}|$ results. The errors on the first group are experimental and theoretical uncertainties, respectively. The second group, adjusted to a common $f(k_+)$, lists the total experimental, $f(k_+)$–related, and Γ_{tot} uncertainties. The two groups are not directly comparable.

	$|V_{ub}|(10^{-3})$	
ALEPH [43]	$4.12 \pm 0.67 \pm 0.76$	neur. net
L3 [44]	$5.70 \pm 1.00 \pm 1.40$	cut, count
DELPHI	$4.07 \pm 0.65 \pm 0.61$	M_X
OPAL [45]	$4.00 \pm 0.71 \pm 0.71$	neur. net
LEP Avg. [13]	$4.09 \pm 0.37 \pm 0.56$	
CLEO [46]	$4.05 \pm 0.61 \pm 0.65$	triple diff.
Belle	$5.00 \pm 0.64 \pm 0.53$	M_X
CLEO	$4.11 \pm 0.34 \pm 0.46 \pm 0.28$	$p_\ell > 2.2$
BaBar	$4.31 \pm 0.28 \pm 0.49 \pm 0.30$	$p_\ell > 2.3$
Belle	$3.99 \pm 0.23 \pm 0.45 \pm 0.27$	$p_\ell > 2.3$
Belle	$4.63 \pm 0.48 \pm 0.48 \pm 0.32$	$M_X < 1.7$ $q^2 > 8$
BaBar	$4.79 \pm 0.40 \pm 0.60 \pm 0.33$	$M_X < 1.55$

0.53 and $f_{qM} = 0.34$ for the CLEO $p_l > 2.2$ GeV/c, the BaBar low M_X, and the Belle $M_X - q^2$ analyses, respectively. The statistical uncertainty derives from the extremes in the fractions found on the contour of Figure 4. The endpoint analysis was symmetrized by taking the average of those extremes, resulting in $f_E = 0.13 \pm 0.02$, $f_M = 0.52 \pm 0.12$ and $f_{qM} = 0.32 \pm 0.06$. The fractions are almost completely correlated as one moves around the contour. A small correction resulting from a re-optimization of background normalization in the $b \rightarrow s\gamma$ analysis [40] (about 1/5 the assigned systematic) is applied. After including the remaining background subtraction systematic and smaller contributions from the α_s uncertainty and from variation of the $f(k_+)$ ansatz [35, 36], the rate fraction results are $f_E = 0.13 \pm 0.03$ (with radiative corrections), $f_M = 0.55 \pm 0.14$ and $f_{qM} = 0.33 \pm 0.07$. A more detailed description is forthcoming [41].

Two alternate approaches to shape function modeling have been taken in experimental studies. The BaBar M_X analysis [17] takes the $f(k_+)$ ansaetze just discussed, but associates $(\Lambda^{SF}, \lambda_1^{SF})$ with the HQET parameters derived from spectral moments of the $b \rightarrow s\gamma$ and $b \rightarrow c\ell\nu$ processes (see Figure 4). Note that while the ellipse appears smaller, the $\overline{\Lambda}^{\overline{MS}}-\lambda_1^{\overline{MS}}$ correlation does not stabilize the average final state energy like the E_γ-based correlation, resulting in f_M uncertainties that are comparable to the E_γ based uncertainties. The Belle M_X-q^2 analysis uses reference [31], which employs a simpler form based on the single parameter $\Lambda^{SF}/\lambda_1^{SF}$ estimated from the $m_b^{(1S)}$ mass and typical estimates for λ_1. These approaches rest on the assumption, now agreed to be incorrect [37, 38], that an arbitrary renormalization scheme can be used in the parameterizations employed. The associated uncertainty is difficult to assess, and has not been included in the $|V_{ub}|$ determinations

Once the renormalization behavior of $f(k_+)$ is understood, the relationship between HQET parameters, kinematic definitions of the b quark mass, and moments of $f(k_+)$ will be better defined. Moments of the $b \rightarrow c\ell\nu$ process and m_b constraints can then inform extraction of $f(k_+)$. Correlations from the m_b dependence in the total rate and in the rate fractions must then be incorporated. Given the complete independence of the kinematic mass determination used for the total rate and the effective quark mass in the E_γ-derived $f(k_+)$, coupled with the large effective mass range sampled in the latter, a linear combination of those two uncertainties [34] seems overly conservative for the present E_γ-derived results.

Table 2, based on the Heavy Flavors Averaging Group (HFAG) summary, [42], summarizes the full set of inclusive $|V_{ub}|$ results. Given the strong correlations among the rate fractions, $|V_{ub}|$ comparisons are meaningful only for results evaluated with common theoretical input. The E_γ-derived shape function is currently the best motivated theoretically and has the most complete categorization of uncertainties. I therefore adjust recent results and $f(k_+)$–related uncertainties to use the rate fractions discussed above. The errors do not include uncertainties from potentially large theoretical corrections that have been categorized but not calculated, as discussed below.

3. COMBINING INCLUSIVE INFORMATION

Evaluation of the total uncertainty on $|V_{ub}|$ remains problematic because of three main theoretical complications ([see, *e.g.*, 47, 48]. The first arises from sub-leading (higher twist) contributions to the OPE re-summation [49–52], which are not universal. Hence with use of $b \to s\gamma$ to constrain $f(k_+)$, there exist subleading contributions both to the use of a shape function in $b \to u\ell\nu$ itself and to the derivation of $f(k_+)$ from $b \to s\gamma$. The subleading contributions, formally of order Λ_{QCD}/m_b, can be as large as $\sim 15\%$. Indeed, a partial estimate [51] for the endpoint region finds corrections that are approximately the same size as the other combined uncertainties.

The second contribution, from "weak annihilation" processes [53, 54], is formally of order $(\Lambda_{QCD}/m_b)^3$ but receives a $16\pi^2$ enhancement. The contribution, which requires factorization violation to be nonzero, is expected to be localized near $q^2 \sim m_b^2$. This results in further enhancement of the effect on $|V_{ub}|$ measurements. For the endpoint region, an effect on the total rate of 2-3% (for factorization violation of about 10%), corresponds to 20-30% on the endpoint rate.

Finally, while global quark hadron duality is well-motivated for spectral moments, the OPE cannot predict the detailed inclusive spectra. The extent of violation of quark–hadron duality locally depends on the size and nature of the region of phase space considered: including a large fraction of the rate is best. The associated uncertainty is difficult to assess.

The problems outlined present a considerable obstacle to a meaningful average of the inclusive results. Results with a potentially large bias will be included with neither correction nor meaningful uncertainty: $|V_{ub}|$ will be biased and have an unreliable uncertainty. As an alternative, we can choose a region of phase space that provides a reasonable compromise among the unknown contributions. The choice is inherently subjective given the different viewpoints within the theory community ([see, *e.g.*, 29, 31]). In this reviewer's opinion, the opportunity to bound experimentally the uncertainties we know of, thereby providing as complete an uncertainty estimate as possible, is more important than achieving the smallest statistical precision. Currently, I find the region restricted to low M_X and higher q^2 the most compelling. It has reduced corrections from $f(k_+)$ and hence from subleading contributions, yet has a sufficient fraction of the spectrum to dilute weak annihilation and local quark hadron duality concerns. The low M_X and p_l endpoint regions, in which one or more of the corrections is more pronounced, then play critical roles in limiting the uncertainty in this region. I stress that estimating a complete

inclusive uncertainty is of fundamental importance, and hence that I consider measurements in all three regions of equal importance. Indeed, I view a single coherent analysis of all three phase space regions simultaneously an important milestone for both B factories, particularly with application of the powerful and clean B tag samples.

For now, however, only Belle has contributed a result for this region of phase space, and I quote that result as my "central value":

$$|V_{ub}|/10^{-3} = 4.63 \pm 0.28_{stat} \pm 0.39_{sys} \pm \qquad (6)$$
$$0.48_{f_{qM}} \pm 0.32_{\Gamma_{thy}} \pm \sigma_{WA} \pm \sigma_{SSF} \pm \sigma_{LQD}.$$

New measurements in this region can be easily combined when available, and will improve the experimental uncertainties. We must determine the uncertainties for weak annihilation (WA), subleading shape function corrections (SSF) and local quark hadron duality (LQD) within this region. Note that the data are not yet precise enough to draw conclusions regarding the presence or absence of these corrections; we use them to provide bounds.

Each phase space region considered should largely contain the WA contribution, which will be most (least) diluted in the low M_X (endpoint) region. For a neglected WA contribution, comparison of $|V_{ub}|$ from these two regions would predict a bias in the M_X, q^2 region to be

$$[(1 - f_{qM})/f_{qM}][f_e f_M/(f_M - f_e)] \approx 0.39 \qquad (7)$$

of the observed difference. The quoted value, which is model dependent, is based on the fractions found for the E_γ–derived $f(k_+)$. Comparison of the CLEO endpoint and BaBar low M_X values, taking into consideration the almost total correlation in the shape function and Γ_{tot} uncertainties, yields $\Delta|V_{ub}|/10^{-3} = 0.69 \pm 0.53$. I take the larger of the central value and error and scale according to Eq. 7 to obtain $\sigma_{WA} \approx 0.27$.

To estimate σ_{SSF}, I assume that subleading corrections scale like the fractional change in the rate prediction ($\Delta\Gamma/\Gamma$) that $f(k_+)$ induces relative to the parton-level calculation. Comparison of the low M_X region ($[\Delta\Gamma/\Gamma]_M \sim 0.15$) to the combined M_X, q^2 region ($[\Delta\Gamma/\Gamma]_{qM} \sim -0.075$) with theory correlations considered, gives $\Delta|V_{ub}|/10^{-3} = 0.16 \pm 0.63$. Scaling by $|(\Delta\Gamma/\Gamma)_{qM}/(\Delta\Gamma/\Gamma)_M| = 0.49$, which is model dependent, we obtain $\sigma_{SSF} \approx 0.31$.

Finally, to bound the local duality uncertainty, I assume that the error scales with the rate fraction f as $(1 - f)/f$ (*ad hoc*, but goes to zero for full phase space and diverges for use of the detailed spectra). To obtain an estimate, I compare the CLEO $p_\ell > 2.2$ GeV/c region to the average of BaBar and Belle in the $p_\ell > 2.3$ GeV/c region and apply the subleading correction estimates [51] ($+0.27 \times 10^{-3}$) to minimize potential cancellation between duality violation and subleading corrections. This

160

yields $(|V_{ub}|^{2.3} - |V_{ub}|^{2.2} + 0.27)/10^{-3} = 0.29 \pm 0.38$. Scaling the uncertainty by

$$\frac{(1 - f_{qM})/f_{qM}}{(1 - f_{2.3})/f_{2.3} - (1 - f_{2.2})/f_{2.2}} \approx 0.29 \qquad (8)$$

based on the fractions in Table 1 gives $\sigma_{LQD} \sim 0.11$.

From this combination of information, we thus find

$$|V_{ub}|/10^{-3} = 4.63 \pm 0.28_{stat} \pm 0.39_{sys} \pm \qquad (9)$$
$$0.48_{f_{qM}} \pm 0.32_{\Gamma_{thy}} \pm 0.27_{WA} \pm 0.31_{SSF} \pm 0.11_{LQD}$$

for a total theory error of 15%. Since experimental uncertainties dominate, quadrature addition seems reasonable. The limits presented here can be improved in robustness (e.g. considering potential cancellations among effects), through more sophisticated scaling estimates, and through improved and additional measurements. Improvement of the $b \to s\gamma$ photon energy spectrum is key. Measurements of D^0 versus D_s^+ semileptonic widths and comparison of neutral and charged B decay can help limit WA contributions [54]. Finally, improved theory for the scaling of the effects over phase space could allow development of a procedure for simultaneous extraction of $|V_{ub}|$ and the corrections, with all experimental information contributing directly to $|V_{ub}|$, or could shift the choice of "preferred" region.

4. EXCLUSIVE MEASUREMENTS OF $b \to u\ell\nu$

I devote considerably less time to the discussion of the exclusive determination of $|V_{ub}|$ – not because it is less important but because the story is simpler. Theoretical issues center on determination of the form factors (FF) involved in the decays. For $B \to \pi\ell\nu$, for example, one has [55]

$$\frac{d\Gamma(B \to \pi\ell\nu)}{dq^2 d\cos\theta_\ell} = |V_{ub}|^2 \frac{G_F^2 p_\pi^3}{32\pi^3} \sin^2\theta_\ell |f_+(q^2)|^2 \qquad (10)$$

with only the single form factor $f_+(q^2)$ for massless leptons. Final states with a vector meson depend on three form factors. The measured rates depend on the variation of the form factors with q^2 ("shape") and their relative normalizations [56], while extraction of $|V_{ub}|$ depends both on their shape and absolute normalization.

A large variety of calculations exist. For extraction of $|V_{ub}|$, the focus has sharpened onto the QCD–based calculations of lattice QCD (LQCD) [57–69] and light cone sum rules (LCSR) [70–78]. Only quenched LQCD FF calculations are available for $b \to u\ell\nu$, and these have sensitivity only in the range $q^2 \gtrsim 16$ GeV^2. Approximations made in the LCSR calculations are only valid for

$q^2 \lesssim 16$ GeV^2. A recent summary [79] shows reasonable agreement where comparison is possible. The FF's have also been evaluated using many quark– or parton–model based techniques [80–94]. Uncertainties in these models are difficult to assess, and they exhibit a broad variation in shape. Finally, various studies of FF's based on constraints from dispersion relations, unitarity or Heavy Quark Symmetry have been made [95–100].

Recent work has helped assessment of the reliability of the FF's based on the available LCSR and quenched (no light quark loops in the propagators) LQCD calculations. Analysis [101–104] of the LCSR approach within the framework of soft collinear effective theory (SCET) has sparked debate regarding potential contributions missing from LCSR. The $B \to \rho\ell\nu$ FF's, in particular, may be overestimated in LCSR, biasing $|V_{ub}|$ low. Unquenched LQCD calculations have begun to appear, and comparison to experiment shows much better agreement with data (few percent) than for quenched results [105]. While work has begun on unquenched FF's, initial results are limited to valence quark masses $\sim m_s$. Initial results [106, 107] are compatible with the $\sim 15\%$ uncertainties assigned for the quenching approximation.

To date, all exclusive measurements employ detector hermeticity to estimate p_ν, which allows full reconstruction of the decays. Two general strategies have been taken. The CLEO [108] and BaBar [109] $B \to \rho\ell\nu$ analyses emphasize higher efficiency and employ relatively loose event cleanliness and ν-consistency criteria. With the resulting background levels, these analyses are primarily sensitive in the region $p_\ell > 2.3$ GeV$/c$. To extract total branching fractions and $|V_{ub}|$, these analyses survey a variety of FF models, including LQCD and LCSR calculations extrapolated over the full q^2 range.

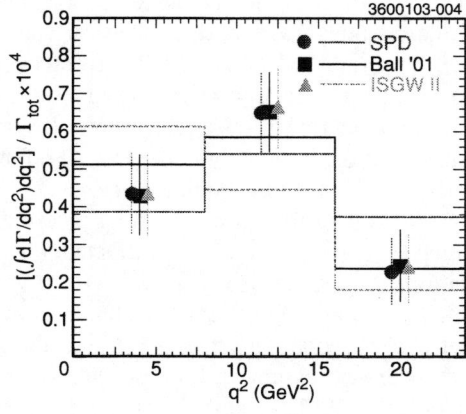

Source: Athar et al. [110]

FIGURE 5. The CLEO $B^0 \to \pi^-\ell^+\nu$ partial branching fractions based on three disparate FF models.

TABLE 3. Summary of exclusive $|V_{ub}|$ measurements. The errors listed are the statistical and experimental systematic uncertainties combined in quadrature, and all form factor uncertainties, respectively. In the CLEO '03 averages, the LQCD and LCSR uncertainties have been treated as correlated.

	$|V_{ub}|(10^{-3})$	q^2 range [GeV^2]	FF
CLEO '00 ρ	$3.23^{+0.33}_{-0.35} \pm 0.58$	all	survey
BaBar '01 ρ	$3.64 \pm 0.33^{+0.39}_{-0.56}$	all	survey
CLEO '03 π	$3.33 \pm 0.28^{+0.57}_{-0.40}$	< 16	LCSR
CLEO '03 π	$2.88 \pm 0.63^{+0.48}_{-0.39}$	> 16	LQCD
CLEO '03 π	$3.24 \pm 0.26^{+0.56}_{-0.40}$	average	
CLEO '03 ρ	$2.67^{+0.47}_{-0.50}{}^{+0.50}_{-0.39}$	< 16	LCSR
CLEO '03 ρ	$3.34^{+0.42}_{-0.48}{}^{+0.69}_{-0.62}$	> 16	LQCD
CLEO '03 ρ	$3.00^{+0.36}_{-0.41}{}^{+0.56}_{-0.47}$	average	
CLEO '03 π,ρ	$3.17^{+0.23}_{-0.24}{}^{+0.53}_{-0.39}$	average	
CLEO '03 π,ρ	$3.26 \pm 0.24^{+0.54}_{-0.39}$	average, no ρ LCSR	

The simultaneous $B \to \pi \ell v$ and $B \to \rho \ell v$ measurement by CLEO [110], on the other hand, applies strict criteria to achieve acceptable background levels over a broad p_ℓ range. With sensitivity down to 1.0 GeV/c (1.5 GeV/c) for $\pi \ell v$ ($\rho \ell v$), this analysis was able to extract independent rates in three q^2 bins. This eliminated model dependence of the measured $\pi \ell v$ rates (Figure 5) and halved that for $\rho \ell v$ rates (see Ref. [56] for a discussion of model dependence). Furthermore, the analysis permits extraction of $|V_{ub}|$ from the LQCD and LCSR FF's within their valid q^2 ranges and so without additional modeling.

Table 3 summarized the results for $|V_{ub}|$ from exclusive measurements. Averages of the CLEO results are given with and without the low q^2 region for $\rho \ell v$ (for which LCSR validity is under debate). Note that FF-related uncertainties have been treated as completely correlated in the LCSR and LQCD averages to remain conservative. The LQCD uncertainties for the π (ρ) modes include 15% (20%) quenching uncertainties, which are called out separately in the LQCD references. Ref. [56] presents a more complete review of recent $B \to X_u \ell v$ branching fractions, including measurements from which $|V_{ub}|$ has not yet been extracted. Of note is evidence for $B^+ \to \omega \ell^+ v$ [111] and $B^+ \to \eta \ell^+ v$ [110].

Potential exists for significant correlation among the dominant experimental systematics [56], so the results have been averaged assuming full correlation. The earlier results [108, 109], which depend more heavily on modeling, are deweighted by 5%. The average yields

$$|V_{ub}| = (3.27 \pm 0.13 \pm 0.19^{+0.51}_{-0.45}) \times 10^{-3} \quad (11)$$

where the errors arise from statistical, experimental systematic and form factor uncertainties, respectively.

Should the LCSR form factors prove to be overestimated, I also average without including information using $q^2 < 16$ GeV^2 in $\rho \ell v$, yielding $|V_{ub}| = (3.26 \pm 0.19 \pm 0.15 \pm 0.04^{+0.54}_{-0.39}) \times 10^{-3}$, where the errors are statistical, experimental systematic $\rho \ell v$ form factor uncertainties, and LQCD and LCSR uncertainties.

The future for exclusive determinations of $|V_{ub}|$ appears promising. The large B tag samples being collected should allow significant improvement in v resolution and reduction of backgrounds and experimental systematics. Unquenched LQCD calculations are underway and will eliminate the primary, poorly controlled, source of uncertainty. Recent advances may also allow use of the full q^2 range for extraction of $|V_{ub}|$ with LQCD [112, 113]. For both LQCD and experiment, $\pi \ell v$ appears to be the golden mode – even the B^* pole now appears manageable [68]. $B \to \eta \ell v$ will provide a valuable cross-check. The $\rho \ell v$ mode will be more problematic for high precision. The broad ρ width leaves experiments open to larger backgrounds, including poorly-understood nonresonant $\pi \pi$ contributions. In unquenched LQCD calculations the ρ is unstable, and methods for accommodating the high energy $\pi \pi$ final state have yet to be developed. The $\omega \ell v$ mode may prove more tractable. Agreement between accurate $|V_{ub}|$ determinations from $\pi \ell v$, $\eta \ell v$ and $\omega \ell v$ will add confidence overall.

5. INCLUSIVE, EXCLUSIVE AVERAGING

Until recently, I have argued against averaging of inclusive and exclusive results because of outstanding uncertainties in the former and no checks on the latter's theory. With the experimental bounds on uncertainties determined above, and some clarification of the reliability of LCSR and of the quenching uncertainties in LQCD, my major concerns are being addressed. I therefore combine the inclusive and exclusive results and find

$$|V_{ub}| = (3.67 \pm 0.47) \times 10^{-3}. \quad (12)$$

Excluding results based on data below $q^2 < 16$ GeV^2 in the $\rho \ell v$ yields a similar result: $|V_{ub}| = (3.70 \pm 0.49) \times 10^{-3}$. Inclusive and exclusive averaging will likely remain controversial in the short term. However, with the progress expected both inclusive and exclusive measurements and theory, I anticipate a noncontroversial value of $|V_{ub}|$, with an uncertainty bettering the 13% presented here, in the next few years.

162

ACKNOWLEDGMENTS

I would like to thank A. Kronfeld, B. Lange, G.P. Lepage, Z. Ligeti, M. Luke and M. Neubert for their input regarding the theory of $|V_{ub}|$ determinations. Special thanks to D. Cronin-Hennessy and E. Thorndike for their analysis of the rate fractions derived from the E_γ spectrum.

REFERENCES

1. Cabibbo, N., *Phys. Rev. Lett.* **10**, 531–532 (1963).
2. Kobayashi, M., and Maskawa, T., *Prog. Theor. Phys.* **49**, 652–657 (1973).
3. Bornheim, A., et al., [CLEO Collaboration], *Phys. Rev. Lett.* **88**, 231803 (2002).
4. Aubert, B., et al., [BaBar Collaboration], arXiv:hep-ex/0207081 (2002).
5. Abe, K., et al., [Belle Collaboration] (2003), BELLE-CONF-0325.
6. Bigi, I. I. Y., Shifman, M. A., Uraltsev, N. G., and Vainshtein, A. I., *Int. J. Mod. Phys.* **A9**, 2467–2504 (1994).
7. Neubert, M., *Int. J. Mod. Phys.* **A11**, 4173–4240 (1996).
8. Bigi, I. I. Y., Shifman, M. A., and Uraltsev, N., *Ann. Rev. Nucl. Part. Sci.* **47**, 591–661 (1997).
9. Hoang, A. H., Ligeti, Z., and Manohar, A. V., *Phys. Rev. Lett.* **82**, 277–280 (1999).
10. Bigi, I. I. Y., arXiv:hep-ph/9907270 (1999).
11. Ligeti, Z., arXiv:hep-ph/9908432 (1999).
12. van Ritbergen, T., *Phys. Lett.* **B454**, 353–358 (1999).
13. Abbaneo, D., Battaglia, M., Gagnon, P., Henrard, P., Lu, J., Mele, S., Piotto, E., and Rosnet, P., [The LEP Vub Working Group Collaboration] (2001), lEPVUB-01/01.
14. El-Khadra, A. X., and Luke, M., *Ann. Rev. Nucl. Part. Sci.* **52**, 201–251 (2002).
15. Bigi, I. I. Y., and Uraltsev, N., *Int. J. Mod. Phys.* **A16**, 5201–5248 (2001).
16. Artuso, M., and Barberio, E., "Determination of $|V_{cb}|$," 2004, to appear in The Review of Particle Properties.
17. Aubert, B., et al., [BaBar Collaboration], arXiv:hep-ex/0307062 (2003).
18. Schwanda, C., "V_{cb}, V_{ub}, HQET at Belle," in *Proceedings of the International Europhysics Conference onă HighăEnergyăPhysics, EPS2003*, Eur. Phys. J. C direct, 2003, digital Object Identifier (DOI) 10.1140/epjcd/s2003-03-109-2.
19. Abreu, P., et al., [DELPHI Collaboration], *Phys. Lett.* **B478**, 14–30 (2000).
20. Neubert, M., *Phys. Rev.* **D49**, 3392–3398 (1994).
21. Neubert, M., *Phys. Rev.* **D49**, 4623–4633 (1994).
22. Dikeman, R. D., Shifman, M. A., and Uraltsev, N. G., *Int. J. Mod. Phys.* **A11**, 571–612 (1996).
23. Aglietti, U., and Ricciardi, G., *Nucl. Phys.* **B587**, 363–399 (2000).
24. Bauer, C. W., Ligeti, Z., and Luke, M. E., arXiv:hep-ph/0007054 (2000).
25. Leibovich, A. K., Low, I., and Rothstein, I. Z., *Phys. Rev.* **D61**, 053006 (2000).
26. Leibovich, A. K., Low, I., and Rothstein, I. Z., *Phys. Lett.* **B486**, 86–91 (2000).
27. Neubert, M., *Phys. Lett.* **B513**, 88–92 (2001).
28. Leibovich, A. K., Low, I., and Rothstein, I. Z., *Phys. Lett.* **B513**, 83–87 (2001).
29. Bigi, I., and Uraltsev, N., *Int. J. Mod. Phys.* **A17**, 4709–4732 (2002).
30. Bauer, C. W., Ligeti, Z., and Luke, M. E., *Phys. Lett.* **B479**, 395–401 (2000).
31. Bauer, C. W., Ligeti, Z., and Luke, M. E., *Phys. Rev.* **D64**, 113004 (2001).
32. Neubert, M., *JHEP* **07**, 022 (2000).
33. Neubert, M., and Becher, T., *Phys. Lett.* **B535**, 127–137 (2002).
34. Kakuno, H., et al., [Belle Collaboration], arXiv:hep-ex/0311048 (2003).
35. Bigi, I. I. Y., Shifman, M. A., Uraltsev, N., and Vainshtein, A. I., *Phys. Lett.* **B328**, 431–440 (1994).
36. Kagan, A. L., and Neubert, M., *Eur. Phys. J.* **C7**, 5–27 (1999).
37. Neubert, M. (2003), private communications, discussed in CLNS-04/1858 (in preparation).
38. Bauer, C. W., and Manohar, A. V., arXiv:hep-ph/0312109 (2003).
39. Cronin-Hennessy, D. (2003), private communication.
40. Chen, S., et al., [CLEO Collaboration], *Phys. Rev. Lett.* **87**, 251807 (2001).
41. Gibbons, L. K., Hennessy, D. C., and Thorndike, E. H. (2004), in preparation.
42. Heavy Flavor Averaging Group (HFAG) (2003), URL: http://www.slac.stanford.edu/xorg/hfag/semi/summer03-lp/summe%r03.shtml.
43. Barate, R., et al., [ALEPH Collaboration], *Eur. Phys. J.* **C6**, 555–574 (1999).
44. Acciarri, M., et al., [L3 Collaboration], *Phys. Lett.* **B436**, 174–186 (1998).
45. Abbiendi, G., et al., [OPAL Collaboration], *Eur. Phys. J.* **C21**, 399–410 (2001).
46. Bornheim, A., et al., [CLEO Collaboration] (2002), CLEO-CONF-02-08.
47. Luke, M., *ECONF* **C0304052**, WG107 (2003).
48. Ligeti, Z., arXiv:hep-ph/0309219 (2003).
49. Leibovich, A. K., Ligeti, Z., and Wise, M. B., *Phys. Lett.* **B539**, 242–248 (2002).
50. Bauer, C. W., Luke, M., and Mannel, T., *Phys. Lett.* **B543**, 261–268 (2002).
51. Neubert, M., *Phys. Lett.* **B543**, 269–275 (2002).
52. Bauer, C. W., Luke, M. E., and Mannel, T., *Phys. Rev.* **D68**, 094001 (2003).
53. Bigi, I. I. Y., and Uraltsev, N. G., *Nucl. Phys.* **B423**, 33–55 (1994).
54. Voloshin, M. B., *Phys. Lett.* **B515**, 74–80 (2001).
55. Gilman, F. J., and Singleton, R. L., *Phys. Rev.* **D41**, 142 (1990).
56. Gibbons, L., *ECONF* **C0304052**, WG105 (2003).
57. Abada, A., et al., *Nucl. Phys.* **B416**, 675–698 (1994).
58. Allton, C. R., et al., [APE Collaboration], *Phys. Lett.* **B345**, 513–523 (1995).
59. Del Debbio, L., Flynn, J. M., Lellouch, L., and Nieves, J., [UKQCD Collaboration], *Phys. Lett.* **B416**, 392–401 (1998).
60. Hashimoto, S., Ishikawa, K.-I., Matsufuru, H., Onogi, T., and Yamada, N., *Phys. Rev.* **D58**, 014502 (1998).
61. Ryan, S., et al., *Nucl. Phys. Proc. Suppl.* **73**, 390–392 (1999).

62. Ryan, S. M., El-Khadra, A. X., Kronfeld, A. S., Mackenzie, P. B., and Simone, J. N., *Nucl. Phys. Proc. Suppl.* **83**, 328–330 (2000).
63. Lellouch, L., arXiv:hep-ph/9912353 (1999).
64. Bowler, K. C., et al., [UKQCD Collaboration], *Phys. Lett.* **B486**, 111–117 (2000).
65. Becirevic, D., and Kaidalov, A. B., *Phys. Lett.* **B478**, 417–423 (2000).
66. Aoki, S., et al., [JLQCD Collaboration], *Nucl. Phys. Proc. Suppl.* **94**, 329–332 (2001).
67. Abada, A., et al., *Nucl. Phys.* **B619**, 565–587 (2001).
68. El-Khadra, A. X., Kronfeld, A. S., Mackenzie, P. B., Ryan, S. M., and Simone, J. N., *Phys. Rev.* **D64**, 014502 (2001).
69. Aoki, S., et al., [JLQCD Collaboration], *Phys. Rev.* **D64**, 114505 (2001).
70. Ball, P., and Braun, V. M., *Phys. Rev.* **D55**, 5561–5576 (1997).
71. Ball, P., and Braun, V. M., *Phys. Rev.* **D58**, 094016 (1998).
72. Khodjamirian, A., Ruckl, R., Weinzierl, S., and Yakovlev, O. I., *Phys. Lett.* **B410**, 275–284 (1997).
73. Khodjamirian, A., Ruckl, R., Weinzierl, S., Winhart, C. W., and Yakovlev, O. I., *Phys. Rev.* **D62**, 114002 (2000).
74. Bakulev, A. P., Mikhailov, S. V., and Ruskov, R., arXiv:hep-ph/0006216 (2000).
75. Huang, T., Li, Z., and Wu, X., arXiv:hep-ph/0011161 (2000).
76. Wang, W. Y., and Wu, Y. L., *Phys. Lett.* **B515**, 57–64 (2001).
77. Wang, W.-Y., and Wu, Y.-L., *Phys. Lett.* **B519**, 219–228 (2001).
78. Ball, P., and Zwicky, R., *JHEP* **10**, 019 (2001).
79. Battaglia, M., et al., arXiv:hep-ph/0304132 (2003).
80. Wirbel, M., Stech, B., and Bauer, M., *Z. Phys.* **C29**, 637 (1985).
81. Korner, J. G., and Schuler, G. A., *Z. Phys.* **C38**, 511 (1988).
82. Isgur, N., Scora, D., Grinstein, B., and Wise, M. B., *Phys. Rev.* **D39**, 799 (1989).
83. Scora, D., and Isgur, N., *Phys. Rev.* **D52**, 2783–2812 (1995).
84. Melikhov, D., *Phys. Rev.* **D53**, 2460–2479 (1996).
85. Beyer, M., and Melikhov, D., *Phys. Lett.* **B436**, 344–350 (1998).
86. Faustov, R. N., Galkin, V. O., and Mishurov, A. Y., *Phys. Rev.* **D53**, 6302–6315 (1996).
87. Demchuk, N. B., Kulikov, P. Y., Narodetsky, I. M., and O'Donnell, P. J., *Phys. Atom. Nucl.* **60**, 1292–1304 (1997).
88. Grach, I. L., Narodetsky, I. M., and Simula, S., *Phys. Lett.* **B385**, 317–323 (1996).
89. Riazuddin, Al-Aithan, T. A., and Gilani, A. H. S., *Int. J. Mod. Phys.* **A17**, 4927–4938 (2002).
90. Melikhov, D., and Stech, B., *Phys. Rev.* **D62**, 014006 (2000).
91. Feldmann, T., and Kroll, P., *Eur. Phys. J.* **C12**, 99–108 (2000).
92. Flynn, J. M., and Nieves, J., *Phys. Lett.* **B505**, 82–88 (2001).
93. Beneke, M., and Feldmann, T., *Nucl. Phys.* **B592**, 3–34 (2001).
94. Choi, H.-M., and Ji, C.-R., *Phys. Lett.* **B460**, 461–466 (1999).
95. Kurimoto, T., Li, H.-N., and Sanda, A. I., *Phys. Rev.* **D65**, 014007 (2002).
96. Ligeti, Z., and Wise, M. B., *Phys. Rev.* **D53**, 4937–4945 (1996).
97. Aitala, E. M., et al., [E791 Collaboration], *Phys. Rev. Lett.* **80**, 1393–1397 (1998).
98. Burdman, G., and Kambor, J., *Phys. Rev.* **D55**, 2817–2826 (1997).
99. Lellouch, L., *Nucl. Phys.* **B479**, 353–391 (1996).
100. Mannel, T., and Postler, B., *Nucl. Phys.* **B535**, 372–386 (1998).
101. Lange, B. O., arXiv:hep-ph/0310139 (2003).
102. Ball, P., arXiv:hep-ph/0308249 (2003).
103. Beneke, M., and Feldmann, T., arXiv:hep-ph/0311335 (2003).
104. Lange, B. O., and Neubert, M., arXiv:hep-ph/0311345 (2003).
105. Davies, C. T. H., et al., [HPQCD Collaboration], arXiv:hep-lat/0304004 (2003).
106. Bernard, C., et al., [MILC Collaboration], arXiv:hep-lat/0309055 (2003).
107. Okamoto, M., et al., arXiv:hep-lat/0309107 (2003).
108. Behrens, B. H., et al., [CLEO Collaboration], *Phys. Rev.* **D61**, 052001 (2000).
109. Aubert, B., et al., [BaBar Collaboration], *Phys. Rev. Lett.* **90**, 181801 (2003).
110. Athar, S. B., et al., [CLEO Collaboration], *Phys. Rev.* **D68**, 072003 (2003).
111. Abe, K., et al., [Belle Collaboration], arXiv:hep-ex/0307075 (2003).
112. Foley, K. M., and Lepage, G. P., *Nucl. Phys. Proc. Suppl.* **119**, 635–637 (2003).
113. Boyle, P. A., arXiv:hep-lat/0309100 (2003).

B LIFETIMES AND MIXING

B Lifetime Measurements at the Tevatron

Daria Zieminska

(*Representing the CDF and DØ Collaborations*)

Indiana University, Physics Department, Bloomington, IN 47405, U.S.A.

Abstract. We present recent measurements of the inclusive b hadron lifetime and the B^+, B^0, B_s^0 and Λ_b lifetimes from exclusive decay channels, performed by the CDF and DØ experiments at the Tevatron.

1. INTRODUCTION

Theoretical predictions of inclusive decay rates of beauty hadrons are based on perturbative calculations of short distance effects and lattice QCD calculations of matrix elements. The latest predictions [1] for the lifetime ratios and for the width differences within the neutral B meson doublets, in next to leading order QCD, are summarized in Table 1, along with the current measurements [2]. The accuracy of the predictions exceeds the precision of the measurements for ratios involving B_s^0 mesons and Λ_b baryons, which are not accessible at the e^+e^- machines running at the $\Upsilon(4S)$. There is a general agreement between data and the predictions, the largest difference (1.5σ) is in the ratio $\tau(\Lambda_b)/\tau(B^0)$ which is found to be smaller than expected.

TABLE 1. *B* hadron lifetime ratios.

	Experiment [2]	Theory [1]
$\tau(B^+)/\tau(B^0)$	1.074 ± 0.014	1.06 ± 0.02
$\tau(B_s^0)/\tau(B^0)$	0.948 ± 0.038	1.00 ± 0.01
$\Delta\Gamma_s/\Gamma_s$	<0.54 (95% CL)	$(7.4\pm 2.4)\%$
$\tau(\Lambda_b)/\tau(B^0)$	0.796 ± 0.052	0.90 ± 0.05

2. TRIGGERS AND DATA SAMPLES

The Fermilab Tevatron Collider Run II started officially in March 2002. By October 2003, the Tevatron has delivered over 200 pb^{-1} of collisions to the CDF and DØ experiments. Present analyses use about half of this data: 115 pb^{-1} (DØ) and 138 pb^{-1} (CDF). The Tevatron produces a full spectrum of heavy flavor hadrons but the abundant b production is overwhelmed by the QCD background. Both experiments use a three-level trigger system to select heavy flavor events. Most of the b hadron lifetime measurements at CDF and DØ are based on decay channels with a J/ψ meson in the final state, and use a trigger requiring the detection of two tracks in the muon system. The acceptance for a muon is $|\eta|<1$ and $p_T > 1.5$ GeV/c (CDF) and $p_T > 1.5$ GeV/c at $1< |\eta| <2$, $p_T > 3.5$ GeV/c at $|\eta| <1$ (DØ). To collect semileptonic B decays, DØ uses a single muon trigger, with a luminosity-dependent prescale. At CDF, a semileptonic sample is obtained by requiring an SVT track emanating from a displaced vertex to accompany a muon at the second level of the trigger.

3. INCLUSIVE B HADRON LIFETIME

The proper time of a B decay is determined from the distance between the primary vertex and the B meson decay vertex in the plane transverse to the beam axis.

$$L_{xy}^B = (\vec{x}_B - \vec{x}_{prim}) \cdot \vec{p}_T / p_T, \qquad (1)$$

where $\vec{p}_T$ is the measured transverse momentum vector and p_T is its value. At DØ, the primary vertex is reconstructed individually for each event. The typical resolution of L_{xy}^B is 40 μm. CDF uses the run-averaged beam position whose contribution to the L_{xy}^B uncertainty is $\approx$30 μm.

In the fully reconstructed B hadron decays, the proper lifetime, τ, is defined by the relation:

$$c\tau = L_{xy}^B \cdot M_B / p_T, \qquad (2)$$

where M_B is the B hadron mass. In the case of the inclusive $B \to J/\psi X$ decay, the boost factor is obtained from a Monte Carlo simulation.

The proper J/ψ decay length distribution measured at DØ, with the result of a binned maximum likelihood fit overlaid, is shown in Fig. 1. The inclusive B hadron lifetime is measured to be 1.562 ± 0.013(stat) ± 0.045(syst) ps. Monte Carlo tests have revealed a bias, not yet fully

CP722, *B Physics at Hadron Machines*, edited by M. Paulini and S. Erhan
© 2004 American Institute of Physics 0-7354-0203-5/04/$22.00

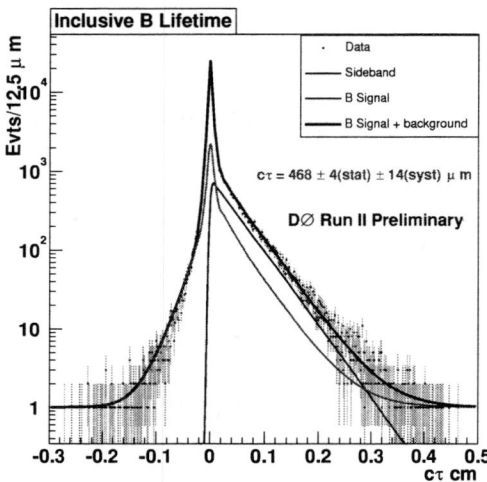

FIGURE 1. $c\tau$ fit result for the inclusive channel $J/\psi X$ at DØ.

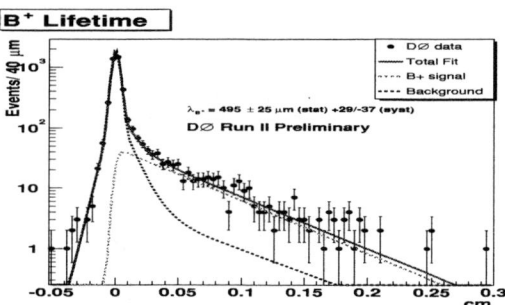

FIGURE 2. Fit result for $c\tau(J/\psi K^{+})$ at DØ.

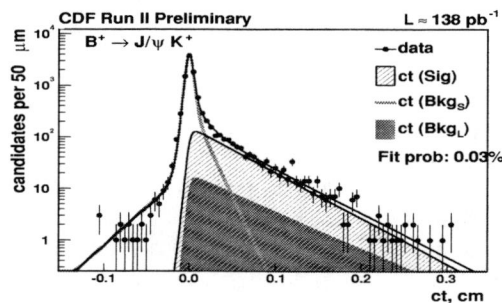

FIGURE 3. Fit projection $c\tau(J/\psi K^{+})$ at CDF.

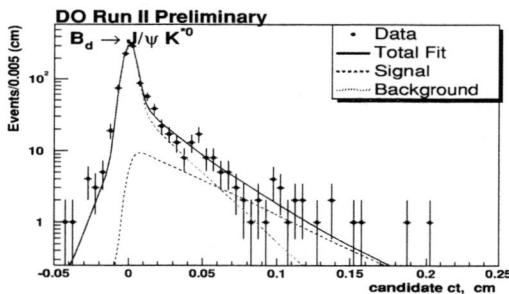

FIGURE 4. Fit projection on $c\tau(J/\psi K^{*0})$ at DØ.

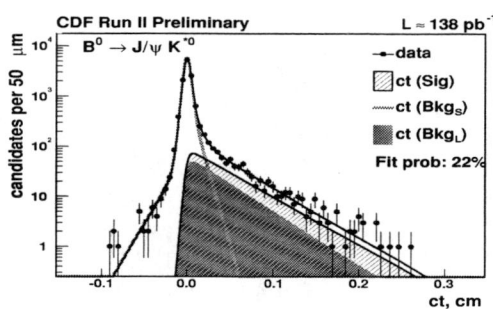

FIGURE 5. Fit projection on $c\tau(J/\psi K^{*0})$ at CDF.

understood, of -0.03 ps. The effect has been accounted for in the estimate of the systematic uncertainty, while the quoted result remains uncorrected. Other sources of the systematic uncertainty are the kinematic correction factor, and the modeling of the background. A similar measurement by CDF, based on 18 pb^{-1}, led to $\tau(b)=1.526 \pm 0.034(\text{stat}) \pm 0.035(\text{syst})$ ps.

4. EXCLUSIVE DECAY CHANNELS

Both experiments have measured the lifetimes of the B^{+}, B^{0} and B_{s}^{0} mesons from the decay channels $B^{+} \rightarrow J/\psi K^{+}$, $B^{0} \rightarrow J/\psi K^{*0}$, and $B_{s}^{0} \rightarrow J/\psi \phi$, respectively. Except for the B^{+} lifetime measurement by DØ, the analyses reported here apply simultaneous unbinned maximum likelihood fits to the candidate mass and proper decay length.

The proper length distributions and fit results are shown in Figs. 2-5 and Figs. 8, 9. The lifetime results are:

$\tau(B^{+})=1.65 \pm 0.08(\text{stat}) {}^{+0.09}_{-0.12}(\text{syst})$ ps (DØ),

$\tau(B^{+})=1.63 \pm 0.05(\text{stat}) \pm 0.04(\text{syst})$ ps (CDF),

$\tau(B^{0})=1.51 \pm 0.18(\text{stat}) \pm 0.20(\text{syst})$ ps (DØ),

$\tau(B^{0})=1.51 \pm 0.06(\text{stat}) \pm 0.02(\text{syst})$ ps (CDF),

$\tau(B_{s}^{0})=1.19 \pm 0.18(\text{stat}) \pm 0.14(\text{syst})$ ps (DØ),

$\tau(B_{s}^{0})=1.33 \pm 0.14(\text{stat}) \pm 0.02(\text{syst})$ ps (CDF).

The B_{s}^{0} lifetime measurements are discussed in more detail in the next subsection.

4.1. $B_s^0 \to J/\psi\phi$

B_s^0 candidates are selected by pairing J/ψ candidates with two oppositely charged tracks, consistent with a ϕ decay. To improve the signal to background ratio, the candidates are required to have $p_T > 6.5$ GeV/c (CDF) and 6 GeV/c (DØ), and the ϕ mesons to have $p_T > 2$ GeV/c. In addition, at DØ each track is required to have $p_T > 1$ GeV/c. In the fit, the signal is modeled by an exponential convoluted with a Gaussian, allowing for an event-by-event variation of the resolution. The background is modeled by a sum of a Gaussian and exponential tails, one at $c\tau < 0$, and two at $c\tau > 0$. The signal is assumed to have a Gaussian mass distribution with a free mean. The width parameter is free at CDF, and at DØ it is fixed at 37 MeV/c^2. A linear mass spectrum is assumed for background. The invariant mass and proper length distributions for the B_s^0 candidates measured at DØ and CDF, with the fit projections overlaid, are shown in Figs. 8 and 9, respectively. There are 120 ± 13 fitted signal events at CDF, and 69 ± 14 events at DØ. The preliminary lifetime results are listed in Fig. 6, along with existing B_s^0 lifetime results. The new measurements are statistics limited. The dominant sources of the systematic uncertainty of the CDF measurement are the alignment and the resolution function. The systematic uncertainty at DØ reflects the current precision of the MC test of the reconstruction and fitting procedure.

The measurements from the $J/\psi\phi$ channel are listed separately from the flavor-specific channels to emphasize the fact that this final state is a mixture of $CP = +1$ and $CP = -1$ states which is different from the CP content of the flavor-specific states. In both cases, all current lifetime measurements are based on a single-exponential fit to the a priori two-exponential time evolution.

In the case of the flavor-specific decay channels, the single-exponential fit determines

$$\Gamma_{fs} = \Gamma_s - (\Delta\Gamma_s)^2/2\Gamma_s + O(\Delta\Gamma_s)^3/\Gamma_s^2), \qquad (3)$$

where $\Delta\Gamma_s$ denotes the width difference $\Gamma_L - \Gamma_H$ between the light and heavy mass eigenstates of the B_s^0 doublet, and $\Gamma_s = (\Gamma_L + \Gamma_H)/2$.

For the $J/\psi\phi$ channel, the time evolution can be expressed in terms of the linear polarization states as [3]:

$$d\Gamma/dt = (|A_0(0)|^2 + |A_\parallel(0)|^2)e^{-\Gamma_L t} + |A_\perp(0)|^2 e^{-\Gamma_H t}. \qquad (4)$$

The $CP = -1$ component, $|A_\perp(0)|^2$, is poorly known, but it is expected to be small [3]. In the case of $A_\perp(0) = 0$, the lifetime measured in the $J/\psi\phi$ channel would be equal to $1/\Gamma_L$.

The measurements presented here are the first step in the program which envisions a study of the time evolution combined with the angular correlation of the two final state vector mesons. The next step is to utilize the fact

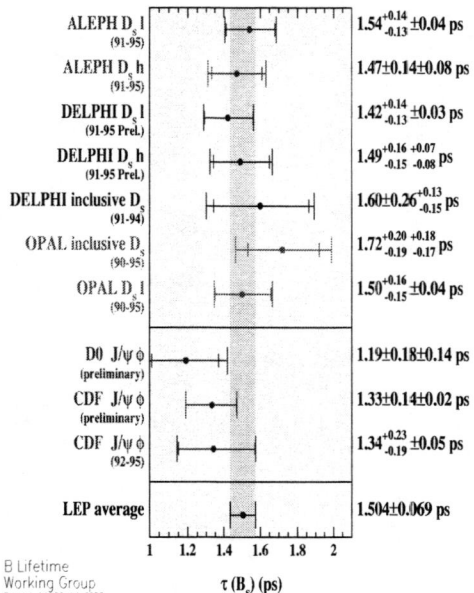

FIGURE 6. Summary of B_s^0 lifetime measurements.

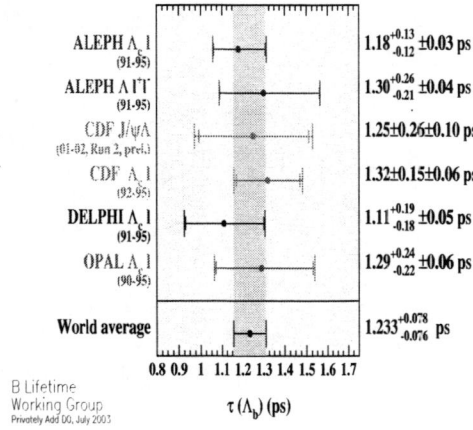

FIGURE 7. Summary of Λ_b lifetime measurements.

that the coefficients of $e^{-\Gamma_L t}$ and $e^{-\Gamma_H t}$ in Eq. (4) have a different dependence on the "transversity" angle (defined as the polar angle of μ^+ in the J/ψ rest frame, where the z-axis is normal to the ϕ decay plane). The extension of the fitting procedure to include the "transversity" angle, and to allow for two decay width parameters is underway.

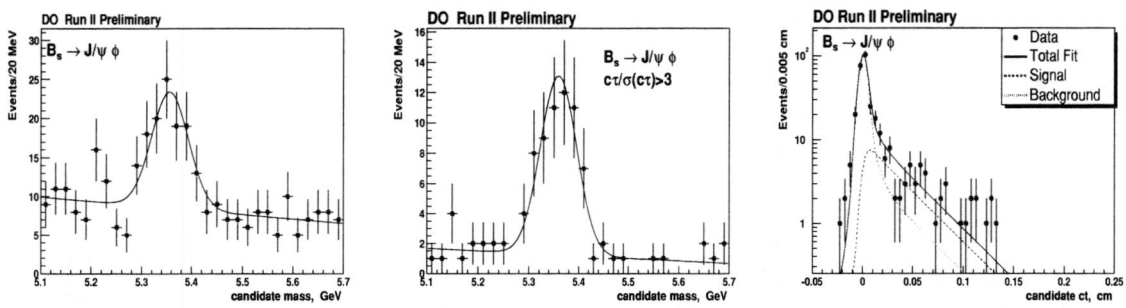

FIGURE 8. Fit projection on $M(J/\psi\phi)$ and $c\tau(J/\psi\phi)$ at DØ.

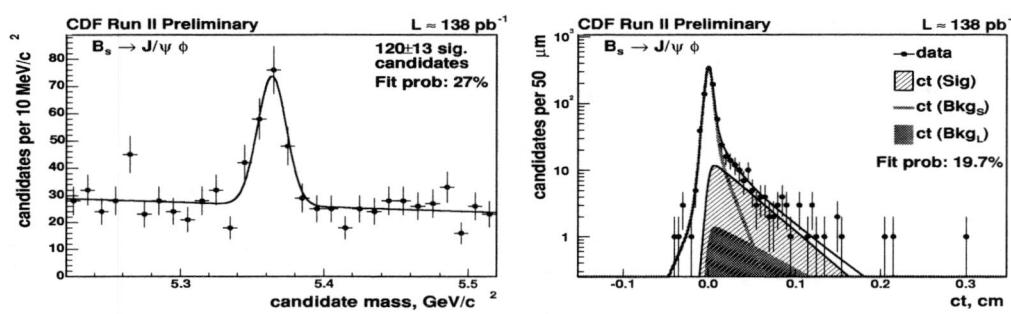

FIGURE 9. Fit projection on $M(J/\psi\phi)$ and $c\tau(J/\psi\phi)$ at CDF.

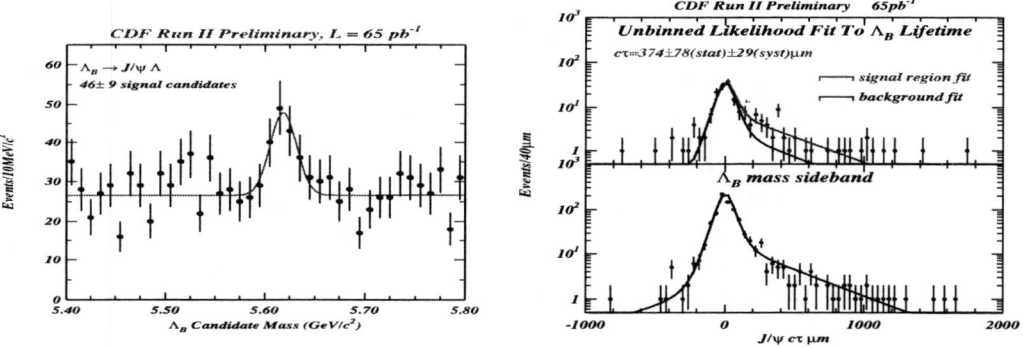

FIGURE 10. Fit projection on $M(J/\psi\Lambda)$ and $c\tau(J/\psi\Lambda)$ at CDF.

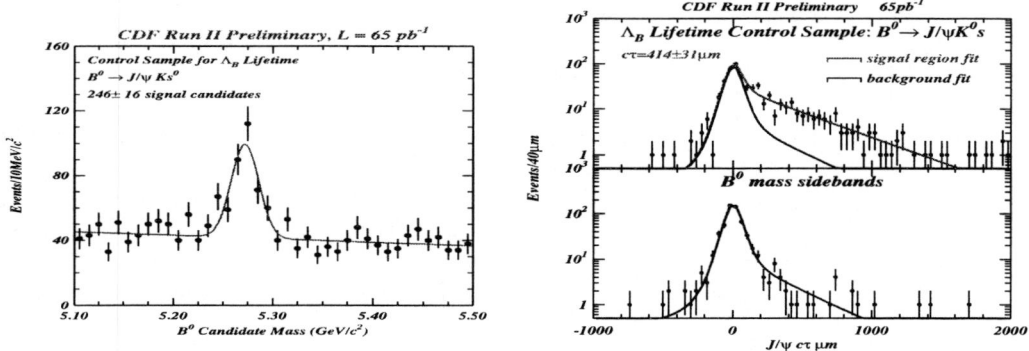

FIGURE 11. Fit projection on $M(J/\psi K_s^0)$ and $c\tau(J/\psi K_s^0)$ at CDF.

4.2. $\Lambda_b \to J/\psi\Lambda$ and $B^0 \to J/\psi K_S^0$

The CDF Collaboration has measured the Λ_b lifetime in the exclusive decay mode $\Lambda_b \to J/\psi\Lambda$, using the $B^0 \to J/\psi K_S^0$ decay as a control process. This preliminary result, based on the exposure of 65 pb^{-1}, is the first measurement of the Λ_b lifetime in the fully reconstructed mode. The measurements thus far have shown some disagreement with the theoretical prediction for $\tau(\Lambda_b)/\tau(B^0)$ – see Table 1. In this analysis, the fits to the proper decay length distributions use background templates determined in the mass sidebands. The mass and decay length distributions for the two channels, along with fit results, are shown in Figs. 10 and 11. The number of fitted Λ_b events and B^0 events is 46 ± 9, and 246 ± 16, respectively. The lifetime results are:

$\tau(B^0)$=1.38 ± 0.10(stat) ± 0.10(syst) ps (CDF),
$\tau(\Lambda_b)$=1.25 ± 0.26(stat) ± 0.10(syst) ps (CDF).

The Λ_b lifetime result is included in Fig. 7.

5. SEMILEPTONIC CHANNELS

Thanks to the "lepton plus displaced track" trigger, CDF is collecting a sample of semileptonic B decays far exceeding in statistics the samples of decays to the exclusive modes. The group has already acquired the world largest sample of semileptonic decays of the B_s^0 and Λ_b hadrons. The data analysis is in progress, the complication arising from the impact parameter cut requiring a realistic simulation of the trigger conditions.

DØ has the information on the track impact parameter available at the second level of the trigger and will implement the impact parameter trigger after the Fall 2003 shutdown. The group has analyzed the abundant inclusive decay channel $B \to D^0\mu X$ as a benchmark for the B_s^0 studies, and determined the lifetime to be 1.460 ± 0.083(stat) ps (see Fig. 12).

6. SUMMARY

After the first year of the Tevatron Run II, CDF and DØ report new measurements of B meson lifetimes in the $B \to J/\psi X$ modes, including a (CP-averaged) lifetime of B_s^0 from the $J/\psi\phi$ channel. CDF reports a unique measurement of the Λ_b lifetime from the fully reconstructed final state $J/\psi\Lambda$. In addition, CDF is collecting a large statistics sample of semileptonic decay modes using a new impact parameter trigger. DØ will add a vertex trigger after the Fall 2003 shutdown. DØ is also increasing the acceptance of the low p_T dimuon trigger by allowing muon tracks that don't penetrate the toroid.

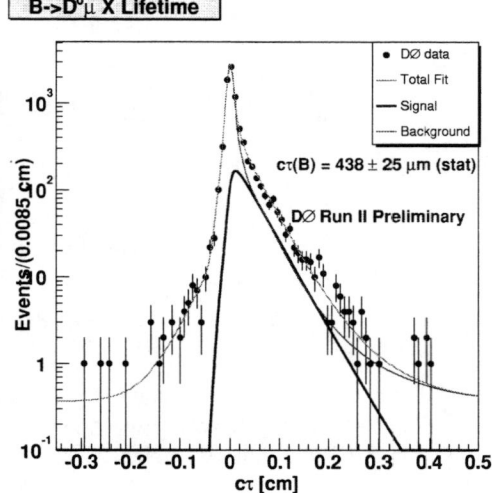

FIGURE 12. $c\tau$ fit for the decay $B \to D^0\mu X$ at DØ.

Fermilab's goal is to collect 2 fb^{-1} in the first phase of Run II (by 2007). Based on the current performance, the projection of the sample size at CDF for the exposure of 2 fb^{-1} is ≈ 2000 B_s^0 and Λ_b events, and a statistical precision of 3% for $\tau(B_s^0)$ and 4% for $\tau(\Lambda_b)$ lifetime, with similar figures for DØ. Note that in the lifetime ratios the systematic uncertainties cancel out to a large degree. Both groups will attempt to measure the lifetime difference in the B_s^0 doublet – see the discussion of future prospects by Ting Miao [4] in these proceedings.

REFERENCES

1. Ciuchini, M., Franco, E., Lubicz, V., Mescia, F., and Tarantino, C., *JHEP* **08**, 031 (2003).
2. Battaglia, M., et al., arXiv:hep-ph/0304132 (2003).
3. Dighe, A. S., Dunietz, I., Lipkin, H. J., and Rosner, J. L., *Phys. Lett.* **B369**, 144–150 (1996).
4. Miao, T., in these proceedings (2004).

B Tagging and Mixing at the Tevatron

Ting Miao

(Representing the CDF and DØ Collaborations)

Fermi National Accelerator Laboratory, Batavia, IL 60510, U.S.A.

Abstract. The measurement of B_s^0 mixing is one of the flagship analyses for the Run II *B* physics program at the Fermilab Tevatron. We report here preliminary results on key elements to this measurement including B_s^0 event reconstruction, proper time resolution and initial *B* flavor tagging. The prospects of B_s^0 mixing with the upgraded CDF and DØ detectors are also discussed.

1. INTRODUCTION

The measurements of the oscillation frequency of neutral *B* mesons, Δm_d for the B^0-$\bar{B}^0$ system and Δm_s for the B_s^0-$\bar{B}_s^0$ system, will be used to precisely determine the length of one side of the CKM unitarity triangle which is shown in Fig. 1. The value of Δm_d is measured to great precision [1], $\Delta m_d = 0.502 \pm 0.006$ ps^{-1}. However, Δm_s for the rapidly oscillating B_s^0-$\bar{B}_s^0$ system has not yet been resolved. Its determination has highest priority for the *B* physics program of the Fermilab Tevatron experiments.

Key elements to the B_s^0 mixing measurement include the ability to collect a large B_s^0 signal sample with small background, the ability of initial *B* flavor tagging and the ability of a precise determination of the proper decay time. We will discuss the status and prospects of these key elements throughout this report.

2. B_S^0 EVENT COLLECTION

Both CDF and DØ started taking Run II data in March 2001. The strengths of the two detectors after upgrade efforts in the 5 year shutdown are somewhat complementary to one another. Here, we only briefly point out components most relevant to *B* physics and refer to details listed in Ref. [2, 3]. Both detectors have axial solenoidal magnetic fields, central tracking and use silicon microvertex detectors. The DØ central tracking volume with the scintillating fibers covers the pseudorapidity region of $|\eta| \leq 1.7$. The DØ silicon detector has a barrel geometry interspersed with disk detectors to extend the tracking volume to $|\eta| \leq 3$. The CDF detector features a 1.4 T solenoid surrounding a silicon microver-

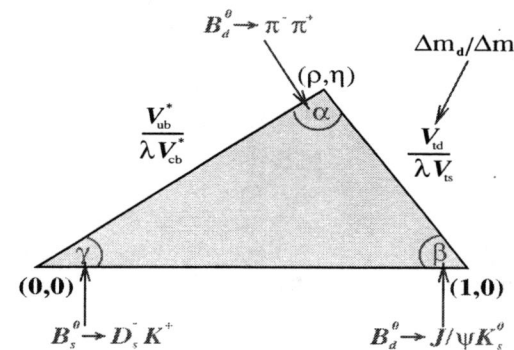

FIGURE 1. Experimental constrains on the CKM unitarity triangle.

tex detector and gas-wire drift chamber to provide excellent mass and vertex resolutions in the central region. The improved tracking coverages of both CDF and DØ, combined with muon detectors and calorimeters, allow for excellent muon and electron identification, as well as precise momentum and vertex measurements which are important for *B* physics. The late addition of a silicon layer very close to the beam spot (L00) and a Time-of-Flight (ToF) system improves dramatically CDF's capability of vertexing and flavor tagging.

Both experiments identify leptons at the trigger level to collect *B* events with leptons in the final state such as semileptonic *B* decays and $B \rightarrow J/\psi X$. DØ has an inclusive muon trigger with large geometry coverage to allow accumulating very large samples of semileptonic decays. Fig. 2 shows the $B_s^0 \rightarrow D_s^- \mu^+ \nu X$ signal collected by DØ with muon triggers. The CDF semileptonic trigger requires an additional displaced track associated with the lepton, providing cleaner samples with a somewhat smaller yields.

CP722, *B Physics at Hadron Machines*, edited by M. Paulini and S. Erhan
© 2004 American Institute of Physics 0-7354-0203-5/04/$22.00

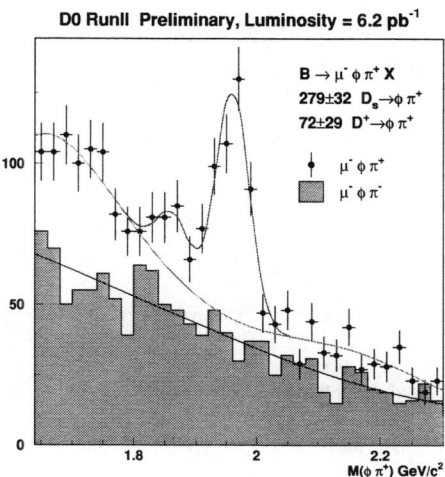

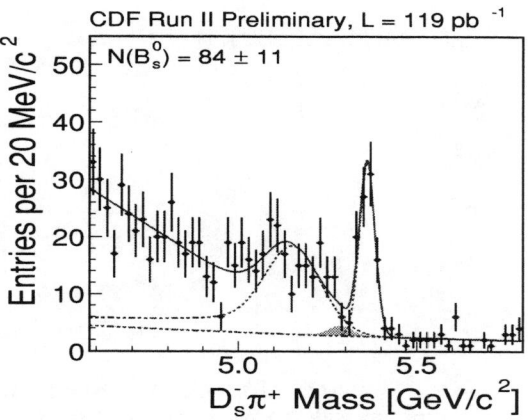

FIGURE 2. B_s^0 events from semileptonic decays of $B_s^0 \to D_s^- \mu^+ \nu X$ with $D_s^- \to \phi \pi^-$ by the DØ experiment. The two peaks correspond to the D^+ and D_s^+ states, which can both decay to $\phi \pi^+$.

FIGURE 3. B_s^0 events from $B_s^0 \to D_s^- \pi^+$ decays with $D_s^- \to \phi \pi^-$ collected with CDF's SVT trigger.

New to experiments at a hadron collider machine is the use of a new Silicon Vertex Tracker (SVT) at CDF to effectively collect events with large track impact parameters. Tracks reconstructed by the eXtremely Fast Tracker (XFT) are passed to the SVT, which appends silicon hits to the track to measure the impact parameter of each track. The SVT has an impact parameter resolution of 47 μm for tracks with $p_T > 2$ GeV/c. With this trigger, CDF collects B_s^0 decays with at least two tracks with impact parameter $d_0 > 120$ μm and $p_T > 2$ GeV/c. Fig. 3 show the B_s^0 signal from $B_s^0 \to D_s^- \pi^+$ with $D_s^- \to \phi \pi^-$ using about 65 pb^{-1} of data collected in the winter of 2002. In this plot, the narrow peak on the right is the B_s^0 signal and the broader bump on its left originates from background contributions from B_s^0 channels other than $B_s^0 \to D_s^- \pi^+$.

There were many improvements on detector coverage and trigger optimization since this early data taking period where CDF was still in the process of commissioning the silicon detector and SVT trigger. The live coverage of the silicon detector has reached above 90%. The SVT efficiency was improved greatly with a change of pattern recognition from requiring 4 hits in 4 silicon layers to 4 out 5 layers. CDF also developed an optimized scheme to handle the L2 bandwidth to maximize the useful trigger rate in different instantaneous luminosity scenarios. With these improvements, CDF was able to gain a factor of two increase for the B_s^0 signal yield which corresponds to a rate of 1600 events per 1 fb^{-1} for $B_s^0 \to D_s^- \pi^+$ with $D_s^- \to \phi \pi^-$. Adding events from channels like $D_s^- \to K^* K^-$, $K_S^0 K^-$ and $B_s^0 \to D_s^- \pi^+ \pi^- \pi^+$, a rate of 2000 B_s^0 events per 1 fb^{-1} is within reach.

The ability of collecting fully reconstructed B_s^0 decays

strengthens our reach for much faster oscillation frequencies and makes the Tevatron the unique place for B_s^0 mixing studies.

3. FLAVOR TAGGING

The initial flavor of a B meson at its production can be determined by using information from particles either produced on the same side together with the B meson of interest or on the opposite of the B meson. Same Side Tagging (SST) uses the charge of an energetic track near the B meson to infer its production flavor. Opposite side tagging algorithms, such as Soft Lepton Tagging (SLT), Jet-Charge Tagging (JetQ) and Opposite Kaon Tagging (OKT), take advantage of the knowledge of correlated $b\bar{b}$ production to infer the flavor of the B meson by the flavor of the other B hadron in the event.

The SST takes advantage of the correlation between the B flavor and the charge of a nearby track produced along with the B during b fragmentation. For a $\bar{b}$ quark to become a B_s^0 meson, it must pair with an s quark from the vacuum. An $\bar{s}$ is produced along with it, which could potentially turn into a K^+ meson in the B_s^0 case. Thus, the correlation of B_s^0-K^+ is the basis for the SST tagging algorithm. The same is true for the B^0-π^+ correlation. B^{**} decays provide a similar correlation.

The SLT algorithm uses the charge information of leptons produced from semileptonic decays, $b \to \ell^- X$ or $\bar{b} \to \ell^+ X$, to identify the B meson flavor. In the JetQ algorithm, the weighted sum of the charges of tracks recoiling against a B meson is used. The OKT uses the fact that it is more likely for a K^- than a K^+ to be produced from a $\bar{B}$ decay in the decay chain $b \to c \to s$. Thus, a K^- is a strong indication that it comes from a $\bar{B}$ meson. This tagging algorithm heavily relies on the

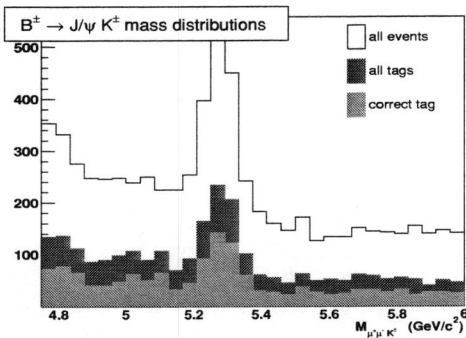

FIGURE 4. $B^+ \to J/\psi K^+$ signal from the DØ experiment. The open histogram with the black line corresponds to all events. The histogram with the blue line are events with a JetQ tag and the histogram with the green line are events which were correctly tagged using JetQ.

kaon identification capability which CDF's ToF system will provide.

The power of a flavor tagging algorithm is determined by its tagging efficiency and its dilution. The tagging efficiency is measured as the fraction of events for which the algorithm returns a decision: $\varepsilon \equiv N_{tag}/N_{total}$. The dilution, defined as $\mathscr{D} \equiv (N_{RS} - N_{WS})/(N_{RS} + N_{WS})$ where N_{RS} and N_{WS} are the respective numbers of correctly and incorrectly tagged events, reflects the purity of the tagging algorithm. The "effective tagging efficiency" is given by the quantity $\varepsilon \mathscr{D}^2$. In a B_s^0 mixing measurement, the statistical power of the sample is directly proportional to $\varepsilon \mathscr{D}^2$. The tagging power is proportional to $\mathscr{D}^2$ because incorrectly measuring the sign removes the event from the "correct" charge bin and puts the event into the "wrong" charge bin.

Using $B^+ \to J/\psi K^+$ events, as shown in Fig. 4, DØ tested the performance of the SST, SLT with muons and JetQ algorithms. The effective tagging efficiencies obtained from DØ are $\varepsilon \mathscr{D}^2 = (5.5 \pm 2.0)\%$ for SST, $(1.6 \pm 1.1)\%$ for SLT with muon, and $(3.3 \pm 1.7)\%$ for JetQ. The combined $\varepsilon \mathscr{D}^2$ from DØ using these three tagging algorithms is estimated to be $\sim 10\%$ [4].

CDF's strategy of flavor tagging for Run II is as follows. The algorithms are to be first optimized using a high statistics and high purity B sample such as semileptonic B decay events collected with the lepton+SVT trigger. Then, the performance of these taggers, in terms of $\varepsilon \mathscr{D}^2$, is measured using independent samples such as $B \to J/\psi K^+$ and fully reconstructed B decays collected with the SVT trigger. Finally, these tagging algorithms will be applied directly to a B_s^0 mixing measurement without retuning to avoid potential bias.

In order to use the high statistics lepton+SVT sample, we need to know its sample composition. First, the B component is enriched by applying kinematic con-

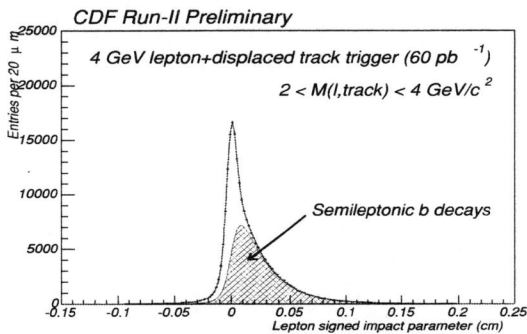

FIGURE 5. Signed impact parameter distribution of leptons with $2 < m(\ell, SVT) < 4$ GeV/c^2.

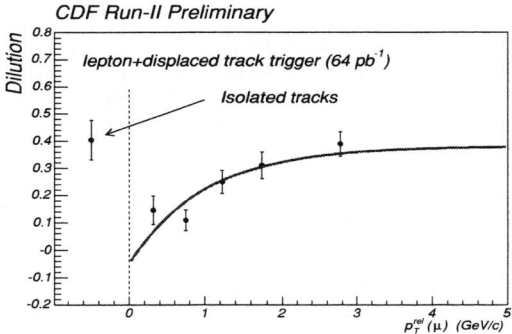

FIGURE 6. Dilution as a function of p_T^{rel} of the tag muon: $\mathscr{D} = A(1 - e^{-p_T^{rel} + B})$. p_T^{rel} is the component of the lepton's momentum that is transverse to the associated jet direction. The data point with $p_T^{rel} < 0$ corresponds to tracks not associated with a jet.

straints such as the requirement of the invariant mass of the lepton and the SVT track to be $2 < m(\ell, SVT) < 4$ GeV/c^2. Then, the remaining backgrounds from charm and prompt decays can be subtracted using the impact parameter distribution of the lepton, shown in Fig. 5. In addition, the B decay on the trigger side can mix or originate from wrong-sign sequential decays, $b \to c \to \ell$. These effects will reduce the 'raw' dilution measured for an opposite side tagging algorithm. CDF estimates the dilution on the trigger side using a Monte Carlo method and applies the correction factors obtained to determine the 'true' dilution of the opposite side tag. The overall correction factors of $\mathscr{D}(trig) = 0.6412 \pm 0.002^{+0.014}_{-0.023}$, and $\mathscr{D}(trig) = 0.6412 \pm 0.002^{+0.022}_{-0.037}$ were obtained for the μ+SVT and e+SVT sample, respectively. Using these samples, CDF obtained a preliminary result on SLT tagging with muons shown in Fig. 6. The performance is a strong function of the lepton relative transverse momentum p_T^{rel}. An averaged effective tagging efficiency $\varepsilon \mathscr{D}^2 = (0.7 \pm 0.1)\%$ was obtained which agrees well with projections using Run I results.

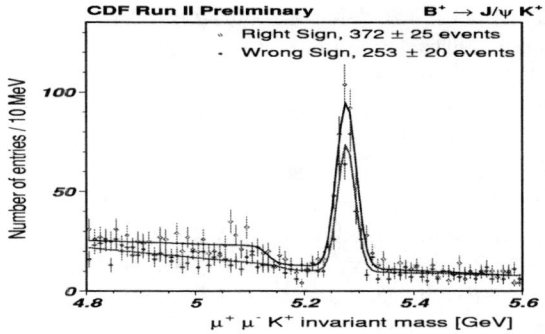

FIGURE 7. The invariant mass distribution of Right Sign (blue line) and Wrong Sign (red line) $B^+ \to J/\psi K^+$ candidates.

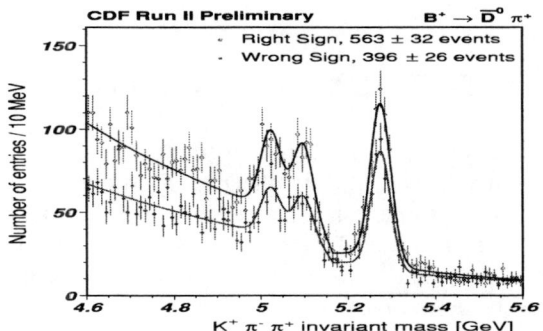

FIGURE 8. The invariant mass distribution of Right Sign and Wrong Sign $B^+ \to \bar{D}^0 \pi^+$ candidates.

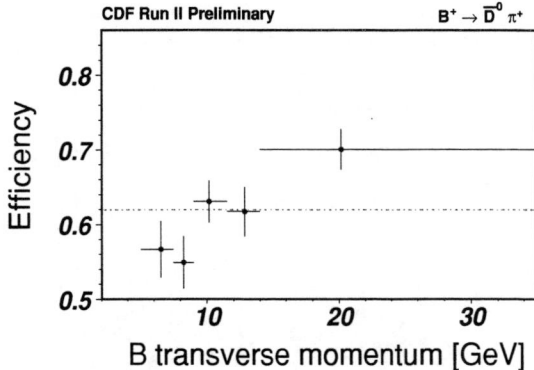

FIGURE 9. SST efficiency as a function of p_T for $B^+ \to J/\psi K^+$; the dot-dashed line denotes the global value.

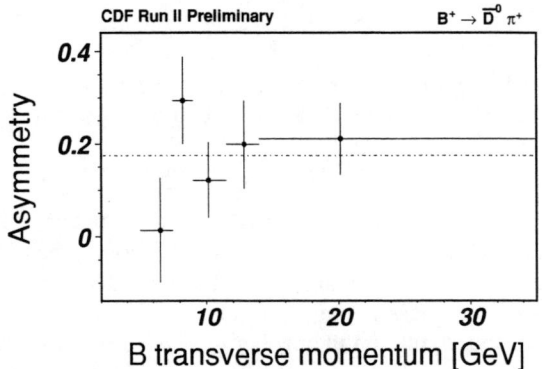

FIGURE 10. SST asymmetry as a function of p_T for $B^+ \to \bar{D}^0 \pi^+$; the dot-dashed line denotes the global value.

4. PROPER TIME RESOLUTION

The B_s^0-$\bar{B}_s^0$ oscillation is expected to have a very high frequency as the current limit, $\Delta m_s > 14.4$ ps^{-1} [1], implies. This means, there are more than 4 oscillations in one lifetime cycle. Thus, the B_s^0 proper time reconstruction resolution plays a key role in an experiment's sensitivity for a Δm_s measurement. At CDF and DØ, the trans-

The performance of the SST algorithm at CDF was tested on two fully reconstructed charged B decay samples from $B^+ \to J/\psi K^+$ and $B^+ \to \bar{D}^0 \pi^+$ as shown in Figs. 7 and 8. The efficiency and dilution, shown in Figs. 9 and 10, is parameterized as a function of the transverse momentum of the B meson. The preliminary result for the tagging efficiency is $\varepsilon \mathscr{D}^2 = (2.4 \pm 1.2)\%$ and $\varepsilon \mathscr{D}^2 = (1.9 \pm 0.9)\%$ independently from the $B^+ \to J/\psi K^+$ and $B^+ \to \bar{D}^0 \pi^+$ samples.

The Run II preliminary results from CDF are summarized in Table 1. In this table, the Run I CDF results and expected Run II performances are also listed. Studies of JetQ and SLT electron algorithms are underway at CDF. Using the preliminary Run II results and projections from Run I, CDF expects a combined efficiency of $\varepsilon \mathscr{D}^2 \approx 4\%$ from SLT, SST and JetQ algorithms. This is an estimation without taking into account potential improvements from particle identification using ToF. The Time-of-Flight system provides a better than 2σ K-π separation for tracks with $p < 1.6$ GeV/c, as shown in Fig. 11. The ToF system is expected to improve both the OKT and SST efficiency dramatically.

TABLE 1. Summary of tagging algorithm performance $\varepsilon \mathscr{D}^2$ [%] at CDF.

	Run I	Run II	Projection (no ToF)	Projection (with ToF)
SST	1.5 ± 0.4	2.1 ± 0.7	2.0	2.0-4.2
SLT-μ	0.6 ± 0.1	0.7 ± 0.1	1.0	1.0
SLT-e	0.3 ± 0.1	-	0.7	0.7
JetQ	1.0 ± 0.3	-	3.0	3.0
OKT	-	-	-	2.4

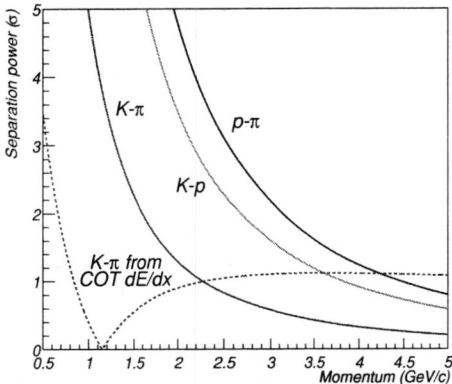

FIGURE 11. K-π separation as function of track momentum from CDF's ToF system.

verse decay length L_{xy} and the transverse momentum p_T of the B meson are used to calculate the proper time

$$t = L/\beta\gamma = L_{xy} \cdot M/p_T.$$

L_{xy} is the two-dimensional flight distance between the beam spot where B is produced and the point where the B decays. p_T is calculated from the B decay daughters and M is the mass of the B_s^0 meson.

The uncertainty on the proper time is derived as

$$\sigma_t = \sqrt{(\sigma_{L_{xy}} \cdot M/p_T)^2 + (\sigma_{p_T}/p_T \cdot t)^2},$$

where $\sigma_{L_{xy}}$ and σ_{p_T} are measurement errors on the decay length and transverse momentum. There are two components to the decay length resolution, the uncertainty of the beam spot position and the vertex fitting errors in determining the decay point. CDF estimates an error of $\sigma_{L_{xy}} \approx 50$ μm in SVT triggered events. The spread of the beam spot contributes an error of 30 μm to the decay length calculation. The beam spot is calculated on a run averaged basis at this time and improvements are under study. The uncertainty on the B transverse momentum depends on the tracking resolution and on details of the event topologies which are very different for semileptonic decays and fully reconstructed hadronic events.

For fully reconstructed hadronic decays, the uncertainty on the B_s^0 transverse momentum is negligible compared to the uncertainty on the decay length due to the excellent tracking resolutions. Therefore, the decay length resolution, which depends on the performance of the silicon system, is the most important issue to a precise measurement of the proper time resolution and the ability for B_s^0 mixing. For $B_s^0 \to D_s\pi$ decays reconstructed by CDF, a proper time resolution of $\sigma_t = 0.067$ ps was obtained. With further improvement on silicon tracking with L00 silicon hits and on the beam spot calculation using tracks

other than the B daughters on an event-by-event basis, CDF believes a resolution of $\sigma_t = 0.05$ ps is achievable.

For B_s^0 semileptonic decays, the uncertainty on the B_s^0 transverse momentum dominates the resolution of proper time due to the unreconstructed neutrino. A Monte Carlo method is used to estimate the B_s^0 meson p_T using kinematic correlations of the momenta of the lepton and charm daughters in the B_s^0 decay. This method results in an error on the order of 10-15%. A method using three-dimensional event topology information is being explored to improve this estimation [5]. In general, a sizable uncertainty introduced from the B_s^0 meson p_T estimate has to be overcome to use B_s^0 semileptonic decays for B_s^0 mixing measurements, especially if Δm_s is indeed very large.

5. B_s^0 MIXING SENSITIVITY

The sensitivity of a B_s^0 mixing measurement is generally quantified as the significance of an observation of B_s^0 mixing at a certain value of Δm_s. Three key numbers, the B_s^0 event yield, the efficiency of flavors tagging and the proper time resolution, determine an experiment's sensitivity for a B_s^0 mixing measurement. To a good approximation, the averaged significance $Sig(\Delta m_s)$, in number of standard deviations σ, is given as

$$Sig(\Delta m_s) = \sqrt{\frac{S\varepsilon\mathscr{D}^2}{2}}e^{-\frac{(\Delta m_s \sigma_t)^2}{2}}\sqrt{\frac{S}{S+B}},$$

where S is the number of signal events, S/B is the signal to background ratio and $\varepsilon\mathscr{D}^2$ and σ_t are the effective tagging efficiency and proper time resolution as discussed in the previous sections.

Both fully reconstructed and semileptonic samples will contribute to the measurement of Δm_s but at different Δm_s regions. For lower values of Δm_s, the high statistics semileptonic sample could play a key role in the measurement. If the value of Δm_s is much higher than the current limit, the proper time resolution σ_t becomes the limiting factor in resolving the oscillations and we need to take advantage of fully reconstructed samples.

Both DØ and CDF are using semileptonic B_s^0 decay samples for the mixing study. DØ predicts a signal yield of 15,000 events from the $B_s^0 \to D_s^- \mu^+ \nu X$ decay with a proper time resolution of 0.15 ps using 500 pb^{-1}. This results in a sensitivity of 1.5 σ for $\Delta m_s = 15$ ps^{-1} [4].

CDF's event rate of reconstructing $B_s^0 \to D_s\pi$ decays is $S = 1600$ events / fb^{-1} with $S/B = 2/1$ and proper time resolution of $\sigma_t = 0.067$ ps. With a tagging efficiency of $\varepsilon\mathscr{D}^2 = 4\%$, CDF will have a 2σ sensitivity for $\Delta m_s = 15$ ps^{-1} using 500 pb^{-1} of data. This will supersede the current world limit which combines contributions from 13 different measurements from LEP, SLD and

CDF [1]. This sensitivity is predicted using numbers based on an already achieved performance of the trigger, reconstruction and flavor tagging.

As discussed in previous sections, improvements are underway to increase the event yield from optimizing the SVT trigger and by including more decay channels in the reconstruction. The proper time resolution is also improving with the usage of hits from the innermost silicon layer and with an event-by-event beam spot calculation method. In addition, the flavor tagging efficiency will double with the help of kaon identification using the ToF system. A higher sensitivity for CDF with fully reconstructed B_s^0 events can be derived taking into account these improvements using a set of numbers of $S = 2000$ events$/\text{fb}^{-1}$, $S/B = 2/1$, $\varepsilon\mathscr{D}^2 = 5\%$ and $\sigma_t = 0.05$ ps. This results in a 5σ sensitivity for $\Delta m_s = 18 \rightarrow 24$ ps^{-1} with $1.7 \rightarrow 3.2$ fb^{-1} of data. This will cover the region of Δm_s that is currently preferred by the indirect fits using the world average [1]. To observe or exclude a value of Δm_s beyond this region will require additional data and improvements along with further progress on triggering, reconstruction and flavor tagging.

6. PROGRESS ON $\Delta\Gamma_s$ MEASUREMENT

A complementary method to a direct Δm_s measurement in a B_s^0 mixing study is the measurement of the width difference $\Delta\Gamma_s$ of the two CP eigenstates of the $B_s^0\bar{B}_s^0$ system. Within the Standard Model, it is predicated that the ratio between $\Delta\Gamma_s$ and Δm_s is large [6]:

$$\frac{\Delta\Gamma_s}{\Delta m_s} = \frac{3\pi}{2} \cdot \frac{m_b^2}{m_t^2} \cdot \frac{\eta(\Delta\Gamma_s)}{\eta(\Delta m_s)} \approx (5.6 \pm 2.6) \times 10^{-3}.$$

If the value of Δm_s is indeed large, the width difference could be as much as $\Delta\Gamma_s/\Gamma_s \sim 15\%$ which will make a precision measurement possible using Run II data.

Experimentally, three methods are being explored to extract the width difference. One can extract it by directly fitting the two lifetime components in events from a decay channel of well defined CP eigenstates, such as $B_s^0 \rightarrow D_s^- \ell^+ v$ which is a 50-50 mixture of CP-even and CP-odd states. A modified method is to separate the CP-even and CP-odd states by a transversity analysis first and then extract the width difference from the two states. The $B_s^0 \rightarrow J/\psi \phi$ channel is expected to be dominated by the CP-even state, since $\Gamma^{CP-even}/\Gamma = 0.778 \pm 0.090 \pm 0.012$ as measured by CDF using Run I data [7]. This channel is being explored for this purpose. Both CDF and DØ are collecting sizable samples of this decay mode using dimuon triggers. CDF's signal is shown in Fig. 12. A precision of 0.05 on $\Delta\Gamma_s/\Gamma_s$ is predicated to be reachable using 4000 $B_s^0 \rightarrow J/\psi \phi$ events [5]. The third method involves B_s^0 decays with a pure CP state such as

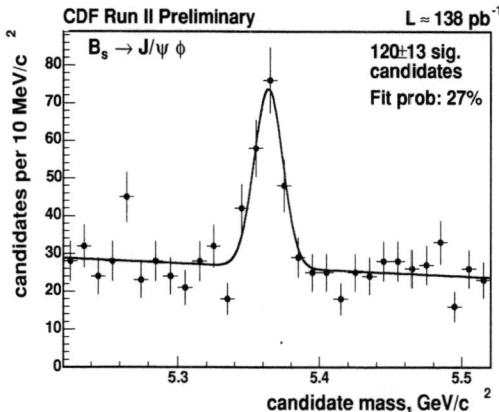

FIGURE 12. Invariant mass spectrum of the $B_s^0 \rightarrow J/\psi \phi$ candidates. The result is using about 138 pb^{-1} CDF data.

$B_s^0 \rightarrow D_s^- D_s^+$ which is a pure CP-even state. The decay branching ratio can be related to $\Delta\Gamma_s$ by [8] $\mathscr{B}(B_s^0 \rightarrow D_s^- D_s^+) = \Delta\Gamma_s/[\Gamma_s(1 + \Delta\Gamma_s/2\Gamma_s)]$. CDF is studying the feasibility of a measurement using $B_s^0 \rightarrow D_s^- D_s^+$ decay events being collected by its SVT trigger.

7. SUMMARY

Both CDF and DØ are collecting large samples of B_s^0 events for challenging B mixing measurements. Much progress has been made on flavor tagging and proper time reconstruction. Over the next few years, both experiments are expected to make significant contributions in the determination of the mixing parameter Δm_s as well as the width difference $\Delta\Gamma_s$ for the $B_s^0\bar{B}_s^0$ system.

REFERENCES

1. Heavy Flavour Averaging Group (2003), URL: http://www.slac.stanford.edu/xorg/hfag/index.html.
2. Blair, R., et al., [CDF Collaboration], The CDF II Detector: Technical Design Report, Tech. Rep. FERMILAB-PUB-96-390-E (1996).
3. Abachi, S., et al., [DØ Collaboration], The DØ Upgrade: The Detector and its Physics, Tech. Rep. FERMILAB-PUB-96-357-E (1996).
4. Jain, V., in these proceedings (2004).
5. Anikeev, K., et al., arXiv:hep-ph/0201071 (2001).
6. Beneke, M., Buchalla, G., and Dunietz, I., *Phys. Rev.* **D54**, 4419–4431 (1996).
7. Affolder, T., et al., [CDF Collaboration], *Phys. Rev. Lett.* **85**, 4668–4673 (2000).
8. Aleksan, R., Le Yaouanc, A., Oliver, L., Pene, O., and Raynal, J. C., *Phys. Lett.* **B316**, 567–577 (1993).

RARE DECAYS

Theory of Radiative and Rare B Decays

Gino Isidori

INFN, Laboratori Nazionali di Frascati, Via E. Fermi 40, I-00044 Frascati, Italy

Abstract. We present a concise theoretical overview of radiative and rare B decays mediated by flavour-changing neutral-current transitions of the type $b \to s(d)\gamma$ and $b \to s(d)\bar{\ell}\ell$.

1. INTRODUCTION

Thanks to the efforts of B factories, the exploration of the mechanism of quark-flavour mixing is now entering a new interesting era. The precise measurements of mixing-induced CP violation and tree-level allowed semileptonic transition have provided an important consistency check of the SM, and a precise determination of the Cabibbo-Kobayashi-Maskawa (CKM) matrix. The next goal is to understand if there is still room for new physics (NP) or, more precisely, if there is still room for new sources of flavour symmetry breaking close to the electroweak scale. From this perspective, radiative and rare B decays mediated by flavour-changing neutral current (FCNC) amplitudes represent a fundamental tool (see e.g. Ref. [1]).

Beside the experimental sensitivity, the conditions which allow to perform significant NP searches in rare decays can be summarized as follows: i) decay amplitude dominated by electroweak dynamics, and thus enhanced sensitivity to non-standard contributions; ii) small theoretical error within the SM, or good control of both perturbative and non-perturbative corrections. In the remainder of this talk, we shall analyze at which level these conditions are satisfied in various decay modes.

2. INCLUSIVE FCNC B DECAYS

Inclusive rare B decays such as $B \to X_s\gamma$, $B \to X_s\ell^+\ell^-$ and $B \to X_s\nu\bar{\nu}$ are the natural framework for high-precision studies of FCNC's in the $\Delta B = 1$ sector [2]. Perturbative QCD and heavy-quark expansion form a solid theoretical framework to describe these processes: inclusive hadronic rates are related to those of free b quarks, calculable in perturbation theory, by means of a systematic expansion in inverse powers of the b-quark mass.

The starting point of the perturbative partonic calculation is the determination of a low-energy effective Hamil-

tonian, renormalized at a scale $\mu = \mathscr{O}(m_b)$, obtained by integrating out the heavy degrees of freedom of the theory. For $b \to s$ transitions –within the SM– this can be written as

$$\mathscr{H}_{\text{eff}} = -\frac{G_F}{\sqrt{2}} V_{ts}^* V_{tb} \sum_{i=1}^{10,\,v} C_i(\mu) Q_i + \text{h.c.} \qquad (1)$$

where $Q_{1...6}$ are four-quark operators, Q_8 is the chromomagnetic operator and

$$
\begin{aligned}
Q_7 &= \frac{e}{4\pi^2}\bar{s}_L \sigma_{\mu\nu} m_b b_R F^{\mu\nu} \\
Q_9 &= \frac{e^2}{4\pi^2}\bar{s}_L \gamma^\mu b_L \bar{\ell}\gamma_\mu \ell \\
Q_{10} &= \frac{e^2}{4\pi^2}\bar{s}_L \gamma^\mu b_L \bar{\ell}\gamma_\mu \gamma_5 \ell \\
Q_v &= \frac{e^2}{4\pi^2 s_w^2}\bar{s}_L \gamma^\mu b_L \, \bar{\nu}_L \gamma_\mu \nu_L
\end{aligned}
\qquad (2)
$$

are the leading FCNC electroweak operators. Within the SM, the coefficients of all the operators in Eq. (2) receive a large non-decoupling contribution from top-quark loops at the electroweak scale. But the m_t dependence is not the same for the four operators, reflecting a different $SU(2)_L$-breaking structure, which can be affected in a rather different way by new-physics contributions [3].

The calculation of the rare decay rates then involves three distinct steps: i) the determination of the initial conditions of the Wilson coefficients at the electroweak scale; ii) the evolution by means of renormalization-group equations (RGE's) of the C_i down to $\mu = \mathscr{O}(m_b)$; iii) the evaluation of the hadronic matrix elements of the effective operators at $\mu = \mathscr{O}(m_b)$, including both perturbative and non-perturbative QCD corrections. Each of the three steps must be taken to matching orders of accuracy in powers of the strong coupling constants α_s and of the large logs generated by the RGE running. The interesting short-distance (electroweak) dynamics that we

CP722, *B Physics at Hadron Machines*, edited by M. Paulini and S. Erhan

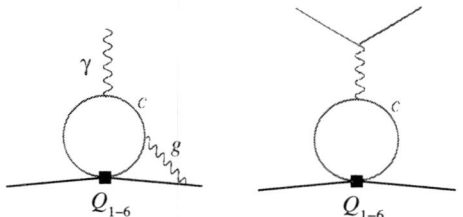

FIGURE 1. Representative diagrams for the mixing of four-quark operators into Q_7 (left) and Q_9 (right).

would like to test enters only in the first step; the following two steps are fundamental ingredients to reduce and control the theoretical error.

The first two steps (initial conditions and RGE's) are process independent and are common also to exclusive modes. Nonetheless, the organization of the leading-log (LL) series is not the same for the three underlying partonic processes, or the four operators in Eq. (2):

$b \to s\gamma$. Here, only Q_7 has a non-vanishing matrix element at the tree level. The large logarithms generated by mixing of four-quark operators into Q_7 (see Fig. 1) play a very important role and enhance the partonic rate by a factor of almost three [4]. Since this mixing vanishes at the one-loop level, a full treatment of QCD corrections beyond lowest order is a rather non-trivial task. This has been achieved already a few years ago, thanks to the joint effort of many authors (see e.g. Ref. [2, 5] and references therein), and is nowadays a rather mature subject. All the ingredients of the partonic calculation have been cross-checked by more than one group. In particular, very recently an independent confirmation of the three-loop mixing of Q_7 and $Q_{1...6}$ [6] – till few months ago the only piece of the calculation performed by one group only – has been presented [7].

$b \to s\ell^+\ell^-$. The three operators with non-vanishing tree-level matrix elements are Q_7, Q_9 and Q_{10}. Similarly to Q_7, QCD corrections are very important also for Q_9. Since Q_9 mixes with four-quark operators already at the one-loop level (see Fig. 1), the organization of the LL series for $b \to s\ell^+\ell^-$ is different than in $b \to s\gamma$: the NLL level is much simpler (no three-loop mixing involved), but less precise. An accuracy below the 10% level on the decay rate (a precision similar to the NLL level in $b \to s\gamma$) is reached here only with a NNLL calculation. All the missing ingredients to reach this goal has finally become available. In particular, NNLL initial conditions and anomalous dimension matrix can be found in [8] and [7], respectively.

It is worth to stress that the impact of QCD corrections is very limited in the axial-current operator Q_{10}.

This operator does not mix with four-quark operators and is completely dominated by short-distance contributions. Together with Q_v, Q_{10} belongs to the theoretically clean $\mathscr{O}(G_F^2)$ hard-GIM-protected part of the effective Hamiltonian. Thus observables more sensitive to Q_{10}, such as the forward-backward (FB) lepton asymmetry in $b \to s\ell^+\ell^-$, have a reduced QCD uncertainty and a stronger sensitivity to possible non-standard phenomena.

$b \to s\nu\bar{\nu}$. In this case only Q_v is involved. Similarly to Q_{10}, QCD corrections play a very minor role since there is no mixing with four-quark operators. As a result, the only non-trivial step of the perturbative calculation for $b \to s\nu\bar{\nu}$ decays is the determination of the initial condition of C_v at the electroweak scale: this is known with a precision around 1% within the SM [9].

These two processes-independent steps of the calculation, namely the determination of the effective Hamiltonian renormalized at a low scale $\mu = \mathscr{O}(m_b)$, can easily be transferred from the $b \to s$ case to the $b \to d$ one. The only difference is the richer CKM structure of the $b \to d$ Hamiltonian, with two independent non-negligible terms ($V_{td}^* V_{tb}$ and $V_{ud}^* V_{ub}$).

The situation is very different for the last step of the calculation, namely the evaluation of the hadronic matrix elements. The latter strongly depend on the specific process and the specific observable we are interested in (e.g. fully inclusive rate or differential distribution). In the following we shall review the results of this step (and thus the final numerical predictions) for some of the most interesting $b \to s$ observables.

2.1. $B \to X_s\gamma$ [*The Most Effective "New Physics Killer"*]

The inclusive $B \to X_s\gamma$ rate is the most precise and clean short-distance information that we have, at present, on $\Delta B = 1$ FCNC's. Combining the precise measurements by ALEPH, BaBar, Belle and CLEO, the world average reads [10]

$$\mathscr{B}(B \to X_s\gamma)^{\exp} = (3.34 \pm 0.38) \times 10^{-4} \qquad (3)$$

On the theory side, the NLL partonic calculation of the matrix elements, performed first in Ref. [11] for the leading terms, has recently been cross-checked and completed in Ref. [12]. Perturbative corrections due to higher-order electroweak effects have also been analyzed (see Ref. [13] and references therein).

Non-perturbative $1/m_b$ corrections are well under control in the total rate. In particular, $\mathscr{O}(1/m_b)$ corrections vanish in the ratio $\Gamma(B \to X_s\gamma)/\Gamma(B \to X_c\ell\nu)$, and the $\mathscr{O}(1/m_b^2)$ ones are known and amount to few per cent [14]. Also non-perturbative effects associated to charm-quark loops have been estimated and found to be very small [15–17]. The most serious problem of non-perturbative origin is related to the (unavoidable) experimental cut in the photon energy spectrum that prevents the measurement from being fully inclusive [16, 18]. With the present cut by CLEO $E_\gamma > 2.0$ GeV [19], this uncertainty is smaller but non-negligible with respect to the error of the perturbative calculation. The latter is around 10% and its main source is the uncertainty in the ratio m_c/m_b that enters through charm-quark loops [20].

According to the detailed analysis of theoretical errors presented in Ref. [20], the SM expectation is

$$\mathscr{B}(B \to X_s\gamma)^{\mathrm{SM}} = (3.73 \pm 0.30) \times 10^{-4} , \quad (4)$$

in good agreement with Eq. (3). It must be stressed that the overall scale dependence is very small: for $\mu \in [m_b/2, 2m_b]$ the central value moves by about 1%. The error in Eq. (4) is an educated guess about the size of possible NNLL terms. In particular, the largest source of uncertainty is obtained by the variation of $\overline{m}_c(\mu)/m_b^{\mathrm{pole}}$ for $\mu \in [m_c, m_b]$.

The comparison between theory and experiments in $\mathscr{B}(B \to X_s\gamma)$ is a great success of the SM and has led us to derive many significant bounds on possible new-physics scenarios. For instance, the $\mathscr{B}(B \to X_s\gamma)$ constraint is one of the main obstacles to build consistent models that predict a sizable difference between $\mathscr{A}_{CP}(B \to \phi K_S^0)$ and $\mathscr{A}_{CP}(B \to \psi K_S^0)$. This constraint can be avoided (see e.g. Ref. [21] and references there in), but the resulting models require a considerable amount of fine tuning. By far more natural are the so-called MFV models [3], where $\mathscr{A}_{CP}(B \to \phi K_S^0) \approx \mathscr{A}_{CP}(B \to \psi K_S^0)$ and deviations from the SM in $\mathscr{B}(B \to X_s\gamma)$ do not exceed the 10%–30% level [3]. Improved measurements of $\mathscr{B}(B \to X_s\gamma)$ are certainly useful to further constrain this possibility. However, since the experimental error has reached the level of the theoretical one, it will be very difficult to clearly identify possible deviations from the SM, if any, in this observable.

Hopes to detect new-physics signals are still open through the CP-violating asymmetry

$$\Delta\Gamma_{CP}(B \to X_s\gamma) = \frac{\Gamma(B \to X_s\gamma) - \Gamma(\bar{B} \to X_s\gamma)}{\Gamma(B \to X_s\gamma) + \Gamma(\bar{B} \to X_s\gamma)} . \quad (5)$$

This is expected to be below 1% within the SM [22], but could easily reach $\mathcal{O}(10\%)$ values beyond the SM, even in the absence of large effects in the total $B \to X_s\gamma$ rate. This is indeed one of the main expectations in models with sizable NP effects in $\mathscr{A}_{CP}(B \to \phi K_S^0)$. The present measurement of $\Delta\Gamma_{CP}(B \to X_s\gamma)$ is consistent with zero [10], but the sensitivity is still one order of magnitude above the SM level.

2.2. $B \to X_s\ell^+\ell^-$ [*The Present Frontier*]

Both Belle [23] and BaBar [24] have recently announced a clear evidence ($\approx 5\sigma$) of the $B \to X_s\ell^+\ell^-$ decay. The two results are compatible and are both based on a semi-inclusive analysis (the hadronic system is reconstructed from a kaon plus 0 to 4 pions, with at most one π^0). Their combination [10]

$$\mathscr{B}(B \to X_s\ell^+\ell^-)^{\mathrm{exp}} = (6.2 \pm 1.1^{+1.6}_{-1.3}) \times 10^{-6} . \quad (6)$$

represents a very useful new piece of information about $\Delta B = 1$ FCNC's, with considerable margin of improvement in the near future.

In principle, these decays offer a phenomenology reacher than $B \to X_s\gamma$, with more than one interesting observable. The joint effort of several groups has recently allowed to evaluate at the NNLL level all the matrix element necessary for the two main kinematical distributions: the dilepton spectrum [25–27] and the lepton FB asymmetry [28, 29].

In addition to the non-perturbative corrections due to the finite b quark mass, $B \to X_s\ell^+\ell^-$ transitions suffer from specific non-perturbative effects due to long-lived $c\bar{c}$ intermediate states ($B \to X_s c\bar{c} \to X_s\ell^+\ell^-$). The heavy-quark expansion, which allow to evaluate the $\Lambda_{\mathrm{QCD}}/m_b$ terms, is rapidly convergent and leads to small corrections for *sufficiently inclusive* observables [30, 31]. A consistent treatment of the second type of effects requires *kinematical cuts* in order to avoid the large non-perturbative background of the narrow $c\bar{c}$ resonances (see Fig. 2). These two requirements are somehow in conflict [26, 31]; nonetheless, we can identify two perturbative windows, defined by:

$$q^2 \equiv M^2_{\ell^+\ell^-} \in [1 \text{ GeV}^2, 6 \text{ GeV}^2] \quad \text{(low)} ,$$
$$q^2 > 14.4 \text{ GeV}^2 \quad \text{(high)} ,$$

where reliable predictions can be performed [26]. It is worth to emphasize that the two regions have complementary virtues and disadvantages:

- *Virtues of the low-q^2 region:* reliable q^2 spectrum; small $1/m_b$ corrections; sensitivity to the interference of C_7 and C_9; high rate.

- *Disadvantages of the low-q^2 region:* difficult to perform a fully inclusive measurement (severe cuts on the dilepton energy and/or the hadronic invariant mass); long-distance effects due to processes of the type $B \to \psi X_s \to X_s + X'\ell^+\ell^-$ not fully under control; non-negligible scale and m_c dependence.

- *Virtues of the high-q^2 region:* negligible scale and m_c dependence due to the strong sensitivity to $|C_{10}|^2$; easier to perform a fully inclusive measurement (small hadronic invariant mass); negligi-

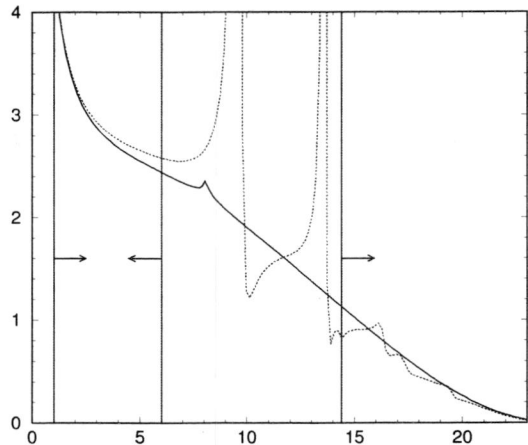

FIGURE 2. Dilepton spectrum of the inclusive $B \to X_s e^+ e^-$ decay within the SM. Vertical axis: $d\mathscr{B}(B \to X_s e^+ e^-)/dq^2$ in units of $10^{-7} \times \text{GeV}^2$; horizontal axis: q^2 in GeV^2. The dotted line denotes the NNLL pure perturbative result, the full line includes an estimates of the non-perturbative $c\bar{c}$ effects [26].

ble long-distance effects of the type $B \to \psi X_s \to X_s + X' \ell^+ \ell^-$.

- *Disadvantages of the high-q^2 region:* q^2 spectrum not reliable (only the integrated rate can be predicted); sizable $1/m_b$ corrections (effective expansion in $1/(m_b - \sqrt{q_{\min}})$ [26]); low rate.

Given this situation, we believe that future experiments should try to measure the branching ratios in both regions and report separately the two results. These two measurements are indeed affected by different systematic uncertainties (of theoretical nature) and provide a different short-distance information. The NNLL SM predictions for the two clean windows [26],

$$\mathscr{B}(B \to X_s \ell^+ \ell^-)^{\text{SM}}_{\text{low}} = (1.63 \pm 0.20) \times 10^{-6} ,$$
$$\mathscr{B}(B \to X_s \ell^+ \ell^-)^{\text{SM}}_{\text{high}} = (4.04 \pm 0.78) \times 10^{-7} , \quad (7)$$

are still affected by a considerable error; however, in both cases the uncertainty is mainly of parametric nature and could be substantially improved in the future. In particular, the large error in the high-q^2 region is mainly due to the uncertainty in the relation between the physical q^2 interval and the corresponding interval for the partonic calculation (i.e. the uncertainty in the relation between m_b and the physical hadron mass), which can be improved with better data on charged-current semileptonic modes. According to the recent analysis of Ref. [27], both results in (7) should be decreased by $\approx 4\%$ to take into account the leading electroweak corrections.

The two results in (7) cannot be directly confronted with (6), which includes an extrapolation to the full q^2 spectrum. The SM expectation for this extrapolated branching ratio is $(4.2 \pm 0.7) \times 10^{-6}$ [32], which is con-

sistent with (6). We stress that this prediction is already saturated by irreducible theoretical errors and, contrary to the results in (7), is very difficult to improve it further.

As anticipated, some of the most interesting short-distance tests in $\mathscr{B}(B \to X_s \ell^+ \ell^-)$ decays can be performed by means of the FB asymmetry of the dilepton distribution:

$$\mathscr{A}_{\text{FB}}(q^2) = \frac{1}{d\mathscr{B}(B \to X_s \ell^+ \ell^-)/dq^2} \int_{-1}^{1} d\cos\theta_\ell$$
$$\frac{d^2\mathscr{B}(B \to X_s \ell^+ \ell^-)}{dq^2 \, d\cos\theta_\ell} \text{sgn}(\cos\theta_\ell) , \quad (8)$$

where θ_ℓ is the angle between ℓ^+ and B momenta in the dilepton centre-of-mass frame. Here the SM predict a zero for $s_0 = q_0^2/m_b^2 = 0.162 \pm 0.008$ [28, 29]: a very precise prediction which could easily be modified beyond the SM, even in absence of significant non-standard effects on the total rate.

3. EXCLUSIVE MODES

On general grounds, theoretical predictions for exclusive FCNC decays are more difficult. The simplest cases are processes with at most one hadron in the final state. Here there has been a substantial progress in the last few years, both by means of analytic approaches [33] and by means of Lattice QCD [34], but still the overall theoretical uncertainty is around 20% at the amplitude level. The largest source of uncertainty is typically the normalization of the hadronic form factors, an error that can be substantially reduced in appropriate ratios or differential distributions. These type of observables become particularly interesting in channels where, because of irreducible experimental problems, the short-distance amplitude cannot be extracted from corresponding inclusive modes. Two of such examples are the ratio

$$R_\gamma(\rho/K^*) = \frac{\mathscr{B}(B \to \rho\gamma)}{\mathscr{B}(B \to K^*\gamma)} , \quad (9)$$

and the normalized FB asymmetry in $B \to K^* \ell^+ \ell^-$.

The ratio $R_\gamma(\rho/K^*)$ is one of the most promising tool to extract short-distance properties about the $b \to s\gamma$ amplitude. On the experimental side, the combination of the bounds on charged and neutral channels, in the isospin limit, leads to $R_\gamma(\rho/K^*) < 0.047$ at 90% C.L. [35]. On the theory side, the $B \to V\gamma$ amplitudes which determine this ratio have been analyzed beyond naive factorization by several authors [36–38]. Within the SM one can write

$$R_\gamma(\rho/K^*) = \left|\frac{V_{td}}{V_{ts}}\right|^2 \frac{(m_B^2 - m_{rho}^2)^3}{(m_B^2 - m_{K^*}^2)^3} \zeta^2 (1 - \Delta R) , \quad (10)$$

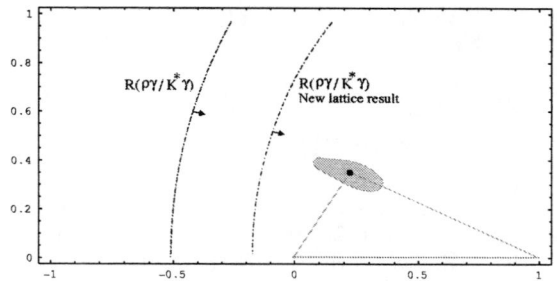

FIGURE 3. Bounds on the ρ–η plane from the BaBar bound on $R_\gamma(\rho/K^*)$: the two curves correspond to different values of ζ [37].

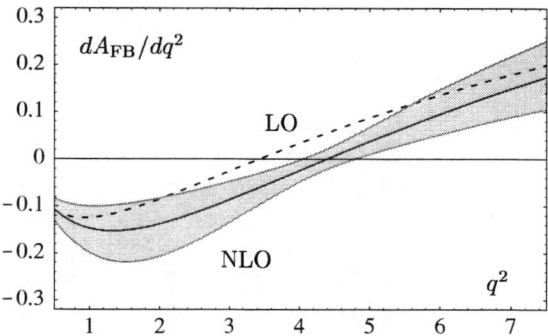

FIGURE 4. Zero of the forward-backward asymmetry in $B^- \to K^{*-}\ell^+\ell^-$ at LO and NLO. The band reflects all theoretical uncertainties from parameters and scale dependence combined [38].

where ζ denotes the ratio of the form factors at $q^2 = 0$ in the $m_b \to \infty$ limit, and ΔR the additional $SU(3)$ (and isospin) breaking due to $1/m_b$ and $\mathcal{O}(\alpha_s)$ corrections. The largest source of uncertainty is ζ: according to the light-cone sum rule estimate $\zeta = 0.76 \pm 0.10$ [37], the present experimental bound on $R_\gamma(\rho/K^*)$ is about twice the SM expectation. However, preliminary Lattice results indicate a larger value, $\zeta = 0.91 \pm 0.08$ [39], which would imply more stringent bounds on $|V_{td}/V_{ts}|$. The two curves in Fig. 3 summarizes the present status.

While the inclusive $B \to X_s\ell^+\ell^-$ rate is already accessible at the B factories, a differential study of the inclusive FB asymmetry is beyond their present and near-future reach. More accessible from the experimental point of view is the FB asymmetry in $B \to K^*\ell^+\ell^-$ (defined as in Eq. (8) with $X_s \to K^*$). Assuming that the leptonic current has only a vector (V) or axial-vector (A) structure (as in the SM), the FB asymmetry provides a direct measure of the A–V interference. Indeed, at the lowest-order one can write

$$\mathscr{A}_{FB}(q^2) \propto \text{Re}\left\{ C_{10}^* \left[\frac{q^2}{m_b^2} C_9^{\text{eff}} + r(q^2) \frac{m_b C_7}{m_B} \right] \right\},$$

where $r(q^2)$ is an appropriate ratio of $B \to K^*$ vector and tensor form factors [40]. There are three main features of this observable that provide a clear and independent short-distance information: 1) The position of the zero of $\mathscr{A}_{FB}(q^2)$ in the low-q^2 region (see Fig. 4) [40]. As shown by means of a full NLO calculation [38], the experimental measurement of q_0^2 could allow a determination of C_7/C_9 at the 10% level. 2) The sign of $\mathscr{A}_{FB}(q^2)$ around the zero. This is fixed unambiguously in terms of the relative sign of C_{10} and C_9 [41]: within the SM one expects $\mathscr{A}_{FB}(q^2 > q_0^2) > 0$ for $|\bar{B}\rangle \equiv |b\bar{d}\rangle$ mesons. 3) The relation $\mathscr{A}[\bar{B}]_{FB}(q^2) = -\mathscr{A}[B]_{FB}(q^2)$. This follows from the CP-odd structure of $\mathscr{A}_{FB}$ and holds at the 10^{-3} level within the SM [41], where C_{10} has a negligible CP-violating phase.

3.1. $B^0/B_s^0 \to \ell^+\ell^-$ [*The Future Frontier*]

The purely leptonic decays constitute a special case among exclusive transitions. Within the SM only the axial-current operator, Q_{10}, induces a non-vanishing contribution to these decays. As a result, the short-distance contribution is not *diluted* by the mixing with four-quark operators. Moreover, the hadronic matrix element involved is the simplest we can consider, namely the B-meson decay constant

$$\langle 0|\bar{q}\gamma_\mu\gamma_5 b|\bar{B}_q(p)\rangle = ip_\mu f_{B_q} \tag{11}$$

Reliable estimates of f_{B_d} and f_{B_s} are obtained at present from lattice calculations and in the future it will be possible to cross-check these results by means of the $B^+ \to \ell^+\nu$ rate. Modulo the determination of f_{B_q}, the theoretical cleanliness of $B_{s,d} \to \ell^+\ell^-$ decays is comparable to that of the *golden modes* $K_L \to \pi^0\nu\bar{\nu}$ and $B \to X_{s,d}\nu\bar{\nu}$.

The price to pay for this theoretically-clean amplitude is a strong helicity suppression for $\ell = \mu$ (and $\ell = e$), or the channels with the best experimental signature. Employing the full NLO expression of C_{10} [9], we can write

$$\mathscr{B}(B_s^0 \to \mu^+\mu^-)^{\text{SM}} = 3.1 \times 10^{-9} \left(\frac{|V_{ts}|}{0.04} \right)^2$$

$$\times \left(\frac{f_{B_s^0}}{0.21 \text{ GeV}} \right)^2 \left(\frac{\tau_{B_s^0}}{1.6 \text{ ps}} \right) \left(\frac{m_t(m_t)}{166 \text{ GeV}} \right)^{3.12}$$

$$\frac{\mathscr{B}(B_s^0 \to \tau^+\tau^-)^{\text{SM}}}{\mathscr{B}(B_s^0 \to \mu^+\mu^-)^{\text{SM}}} = 215 .$$

The corresponding B^0 modes are both suppressed by an additional factor $|V_{td}/V_{ts}|^2 = (4.0 \pm 0.8) \times 10^{-2}$. The present experimental bound closest to SM expectations

185

is the one obtained by CDF on $B_s^0 \to \mu^+\mu^-$ [10, 42]:

$$\mathscr{B}(B_s^0 \to \mu^+\mu^-) < 9.5 \times 10^{-7} \quad (95\% \text{ CL}),$$

which is still very far from the SM level. The latter will certainly not be reached before the LHC era.

As emphasized in the recent literature [43–45], the purely leptonic decays of B_s^0 and B^0 mesons are excellent probes of several new-physics models and, particularly, of scalar FCNC's. Scalar FCNC operators, such as $\bar{b}_R s_L \bar{u}_R \mu_L$, are present within the SM but are absolutely negligible because of the smallness of down-type Yukawa couplings. On the other hand, these amplitudes could be non-negligible in models with an extended Higgs sector. In particular, within the MSSM, where two Higgs doublets are coupled separately to up- and down-type quarks, a strong enhancement of scalar FCNC's can occur at large $\tan\beta = v_u/v_d$ [43]. This effect is very small in non-helicity-suppressed B decays and in K decays (because of the small Yukawa couplings), but could enhance $B \to \ell^+\ell^-$ rates by orders of magnitude. As pointed out in Ref. [46], $\mathscr{O}(100)$ enhancements in $\mathscr{B}(B \to \ell^+\ell^-)$ correspond to $\mathscr{O}(10\%)$ breaking of universality in $\mathscr{B}(B \to K\mu^+\mu^-)$ vs. $\mathscr{B}(B \to Ke^+e^-)$. Therefore, the present search for $B \to \ell^+\ell^-$ at CDF is already quite interesting, even if the sensitivity is well above the SM level. In a long-term perspective, the discovery of such processes is definitely one of the most interesting items in the B-physics program of hadron colliders.

4. CONCLUSIONS

Rare FCNC decays of B mesons provide a unique opportunity to perform high-precision studies of quark-flavour mixing. The $B \to X_s\gamma$ rate, where both experimental and theoretical errors have reached a comparable level around 10%, represents the highest peak in our present knowledge of FCNC's. The lack of deviations from SM expectations in $\Gamma(B \to X_s\gamma)$ should not discourage the measurement of other clean and independent FCNC observables, such as the forward–backward asymmetry in $B \to X_s\ell^+\ell^-$ or the $B \to \ell^+\ell^-$ rates. Even if new physics will first be discovered elsewhere, the experimental study of these theoretically-clean observables would still be very useful to investigate the flavour structure of any new-physics scenario.

Acknowledgments

I am grateful to the organizers of Beauty 2003 for the invitation and the financial support that allowed me to attend this interesting conference. This work is partially supported by the EC-Contract HPRN-CT-2002-00311 (EURIDICE).

REFERENCES

1. A. J. Buras, hep-ph/0307203; Y. Nir, *Nucl. Phys. Proc. Suppl.*, **117**, 111 (2003) [hep-ph/0208080]; G. Isidori, *Int. J. Mod. Phys. A*, **17**, 3078 (2002) [hep-ph/0110255].
2. For a recent review see T. Hurth, *Rev. Mod. Phys.*, **75**, 1159 (2003) [hep-ph/0212304].
3. G. D'Ambrosio *et al.*, *Nucl. Phys. B*, **645**, 155 (2002) [hep-ph/0207036]; A. J. Buras *et al.*, *Phys. Lett. B*, **500**, 161 (2001) [hep-ph/0007085]; A. J. Buras, *Acta Phys. Polon. B*, **34**, 5615 (2003) [hep-ph/0310208].
4. S. Bertolini, F. Borzumati, A. Masiero, *Phys. Rev. Lett.*, **59**, 180 (1987); N. G. Deshpande *et al.*, *Phys. Rev. Lett.*, **59**, 183 (1987).
5. A. J. Buras and M. Misiak, *Acta Phys. Polon. B*, **33**, 2597 (2002) [hep-ph/0207131].
6. K. Chetyrkin, M. Misiak and M. Munz, *Phys. Lett. B*, **400**, 206 (1997); **425**, 414 (1997) (E) [hep-ph/9612313].
7. P. Gambino, M. Gorbahn and U. Haisch, *Nucl. Phys. B*, **673**, 238 (2003) [hep-ph/0306079].
8. C. Bobeth, M. Misiak and J. Urban, *Nucl. Phys. B*, **574**, 291 (2000) [hep-ph/9910220].
9. G. Buchalla and A. J. Buras, *Nucl. Phys. B*, **548**, 309 (1999) [hep-ph/9901288]; M. Misiak and J. Urban, *Phys. Lett. B*, **451**, 161 (1999) [hep-ph/9901278].
10. M. Nakao, hep-ex/0312041.
11. C. Greub, T. Hurth and D. Wyler, *Phys. Rev. D*, **54**, 3350 (1996) [hep-ph/9603404].
12. A. J. Buras, A. Czarnecki, M. Misiak and J. Urban, *Nucl. Phys. B*, **631**, 219 (2002) [hep-ph/0203135]; **611**, 488 (2001) [hep-ph/0105160].
13. P. Gambino and U. Haisch, *JHEP*, **0110**, 020 (2001) [hep-ph/0109058].
14. A.F. Falk, M. Luke and M.J. Savage, *Phys. Rev. D*, **49**, 3367 (1994) [hep-ph/9308288].
15. M. B. Voloshin, *Phys. Lett. B*, **397**, 275 (1997) [hep-ph/9612483]; A. Grant, A. Morgan, S. Nussinov and R. Peccei, *Phys. Rev. D*, **56**, 3151 (1997) [hep-ph/9702380].
16. Z. Ligeti, L. J. Randall and M. B. Wise, *Phys. Lett. B*, **402**, 178 (1997) [hep-ph/9702322].
17. G. Buchalla, G. Isidori and S. J. Rey, *Nucl. Phys. B*, **511**, 594 (1998) [hep-ph/9705253].
18. A. L. Kagan and M. Neubert, *Eur. Phys. J. C*, **7**, 5 (1999) [hep-ph/9805303].
19. D. Cassel, these proceedings; S. Chen *et al.* [CLEO Collab.], hep-ex/0108032.
20. P. Gambino and M. Misiak, *Nucl. Phys. B*, **611**, 338 (2001) [hep-ph/0104034].
21. M. Ciuchini *et al.*, eConf **C0304052** (2003) WG307 [hep-ph/0308013].
22. A. L. Kagan and M. Neubert, *Phys. Rev. D*, **58**, 094012 (1998) [hep-ph/9803368]; T. Hurth, E. Lunghi and W. Porod, hep-ph/0312260.
23. J. Kaneko *et al.* [Belle Collaboration], *Phys. Rev. Lett.*, **90**, 021801 (2003) [hep-ex/0208029].
24. B. Aubert *et al.* [BaBar Collaboration], hep-ex/0308016.
25. H. H. Asatrian, H. M. Asatrian, C. Greub and M. Walker,

Phys. Lett. B, **507**, 162 (2001) [hep-ph/0103087]; *Phys. Rev. D*, **65**, 074004 (2002) [hep-ph/0109140]; *Phys. Rev. D*, **66**, 034009 (2002) [hep-ph/0204341].

26. A. Ghinculov, T. Hurth, G. Isidori and Y. P. Yao, hep-ph/0312128; hep-ph/0310187; *Nucl. Phys. Proc. Suppl.*, **116**, 284 (2003) [hep-ph/0211197].

27. C. Bobeth, P. Gambino, M. Gorbahn and U. Haisch, hep-ph/0312090.

28. A. Ghinculov, T. Hurth, G. Isidori and Y. P. Yao, *Nucl. Phys. B*, **648**, 254 (2003) [hep-ph/0208088].

29. H. M. Asatrian, K. Bieri, C. Greub and A. Hovhannisyan, *Phys. Rev. D*, **66**, 094013 (2002) [hep-ph/0209006]; H. M. Asatrian, H. H. Asatryan, A. Hovhannisyan and V. Poghosyan, hep-ph/0311187.

30. A. Ali, G. Hiller, L. T. Handoko and T. Morozumi, *Phys. Rev. D*, **55**, 4105 (1997) [hep-ph/9609449].

31. G. Buchalla and G. Isidori, *Nucl. Phys. B*, **525**, 333 (1998) [hep-ph/9801456].

32. A. Ali, E. Lunghi, C. Greub and G. Hiller, *Phys. Rev. D*, **66**, 034002 (2002) [hep-ph/0112300].

33. I. Stewart, these proceedings.

34. A. El-Khadra, these proceedings.

35. B. Aubert *et al.* [BaBar Collaboration], hep-ex/0207073.

36. B. Grinstein and D. Pirjol, *Phys. Rev. D*, **62**, 093002 (2000) [hep-ph/0002216]; A. Ali and A. Y. Parkhomenko, *Eur. Phys. J. C*, **23**, 89 (2002) [hep-ph/0105302]. S. W. Bosch and G. Buchalla, *Nucl. Phys. B*, **621**, 459 (2002) [hep-ph/0106081].

37. A. Ali and E. Lunghi, *Eur. Phys. J. C*, **26**, 195 (2002) [hep-ph/0206242]; T. Hurth and E. Lunghi, eConf **C0304052** (2003) WG206 [hep-ph/0307142].

38. M. Beneke, T. Feldmann and D. Seidel, *Nucl. Phys. B*, **612**, 25 (2001) [hep-ph/0106067].

39. D. Becirevic and J. Reyes, hep-lat/0309131.

40. G. Burdman, *Phys. Rev. D*, **57**, 4254 (1998) [hep-ph/9710550].

41. G. Buchalla, G. Hiller and G. Isidori, *Phys. Rev. D*, **63**, 014015 (2001) [hep-ph/0006136].

42. C.-J. Lin, these proceedings.

43. C. Hamzaoui, M. Pospelov and M. Toharia; *Phys. Rev. D*, **59**, 095005 (1999) [hep-ph/9807350]; K. S. Babu and C. Kolda, *Phys. Rev. Lett.*, **84**, 228 (2000) [hep-ph/9909476]. G. Isidori and A. Retico, *JHEP*, **0111**, 001 (2001) [hep-ph/0110121]. A. J. Buras *et al.*, *Nucl. Phys. B*, **659**, 3 (2003) [hep-ph/0210145]. A. Dedes, *Mod. Phys. Lett. A*, **18**, 2627 (2003) [hep-ph/0309233].

44. C. S. Huang *et al.* *Phys. Rev. D*, **59**, 011701 (1999) [hep-ph/9803460]; *Phys. Rev. D*, **63**, 114021 (2001) [hep-ph/0006250]; S. R. Choudhury and N. Gaur, *Phys. Lett. B*, **451**, 86 (1999) [hep-ph/9810307]; P.H. Chankowski and L. Slawianowska, *Phys. Rev. D*, **63**, 054012 (2001) [hep-ph/0008046]; C. Bobeth *et al.*, *Phys. Rev. D*, **64**, 074014 (2001) [hep-ph/0104284]; *Phys. Rev. D*, **66**, 074021 (2002) [hep-ph/0204225]; Z. Xiong and J. M. Yang, *Nucl. Phys. B*, **628**, 193 (2002) [hep-ph/0105260]; A. Dedes *et al.*, *Phys. Rev. Lett.*, **87**, 251804 (2001) [hep-ph/0108037]; S. w. Baek, P. Ko and W. Y. Song, *Phys. Rev. Lett.*, **89**, 271801 (2002) [hep-ph/0205259]; *JHEP*, **0303**, 054 (2003) [hep-ph/0208112]; J. K. Mizukoshi, X. Tata and Y. Wang, *Phys. Rev. D*, **66**, 115003 (2002) [hep-ph/0208078]; G. L. Kane, C. Kolda and J. E. Lennon, hep-ph/0310042.

45. G. Isidori and A. Retico, *JHEP*, **0209**, 063 (2002) [hep-ph/0208159]; R. Fleischer, G. Isidori and J. Matias, *JHEP*, **0305**, 053 (2003) [hep-ph/0302229]; A. J. Buras, *Phys. Lett. B*, **566**, 115 (2003) [hep-ph/0303060].

46. G. Hiller and F. Kruger, hep-ph/0310219.

Radiative and EW Penguin *B* Decays with the Belle Detector

Thomas Ziegler

(Representing the Belle Collaboration)

Princeton University, Department of Physics, Jadwin Hall, Princeton, NJ 08540, U.S.A.

Abstract. Radiative *B* decays which are described by electroweak penguin or box diagrams are an extremely interesting field to test the Standard Model (SM) and search for physics beyond the Standard Model. Many modes were added to the list of radiative *B* decays in the last years and in some channels the statistics is sufficient to allow for measurements of e.g. *CP* and isospin asymmetries. Theoretical and systematic errors cancel largely in these ratios and measurements represent a sensitive tool for tests of the Standard Model. In purely leptonic *B* decays, the observation of certain decay modes would hint directly to new physics processes, as most of the branching fractions in the Standard Model are beyond the reach of current statistics in the *B* factories. This paper will summarize recent results from the Belle collaboration and will give a status of the current situation and future prospects of radiative and rare *B* decays.

1. INTRODUCTION

Rare radiative decays $b \to s\gamma$ or $b \to s\ell^+\ell^-$ can proceed in lowest order only by penguin or box diagrams and are therefor highly suppressed. In the last years, the increasing integrated luminosities of the *B* factories allowed the observation of many decay modes which can be used to test the predictions of the Standard Model and search for new physics. Absolute branching ratios suffer in the Standard Model calculations from large hadronic uncertainties, which cancel in ratios like e.g. *CP*- or isospin asymmetries. In addition, systematic errors of measurements can be eliminated to a large degree in these ratios which therefore provide an excellent tool to test the Standard Model and its predictions. Various extensions of the Standard Model like SUSY or 2-Higgs-Double models (2HDM) might change these variables significantly without necessarily changing the absolute branching ratios due to different couplings of new particles.

The same holds true for purely leptonic *B* decays $B \to \ell\ell(\gamma)$, where the ℓ stands for a charged lepton or a neutrino. None of these decays were observed yet and the Standard Model branching ratios are too low to be accessible with the current data statistics in the *B* factories for most of the channels. However, various models beyond the SM would increase the branching ratios by up to three orders of magnitude, so that any observation of these channels could hint directly to new physics.

The results presented in this paper were obtained from data taken with the Belle detector at the asymmetric e^+e^- *B* factory KEKB in Tsukuba, Japan. In summer 2003 in total about 158 fb^{-1} of data were written to disk, of which about 140 fb^{-1} were taken on the $\Upsilon(4S)$ resonance, which decays almost exclusively into neutral or charged *B* pairs. Details on the machine and the Belle detector can be found in [1] and [2], respectively.

Common to all presented decay channels is the background of continuum events $e^+e^- \to q\bar{q}$. The cross section for this process is roughly 3 times higher than for *B* pair production and has to be suppressed. As the *B* mesons decay almost at rest, the decays are rather spherical in contrast to the more 2-jet like continuum background. Variables that are used to suppress the background are e.g. modified Fox-Wolfram moments, sphericity tensor and axis, thrust and its axis. These variables are correlated and are combined in a Fisher discriminant. The output of the Fisher variable is then combined with $\cos\theta_B^*$ into a likelihood ratio $LR = \mathscr{L}_B/(\mathscr{L}_B + \mathscr{L}_{q\bar{q}})$, where θ_B^* represents the flight direction of the *B* candidate with respect to the beam pipe in the center-of-mass frame of the $\Upsilon(4S)$. Details can be found in the respective papers for each analysis.

Common kinematic variables that are used in all analyses are the beam constraint mass $M_{bc} = \sqrt{E_{beam}^{*2} - |P_B^*|^2}$ and the energy difference of the *B* candidate and the beam energy $\Delta E = E_B^* - E_{beam}^*$. The "*" indicates that all variables are estimated in the $\Upsilon(4S)$ rest frame.

2. RADIATIVE *B* DECAYS $b \to s\gamma$

The first observation of radiative *B* decays was reported by CLEO 10 years ago in the channel $B \to K^*\gamma$ [3]. The lowest order Feynman diagram for this decay is a pen-

CP722, *B Physics at Hadron Machines*, edited by M. Paulini and S. Erhan
© 2004 American Institute of Physics 0-7354-0203-5/04/$22.00

guin diagram which is dominated by a top quark in the loop due to the large mass of the top quark compared to the other up-type quarks. It turns out that the total inclusive branching ratio for $B \to X_S \gamma$ is relatively large $\mathcal{B}(B \to X_S \gamma) \approx 3.5 \cdot 10^{-4}$. In recent years many additional exclusive decay channels could therefor be observed and the statistics in some channels is large enough to allow for measurements of various asymmetries and photon spectra.

2.1. The Decay $B \to K^* \gamma$

In Belle this mode is reconstructed by identifying K^* in the 4 final states $K^* \to K^+ \pi^-, K_S^0 \pi^0, K^+ \pi^0$ and $K_S^0 \pi^0$. A high energetic photon with an energy of 1.8 GeV $<$ $E_\gamma^* <$ 3.4 GeV is required. Photons are rejected if they come from a π^0 or a η (π^0/η veto). The π^0 is reconstructed from photon pairs, the K_S^0 from two charged pions. K^*'s are identified if the invariant mass of the K-π-pair is $|M_{inv}(K\pi) - M(K^*)| <$ 75 MeV/c^2. Tests with the helicity distribution of the K^* show, that there is practically no background from non-vector mesons in the selection. The yield of the selection is estimated from a fit to the beam constrained mass distribution and results in the branching ratios shown in Table 1 along with the Standard Model predictions. Details of the analysis can be found in [4].

The measurements are slightly higher than the theoretical calculations, but the latter have quite large errors due to hadronic uncertainties. The difference is about 1.5 standard deviations and not significant.

In this decay channel the isospin asymmetry has been investigated between the neutral and charged B decays:

$$\Delta_{0+} = \frac{(\tau_{B^+}/\tau_{B^0})\mathcal{B}(B^0 \to K^{*0}\gamma) - \mathcal{B}(B^+ \to K^{*+}\gamma)}{(\tau_{B^+}/\tau_{B^0})\mathcal{B}(B^0 \to K^{*0}\gamma) + \mathcal{B}(B^+ \to K^{*+}\gamma)}.$$

The analysis uses $\tau_{B^+}/\tau_{B^0} = 1.083 \pm 0.017$ taken from [10]. Furthermore it assumes the ratio of the charged-to-neutral branching ratio of the $\Upsilon(4S)$ to be $f_+/f_0 = 1$. The measurement shows $f_+/f_0 = 1.072 \pm 0.057$ [10] or a more recent result from Belle $f_+/f_0 = 1.01 \pm 0.03 \pm 0.09$ [11].

The measured isospin asymmetry results in $\Delta_{0+} = (+0.3 \pm 4.5 \pm 1.8)\%$ which compares to the SM prediction of about 5-10%. Both numbers are in good agreement as the statistical error is still large.

In addition, the CP asymmetry was studied, which is expected to be smaller than 1% in the Standard Model:

$$\mathcal{A}_{CP} = \frac{1}{1 - 2w} \cdot \frac{N(\bar{B} \to \bar{K}^*\gamma) - N(B \to K^*\gamma)}{N(\bar{B} \to \bar{K}^*\gamma) + N(B \to K^*\gamma)},$$

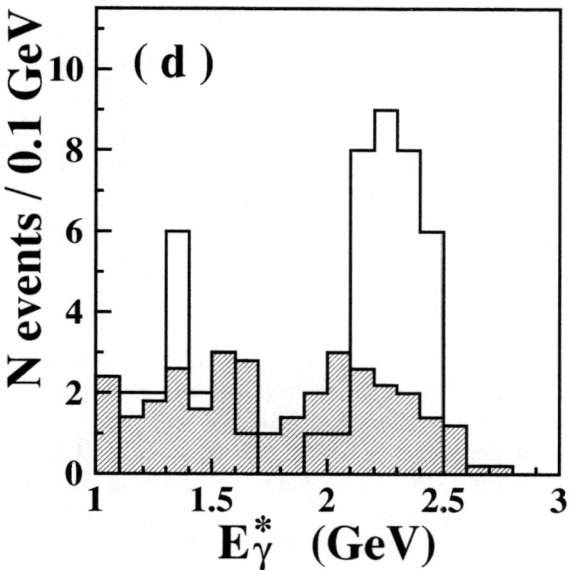

FIGURE 1. Energy spectrum of the photon in the decay $B \to \phi K \gamma$ [7].

where w represents the wrong tag fraction. The result of the measurement is $\mathcal{A}_{CP} = (-0.1 \pm 4.4 \pm 0.8)\%$ and in good agreement with the Standard Model.

The measurements of both asymmetries are highly statistically dominated due to the cancellation of most systematic uncertainties and can be significantly improved by the increasing integrated luminosities in the next years.

2.2. First Observation of the Decay $B \to \phi K \gamma$

High energetic photons are selected with energies of 2.0 GeV $< E_\gamma^* <$ 2.7 GeV. As before, a π^0/η veto is applied. The ϕ is reconstructed from two charged kaons where the invariant mass of the kaon pair is required to be within $|M(K^+K^-) - M(\phi)| <$ 10 MeV/c^2. The photon and ϕ are combined with a charged kaon or a K_S^0 to reconstruct the B candidate. The yield of the event selection is estimated from a fit to the beam constrained mass and the branching ratio for $B^+ \to \phi K^+ \gamma$ is shown in Table 1. The branching ratio for the neutral mode is not yet significant and a limit was calculated

$$\mathcal{B}(B^0 \to \phi K^0 \gamma) = (4.6 \pm 2.4 \pm 0.6) \cdot 10^{-6}$$
$$< 8.3 \cdot 10^{-6} \quad @90\% \text{ CL.}$$

The details of the analysis are described in [7]. First measurements of the energy distribution of the photon have been performed but are not significant yet as the

TABLE 1. Summary of exclusively reconstructed $b \to s\gamma$ transitions at Belle.

Channel	$\mathscr{B}$ [10^{-6}]	$\int \mathscr{L} dt$	$\mathscr{B}$(theor.) [10^{-6}]
$B \to X_s\gamma$	$336 \pm 53 \pm 42 + 50/ - 54$ [5]	6 fb^{-1}	
$B^0 \to K^{*0}\gamma$	$40.9 \pm 2.1 \pm 1.9$ [4]	78 fb^{-1}	69 ± 21 [6]
$B^+ \to K^{*+}\gamma$	$44.0 \pm 3.3 \pm 2.4$ [4]	78 fb^{-1}	74 ± 23 [6]
$B^+ \to \phi K^+\gamma$	$3.4 \pm 0.9 \pm 0.4$ [7]	90 fb^{-1}	
$B^+ \to K^+\pi^+\pi^-\gamma$	$24 \pm 5^{+4}_{-2}$ [8]	29 fb^{-1}	
$B^0 \to K_2^{*0}(1430)\gamma$	$13 \pm 5 \pm 1$ [8]	29 fb^{-1}	17.3 ± 8.0 [9]

statistics is too low to draw any definite conclusions. However, the energy spectrum fits well with the expectation from the Standard Model where energies of the photon are expected to be higher than 2 GeV. This can be seen in Fig. 1. As the photon is emitted from the penguin loop, the energy spectrum will give important information on the transition dynamics in this decay mode and will be one of the important measurements in the future.

2.3. Summary and Discussion of Radiative $b \to s\gamma$ Decays

In Table 1 all currently available results on branching ratios of radiative decays from Belle are shown. The accuracy of most measurements are already limited by the systematics and due to the fact, that the theoretical predictions on absolute branching ratios underly large hadronic uncertainties, tests on the Standard Model are quite limited. The explicitly reconstructed decay channels cover about $35 \pm 6\%$ of the total inclusive branching ratio for $B \to X_s\gamma$. Analysis are currently under way to add more explicit decay channels as e.g. $K\pi^n\gamma$ with $n > 2$, baryons and a photon, modes with resonances like e.g. η or η' and $B \to KKK\pi^n\gamma$.

As described above in measurements of various asymmetries, the systematic errors in both, theoretical calculations and experimental observables, are largely decreased and provide a powerful test on physics beyond the Standard Model. However, these measurements are still statistically limited. Table 1 also shows the size of the data set that was used for each analysis, which is in many cases much lower than the currently available data of about 140 fb^{-1} and leaves a lot of room for improvement in the near future.

Measurements of the photon spectrum will provide further handles on the penguin loop dynamics and the explicit decay $B \to \phi K_S^0\gamma$ will be one of the most promising candidates for future measurements on time-dependent CP violation.

3. THE DECAYS $B \to K^{(*)}\ell^+\ell^-$

The lowest order Feynman diagram for these decays are similar to the processes described above. Here, a virtual photon decays into the charged lepton pair. In addition a box diagram and a penguin diagram where the photon is replaced by a Z^0 have to be taken into account in lowest order. In Belle the decay $B \to K\ell^+\ell^-$ is reconstructed using neutral or charged kaons, for the decay $B \to K^*\ell^+\ell^-$ the K^* is reconstructed in the three decay modes $K^+\pi^-$, $K_S^0\pi^+$ and $K^+\pi^0$, where the invariant mass of the kaonpion pair is required to be within $|M(K\pi) - M(K^*)| < 75$ MeV$/c^2$. The leptons are identified with a likelihood probability function and minimal momenta are required $p_e > 0.4$ GeV$/c$ and $p_\mu > 0.7$ GeV$/c$ for electron and muon, respectively. In both cases a veto on the invariant mass of the lepton pair is applied not to originate from a J/ψ or $\psi(2S)$. The yield of the selection is extracted from a fit to the beam constrained mass and results in the branching ratios

$$\mathscr{B}(B \to K\ell^+\ell^-) = (4.8^{+1.0}_{-0.9} \pm 0.3 \pm 0.1) \cdot 10^{-7}$$
$$\mathscr{B}(B \to K^*\ell^+\ell^-) = (11.7^{+2.6}_{-2.4} \pm 0.8 \pm 0.4) \cdot 10^{-7}.$$

The SM predictions for these branching ratios are [12]:

$$\mathscr{B}(B \to K\ell^+\ell^-) = (3.5 \pm 1.2) \cdot 10^{-7}$$
$$\mathscr{B}(B \to K^*\ell^+\ell^-) = (11.9 \pm 3.9) \cdot 10^{-7}.$$

Due to the large errors on the theoretical prediction the measurement is well compatible with the calculation. Details on the analysis can be found in [13].

However, more information about this decay can be obtained from the invariant mass spectrum $q^2 = m_{inv}(\ell^+\ell^-)$ of the lepton pair, which describes the 'virtuality' of the photon and carries information about the dynamics in the penguin transition. The measurement in the two decay channels is presented in Fig. 2 and shows that the theoretical predictions in some bins are rather accurate. It also can be seen, that data and theoretical calculations are very well compatible, except for maybe the first bin in Fig. 2(a), which could very easily be a statistical fluctuation. In this comparison the experimental statistical error is the limiting factor and it will be signif-

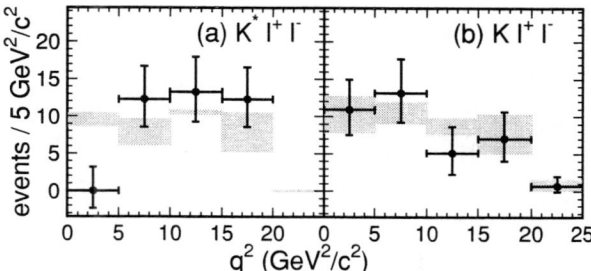

FIGURE 2. The invariant mass distribution of the lepton pair in the decays $B \to K^{(*)}\ell^+\ell^-$. The yellow shaded areas are SM predictions [13].

icantly improved by the increasing data set available in the future.

The forward-backward asymmetry $\mathscr{A}_{FB}$ of the lepton pair will deliver additional information about the transition dynamics and will be one of the important measurements for these decay modes.

4. THE PURELY (RADIATIVE) LEPTONIC B DECAYS

These decay modes are an interesting class of transitions, as the absolute branching ratios can be calculated much cleaner than for hadronic decays. The decay $B \to \ell \nu$ would be e.g. an interesting alternative for a clean measurement of V_{ub}. Unfortunately, the purely leptonic decays are highly helicity suppressed and the branching ratios in SM calculations are too low to be seen in the B factories.

However, for the purely leptonic decay modes the branching ratios might be enhanced of up to 3 orders of magnitude in various extensions of the Standard Model (e.g. 2-Higgs-Doublet models or Z-mediated flavour changing neutral currents (FCNC)). The purely leptonic decay modes are therefor an ideal testing ground for new physics beyond the Standard Model as they might increase the branching ratios to a level accessible to the B factories. In the Standard Model mixed leptonic decays like e.g. $B \to e^+ \mu^-$ are only allowed due to neutrino oscillations and therefor completely negligible. Again, various models like e.g. SUSY or Lepto-Quark models allow these decays and they could be a sensitive tool to search for these physics processes.

Another interesting aspect are radiative leptonic decays where the emission of a photon avoids the helicity suppression. In SM calculations, the branching ratios are expected to be of the order of 10^{-6} for $B \to \ell \nu \gamma$ and could be possibly reconstructed at the B factories. It is not clear however, if they could be used, e.g., for measurements of V_{ub} as the theoretical calculations suffer from

new uncertainties.

Belle performed a search for the decay $B \to \ell^+ \ell^-$ where ℓ could be either an electron or a muon. The lepton identification was done with rather tight cuts on a likelihood probability function ($\mathscr{L} > 0.9$). The signal box was defined as $5.27 < M_{bc} < 5.29$ GeV$/c^2$ and $|\Delta E| < 0.05$ GeV. No events were found in the box and the obtained limits are

$$
\begin{aligned}
\mathscr{B}(B \to ee) &< 1.9 \cdot 10^{-7}, \\
\mathscr{B}(B \to \mu\mu) &< 1.6 \cdot 10^{-7}, \\
\mathscr{B}(B \to e\mu) &< 1.7 \cdot 10^{-7},
\end{aligned}
$$

The analysis was based on a data set of 78 fb^{-1}.

5. SUMMARY

With the B factories our knowledge of radiative $b \to s\gamma$ decays improved quite dramatically in the last three years. Up to then the only observed radiative B decay was the decay $B \to K^*(892)\gamma$ which was observed by CLEO ten years ago. This mode turned into a precision measurement with Belle and BaBar. Now, about 36% of the total branching ratio $B \to X_s \gamma$ are explicitly measured and recently the decay modes $B \to \phi K \gamma$ and $B \to K^*\ell^+\ell^-$ were added by Belle. The measurements of all branching ratios agrees well with the Standard Model, however the theoretical calculations suffer from large hadronic uncertainties. These can be largely avoided in ratios where also experimental systematic uncertainties largely cancel out. Various measurements on e.g. CP- and isospin-asymmetries agree well with the SM, but are limited by the experimental statistical error.

Measurements of the real or virtual (in the case of $B \to K^{(*)}\ell\ell$) photon in the decay deliver information on the dynamics in the penguin loop but are also still limited by the statistical error of the data. All these measurements will improve significantly in the near future with the increasing collected luminosity of the B factories.

The decay mode $B \to \phi K_0 \gamma$ is one of the most promising candidates for a time-dependent CP analysis and will be one of the important measurements for radiative B decays.

The purely leptonic decay channels are an ideal testing ground to search for various models beyond the SM as 2-Higgs-Doublet and Lepto-Quark models as well as SUSY and Z-mediated flavour changing neutral currents. None of the decays has been observed yet which is compatible with the SM in which branching ratios are predicted to be too low to be observed with the available data sets yet. Radiative leptonic decays are not helicity suppressed and their observation should be within reach of the B factories. They might be an interesting alternative approach for a measurement of V_{ub}, but theoretical uncertainties still have to be understood.

Many measurements are on its way or planned and the next years will bring many more interesting results from rare radiative B decays.

ACKNOWLEDGMENTS

We wish to thank the KEKB accelerator group for the excellent operation of the KEKB accelerator. We acknowledge support from the Ministry of Eduction, Culture, Sports, Science and Technology of Japan and the Japan Society for the Promotion of Science; the Australian Research Council and the Australian Department of Education, Science and Training; the National Science Foundation of China under contract No. 10175071; the Department of Science and Technology of India; the BK21 program of the Ministry of Education of Korea and the CHEP SRC program of the Korea Science and Engineering Foundation; the Polish State Committee for Scientific Research under contract No. 2P03B 01324; the Ministry of Science and Technology of the Russian Federation; the Ministry of Education, Science and Sport of the Republic of Slovenia; the National Science Council and the Ministry of Education of Taiwan; and the U.S. Department of Energy.

REFERENCES

1. Kurokawa, S., and Kikutani, E., *Nucl. Instrum. Meth.* **A499**, 1–7 (2003).
2. Abashian, A., et al., [Belle Collaboration], *Nucl. Instrum. Meth.* **A479**, 117–232 (2002).
3. Ammar, R., et al., [CLEO Collaboration], *Phys. Rev. Lett.* **71**, 674–678 (1993).
4. Belle Collaboration, Abs. 537, BELLE-CONF-0319, Preliminary (2003).
5. Abe, K., et al., [Belle Collaboration], *Phys. Lett.* **B511**, 151–158 (2001).
6. Bosch, S. W., and Buchalla, G., *Nucl. Phys.* **B621**, 459–478 (2002).
7. Drutskoy, A., et al., [Belle Collaboration], *Phys. Rev. Lett.* **92**, 051801 (2004).
8. Nishida, S., et al., [Belle Collaboration], *Phys. Rev. Lett.* **89**, 231801 (2002).
9. Veseli, S., and Olsson, M. G., *Phys. Lett.* **B367**, 309–316 (1996).
10. Hagiwara, K., et al., [Particle Data Group Collaboration], *Phys. Rev.* **D66**, 010001 (2002).
11. Hastings, N. C., et al., [Belle Collaboration], *Phys. Rev.* **D67**, 052004 (2003).
12. Ali, A., Lunghi, E., Greub, C., and Hiller, G., *Phys. Rev.* **D66**, 034002 (2002).
13. Choi, S. K., et al., [Belle Collaboration], *Phys. Rev. Lett.* **91**, 262001 (2003).

Spectroscopy and Rare Decays at CDF

Cheng-Ju S. Lin

(Representing the CDF Collaboration)

Fermi National Accelerator Laboratory, Batavia, IL 60510, U.S.A.

Abstract. We report recent results on B hadron mass and rare decay measurements using data collected by the CDF detector in Run II.

1. INTRODUCTION

The CDF Run II detector [1, 2] has been collecting physics quality data since the beginning of 2002. The upgraded detector and trigger/DAQ system significantly enhanced the physics capability of the experiment. Some of the detector upgrades that are important to heavy flavor physics include the extended silicon vertex detector (SVX II), a time-of-flight (ToF) system for particle identification and a new layer of silicon detector (L00) mounted directly on the beam pipe (r ~ 1.5 cm) for precision vertexing near the interaction point. The brand new trigger system is designed to handle higher beam crossing rate and instantaneous luminosity. One of the new additions to the trigger system is the Silicon Vertex Trigger (SVT). The SVT is able to trigger on hadronic B and charm decays by selecting tracks with large impact parameters. The typical impact parameter resolution for an SVT track is about 48 μm (see Fig. 1).

With over 200 pb^{-1} of physics data now on tape, CDF is well positioned to make substantial contribution to the heavy flavor physics community. In this paper, we will report recent B hadron mass measurements ($B^+, B^0, B_s^0, \Lambda_b$) using exclusively reconstructed decay channels. We will also report on the search for flavor changing neutral current (FCNC) decays of $D^0 \to \mu^+\mu^-$ and $B_s^0 \to \mu^+\mu^-$.

2. SPECTROSCOPY

A good understanding of track parameter corrections due to energy loss is a prerequisite for precision mass measurements. Without proper correction, the measured mass would exhibit a strong p_T dependence. At CDF, a sample of muons from J/ψ decays are used to calibrate the energy loss correction. Figure 2 shows the J/ψ mass p_T dependence for the various scenarios. The lowest

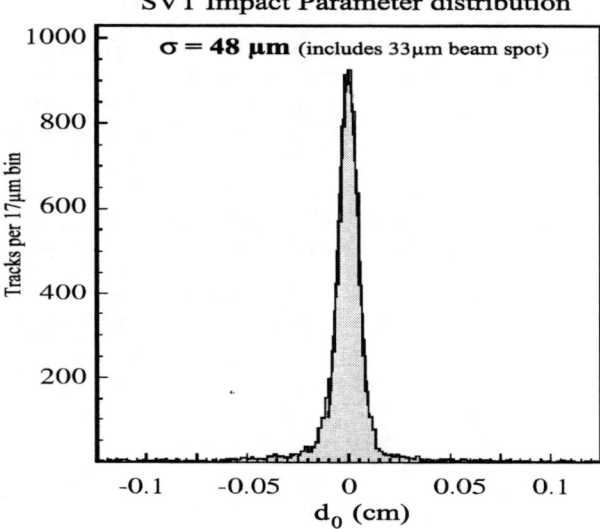

FIGURE 1. Impact parameter distribution of SVT tracks. The fitted impact parameter resolution is 48 μm (including 33 μm beam spot uncertainty). The typical SVT d_0 cut for displaced track is ± 120 μm.

curve shows the p_T dependence without any correction. As expected, the fitted J/ψ mass shows a strong p_T dependence. With the nominal energy loss correction (open triangle), there is still a residual p_T dependence. This residual effect is removed by iteratively tuning the GEANT material description used by the track fitter. Lastly, a global magnetic field correction is applied to shift the J/ψ mass to the PDG value (shown in solid circles). The momentum calibration procedure is cross-checked by measuring the mass of K_s^0, D, Υ and ψ'. The measured masses, after applying all corrections, are consistent with the PDG values.

Using ~ 80 pb^{-1} of data, we have measured the mass of various B hadrons using fully reconstructed J/ψ decay

CP722, B Physics at Hadron Machines, edited by M. Paulini and S. Erhan
© 2004 American Institute of Physics 0-7354-0203-5/04/$22.00

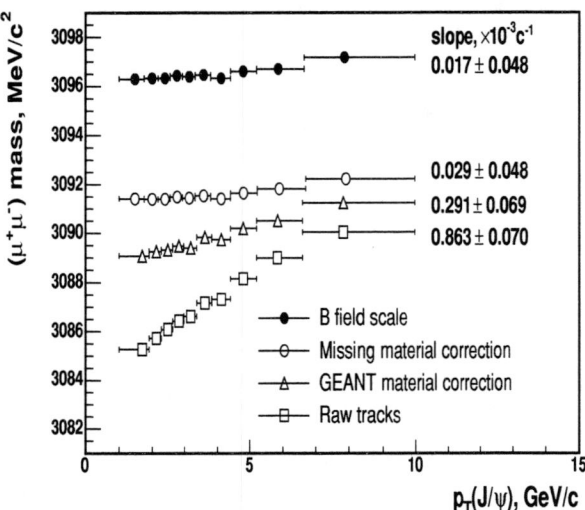

FIGURE 2. The J/ψ mass as a function of p_T. The four distributions are: no correction (square box), default energy loss correction (triangle), full energy loss correction (open circle), full energy loss and magnetic field corrections (filled circle).

TABLE 1. Summary of CDF B hadron mass measurements. The first quoted error is statistical and the second is systematic.

	Mass (MeV/c^2)	Data (pb^{-1})
$B^+ \to J/\psi K^+$	$5279.32 \pm 0.68 \pm 0.94$	80
$B^0 \to J/\psi K^{*0}$	$5280.30 \pm 0.92 \pm 0.96$	80
$B_s^0 \to J/\psi \phi$	$5365.50 \pm 1.29 \pm 0.94$	80
$\Lambda_b \to J/\psi \Lambda$	$5280.30 \pm 0.20 \pm 0.96$	70

channels. The reconstructed channels are: $B^+ \to J/\psi K^+$, $B^0 \to J/\psi K^{*0}$, $B_s^0 \to J/\psi \phi$ and $\Lambda_b \to J/\psi \Lambda$. The measured values are shown in Table 1. The invariant mass distributions for B_s^0 and Λ_b candidates are shown in Figs. 3 and 4, respectively. The new B_s^0 and Λ_b mass measurements from CDF are the world's best.

At CDF, we have also reconstructed a sample of charged hyperons, Ξ and Ω. The long lived charged hyperons, which have proper lifetimes ($c\tau$) on the order of centimeters, could leave tracks in the inner layers of the CDF silicon vertex detector before decaying. By requiring the pseudo hyperon tracks reconstructed from the decay products to be matched to the silicon track from the parent hyperon, the combinatorial background can be significantly reduced. Figures 5 and 6 show the invariant mass distribution of $\Xi^- \to \Lambda \pi^-$ and $\Omega \to \Lambda K^-$ candidates with the silicon track requirement, respectively. The sample of clean hyperons will be used to reconstruct the bottom-strange baryons: Ξ_b^0, Ξ_b^- and Ω_b^- to study their production properties.

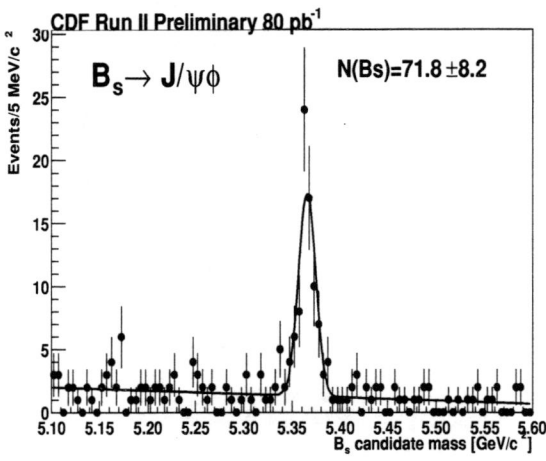

FIGURE 3. The invariant mass distribution for $B_s^0 \to J/\psi \phi$ candidates. The B_s^0 candidate is reconstructed with J/ψ decaying to $\mu^+\mu^-$ and ϕ to K^+K^-.

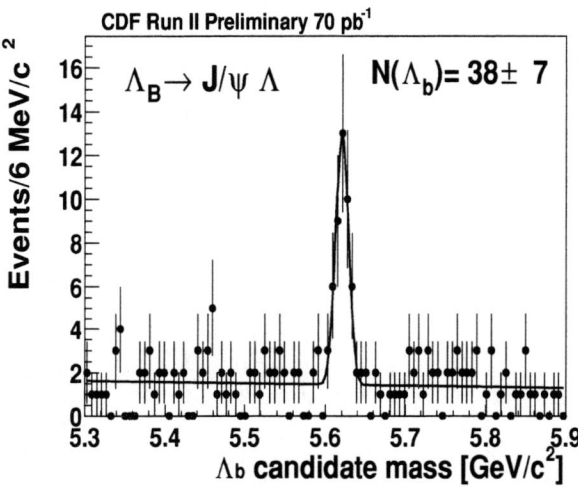

FIGURE 4. The invariant mass distribution for Λ_b candidates. The Λ_b candidate is reconstructed with J/ψ decaying to $\mu^+\mu^-$ and Λ to $p^+\pi^-$.

3. RARE DECAY SEARCH: $D^0 \to \mu^+\mu^-$

In the Standard Model, the FCNC decay $D^0 \to \mu\mu$ is heavily suppressed, with an expected branching ratio on the order of 10^{-13}. However, in some R-parity violating SUSY models, the branching ratio could be as large as 10^{-6} [3]. Thus, this decay channel provides a window of opportunity to search for new physics beyond the Standard Model.

The $D^0 \to \mu^+\mu^-$ decays are searched for in the hadronic data sample triggered by the SVT. The measurement is normalized to the decay of $D^0 \to \pi^+\pi^-$, which has similar kinematics as the signal mode and a

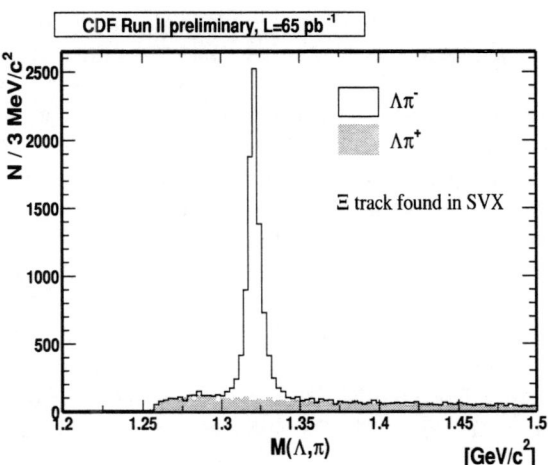

FIGURE 5. The invariant mass distribution for $\Xi \to \Lambda\pi^-$ candidates ($\Lambda \to p^+\pi^-$) with Ξ tracked in the silicon vertex detector. The distribution for the wrong sign candidates is shown in yellow.

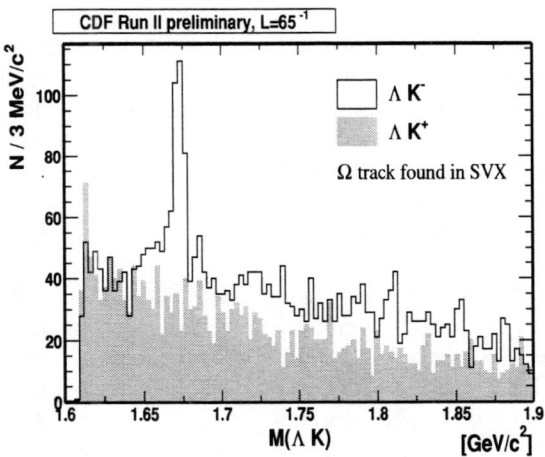

FIGURE 6. The invariant mass distribution for $\Omega \to \Lambda K^-$ candidates ($\Lambda \to p^+\pi^-$) with Ω tracked in the silicon vertex detector. The distribution for the wrong sign candidates is shown in yellow.

well measured branching ratio. To reduce combinatoric backgrounds, the D^0 for both $\mu\mu$ and $\pi\pi$ channels are required to come from $D^{*\pm}$ decay. By using the same track-based trigger for both signal and normalization modes, various efficiencies cancel in the relative branching ratio measurement. The remaining main ingredients needed for the measurement are: (1) the number of reconstructed $D^0 \to \pi^+\pi^-$ in the fiducial muon region, (2) the number of signal events (or an upper limit), (3) relative acceptance and reconstruction efficiency of $D^0 \to \mu^+\mu^-$ to $D^0 \to \pi^+\pi^-$, and (4) the expected background.

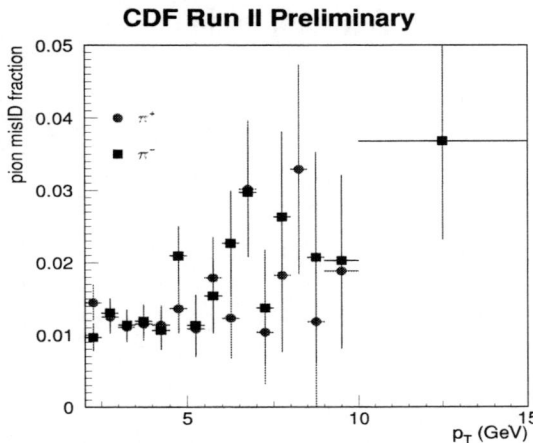

FIGURE 7. The fraction of $\pi^{\pm}$ misidentified as muons vs p_T.

The overall reconstruction efficiency ratio, $\varepsilon(\pi\pi)/\varepsilon(\mu\mu)$, is estimated from a combination of measurements and Monte Carlo simulation, as follows. First, $\varepsilon(\mu\mu)$ is obtained by convoluting the p_T spectrum of pions from $D^0 \to \pi^+\pi^-$ with the measured muon reconstruction efficiency. To properly account for the effect of hadronic interactions with detector material, $\varepsilon(\pi\pi)$ was estimated from a detailed GEANT detector simulation [4]. Combining the two values, the efficiency ratio is found to be 1.13±0.04. The same Monte Carlo simulation used in deriving the reconstruction efficiency ratio is also used to determine the geometric acceptance ratio. The geometric acceptance ratio, $\alpha(\pi\pi)/\alpha(\mu\mu)$, is found to be 0.96±0.02. The expected background in the signal mass window has two contributions. The first contribution is combinatoric, which can be estimated from the high-mass side-band region. The expected background from this source is 1.6±0.7 events. The second contribution comes from $D^0 \to \pi^+\pi^-$ with both tracks misidentified as muons. The latter contribution can be estimated as the number of $D^0 \to \pi^+\pi^-$ events in the signal mass region (using μ mass hypothesis) times the square of the misidentification probability. The misidentification probability is measured from a sample of $D^0 \to K\pi$ events from tagged $D^{*\pm}$. The $\pi \to \mu$ misidentification probability as a function of p_T is shown in Fig. 7. The average $\pi \to \mu$ probability (convoluted with the track p_T spectrum) is about 1.3%. This translates to 0.22±0.02 expected misidentification events. The combined background from the two sources is 1.8±0.7 events.

Based on 69 pb^{-1} of data, we found no events remaining in the signal region (Fig. 8). Using the prescription of Cousins and Highland [5], the limit on the branching ra-

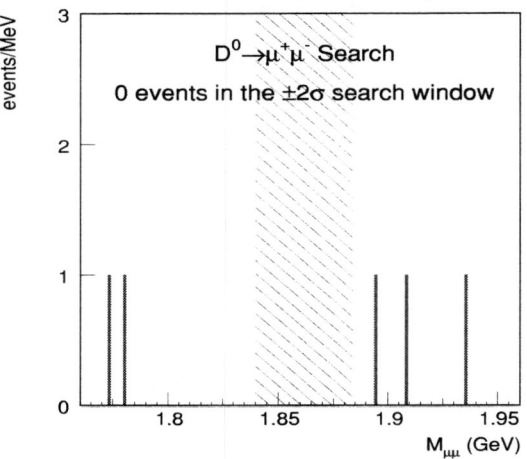

CDF Run II Preliminary

$D^0 \to \mu^+\mu^-$ Search

0 events in the $\pm 2\sigma$ search window

FIGURE 8. The di-muon mass spectrum with $D^{*\pm}$ tag.

tio is:

- $\mathcal{B}(D^0 \to \mu^+\mu^-) < 2.4 \times 10^{-6}$ at 90% CL,
- $\mathcal{B}(D^0 \to \mu^+\mu^-) < 3.1 \times 10^{-6}$ at 95% CL.

This result improves the previous published limits from BEATRICE and E771 [6, 7] by about a factor of two.

4. RARE DECAY SEARCH: $B_s^0 \to \mu^+\mu^-$

Similar to $D^0 \to \mu^+\mu^-$, the FCNC decay of $B_s^0 \to \mu^+\mu^-$ is suppressed at tree level in the Standard Model. The expected branching ratio is $\mathcal{B}(B_s^0 \to \mu^+\mu^-) = 3.8 \times 10^{-9}$. In many SUSY models, this branching ratio can be enhanced to as large as $\sim 10^{-6}$ [8–10], which would make the decay observable in Run II.

The $B_s^0 \to \mu^+\mu^-$ decays are searched for in the dataset collected by the di-muon trigger. The B_s^0 candidates are required to be in the kinematic fiducial box of $p_T > 6$ GeV/c and rapidity $|y_{B_s}| < 1$. This kinematic box is chosen so that we can readily normalize the result to the measured B production cross-section at the Tevatron. The branching ratio is obtained from the expression:

$$\mathcal{B}(B_s^0 \to \mu^+\mu^-) = \frac{N(B_s^0 \to \mu^+\mu^-)}{2 \cdot \sigma_{B_s} \cdot \mathcal{L}_{total} \cdot \alpha \cdot \varepsilon_{total}}, \quad (1)$$

where $N(B_s^0 \to \mu^+\mu^-)$ is the number of observed signal events (or upper limit if no signal is observed), σ_{B_s} is the B_s^0 production cross section. The B_s^0 production cross section is obtained by multiplying the measured B^0 production cross section [11] by the world average value of f_s/f_d [12]. The uncertainty on σ_{B_s} is the dominant source of systematic uncertainty in this analysis.

$\mathcal{L}_{total}$ is the total integrated luminosity. The geometric and trigger acceptance, α, is estimated from the Monte Carlo. The total efficiency, ε_{total}, includes tracking, trigger, reconstruction and analysis cut efficiencies. The first three efficiencies (tracking, trigger and reconstruction) are measured directly from the data. The analysis cut efficiency accounts for the efficiency of the offline analysis cuts used in the optimization. The four primary discriminating variables that we use to optimize the analysis are: the invariant mass of the muon pair ($M_{\mu\mu}$), the proper lifetime of the B_s^0 candidate ($c\tau$), the 2-D (transverse plane) opening angle between the B_s^0 momentum vector and the decay vertex axis ($\Delta\phi$), and the isolation cut (Iso). The isolation variable is defined as $Iso = p_T(B_s^0)/(p_T(B_s^0) + \Sigma p_T(trk_i))$, where the summation is over all tracks within a cone of $\sqrt{\Delta\eta^2 + \Delta\phi^2} < 1$ about the B_s^0 momentum direction. The analysis cut efficiency is estimated from the Monte Carlo and cross-checked using a sample of $B^+ \to J/\psi K^+$ events from the data.

The number of background events remaining in the signal region is estimated from Monte Carlo and data. From the Monte Carlo study, we concluded that contamination from resonant decays (e.g. $B \to h^+h^-$) are negligible at the current level of sensitivity. The dominant source of background comes from combinatorics. We have estimated the combinatorial background from mass side-bands. To improve the estimate, we factorize the expected rejection for each (non-correlated) group of cuts separately. The total background is obtained from the expression: $N_{bkg} = N_{SB}(c\tau, \Delta\phi) \cdot R_{Iso} \cdot R_{mass}$, where $N_{SB}(c\tau, \Delta\phi)$ is the number of sideband events passing a given set of $c\tau$ and $\Delta\phi$ cuts, R_{Iso} is the expected rejection for a given Iso cut and R_{mass} is the expected rejection for a given mass window cut. The variables $c\tau$ and $\Delta\phi$ are correlated and therefore cannot be treated separately. The resulting total background is estimated to be 0.54 ± 0.2 events.

The di-muon invariant mass spectrum (using 113 pb^{-1} of data) with the final selection cuts are shown in Fig. 9. There is one event within the B_s^0 search window. Given the observation is consistent with background expectation, we calculate the upper limit on the branching ratio using the method of Ref. [13]. The resulting limits are:

- $\mathcal{B}(B_s^0 \to \mu^+\mu^-) < 9.5 \times 10^{-7}$ at 90% CL,
- $\mathcal{B}(B_s^0 \to \mu^+\mu^-) < 1.2 \times 10^{-6}$ at 95% CL.

This result is a factor of two better than the published limit [14].

We have extended the analysis to search for $B^0 \to \mu^+\mu^-$ decays. The analysis procedure is identical to the B_s^0 search. The event in the B_s^0 mass window also falls in the B^0 search window. Given the observation is also consistent with background expectation ($N_{bkg}^{B^0} =$

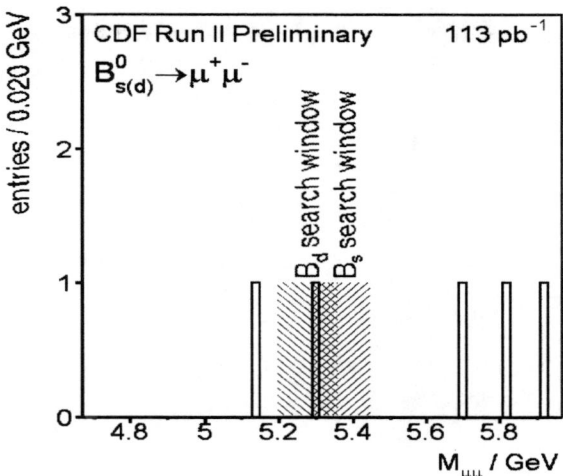

FIGURE 9. The di-muon mass spectrum. The 3σ mass search window for B_s^0 is shown in the blue hatched region. The 3σ mass search window for B^0 is shown in the red hatched region. There is one event overlapping both search windows.

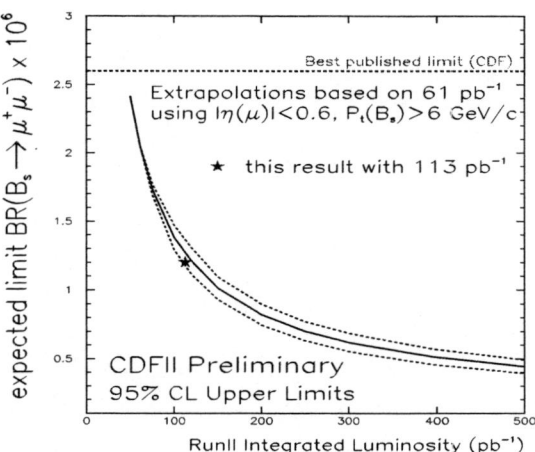

FIGURE 10. The expected 95% CL upper limit on $\mathcal{B}(B_s^0 \to \mu^+\mu^-)$ as a function of integrated luminosity.

0.59 ± 0.22), we set an upper limit on the branching ratio:

- $\mathcal{B}(B^0 \to \mu^+\mu^-) < 2.5 \times 10^{-7}$ at 90% CL,
- $\mathcal{B}(B^0 \to \mu^+\mu^-) < 3.1 \times 10^{-7}$ at 95% CL.

Figure 10 shows the expected sensitivity of our $B_s^0 \to \mu^+\mu^-$ search as a function of integrated luminosity. In the absence of a signal, with twice the data and no improvement in the analysis technique, the limit is expected to improve by another factor of two. To push the sensitivity beyond the 10^{-7} level would require re-optimizing the analysis.

5. SUMMARY

The Run II physics program is now well underway. Using only a fraction of the available data, CDF has already made world class measurements in the area of heavy flavor physics. In this paper, we presented some preliminary measurements on B hadron masses and rare decay searches, some of which are already the best in the world. With more data on the way, the prospect of performing precision spectroscopy on the broad spectrum of hadrons produced (including B_c, charm-strange and bottom-strange baryons, *etc.*) at the Tevatron is very promising. CDF will also remain competitive in the area of rare decay measurements.

REFERENCES

1. Abe, F., et al., [CDF Collaboration], *Nucl. Instr. Meth.* **A271**, 387–403 (1988).
2. Blair, R., et al., [CDF Collaboration], The CDF II Detector: Technical Design Report, Tech. Rep. FERMILAB-PUB-96-390-E (1996).
3. Burdman, G., Golowich, E., Hewett, J., and Pakvasa, S., *Phys. Rev.* **D66**, 014009 (2002).
4. Brun, R., Hagelberg, R., Hansroul, M., and Lassalle, J. C., GEANT: Simulation Program for Particle Physics Experiments. User Guide and Reference Manual., Tech. Rep. CERN-DD-78-2-REV and CERN-DD-78-2 (1978).
5. Cousins, R. D., and Highland, V. L., *Nucl. Instrum. Meth.* **A320**, 331–335 (1992).
6. Adamovich, M., et al., [BEATRICE Collaboration], *Phys. Lett.* **B408**, 469–475 (1997).
7. Alexopoulos, T., et al., [E771 Collaboration], *Phys. Rev. Lett.* **77**, 2380–2383 (1996).
8. Dedes, A., Dreiner, H. K., Nierste, U., and Richardson, P., arXiv:hep-ph/0207026 (2002).
9. Baer, H., et al., *JHEP* **07**, 050 (2002).
10. Arnowitt, R., Dutta, B., Kamon, T., and Tanaka, M., *Phys. Lett.* **B538**, 121–129 (2002).
11. Acosta, D., et al., [CDF Collaboration], *Phys. Rev.* **D65**, 052005 (2002).
12. Heavy Flavour Averaging Group (2003), URL: http://www.slac.stanford.edu/xorg/hfag/osc/PDG_2003/.
13. Feldman, G. J., and Cousins, R. D., *Phys. Rev.* **D57**, 3873–3889 (1998).
14. Abe, F., et al., [CDF Collaboration], *Phys. Rev.* **D57**, 3811–3816 (1998).

B TRIGGER AT FUTURE EXPERIMENTS

The ATLAS *B* Physics Trigger

Simon George

(Representing the ATLAS Collaboration)

Department of Physics, Royal Holloway, University of London, Surrey TW20 0EX, U.K.

Abstract. This paper gives an overview of the ATLAS Trigger and Data Acquisition (T/DAQ) system with an emphasis on *B* physics capabilities. It describes recent work on how to maintain the *B* physics program within some constraints that have arisen: a higher target start-up luminosity, an incomplete configuration of the detector at start up, and cost constraints for the T/DAQ system. It also shows how the High Level Trigger (HLT) software has advanced and gives some results of new performance measurements.

1. INTRODUCTION

ATLAS is a general-purpose experiment for the Large Hadron Collider (LHC), currently under construction at the European Laboratory for Particle Physics (CERN). The LHC will collide protons with 14 TeV centre-of-mass energy within bunches crossing at 40 MHz. At start up in 2007, the target peak luminosity is $2 \times 10^{33}\,\mathrm{cm^{-2}s^{-1}}$, rising to the full design luminosity of $1 \times 10^{34}\,\mathrm{cm^{-2}s^{-1}}$ after a few years' running. There will be an average of 4.6 and 23 interactions per bunch crossing for initial and design luminosity, respectively.

The ATLAS first-level trigger has to make a decision every 25 ns. The final event rate for mass storage must be reduced to ~ 200 Hz in view of the large event size ($O(1)$ MByte) and the available offline storage and processing resources. About 1% of collisions produce a $b\bar{b}$; the challenge for the trigger is to select those events of most interest for the ATLAS *B* physics program, within the limited trigger resources (processing power and network bandwidth) available. The ATLAS *B* physics trigger is based on a lepton signature at the first level, which can optionally be accompanied by additional, lower transverse-energy signatures of leptons and jets. These signatures are refined in the higher trigger levels where specific decays are reconstructed. The ATLAS physics program is described in Ref. [1]. More details on the design of the ATLAS T/DAQ system can be found in Refs. [2, 3].

The whole ATLAS detector is described in Ref. [4] and the inner detector in Ref. [5]. The largest volume part of the detector is the muon spectrometer incorporating toroid magnets with monitored drift tubes (MDT), resistive plate chambers (RPC) and thin gap chambers (TGC). Inside this is the calorimeter system, comprising

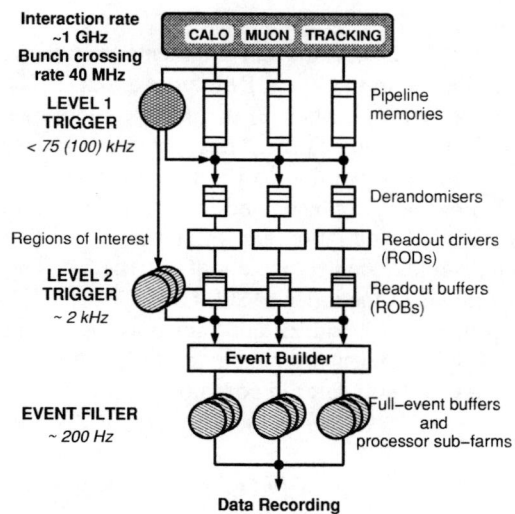

FIGURE 1. Overview of the ATLAS T/DAQ System.

a liquid-argon electromagnetic and hadronic endcap and electromagnetic barrel, and a scintillating-tile hadronic barrel. Within this is a solenoid providing a 2 T field and housing the inner detector (ID), which consists of silicon pixels nearest the beam pipe, then silicon microstrip detectors (SCT) at intermediate radii, and a straw-tube transition-radiation tracker (TRT) as the outermost part.

2. THE ATLAS T/DAQ SYSTEM

2.1. Overview

ATLAS has a three-level trigger, shown in Fig. 1. The first-level (LVL1) has to operate at the bunch-crossing

CP722, *B Physics at Hadron Machines*, edited by M. Paulini and S. Erhan

rate of 40 MHz, and takes a decision based on coarse-granularity calorimeter towers and muon trigger stations in less than 2.5 μs, while detector data are held in pipeline memories. It reduces the event rate to less than 75 kHz, although this will be limited to 25 kHz LVL1 output for initial data taking due to resource limitations in the HLT/DAQ system. Accepted events are transferred from the pipeline memories of the detector to readout buffers (ROB) where they are held pending a second-level (LVL2) decision.

LVL2 has access to full-granularity data from all detectors (including the ID) by requesting them from the ROB's over the network. Information on the Regions of Interest (RoI) identified by LVL1 is transferred to LVL2, where it is used to reduce the amount of data requested to a few percent of the full event. Specialized trigger algorithms are used with an emphasis on fast rejection, to achieve the target average decision time of about 10 ms. Accepted events are classified by the signature(s) they match.

After a LVL2 accept, event fragments from the ROB's are transferred to the Event Builder at about 2 kHz. Full events are then distributed to the third-level, or event filter (EF), processing farms. Here the event selection is further refined according to the LVL2 classification by performing more sophisticated reconstruction and using more detailed calibration and alignment data. The processing can take on average about one second. The final output rate is of the order of 200 Hz.

Together, LVL2 and EF are known as the High Level Trigger (HLT) and have a large part of their infrastructure (hardware and software) in common.

2.2. Region of Interest Mechanism

The Region of Interest mechanism is a fundamental architectural principle of the ATLAS trigger.

The LVL1 event selection is based mainly on local signatures, identified at coarse granularity in the muon detectors and calorimeter. Significant further rejection can be achieved by examining the full granularity data from muon, calorimeter and inner detectors in the same localities.

The RoI is the geographical location of the LVL1 signature. It is passed to LVL2 where it is quickly translated into a list of corresponding ROB's. The data are requested sequentially, one detector at a time, so that only as much data are transferred as are needed. In most cases events are rejected and the data from all detectors in the RoI are not needed.

The RoI mechanism is a powerful way to gain additional rejection before event building. The benefit is an order of magnitude reduction in the dataflow bandwidth,

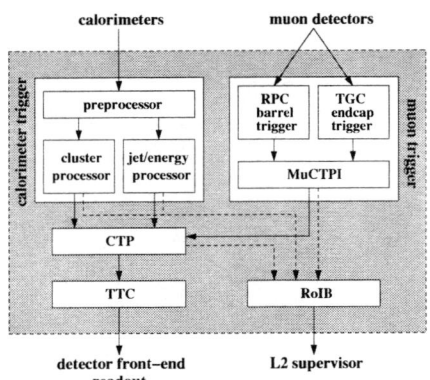

FIGURE 2. The LVL1 Trigger system.

at the small cost of more control traffic.

Simulations of the LVL1 trigger show that there will be about 1.6 RoI's from triggers per event. These can be accompanied by RoI's with lower thresholds to give additional guidance to the HLT.

2.3. LVL1 Trigger

The LVL1 trigger works by identifying basic signatures that are consistent with interesting physics: muon tracks in the muon spectrometer, electromagnetic/hadronic/jet-like calorimeter clusters, and missing/sum transverse energy (E_T). The trigger decision is based on multiplicities and thresholds of these signatures.

LVL1 is a hardware trigger. The logic is implemented in programmable and custom electronics (FPGA's and ASIC's). These will have programmable thresholds which can be configured to suit the expected conditions of a data-taking run. As shown in Fig. 2, there are separate muon and calorimeter triggers, whose results are combined in the Central Trigger Processor (CTP). If the event is accepted, signals from the CTP, muon and calorimeter triggers are sent to the Region of Interest Builder (RoIB) where they are prepared for LVL2. The CTP also signals the detector front-end readout via the Timing, Trigger and Control (TTC).

2.4. High Level Trigger

The HLT is built mainly from commodity equipment (PCs, Ethernet switches), with a few custom components such as the RoIB and ROB's in the DAQ system. The majority of the implementation is in software; indeed the HLT is a large and complex software-engineering project. Fig. 3 shows the main components and high-level interactions of the system that has been designed.

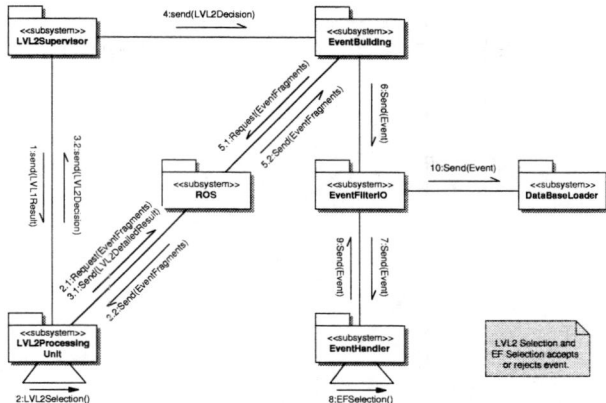

FIGURE 3. Overview of HLT components and the messages they exchange.

Note that the Read Out Subsystem (ROS) contains the ROB's.

To optimize processor usage, the LVL2 processing unit can run several worker threads at once on a dual-processor computer. Each worker thread will run a copy of the event-selection software which processes events independently. This allows idle cycles to be used to process another event while waiting for the ROS to return event data.

2.4.1. Event Selection Software

LVL2, EF and the offline software ('Athena') share a common event-selection software suite. It is used both to perform the real event selection online, and to emulate the HLT offline. The latter is also used as a development environment for trigger algorithms.

The necessary differences between these environments are hidden behind interfaces, so for example trigger algorithms have no knowledge of whether event data are being accessed from a file, database, memory, or ROB's over a network. Thus the principle of re-use of offline software as far as possible has been established. In current prototypes, offline software provides the basic framework and services, the data unpacking mechanism, part of the event data model and detector description. LVL2 has its own specialized algorithms and data preparation. All software used in LVL2 must meet the additional constraints of being multi-thread safe and efficient (i.e. while there are available processor cycles, the event input rate that can be sustained by one processor should scale more or less linearly with the number of threads, showing that performance is not limited due to locks). The EF re-uses offline reconstruction algorithms; to facilitate this they are currently being redesigned in the form of a 'toolkit' as described in Ref. [6].

A tool to aid integration of offline software and trigger algorithms with the LVL2 test beds has been developed: 'AthenaMT' provides a single-PC test environment and automatically sets up emulation of the full LVL2 system including the multiple worker threads. Simulated data for tests have a fully simulated detector response and the data are formatted as they are expected to come from the readout electronics. This ensures that the additional processing time to deal with data in this format is taken into account.

Both LVL2 and EF use a common 'steering'. This is a RoI-driven framework which calls the algorithms necessary to find features in subdetectors and combine these to establish hypotheses. Early rejection of events is achieved by comparing the event with a Trigger Menu (lists of signatures) after each step in the processing: the event is rejected if it is no longer consistent with any signature in the menu.

3. *B* PHYSICS TRIGGER STRATEGY

At the peak target luminosity of $2 \times 10^{33}\,\mathrm{cm^{-2}s^{-1}}$ in the initial running (and the eventual target of $1 \times 10^{34}\,\mathrm{cm^{-2}s^{-1}}$), the *B* physics trigger will be limited to a di-muon trigger. This is based on muons found in LVL1 with a p_T threshold of around 6 GeV/c. At LVL2 and EF, the regions of interest will be confirmed in the precision muon chambers, the tile calorimeter, then extrapolated to the ID. The trigger selection is based on identifying candidates for specific exclusive or semi-inclusive decays, for example $J/\psi \to \mu^+\mu^-$ and $B^0/B_s^0 \to \mu^+\mu^-(X)$.

As the luminosity falls to around $1 \times 10^{33}\,\mathrm{cm^{-2}s^{-1}}$, further *B* triggers can be added based on a single muon trigger plus at least one additional trigger from the semi-inclusive reconstruction of specific decay candidates, for example $J/\psi \to e^+e^-$, $B \to h^+h^-$ and $D_s^+ \to \phi\pi^\pm$. Two strategies have been investigated for these additional triggers: 'RoI-guided' and 'full-scan'. In both cases the LVL1 muon is confirmed at LVL2 using the tile calorimeter and ID.

The baseline, RoI-guided strategy extends the fundamental architectural principle of the ATLAS trigger to *B* physics. The LVL1 trigger provides information on low-E_T EM and jet RoI in addition to the muon trigger. At LVL2, tracks are reconstructed in the ID within these RoI and the reconstructed track information is used to search for candidates with specific decays. Track reconstruction and *B* decay candidate selection is repeated in the EF, using the LVL1 RoI's again or the LVL2 tracks as seeds. The EF selection is refined by fitting tracks to include secondary vertex information.

This approach significantly reduces the resources

needed for the B physics trigger compared to the second approach. However, there could prove to be too many low-threshold jet and EM RoI's, and/or the inefficiency for identifying RoI's corresponding to the B decays of interest could be unacceptably low, although initial studies based on fast simulation are encouraging. Studies based on full simulation and the full-scan strategy show that, for example, tracks from $D_s^+ \to \phi \pi^\pm$ are contained within a cone of about 1.2 radians by 1 unit of pseudorapidity.

The full-scan strategy does not use RoI's to seed searches for B decay products. Instead, LVL2 track reconstruction is performed in the full acceptance of the SCT and pixels and the resulting tracks are used to form $B \to h^+ h^-$ and $D_s^+ \to \phi \pi^\pm$ candidates. Given additional resources, the TRT can be scanned for tracks in order to identify $J/\psi \to e^+ e^-$ candidates. The EF can perform another full scan or use the LVL2 tracks as seeds. This approach is expected to be more efficient than the baseline above, but at a significantly higher resource cost.

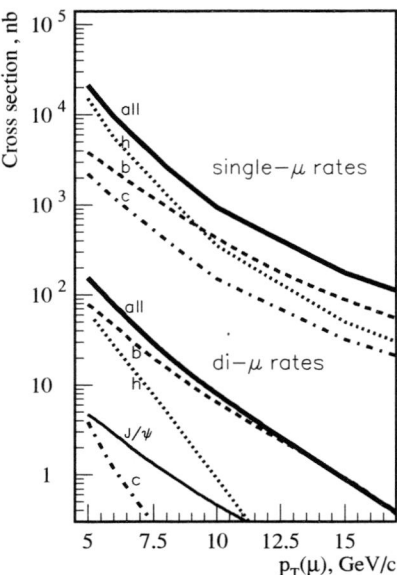

FIGURE 4.   Single and di-muon cross sections.

3.1. Start-up Conditions

When the LHC starts up, the initial beam conditions and the state of the ATLAS detector will not be optimal. Anticipated problems include: luminosity will vary somewhat from fill to fill, variable beam-related background, the ATLAS detector will be incomplete (see Ref. [5]), the detector will need understanding and tuning, and the T/DAQ processing capacity and bandwidth may be limited.

The target peak luminosity for the LHC initial running is 2×10^{33} cm^{-2}s^{-1}. However, luminosity will fall by a factor of about two during the coast. The T/DAQ system will clearly be built with sufficient capacity to meet the requirements of triggering on 'discovery' physics at the target peak luminosity, which will in particular require more processing power. When the luminosity drops, demand on trigger resources will also fall and some spare capacity can be used for B physics triggers.

To cope with the initial running conditions, trigger algorithms must be robust with respect to noise, misalignment, and be able to operate effectively with the reduced detector that will be available at start-up time. The configuration must be flexible to adapt thresholds, pre-scales and other parameters to cope with varying noise, luminosity, etc.

3.2. B Physics Trigger Rates

The estimated B physics trigger rates for the signatures discussed in this paper are shown in Table 1. The left column applies to higher luminosities where the strategy is

based solely on di-muon triggers, while the right column shows the increased coverage planned for lower luminosities. Further details can be found in Refs. [2, 7]. Note that these are only a small fraction of the total LVL2 and EF accept rates.

3.3. Muon Triggers

The LVL1 muon trigger has logic designed for both low and high-p_T thresholds; low-p_T is the one used for B physics. It uses data from the two inner RPC stations (barrel) and two outer TGC stations (endcap); each station contains two layers and each layer has two projections. The algorithm starts with hits in the outer station and looks for coincident hits within a window (determined by the p_T threshold) in the inner station. Three out of four layers must contain hits in each projection.

The main single-muon background comes from π/K decays in flight. These can mostly be rejected at LVL2, by matching muon tracks to ID tracks and applying sharp p_T cuts. Fig. 4 shows how the prompt-muon cross section falls steeply with increasing muon p_T (and the background drops off even more quickly), which means that the trigger rate can be controlled by fine-tuning of the threshold. Simulations predict that a p_T threshold of 6 GeV/c has a LVL1 rate of about 20 kHz at 1×10^{33} cm^{-2}s^{-1}.

For di-muon triggers, lower thresholds are possible if the rates are low enough: $p_T > 5$ GeV/c in the barrel, 3 GeV/c in the endcaps. At 1×10^{34} cm^{-2}s^{-1} the di-

TABLE 1. Estimated B physics Trigger Rates.

Trigger	$2 \times 10^{33}\,\mathrm{cm^{-2}s^{-1}}$		$1 \times 10^{33}\,\mathrm{cm^{-2}s^{-1}}$	
	LVL2	EF	LVL2	EF
$B^0/B_s^0 \to \mu^+\mu^-(X)$	200 Hz	small	100 Hz	small
$J/\psi \to \mu^+\mu^-$		10 Hz		5 Hz
$D_s^+ \to \phi\pi^\pm$	–	–	60 Hz	9 Hz
$B \to \pi^+\pi^-$	–	–	20 Hz	3 Hz
$J/\psi \to e^+e^-$	–	–	10 Hz	2 Hz
Total	**200 Hz**	**10 Hz**	**190 Hz**	**20 Hz**

muon trigger rate should be below 1 kHz for a 6 GeV/c threshold, only a small fraction of the total LVL1 rate. This is dominated by heavy-flavour decays but is subject to large uncertainties due to its sensitivity to very low p_T muons in the endcap and a small fraction of single muon events that suffer from double counting. Di-muon triggers are used to select decays such as $B^0 \to J/\psi(\mu^+\mu^-)K_S^0$ and $B^0/B_s^0 \to \mu^+\mu^-(X)$.

At LVL2, muon triggers are first confirmed using the precision chambers (MDT). Better track measurement gives a tighter threshold which, due to the steeply falling muon p_T spectrum, significantly reduces the trigger rate. Further rejection is obtained by extrapolating tracks to the ID, requiring a track with closely matching z, ϕ, and p_T. Track matching helps to reject muons from π/K decays and using the ID further improves p_T resolution for prompt muons. In the barrel this has been fully simulated and studied in detail. At luminosity of $1 \times 10^{33}\,\mathrm{cm^{-2}s^{-1}}$, the LVL2 rate for a single-muon trigger with a threshold of 6 GeV/c and $|\eta| < 1$ is about 2 kHz, about half of which is remaining π/K decays and the rest heavy-flavour decays. At $2 \times 10^{33}\,\mathrm{cm^{-2}s^{-1}}$, a higher threshold of about 8 GeV/c gives similar rates. Extrapolated to the full detector, the resulting rate is estimated to be about 5 kHz. The EF will perform near offline-quality track reconstruction, vertex fit and mass cuts, for example to select $J/\psi \to \mu^+\mu^-$.

3.4. LVL1 Jet and EM Cluster Trigger

The LVL1 calorimeter trigger uses ~ 7000 dedicated, relatively coarse calorimeter 'towers' of 0.1 units in pseudorapidity by 0.1 radians in azimuthal angle. The towers have two layers: the electromagnetic and hadronic calorimeters. A sliding window algorithm is used to find clusters which satisfy criteria for EM, tau/hadron and jet selections. A E_T threshold is applied, with isolation in both layers to provide powerful jet rejection for EM/tau/hadron triggers.

The average multiplicity of LVL1 EM and jet RoI in events with a muon trigger is important as it determines

the fraction of the detector that must be read out in order to make the B trigger decision. This has been studied using a sample of $b\bar{b}$ events containing a muon with $p_T > 6$ GeV/c, produced by a fast simulation which contains a detailed simulation of the calorimeter and LVL1 electronics. With a jet E_T threshold of 5 GeV there are on average about 2 RoI's per event. An EM cluster threshold of 2 GeV yields about 1 RoI per event.

3.5. Hadronic Final States

Tracks, found in the SCT and pixels by either by the RoI-guided or full-scan strategies, are used to reconstruct the B decays $B \to h^+h^-$ and $D_s^+ \to \phi\pi^\pm$. Studies show that the RoI-guided method would be efficient for a B hadron with $p_T > 15$ GeV/c. At LVL2, kinematic and topological cuts are made to reduce combinatorial background. At EF there is more processing time so the resolutions of reconstructed track parameters are better and a vertex fit is available. The rate can therefore be further reduced by tighter mass cuts, and using decay-length and vertex-quality cuts. Studies show that the tracking algorithms are very robust with respect to a missing middle pixel layer (as in the initial detector configuration) and anticipated levels of misalignment at LVL2.

3.6. Muon-Electron Final States

To select channels such as $B^0 \to J/\psi K_S^0$ with $J/\psi \to \mu^+\mu^-$ or $J/\psi \to e^+e^-$ with opposite side electron or muon tag, respectively, electron identification is required. ATLAS provides two options for this: use the RoI-guided strategy to find silicon tracks guided by EM RoI's, or perform full, unguided reconstruction of tracks in the TRT. The RoI-guided method is much faster, since typically only about 0.3% of the full ID is reconstructed. However, the lowest nominal threshold possible in the calorimeter with acceptable RoI multiplicity is 2 GeV and this is not efficient until a higher energy, significantly higher than the minimum threshold possible with a full

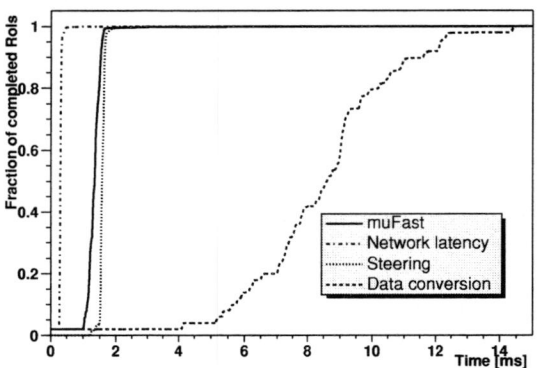

FIGURE 5. Testbed measurements of the LVL2-muon trigger processing time.

scan of the TRT. The efficiency to find a separate RoI for both e^- and e^+ with $p_T > 3$ GeV/c is about 80%.

4. PERFORMANCE

4.1. Testbed Performance Measurements

Software to perform the LVL2 muon reconstruction in the barrel of the muon spectrometer was set up on a LVL2 farm node, a PC with dual 2.2 GHz Xeon processors, with separate PCs running the LVL2 supervisor process and ROS emulation.

As shown in Fig. 5, the algorithm time itself is very small, and the time to access data over the network is also modest, since typically data are fetched from 1-2 ROB's. Data preparation time dominates the processing time despite the fact that only data within the RoI's need to be prepared – in the absence of the RoI mechanism the time would be much larger. This time is spent decoding binary data from the RPC and MDT detectors, converting logical addresses to geometrical positions taking into account the complex geometry of the muon detectors, and instantiating collections of objects, using the 'offline' framework and services. There is still considerable scope for code optimization. The results are pessimistic because of a conservative level of radiation in the cavern and the fact that only muon events, not background events, have been simulated. Even so, these results already show adequate performance, since computers will be bought in 2006 when 8 GHz is expected to be the commodity processor speed.

4.2. Resource Estimates

A lot of care has been taken to ensure that the studies provide a realistic basis for resource calculations, such as simulation of realistic raw data, timing of network requests and data preparation, and the reduced rate of later steps in the sequential-processing scheme. This has lead to the conclusion that the LVL2 budget of ~10 ms will be achievable for high-p_T 'discovery' physics. The time for muon and calorimeter triggers, for which studies are most advanced, is already acceptable; this work will be extended to cover all detectors and triggers.

The target time of 10 ms at 25 kHz for LVL2 yields a requirement of 250 processors, scaling to 750 for the full 75 kHz LVL1 rate. Note that the number of EF processors required is much more.

5. CONCLUSIONS

The latest picture of ATLAS and LHC at start up is less favourable for B physics than in the past, but ATLAS has responded with a variety of flexible trigger schemes to make the most of it. The RoI-based strategy allows a full program at modest resource cost, with slightly reduced trigger efficiency. ATLAS will take advantage of beam-coast and lower-luminosity fills to increase the coverage of B physics at lower luminosities. Algorithms are robust enough for the initial detector and conditions. For the muon spectrometer and calorimeter, there is now a full simulation of realistic raw data and a full chain of algorithms to retrieve, unpack and process them.

Further studies of the RoI strategy will be made and more of the software will be tested and optimized through test bed and test beam deployment. Thus, the T/DAQ system will be prepared for commissioning in 2006 and ready for first collisions in 2007 when it is hoped to record B physics data from ATLAS.

The ATLAS HLT, DAQ and Controls TDR (Ref. [2]) was recently recommended for approval by the LHCC.

REFERENCES

1. Eerola, P., in these proceedings (2004).
2. ATLAS Collaboration, ATLAS High Level Trigger, Data Acquisition and Controls Technical Design Report, Tech. Rep. CERN/LHCC/2003-22 (2003).
3. ATLAS Collaboration, ATLAS Level-1 Trigger Technical Design Report, Tech. Rep. CERN/LHCC/98-14 (1998).
4. ATLAS Collaboration, ATLAS Detector and Physics Performance Technical Design Report, Tech. Rep. CERN/LHCC/99-14, 99-15 (1999).
5. Moser, H.-G., in these proceedings (2004).
6. Parodi, F., in these proceedings (2004).
7. Baines, J., [ATLAS Collaboration], *Nucl. Phys. Proc. Suppl.* **120**, 139–144 (2003).

Online Event Selection at the CMS Experiment

Marcin Konecki

(Representing the CMS Collaboration)

University of Basle, Switzerland

Abstract. Triggering in the high-rate environment of the LHC is a challenging task. The CMS experiment has developed a two-stage trigger system. The Level-1 Trigger is based on custom hardware devices and is designed to reduce the 40 MHz LHC bunch-crossing rate to a maximum event rate of ~ 100 kHz. The further reduction of the event rate to $\mathcal{O}(100\,\text{Hz})$, suitable for permanent storage, is performed in the High-Level Trigger (HLT) which is based on a farm of commercial processors. The methods used for object identification and reconstruction are presented. The CMS event selection strategy is discussed. The performance of the HLT is also given.

1. INTRODUCTION

The Large Hadron Collider (LHC) is the next accelerator at CERN. The main motivation [1] for building the LHC is to explore the energy region up to a few TeV range, in view of improving the understanding of the Electroweak Symmetry Breaking mechanism, and of searching for Supersymmetry and any kind of New Physics. Precise tests of the Standard Model are also important topics.

At the LHC, protons will collide with centre-of-mass energy $\sqrt{s}=14$ TeV, about seven times higher than at the Tevatron II at Fermilab. The total inelastic, non-diffractive $p\text{-}p$ cross section at LHC energies exceeds 50 mb. On the other hand the effective cross section in some Higgs boson discovery channels can be as low as the fb-pb level. A comparison shows that an event selection of the order of 1 in $10^{\sim 13}$ is necessary. In order to collect significant statistics, a large number of $p\text{-}p$ collisions is essential. It is obtained by a strong beam focusing, a large number of protons per bunch and high bunch collision frequency (40 MHz).

It is expected that soon after the startup (first physics runs are planned in year 2007) the luminosity will reach $2 \cdot 10^{33}\text{cm}^{-2}\text{s}^{-1}$ (so called *low luminosity* mode) and will be increased later to the designed luminosity of $10^{34}\text{cm}^{-2}\text{s}^{-1}$ (*high luminosity* mode).

The Compact Muon Solenoid (CMS) is a general purpose experiment for physics discoveries at the highest luminosities of LHC. The present construction status is described in Ref. [2]. The CMS design principles [3] include efficient lepton and photon selection and measurements. These particles are of key importance for low statistics discovery physics. The event selection system has to be flexible to preserve signal signatures and effi-

ciently reduce the background.

The CMS is capable to handle these requirements keeping the output data volume rate at the level of $\mathcal{O}(100\,\text{MB/s})$, suitable for permanent storage and further, offline analysis. It has been demonstrated and documented in the CMS Trigger and Data Acquisition reports [4] and [5]. The key elements of the CMS event selection strategy are presented in this paper.

2. THE CMS TRIGGER AND DATA ACQUISITION SYSTEM

Each LHC beam crossing at high luminosity produces about 20 $p\text{-}p$ collisions. Each event recorded in the CMS detector results in approximately 1 MB of zero suppressed data. This limits the final data storage rate to about 100 Hz. The event selection at CMS is done in two triggering steps only.

The Level-1 Trigger reduces the 40 MHz beam collision rate below 100 kHz. It is based on custom, partially programmable hardware devices (mostly special-purpose ASIC's but also FPGA's where appropriate). It analyzes coarsely segmented data from the calorimeter and muon systems only. During the Level-1 Trigger latency the full granularity data are stored in detector front-end pipelines.

The 100 kHz event rate of 1 MB event data is a design constraint for the CMS Data Acquisition (DAQ) and the High-Level Trigger (HLT) system. A schematic view of the central part of CMS DAQ is shown in Fig. 1.

Data from the detectors are stored in modules of the detector Front-End system upon the Level-1 accept is issued by the Global Trigger. Then data are read out and

CP722, *B Physics at Hadron Machines,* edited by M. Paulini and S. Erhan
© 2004 American Institute of Physics 0-7354-0203-5/04/$22.00

buffered in Readout Units. At this stage data from each event are spread out over 64 units. The Builder Network is responsible for collecting data belonging to one event and providing it to the Filter Systems. The Builder Network itself is a large switching fabric with 100 GB/s bandwidth, imposed by the Level-1 rate and event size. In the Filter Systems each event is buffered in one Builder Unit and the data from a complete event is provided to a Filter Unit. Each Filter Unit contains a set of commercial CPU units and the Filter Units form a Filter Farm. Each event is assigned to a single processor where the HLT algorithms are executed and the actual event selection takes place. The HLT reduces the Level-1 output rate below the level of the maximal event storage rate of the order of 100 Hz. The final treatment of the events selected at HLT is addressed to Computing Services that forward the data to mass storage and perform some monitoring tasks. The DAQ system is completed with the Event Manager, responsible for the data flow control, and Control System which takes care of configuration, DAQ monitoring and various controlling tasks.

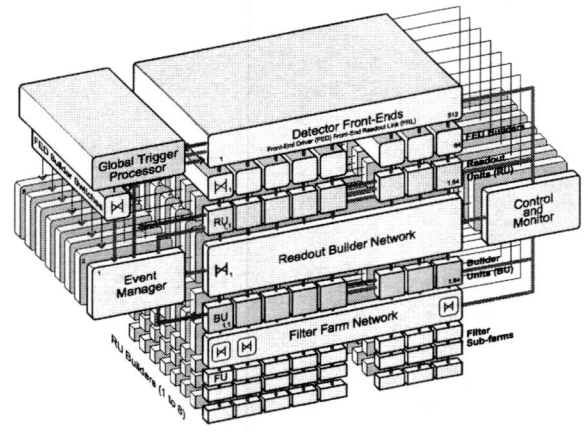

FIGURE 1. Schematic view of the central part of the CMS Data Acquisition System.

The CMS Trigger system does not have any intermediate step between Level 1 and HLT. The traditional "Level 2" trigger is often based on dedicated hardware and uses limited data granularity. The CMS choice to skip this intermediate filter and execute HLT algorithms directly after Level 1 provides a lot of advantages. This design benefits maximally from the computer technology and its developments; it has maximal flexibility without design and architectural limitations of dedicated hardware; there is no limitation on data accessing type and granularity; there is maximal freedom in the choice of the selection algorithms. On the other hand the maintenance of a huge amount of data makes the system challenging and the HLT algorithms are required to be efficiently performing.

Another important feature of the CMS DAQ is that the system is build out of eight 12.5 kHz units (Fig. 1). This modularity makes the DAQ system staging possible. It is especially important because the expected initial event rate at LHC will be substantially lower than the design rate. Only a part of the system has therefore to be ready at startup, allowing overall cost optimization. The CMS choice, called 50 kHz *DAQ scenario* is to build only four DAQ units at startup, providing only half of the design bandwidth.

3. LEVEL-1 TRIGGER

The Level-1 Trigger is based on coarse-grained trigger data from the muon system and calorimeters. The main components and their dependence are shown in Fig. 2.

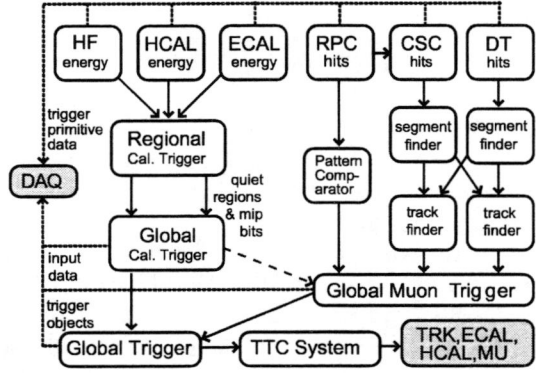

FIGURE 2. The collaboration between the Level-1 Trigger components.

The Level-1 Calorimeter Trigger uses the information from the electromagnetic calorimeter (ECAL), the hadron calorimeter (HCAL) and the hadron forward calorimeter (HF). The following categories of Level-1 objects are delivered: a maximum of four isolated and non-isolated electrons/photons; at most four taus, central and forward jets; total and missing transverse energy and jet multiplicity. The electron/photon, jet and tau candidates are found locally by the Regional Calorimeter Trigger and combined by the Global Calorimeter Trigger that also make use of energy sums in the calorimeter regions to define global objects. The Global Calorimeter Trigger calculates also additional information which can be used for muon isolation.

The Level-1 Muon Trigger provides up to four Level-1 muons. Trigger data are provided by Resistive Plate Chamber (RPC), Drift Tubes (DT) and Cathode Strip Chamber (CSC) detectors. Each of these detectors has a dedicated triggering system which builds and defines muon candidates. The final Level-1 muons are selected by the Global Muon Trigger supplied by RPC, DT and

CSC subtriggers and calorimeter information needed for optional isolation conditions.

The outputs of the Global Calorimeter Trigger and the Global Muon Trigger are sent to the Level-1 Global Trigger, where the decision whether to accept or reject the event is made. The Global Trigger comprises a novel concept of selection based not only on objects exceeding energy or momentum thresholds but also complex event topology. This is made possible by the availability of candidate position, charge and quality measurement information in the Global Trigger algorithms calculations.

The Global Trigger accept bit pattern is forwarded to the DAQ/HLT system and to the Trigger Timing and Control system in order to signal all detector front-end and readout subsystems. The overall Level-1 Trigger latency is 3.2 μs and is dominated by the signal propagation from the detector to the control room and back.

The Calorimeter Algorithms. The electron/photon identification algorithms are defined in a sliding window of 3×3 towers. Trigger towers in the central part of the detector have a $\Delta\eta \times \Delta\varphi$ size approximately equal to 0.087×0.087 following the HCAL segmentation. The ECAL segmentation is much tighter and multiple crystals (5×5 in barrel) are assigned to one tower. The size of the trigger towers varies with the pseudorapidity in the endcaps.

The transverse energy of the electron/photon candidate is defined as the transverse energy deposit in the central tower of the sliding window summed with the largest deposit in one of its four nearest neighbors. In order to reduce the background (Fig. 3) additional identification criteria are introduced. The object is classified as non-isolated electron/photon if, in the central tower, most of the energy is deposited in the narrow strip ($n_\eta \times n_\varphi = 2 \times 5$) of crystals (Fine-Grain energy profile) and the HCAL contribution is below a given threshold. The isolated electron/photon has to fulfill the above criteria in all towers of the sliding window supplemented with additional isolation cut. This isolation cut requires that at least one corner of the sliding window to which five towers are associated is quiet, *i.e.* all the associated towers are below a (programmable) threshold.

The Jet and Tau identification algorithms work in the sliding window made of 3×3 trigger regions where the trigger region is made of 4×4 towers, except for HF where the region corresponds just to a tower. The sliding window is centered on the region with the largest transverse energy, required to be above a fixed value threshold. This central region defines the position of the jet. The jet transverse energy is a sum of transverse energies in a sliding window. Additional requirements are necessary for the identification of jets from hadronic decays of taus (tau-jets). The hadronic decays of tau are dominated by modes with one or three charged particles

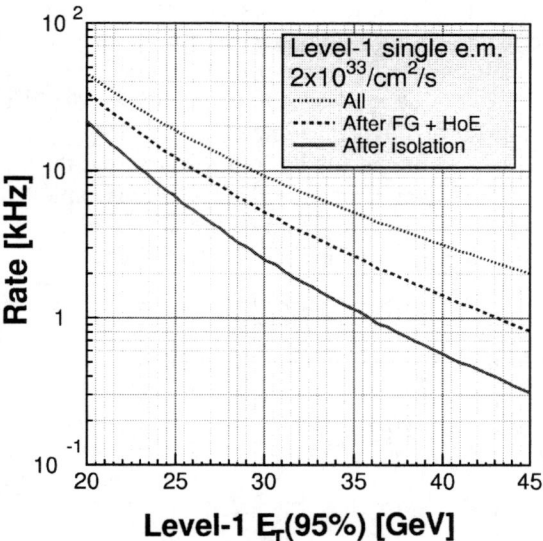

FIGURE 3. Level-1 electron/photon rates. The candidates are selected be energy thresholds, non-isolated electron/photon identification and isolation criteria.

accompanied by a small number of neutral particles. The products of decay form a low-mass, thin jet with the energy deposited both in ECAL and HCAL. Since the calorimeter response to single particles is often confined in a small cluster extending to no more than 2×2 towers (tau-jet patterns) the sliding window provides a tau-jet tag if all its active regions (ECAL or HCAL energy above a threshold) contain only tau-jet patterns.

On top of the described jet objects the Level-1 Calorimeter Trigger computes, in the pseudorapidity region $|\eta| < 5$, the sum of the transverse energy (E_T) and total missing transverse energy (E_T^{MISS}) by summing E_T components and the information on jet multiplicity (to allow triggering on events with a large number of low energetic jets).

The Muon Algorithms. Each of the CMS muon subtriggers (RPC, DT, CSC) has its own trigger algorithm. The RPC trigger is based on a Pattern Comparator. The hit positions from all RPC detectors in a certain η and φ range are assigned to a single comparator to form a pattern. It is compared with a set of predefined patterns corresponding to tracks with different momenta. In the case of the DT and the CSC's the multi-hit information in one chamber is used by the local triggers to deliver a track segment. The segments are then combined by a track finder and muon candidates are formed. The three trigger systems deliver the charge, the position and the estimate of the maximal transverse momentum of each muon candidate, as well as the quality of the measurement.

209

The role of the Global Muon Trigger is to combine the information from its suppliers in the optimal way. The DT are located in the barrel part of the CMS detector, the CSC's in the forward region while RPC's extend[1] over the full Level-1 Trigger pseudorapidity range $|\eta| < 2.1$. They have different trigger logics so that the three sub-triggers are complementary. This feature is exploited by the Global Muon Trigger to gain both in rate reduction (shown in Fig. 4) and in increase of efficiency.

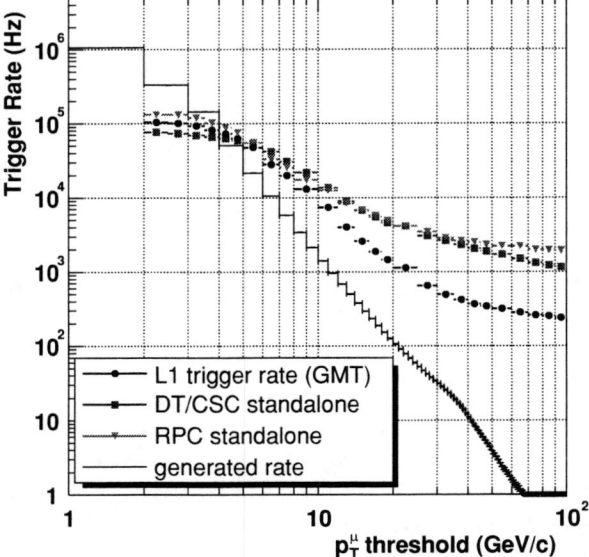

FIGURE 4. The Level-1 single muon rate (low luminosity) provided by the Global Muon Trigger (GMT). The standalone RPC trigger rate and the combined DT and CSC trigger rates are shown separately. The simulated rate is also shown.

The Level-1 Selection. The trigger thresholds are defined by the total available bandwidth and the allocation to different trigger streams which has to compromise various discovery scenarios. The CMS DAQ system is designed to accept 100 kHz Level-1 output rate, but due to staging at startup the maximum input rate is limited to 50 kHz. The CMS strategy is to apply a safety factor of three on the available bandwidth to take into account simulation uncertainties and non-simulated backgrounds. Thus the trigger tables are designed for the output rate of 16 kHz and 33 kHz for low and high luminosity mode, respectively. Since the bandwidth allocation cannot be optimized without data, it has been divided in equal parts to electron/photon triggers, muons, taus and jet plus combined triggers (as shown in Table 1). This

strategy provides inclusive selection optimal for discoveries.

TABLE 1. The Level-1 Trigger table for $2 \cdot 10^{33} \mathrm{cm}^{-2}\mathrm{s}^{-1}$ luminosity and 50 kHz DAQ scenario with safety factor three superimposed. The listed thresholds correspond to 90-95% efficiency.

	Threshold [GeV, GeV/c]	Individual rate [kHz]	Cumulative rate [kHz]
$1 e/\gamma, 2 e/\gamma$	29, 17	4.3	4.3
$1\mu, 2\mu$	14, 3	3.6	7.9
$1\tau, 2\tau$	86, 59	3.2	10.9
1,3,4-jet(s)	177, 86, 70	3.0	12.5
Jet $\oplus E_T^{MISS}$	$88 \oplus 46$	2.3	14.3
$e/\gamma \oplus$ Jet	$21 \oplus 45$	0.8	15.1
min. bias		0.9	16.0

4. HIGH-LEVEL TRIGGER

The HLT algorithms run on a computer Filter Farm, where each event is processed by a single processor. The reconstruction and selection code is written in Object Oriented C++. To cope with the constraints on the event storage resources, the HLT algorithms must achieve a factor ~ 1000 reduction on the event rate. On top of the hardware constraints the HLT selection must satisfy physics requirements to ensure the most inclusive selection and the highest efficiency for "signal" signatures.

The strategy to eliminate unwanted events may explore fast, simplified but not accurate reconstructions or use of minimal amount of precise information. Both ways are used to optimize the event rejection speed although the latter is preferred. The CMS guideline is to keep the HLT reconstruction code as close as possible to that used for the offline reconstruction. The only restriction is that the HLT reconstruction can not rely on the detailed knowledge of the calibration constants which are not available online. The following features are included in the HLT reconstruction code:

- **Reconstruction on Demand ("when needed"):** The reconstruction is started only if it is required by the trigger decision, rather than always reconstruct particular objects,
- **Regional Reconstruction ("where needed"):** Data are reconstructed only in the region of interest, defined by a trigger condition. Data from other "partitions" are not accessed. This strategy does not hold for non-seeded objects (for example E_T^{MISS}) where the **global** reconstruction is necessary.
- **Partial and Conditional Track Reconstruction ("if needed"):** Tracking is stopped if the combinatorial ambiguities are solved and/or the precision is sufficient for further triggering steps. The recon-

---

[1] The initial pseudorapidity coverage of RPC trigger has been recently limited to approximately $|\eta| < 1.6$ (staging). This change is not included in the results presented here.

struction is abandoned if the track fit parameters disagree with the kinematical requirements (for example momentum below threshold).

The event reconstruction and selection takes place in several steps. Because the reconstruction is triggered by Level-1 Trigger based on calorimetry and muon system, the natural way is to fully explore the provided information with full granularity reconstruction therein. The calorimeter and muon system based reconstruction is called *Level 2*. The reconstructed objects are supplemented with limited information from the Tracker System (usually by Pixel detector hits) at *Level 2.5* and finally reconstructed using more Tracker data at *Level 3*. The *virtual trigger levels* introduced are a matter of convention. The HLT algorithms have no data access restrictions and the order of their execution can be optimized to follow the master principle to reject the event as soon as possible. The description of main algorithms used at HLT is given below.

Electron and Photon Reconstruction. The Level-1 electron/photon trigger rate is dominated by fake candidates caused by jets in which a single π^0 carries a substantial fraction of energy. The first step at HLT is to find the ECAL clusters and confirm the Level-1 candidates using the full granularity. Because of the sizable amount of material ($0.4 - 1.4 X_0$) in front of the electromagnetic calorimeter, electrons radiate a large fraction of their energy. The recovery of the energy lost in Bremsstrahlung photons is done by measuring the energy in cluster of clusters (*super clusters*) spread in φ due to electron bending in magnetic field. The rejection of fake electrons is done at Level 2.5, where the electron candidate is propagated back to the Pixel detector layers and hit matching is performed. The rejection power is shown in Fig. 5. This selection does not affect photons due to much higher energy thresholds required for them. At Level 3 the electron track is reconstructed in the tracker. The cuts on the ratio of the measured energy and the measured momentum as well as track-cluster matching are applied.

Muon Reconstruction. In contrast with the electrons/photons the Level-1 muon candidates are real muons and the high rate is caused by momentum mismeasurement. The Level-2 muon algorithm works in a region of interest defined by the Level 1. It uses DT segments, CSC and RPC hits reconstructed locally on chambers. The muon trajectory is built and updated by collecting measurements using the Kalman Filter Fit. Since the most interesting muons originate from the beam spot or nearby, the momentum measurement is improved by updating the muon state with the vertex constraint.

The Level-3 muon reconstruction starts with the determination of a region of interest, with parameters and

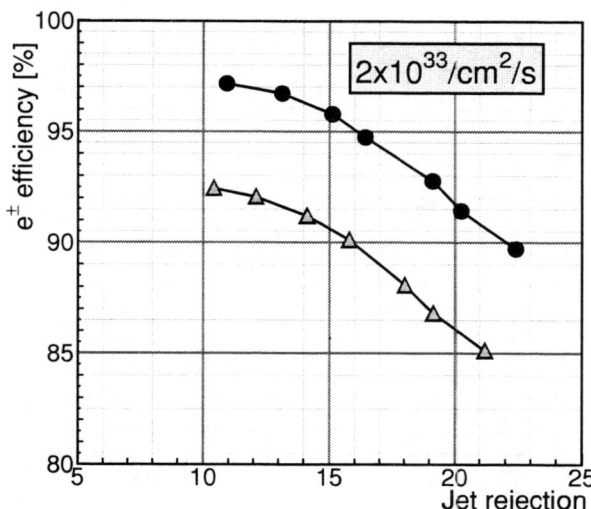

FIGURE 5. The electron selection at Level 2.5 achieved by matching the electron candidate with Pixel detector hits. The upper curve corresponds to the baseline Pixel detector layout with three layers on pixels and two disks. The bottom curve corresponds to only two layers of pixels available.

uncertainties defined by the Level-2 muon state at the vertex. The seeds for the regional reconstruction are given by Pixel hit pairs. The Kalman Fitting technique is employed to build a Tracker based muon state which is then updated with muon system measurements (collected by original Level-2 muon). The typical resolution of $(p_T{}^{gen} - p_T{}^{rec})/p_T{}^{rec}$ is 10-16% at Level 2 improved to 1.0-1.7% at Level 3.

The muon rate in the p_T range below 30 GeV/c is dominated by muons from b, c and K, π decays. In order to separate them from those produced in heavy object decays, isolation algorithms are used (Fig. 6). Three techniques were studied: *calorimeter isolation*, with combined ECAL and HCAL energies, *pixel isolation*, with tracks from simplified standalone reconstruction in the Pixel detector and *tracker isolation*, with tracks reconstructed by standard regional reconstruction algorithm. The isolation algorithms apply a threshold (which may depend on η) to the deposits ($\sum E_T$ or $\sum p_T$) in a cone around the muon.

Jet and Missing Energy Reconstruction. The HLT jet finding is a global algorithm based on iterative grouping of calorimeter towers. Initially, the most energetic tower defines the direction of a "protojet". The towers in a fixed size cone around a protojet direction are assigned to the protojet. The size of the jet cone is chosen to be $\Delta R = \sqrt{\Delta \eta^2 + \Delta \varphi^2} \approx 0.5$. The protojet energy and position are recalculated providing a new protojet definition. The protojet position is determined by weighting

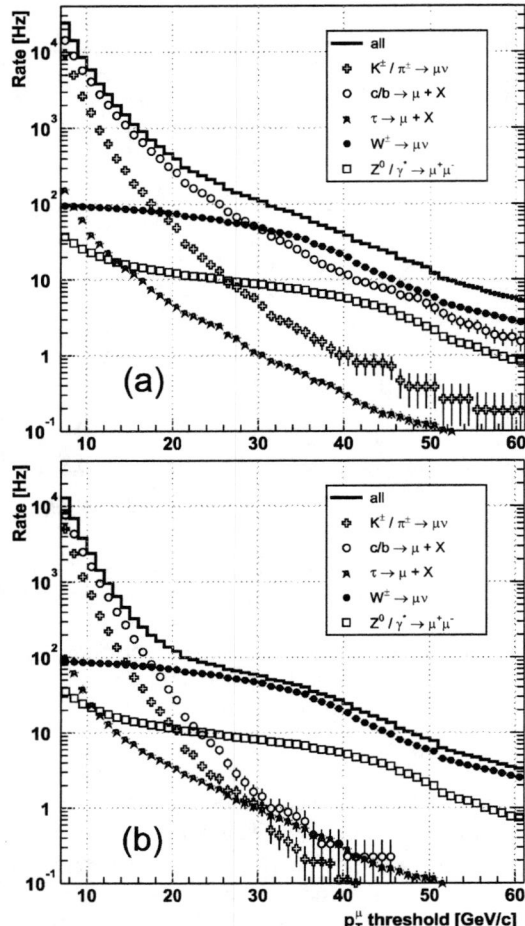

FIGURE 6. Contribution to the Level-3 muon rate (high luminosity) before (a) and after (b) isolation algorithms applied. The b, c and K, π components are reduced while W, Z well preserved. The typical isolation cone has a size ΔR =0.2-0.3 with ~ 3 GeV $\sum p_T$ cut.

the tower positions with their energies. The procedure is repeated until the variation of the protojet energy and position between two iteration steps is small enough. Subsequently the towers contributing to a jet are removed from the list of active towers and the procedure is repeated. The simple calculation of the vector sum of the tower energies is used to calculate the total missing transverse energy E_T^{MISS}, useful for neutrinos identification.

Tau-Jet Reconstruction. The CMS tau-jet finding algorithm is optimized for the identification of isolated high energetic taus (*e.g.* from SUSY Higgs decays). At Level 2 a jet is tagged as a tau jet if the jet energy is concentrated in a narrow cone ($\Delta R = 0.13$). The fake tau-jet rejection is improved at Level 2.5 and Level 3 by improving the isolation with reconstructed tracks.

Primary Vertex Reconstruction. Because of the large number of collisions in a single bunch crossing the knowledge of the position of the signal vertex is useful. Recently updated (Ref. [6]) HLT primary vertex finding algorithms are based on fast, three-hit track reconstruction in the Pixel detector. Since the Pixel hit position measurement is very accurate and the track bending is only in transverse plane, the pixel tracks provide a good measurement of the track intercept with the beam axis (Fig. 7) on which the primary vertex finding algorithms rely. The vertex finding efficiency is 95-100%. There are,

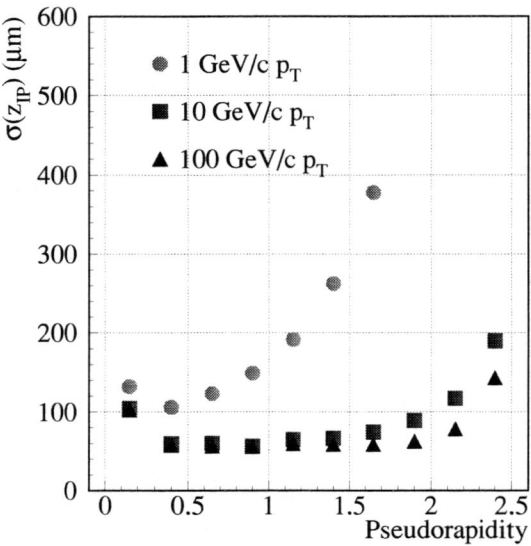

FIGURE 7. The resolution of impact parameter along beam line for single track measured in pixel detector only.

however several vertices found per event, due to event pileup. The efficiency to select correctly the primary vertex is thus lower (50-97%, event type and luminosity dependant) for general algorithms. It can be however optimized for a particular channel.

HLT Selection and Efficiency Performance. The HLT threshold values for the low luminosity case are presented in Table 2. The thresholds are optimized to achieve the 100 Hz output rate. The initial inclusive event selection principle is followed. The efficiency for some signals, obtained using the above triggering conditions, are given in Table 3.

The Timing Performance. Due to limited computing resources the execution time of the HLT algorithms is a crucial parameter. The CPU time needed to select HLT physics objects was measured on 1 GHz Pentium III machines (Table 4). The average selection time, defined as the object selection time weighted by its Level-1 input rate, is approximately 300 ms/event.

TABLE 2. High-Level Trigger table for the $2 \cdot 10^{33} \mathrm{cm}^{-2}\mathrm{s}^{-1}$ luminosity case.

	Threshold [GeV, GeV/c]	Individual rate [Hz]	Cumulative rate [Hz]
$1\,e, 2\,e$	29, 17	34	34
$1\,\gamma, 2\,\gamma$	80, (40$\oplus$25)	9	43
$1\,\mu, 2\,\mu$	19, 7	29	72
$1\tau, 2\tau$	86, 59	4	76
Jet $\oplus E_T^{MISS}$	180 $\oplus$ 123	5	81
1,2,3-jet(s)	657, 247, 113	9	89
$e/\gamma \oplus$ Jet	19 $\oplus$ 52	1	90
incl. b-jets	237	5	95
others		10	1005

TABLE 3. HLT selection efficiencies for some types of signal. The numbers are given for events in detector fiducial volume only.

Channel	Efficiency
$H(115\ \mathrm{GeV}) \to \gamma\gamma$	77%
$H(160\ \mathrm{GeV}) \to WW^* \to 2\mu$	92%
$H(150\ \mathrm{GeV}) \to ZZ \to 4\mu$	98%
$A/H(200\ \mathrm{GeV}) \to 2\tau$	45%
SUSY (~ 0.5 TeV sparticles)	$\sim 60\%$
with R_p - violation	$\sim 20\%$
$W \to e\nu$	42%
$W \to \mu\nu$	69%
$t \to \mu X$	72%

The CMS Collaboration has assumed that the computer CPU power will follow the Moore's law and double the CPU power each 18 months. This assumption implies that in 2007 eight times[2] faster machines will be widely available on the market. Currently available algorithms will be executed in $\simeq$ 40 ms at LHC startup. Thus, for the 50 kHz DAQ scenario the required number of processors is about 2000 corresponding to a well affordable Filter Farm of about 1000 dual CPU boxes.

This estimate is not accurate. The uncertainties are coming from the following sources: not yet estimated delay time caused by raw data accessing and formatting; assumption that the superimposed DAQ safety factor does not change CPU average; non simulated sources of backgrounds; possible underestimated occupancies in detector that may slow down for example pattern recognition. On the other hand the performance of the algorithms is expected to improve.

---

[2] The CPU time measurements with 1 GHz Pentium III machines were done in the year 2002. These machines are already obsolete.

TABLE 4. The average CPU time (1 GHz Pentium III) for the HLT algorithms.

	CPU [ms]	Rate [kHz]	Time [s]
$1e/\gamma, 2\,e/\gamma$	160	4.3	688
$1\mu, 2\mu$	710	3.6	2556
$1\tau, 2\tau$	130	3.0	390
Jets, Jet $\oplus E_T^{MISS}$	50	3.4	170
$e/\gamma \oplus$ Jet	165	0.8	132
b-jets	300	0.5	150

5. SUMMARY

The CMS experiment has two trigger levels to reduce the 40 MHz LHC bunch crossing rate to 100 kHz (50 kHz for initial running conditions) in the first step and to the order of 100 Hz appropriate for mass storage, in the second step. The algorithms for the event selection for both trigger stages were developed. The current average execution time of the HLT algorithms is 300 ms/event (1 GHz Pentium III). This execution time is adequate provided that the progress in computer technology arise as expected. The CMS Trigger table is initially optimized for the most inclusive selection allowing powerful event rejection while preserving signatures of possible discovery channels. The current trigger table, however, does not include some Standard Model physics selections relevant to CMS. An example is the B physics selection for which highly efficient and low CPU cost HLT algorithms are under development [7].

REFERENCES

1. Jarlskog, C., and Rein, G., editors, *Proceedings of the Large Hadron Collider Workshop*, Aachen, 4-9 October 1990, CERN 90-10, ECFA 90-133.
2. Neal, H., in these proceedings (2004).
3. CMS Collaboration, CMS: The Compact Muon Solenoid, Technical Proposal, Tech. Rep. CERN/LHCC 94-38 (1994).
4. CMS Collaboration, CMS: The Trigger and Data Acquisition Project, Volume I: The Level-1 Trigger, Tech. Rep. CERN/LHCC 2000-038 (2000).
5. CMS Collaboration, CMS: The Trigger and Data Acquisition Project, Volume II: Data Acquisition & High-Level Trigger, Tech. Rep. CERN/LHCC 2002-26 (2002).
6. Cucciarelli, S., et al.,, Track-Parameter Evaluation and Primary-Vertex Finding with the Pixel Detector, Tech. Rep. CMS NOTE-2003/026 (2003).
7. Marinelli, N., in these proceedings (2004).

The LHCb Trigger System: Implementation & Performance

Olivier Callot

(Representing the LHCb Collaboration)

Laboratoire de l'Accélérateur linéaire, CNRS-IN2P3, Orsay, France and CERN, Geneva, Switzerland

Abstract. The LHCb trigger has been recently described in a TDR [1] submitted to the LHCC. This presentation describes shortly the main components of this three-level system, a synchronous hardware Level-0 followed by a software Level-1 running at 1 MHz, and finally a High Level Trigger to bring the rate down to 200 Hz. A short summary of the expected performance on several benchmark channels is also given.

1. THE TRIGGERING CHALLENGE

The LHCb experiment [2] will run at the LHC, with 40 MHz crossing rate and a moderate luminosity of $2 \times 10^{32}\,\mathrm{cm}^{-2}\,\mathrm{s}^{-1}$, resulting in 12 million events per second visible in the detector. But only a few of these events contain an interesting B decay mode, useful for studying the CP asymmetries. These are the main goal of the experiment, together with the search for rare decays. The trigger should have a rejection power around 50,000 while retaining efficiency on interesting B decay modes.

The trigger system of LHCb is made of three levels:

- The Level-0 is a hardware system, with fixed latency of 4 μs, with custom electronics working in pipe-line mode. The output rate is 1 MHz.
- The Level-1 is a software trigger, running on a small part of the information, packed in a dense format to reduce the required bandwidth. On average, the processing takes one millisecond with a absolute maximum of 58 ms, determined by the limited buffer space in the FE boards.
- The Hight Level Trigger (HLT) accesses the complete information of the event and perform the selection of interesting decay modes.

Last, the Timing and Fast Control system (TFC) coordinates the distribution of trigger signals and ensures the proper rate limitations when appropriate.

2. THE LEVEL-0 SYSTEM

The basic idea of Level-0 is to select events containing a B decay by the presence of a large p_T or E_T object. This object can be a hadron, a muon, an electron, a photon or

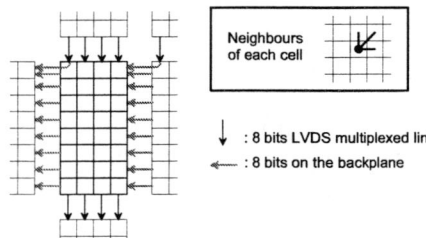

FIGURE 1. Links between front-end cards.

a π^0. The Level-0 is also rejecting too complex events, since they fulfill more easily the trigger requirements but are not usable for final physics analysis. Rejecting them earlier allows having more bandwidth for useful events. The Level-0 is made of four sub-systems.

2.1. The Level-0 Calorimeter Trigger

The calorimeter trigger searches for high E_T local deposits in each possible group of 2×2 cells. The electromagnetic calorimeter ECAL has about 6000 square cells, with a size from 4 cm to 12 cm. For the hadronic calorimeter HCAL, there are only 1500 cells, either 13 cm or 26 cm. A first operation converts the 12 bits ADC output to E_T coded on 8 bits, with a saturation at 5.1 GeV of p_T. The second operation is to collect the of each cell, which is easy when on the same front-end board, but requires many high speed connections with a dedicated backplane or LVDS links, as depicted on Fig. 1 when on a different board.

As only the highest candidate is interesting for triggering, the highest E_T sum on the front-end board is selected, and transmitted to the next stage. This reduces the number of connections by a large factor.

CP722, *B Physics at Hadron Machines,* edited by M. Paulini and S. Erhan

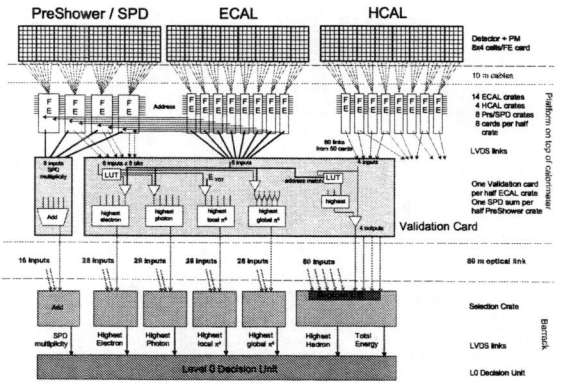

FIGURE 2. Overall view of the Level-0 Calorimeter Trigger.

On the next stage, called Validation and shown on Fig. 2, the information from ECAL is validated by the information from the SPD and the PreShower. Both detectors have the same geometry as ECAL, projective from the interaction point. The SPD is in front, and the PreShower after, a lead sheet of 2.5 X_0. The cells with the same ID are retrieved and used to validate the candidate as electron or photon. π_0 candidates are also built, either by combining photons on two neighboring cards, or by looking at the total energy in a single card, for the configuration where the two photons are in the same card. A validation card processes 8 front-end cards, and only the candidate with the highest E_T in each category is output to the Selection crate. Similarly, the ECAL energy in front of every HCAL candidate is added to it. This is somewhat more complex as the geometries are not identical, and only the ECAL candidates are available, not all cells.

The Selection crate finishes the process of selecting the candidate with highest E_T in each category and sends them to the Level-0 Decision Unit. It also produces two global variables, the total E_T in HCAL, useful to reject empty events that could be triggered by a halo muon from the machine, and the SPD multiplicity, summing local multiplicities counted on the front-end cards.

The front-end electronics and the Validation cards are located on the Calorimeter platform, inside the cavern, and will be exposed to non-negligible amount of radiation when the accelerator is operating. This implies using anti-fuse PGA and triple voting technique for all control information. The processing is fully synchronous and pipe-lined, thus independent of history and occupancy.

2.2. The Level-0 Muon Trigger

The muon trigger is searching for straight lines in the four pad chambers M2-M5 located behind the calorimeter. The geometry of the chambers is projective, meaning that an infinite momentum track hits cells with the same number in the four chambers. A Field Of Interest (FOI) of ± 3 to ± 5 pads in the bending direction, and one pad in the non-bending direction allows for finite momentum and multiple scattering. Once a line is found, its extrapolation in the M1 chamber located in front of the calorimeter system is searched for, and used to determine the p_T of the muon. This is shown on Fig. 3.

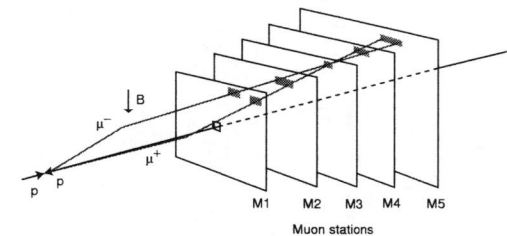

FIGURE 3. Overall view of the Level-0 Muon trigger.

Contrary to the Calorimeter trigger, the electronics is entirely placed in the electronics barracks, which permits using normal components. A bit map of the detector is sent at 40 MHz by optical fibers from the detector to the barracks, 148 ribbons of 12 fibers. For each quarter of the detector, 15 processing boards are handling a part of the detector. Neighbor exchange is also implemented on a dedicated high speed backplane. The two highest p_T candidates of each board are combined to get the two highest p_T candidates of the quarter, that are sent to the Decision Unit. The processing is completely synchronous, independent of the history and occupancy.

2.3. The Level-0 Pile-Up System

The pile-up system identifies the presence of multiple interactions in the same bunch crossing, using dedicated R sensors of the Velo (VErtex LOcator) detector looking at backward tracks. In each 45° sector, the sensor measures the radius with an effective pitch (real strips are OR-ed 4 by 4) from 160 to 400 μm. Measuring the radius at two positions allows computing the Z of intercept with the axis. Histogramming all combinations enables to find the position of vertices as shown in Fig. 4

In fact this is a two pass process: first the main peak is found, all hits contributing to it are removed and the search for peak is performed again, on a cleaner set of points.

The detector electronics provides a discriminated output, which is combined 4 by 4 and sent via optical links to the barracks. Four processing boards are fed in turn by the complete data of one event, thus processing one event every 100 ns, and producing the peak analysis in a bit more than 1 μs, which is also sent to the Decision Unit.

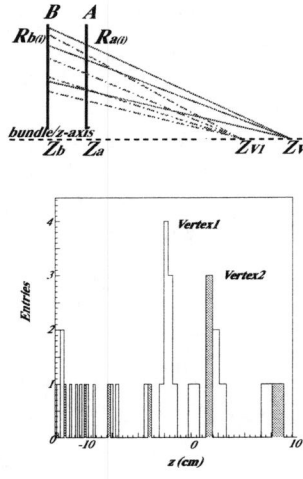

FIGURE 4. Principle of the Pile-Up system.

2.4. The Level-0 Decision Unit

As already mentioned, this board receives the information from the three other sub-systems, and takes the decision for each crossing. In a first step, the various inputs are synchronized, then thresholds are applied on each input to see if one type of candidate (calorimeter or muon) is in the acceptable range. Pile-up veto and multiplicity cut are applied for most decisions, keeping very clear signature, like di-muon, unaffected. A decision is transmitted to the TFC system at 40 MHz, with a nominal accept rate of 1 MHz.

3. THE LEVEL-1 AND HLT SYSTEM

The Level-1 trigger is a software algorithm operating on a farm of general purpose CPUs. The same farm is also used for the HLT. The aim of the algorithm is to identify displaced vertices by performing a fast tracking in the Velo, in order to identify large impact parameter tracks and confirm the significance by measuring their p_T in the fringe field of the magnet.

3.1. The Level-1 Hardware

The main difficulty is to collect data from about 130 sources and dispatch them to the proper CPU in a farm of more than 1000 nodes, at 1 MHz. First, the worker nodes are grouped in about 100 sub-farms, with a Sub-Farm Controller (SFC) such that the event fragments are sent to the SFC which is in charge of assembling the event and distributing it to a free node, with local resource management. The second simplification is to group events

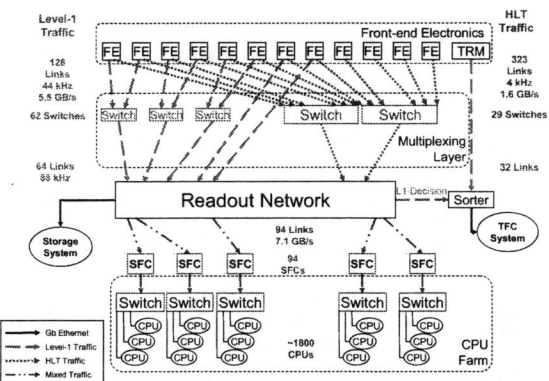

FIGURE 5. Architecture of the L1 and HLT system.

by Multi Event Packet (MEP) at the source, that is in the LHCb common readout board TELL1 [3]. About 25 events are buffered, and sent together to the same SFC node. This reduces the Ethernet frame overhead as the payload per event is on average 30 bytes per source. This also minimizes the number of interrupts to handle in the SFC, making the system already implementable today with commercial components. The big IP router connecting the sources to the destination is also available today, even if quite expensive. A few Ethernet switches allow merging the traffic of some input sources to fully use the bandwidth of this router. One should note that the very same infrastructure is used for the HLT, where a dedicated output of the TELL1 board handles this lower rate traffic, with more sources and bigger event fragments, but still packed in a MEP with only 10 events. The whole architecture is shown on Fig. 5

3.2. The Level-1 Software

As mentioned, the Level-1 performs a fast pattern recognition in the Velo to identify large impact parameter tracks. A first step is to perform the tracking in the R-Z view, where tracks are (almost) straight lines. The search is performed independently in each 45° sector of the R sensors. An example for a busy sector is shown on Fig. 6.

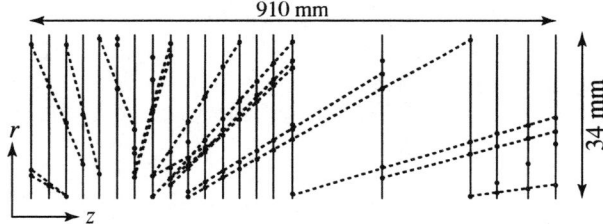

FIGURE 6. Event display of the 2D tracking in one 45° sector.

On average, 58 forward and 30 backward tracks are found. A vertex search is performed, using the sector number as ϕ information. This gives a resolution on the vertex around 60 μm along the beam line and 25 μm in the transverse direction. Tracks with an impact parameter between 150 μm and 3 mm are selected, and reconstructed in 3 dimensions using the ϕ sensors of the Velo. These space tracks are extrapolated to the TT tracking station, which is in the fringe field of the magnet. The observed track's displacement allows estimating the momentum within 10-20% accuracy, enough to reject the low momentum tracks which have an impact parameter due to multiple scattering before the Velo measurements.

The selection of interesting events is based on the p_T product of the two highest p_T tracks reconstructed in space. But some clean topologies receive a bonus, like large mass dimuon from a J/ψ decay or events with a high E_T photon. Only 4% of the events are kept, to reach 40 kHz output rate. The processing time takes now on average 8 ms per event passing Level-0. We know already how to gain 20% on that, and we expect the computer industry to deliver faster CPU according to Moore's law, so that we will need only 1 ms in 2007, which means 1000 CPUs for Level-1.

3.3. The High Level Trigger

Starting from the full information on accepted events, the HLT performs the full tracking, first on selected tracks to confirm Level-1, then on the whole event to select the B decay modes of interest. The first part has already be implemented, a fast tracking in the Velo and then in the whole tracking system, giving a momentum accuracy around 0.6%. About half the events can be immediately rejected, in less than 25 ms as they have passed Level-1 due to wrongly measured tracks. The complete event reconstruction takes now around 50 ms and will be followed by a selection of the interesting decays. This last part is not yet coded, but prototypes show that a few ms per decay mode family is reachable. This indicates that the size of the HLT processing part of the farm can be around 400 CPU nodes.

3.4. Timing and Fast Control

Based on the RD45 TTC system, out Timing and Fast Control is in charge of distributing Level-0 and Level-1 decisions at the rate of 40 MHz and 1 MHz, respectively. The destination assignment of the SFC node is also distributed by the TFC system in a simple round-robin scheme, at 40 kHz and 4 kHz, respectively. Another important component is the Readout Supervisor, in

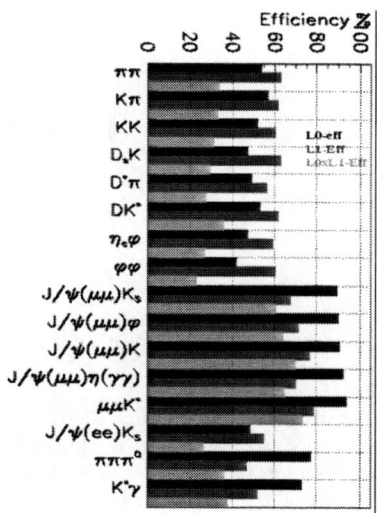

FIGURE 7. Performance of the Level-0 and Level-1 trigger on several decay modes.

charge of moderating the rates when needed, either by emulating the occupancy of the Level-0 de-randomizer buffers, or by handling throttle signals.

4. PERFORMANCE AND FUTURE DEVELOPMENTS

The settings of the Level-0 and Level-1 systems have been extensively studied this year and are reported in the Trigger TDR [1]. A short summary is depicted on Fig. 7. In particular the relative rate of the various Level-0 components has been tuned for the best average efficiency on many channels.

The LHCb trigger TDR has been submitted in September to the LHCC, approval is expected soon. Level-0 is now mature, and hardware implementation has started. Level-1 and HLT are based on commercial components. The system integration and the control of such a large size system of over 1500 nodes and hundreds of Giga-Bit Ethernet routing connections requires intense work. The algorithms are being developed and improved, with all indications that LHCb will be ready in 2007 to efficiently trigger on interesting physics events.

REFERENCES

1. Antunes Nobrega, R., et al., [LHCb Collaboration], LHCb Trigger System Technical Design Report, Tech. Rep. CERN/LHCC 2003-031 (2003).
2. Raven, G., in these proceedings (2004).
3. Bay, A., et al., Common L1 Read-out Board for LHCb: Specification, Tech. Rep. LHCb 2003-007 (2003).

The BTeV Trigger System

Michael H.L.S. Wang

(Representing the BTeV Collaboration)

Fermi National Accelerator Laboratory, Batavia, IL 60510, U.S.A.

Abstract. BTeV is a high-statistics B physics experiment that will achieve new levels of sensitivity in testing the Standard Model explanation of CP violation, mixing and rare decays in the b and c quark systems by operating in the unique environment of a hadron collider. In order to achieve its goals, it will make use of a state-of-the-art Si-pixel vertex detector and a novel 3-level hierarchical trigger that will look at every single beam crossing to detect the presence of heavy quark decays. This talk will focus on the Level-1 vertex trigger describing how it fits into the overall design of the BTeV trigger and data-acquisition system.

1. INTRODUCTION

BTeV is a collider B physics experiment that will begin taking data in 2009 at the C0 interaction region of the Fermilab Tevatron. In addition to the large $b\bar{b}$ cross sections at hadron colliders ($\sigma(b\bar{b}) \sim 100~\mu$b), it will exploit the unique characteristics of hadronic b production in the forward region which includes the large production cross sections for highly correlated $B\bar{B}$ pairs and the large boost received by B mesons that result in longer observed decay lengths and reduced effects due to multiple scattering [1, 2].

To take advantage of these features, it will employ a unique forward spectrometer consisting of a muon detector, a Rich Imaging Cherenkov detector, and an electromagnetic calorimeter for particle ID, as well as a combination of straw tubes and Si-microstrips for charged particle tracking [2]. The real strength of the BTeV detector, however, is the 30 station Si-pixel vertex detector spanning $\sim$120 cm, centered at the C0 collision point, and immersed in the 1.5 Tesla dipole field of the SM3 magnet [3]. Each pixel station has over 7.6×10^5 rectangular pixels measuring $50 \times 400~\mu$m^2 distributed over bend and non-bend views for a total of over 22×10^6 pixels in the full detector. The long dimension of the pixels in the bend view is oriented parallel to the dipole field, while those in the non-bend view are perpendicular to the field.

In order to deal with the rare occurrence of a b event (about 1/1000 $p\bar{p}$ collisions at the Tevatron) and the high track multiplicities in these events, BTeV will couple the superior pattern recognition capability of the Si-pixel detector with a unique online trigger system to acquire data at the full crossing[1] rate of 7.6 MHz. With an estimated event size of 100 kB, data will come out of the detector front-end at an extremely high rate of 800 GB/sec. To handle this, BTeV will employ a three-level hierarchical trigger architecture which will be described in some detail below.

2. LEVEL-1 TRIGGER ALGORITHM

We begin our discussion of the trigger architecture by briefly describing the two stage algorithm employed in the first level trigger (L1) [4, 5]. The first stage, referred to as the segment finding stage, searches for track segments using hit clusters spanning three adjacent pixel stations. Its search is confined to two separate regions of the pixel detector, an inner region close to the beam axis and an outer region close to the edge of the pixel planes. Segments found in the inner region are called inner triplets which represent the beginning of a track. Segments found in the outer region are called outer triplets which represent the end of a track just before exiting the pixel detector volume.

The second stage consists of two phases, a track finding phase followed by a vertex finding phase. The track finding phase projects the inner triplets downstream based on their slopes in the non-bend plane and their "curvature" (slope change) in the bend plane. Outer triplets within a certain distance of the projected trajectories are matched to the inner triplets to form com-

---

[1] In this paper, the terms crossing and event will be used interchangeably.

CP722, *B Physics at Hadron Machines*, edited by M. Paulini and S. Erhan
© 2004 American Institute of Physics 0-7354-0203-5/04/$22.00

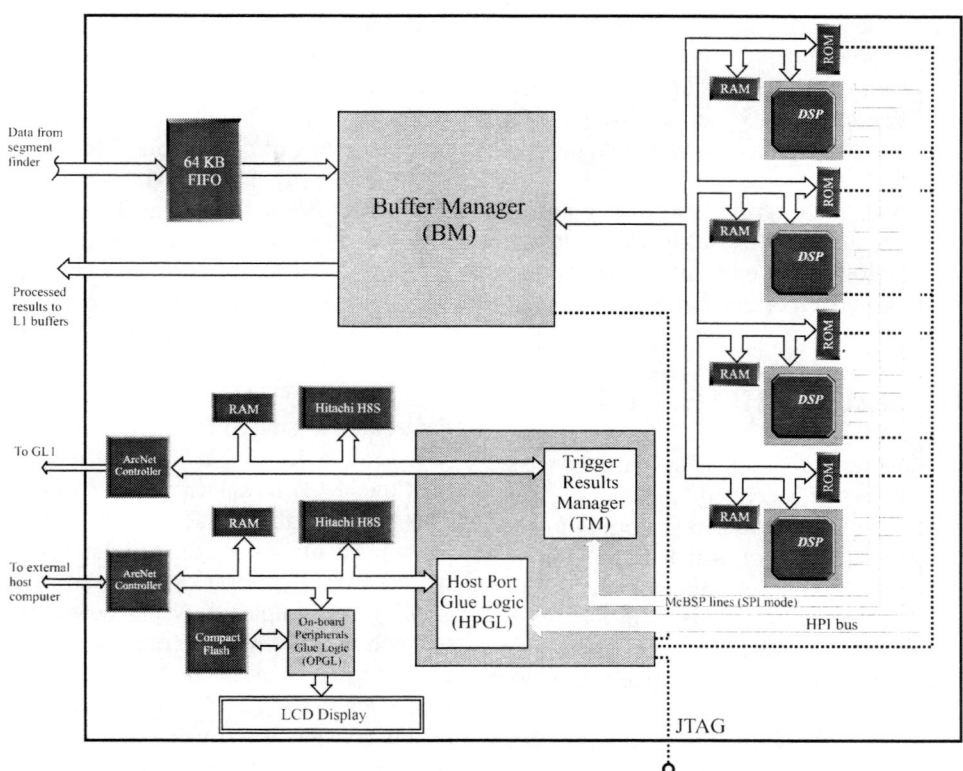

FIGURE 1. Block Diagram of Pre-prototype Level-1 Track and Vertex Hardware.

plete tracks. The vertex finding phase uses these found tracks to locate the primary interaction vertices. Remaining tracks not belonging to these primary interaction vertices are then tested for their detachment from these vertices to see if they belong to candidate B decay products. A L1 trigger accept is generated if there are at least 2 tracks in the instrumented arm of the BTeV detector meeting the following criteria: $p_T^2 \geq 0.25 \ (\text{GeV}/c)^2$, $b/\sigma \geq n$, and $b \leq 2$ mm where b is the impact parameter, σ is the error in b, and n is a preset value ranging from 4-7.

3. LEVEL-1 TRIGGER HARDWARE

The first stage of the L1 trigger algorithm (segment finding stage) has been implemented in VHDL and is currently being tested on a modified PCI Test Adapter (PTA). The PTA is a PCI-based, general purpose I/O card developed at Fermilab primarily for use in detector development test stands. The modified version of the board uses an Altera EPC20K1000 FPGA, roughly 80% of whose resources are used to implement the algorithm for 3 adjacent pixel detector stations. Several PTA cards can be used together on a PC motherboard to test a larger portion of the full segment finding hardware.

Pre-prototype hardware has also been developed to serve as a baseline design for implementing the second stage of the L1 trigger algorithm (track/vertex finding). A block diagram of this hardware is shown in Figure 1. It consists of a motherboard with mezzanine sites for 4 Texas Instruments (TI) TMS320C6711 digital signal processors (DSP). Segment data from the first stage of the algorithm is received by the motherboard through a 32-bit input port and buffered in a 64kB FIFO. A buffer manager (BM) implemented on a Xilinx Virtex II FPGA makes the data for each crossing available to one of the 4 DSP's. This data is transferred via direct memory access (DMA) into the internal memory of the DSP which executes the track and vertex finding algorithm coded in C. Processed data is sent via DMA back to the BM and on to a 32-bit output port. The DSP sends the trigger summaries destined for Global Level-1 (GL1) through its multi-channel buffered serial port (McBSP) to the trigger results manager (TM) implemented on part of a second Xilinx FPGA. The TM makes this data available to a Hitachi H8S micro-controller which reads this data and sends it to GL1 via an ArcNet controller.

A second ArcNet controller allows an external host PC to communicate with a second H8S micro-controller that is connected to the host port interface bus (HPI) of each DSP. This allows the host PC to initialize, control, and communicate with each DSP via the micro-controller. Additional logic provides the micro-controller access to a compact flash memory card for file storage and an LCD display for status messages. JTAG ports are available for in-circuit debugging and for downloading code into the DSP's and FPGA's.

Having built, tested, and verified the pre-prototype track/vertex hardware, the latest serial link and alternative processor technologies are currently under evaluation for the prototype version of the hardware.

4. TRIGGER ARCHITECTURE

The three-level trigger system shown in Fig. 2 consists of eight parallel data pathways called "highways". Data from each beam crossing is distributed uniformly to one of these eight highways, each of which forms a complete and independent three-level trigger system. This reduces the full data rate of 800 GB/s from the detector to 100 GB/s into each highway allowing the use of cheap components such as commercially available ethernet switches.

Starting from the left hand side of Fig. 2, digitized data packets from the front-end boards are sent via high-speed copper links to data-combiner boards (DCB) located in the collision hall [6]. DCB's merge multiple input streams to fewer output streams by combining smaller packets from many input channels into larger packets. Each DCB interfaces with eight optical transmitters, each connected to one of the eight highways. Data packets from the output stream for one crossing are sent by one of these transmitters via optical fibers over a distance of about 30 m from the collision hall to the counting room. Optical receivers in the counting room then route the data packets to the appropriate highway. Three data-combiner boards are grouped together to form a DCB subsystem.

For each highway, data from all detector components is sent from the optical receivers to L1 buffers for temporary storage as trigger decisions are made. An exception is the vertex detector whose data goes through the pixel-processors before being stored in L1 buffers. Data from the pixel processors are also sent to an FPGA based segment tracker that executes the segment finding stage of the algorithm. Inner and outer triplets belonging to the same crossing are then routed by a network switch from the segment tracker to one node in a farm of over 300 programmable embedded processors that execute a C version of the track and vertex finding stage of the algorithm. For each node, complete processed results are routed to the L1 buffers while summarized trigger results are sent to a Global Level-1 (GL1) processor responsible for the ultimate trigger decision. Crossings accepted by GL1 are then maintained as a list in the Information

Transfer Control Hardware (ITCH) which also broadcasts accept messages to all L1 buffers indicating which crossings are to be saved.

The basic building blocks of the L1 buffers are modules consisting of commercial DRAM configured as circular buffers [6]. Three modules are grouped into an L1 buffer subsystem for a total of 32 subsystems in each highway. Each of these subsystems has 24 input buffers serving 8 DCB subsystems in the collision hall. A Level-1 accept concatenates the data from all 24 input buffers copying them to an output buffer where it remains until transferred to a Level-2/3 (L2/3) node. The use of low-cost DRAM allows enough memory to buffer ~300 thousand crossings in each highway corresponding to ~300 ms of L1 trigger decision time (since the crossing time of 132 ns is increased $8\times$ to 1 μs per highway). This is nearly three orders of magnitude more than the average L1 processing time of 330 μs. The use of large buffers with circular access is far more cost effective than smaller ones employing sophisticated memory management. It also allows the system to handle the long processing times of events in the tail of the L1 time distribution. Each of the 32 L1 buffer subsystems feeds a port in a highway switch consisting of a commercial 64-port gigabit ethernet switch. 12 pairs of ports on this switch feed the paired gigabit uplink ports of a dozen 24-port fast-ethernet fanout switches. These fanouts, in turn, feed data to nearly 300 commodity Linux-PC's that make up the L2/3 processor farm in each highway where the DRAM in each of the nodes functions as a Level-2/3 buffer.

After receiving a request from an idle node in the L2/3 processor farm for an event, the ITCH responds by assigning an accepted crossing number to that node. Once it receives its assignment, the L2/3 node sends a request to a subset of the L1 buffers which respond by sending their data to that node. All requests and data transfers between the L2/3 farm and the L1 buffers and the ITCH are routed through the highway and fanout switches. Upon receiving the data, the L2/3 node executes the L2 trigger algorithm which now has the option of using additional information from the first few stations of the straw and Si-microstrip trackers in doing a more refined analysis of the event. If the event passes L2, data from the rest of the L1 buffers is transferred to the same node to execute L3. At this stage, the processing node has, at its disposal, particle ID information in addition to that from the rest of the forward tracking stations to further improve upon the L2 results.

In practice, each L2/3 node issues multiple event requests, storing data for ~16-32 events in the L2/3 buffers at any given instant. When a Level-2/3 processor completes executing the L2 trigger algorithm on an event, it performs a context switch to process one of the other events in the buffer while new data is requested to fill the buffers. Processing of events that pass L2 is temporarily

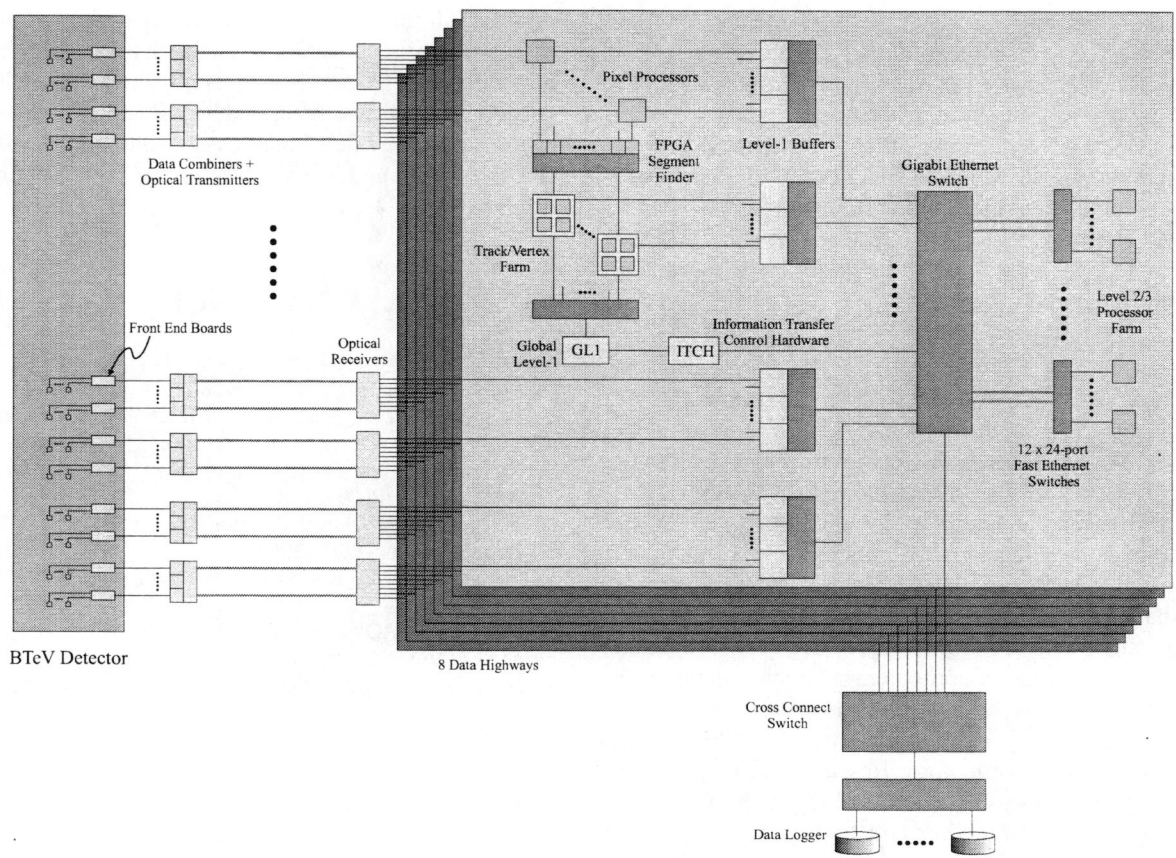

FIGURE 2. BTeV Three-Level Eightfold-way Trigger Architecture.

suspended while the L2/3 processor switches to another event, resuming L3 processing on the L2-accepted event after additional data has been transferred. This way, no dead-time is incurred between events due to data transfers between the L1 and L2/3 buffers.

If the event passes L3, the processed results are propagated back up the fanout and highway switches to an external cross-connect switch that routes accepted events from all 8 highways to a small cluster of data-logging nodes for archival.

5. DATA RATES

Each of the 256 DCB subsystems serving the detector front-end interfaces with two dozen 2.5 GB/s optical links to bring the data from the collision hall to the optical receivers in the counting room [6]. This makes it possible to achieve a peak design rate of 1.5 TB/s providing sufficient headroom for the estimated data rate of 800 GB/s. The data rate going into the L1 buffers in each highway is reduced 8× to 100 GB/s through the use of the parallel highway architecture described

above. The data coming out of the pixel detector front-ends is on the order of 40 GB/s which is increased to about 70 GB/s after additional information, such as the ID's of the pixel read-out chips, is inserted into the data stream by the pixel DCB's. This means the data rate going into the pixel processors in each highway is ~9 GB/s. Assuming an average size of 4 kB for the total number of triplets found by the L1 segment tracker for each crossing, the data rate going into the L1 track/vertex processor farm is ~4 GB/s. The resulting data rate into each node in a farm of over 300 processors is an easily manageable ~13 MB/s. If each node produces 20 bytes of summarized trigger results, the total data rate going from the track/vertex farm to the GL1 processor in each highway is ~20 MB/s.

The L1 trigger rejects 99% of the input to the L1 buffers reducing the output from the buffers by 100× to 1 GB/s (for simplification, we will treat L2 and L3 as a single trigger stage in this discussion). This means the data rate coming out of each of the 32 L1 buffer subsystems is ~31 MB/s which can readily be handled by a single gigabit ethernet port on the highway switch. Since there are 12 fanout switches, the data rate into each fanout is ~83 MB/s which can easily be accommodated by the pair of gigabit uplink ports available on the fanouts. Data is distributed to each node in the L2/3 processor farm at ~3.5 MB/s using one of the 24 fast ethernet ports on the fanout switches. The L2/3 trigger rejects 95% of the incoming data reducing this by $20\times$ to 50 MB/s. A $2\times$ data compression further reduces this to 25 MB/s. The resulting data rate from all 8 highways into the cross-connect switch and the data-logging cluster is a mere 200 MB/s which can easily be handled by commercially available storage technology.

6. FAULT-ADAPTIVE TRIGGER

The BTeV trigger is a highly complex system involving a heterogeneous mix of several thousand FPGA's, embedded processors, and commodity PC's. A system of this complexity is bound to experience hardware and software failures. Such failures, if not dealt with properly, can have a detrimental effect on the trigger's ability to accomplish its task. To address this issue, BTeV is collaborating with a group of computer scientists and engineers from 5 institutions in the Real Time Embedded Systems (RTES) project [7]. RTES is funded by the NSF through the Information Technology Research (ITR) grant. Its primary goal is to address the issue of reliability in large scale clusters with real time constraints.

Within the RTES framework, the BTeV trigger is organized hierarchically into global and regional management nodes operating worker nodes that represent trigger nodes. The main technology employed by RTES are software components called Adaptive, Reconfigurable, and Mobile Objects for Reliability (ARMOR) and Very Lightweight Agents (VLA) that reside on each node. These agents gather data on the status and condition of the software and hardware on each node sending this data to entities above it in the hierarchy. Depending on the severity of the problem, these agents may attempt a local recovery upon detecting an error condition. For more severe problems, it will rely on higher level entities for a more complex analysis of the problem and to take necessary measures.

By serving as a test-bed on which the RTES collaboration can conduct their investigations and apply their solutions, BTeV hopes to obtain the tools and technology necessary to build a reliable fault-adaptive trigger system.

7. CONCLUSION

We have provided an overview of the software and hardware aspects of BTeV's Level-1 vertex trigger. By showing how it works with the pixel detector and how it fits into the full three-level, eightfold-way trigger and data-acquisition system, we have described, in concrete terms, how BTeV intends to achieve its physics goals with this trigger. We have also described how BTeV will make use of the technology developed by RTES to address the issue of software and hardware failures in its sophisticated trigger system.

REFERENCES

1. Kulyavtsev, A., et al., [BTeV Collaboration], BTeV Proposal, Tech. Rep. BTeV-doc-66 (2000).
2. Drobychev, G. V., et al., [BTeV Collaboration], Update to BTeV Proposal, Tech. Rep. BTeV-doc-316 (2002).
3. Kwan, S., [BTeV Collaboration], *Nucl. Instrum. Meth.* **A511**, 48–51 (2003).
4. Wang, M., BTeV Level 1 Vertex Trigger Algorithm, Tech. Rep. BTeV-doc-1179 (2003).
5. Gottschalk, E. E., *Nucl. Instrum. Meth.* **A473**, 167–173 (2001).
6. The BTeV DAQ Group, Overview of the BTeV Readout and Controls System, Tech. Rep. BTeV-doc-1091. (2002).
7. Kowalkowski, J., *ECONF* **C0303241**, THGT001 (2003).

B PHYSICS AT FUTURE HADRON MACHINES

B Physics at LHC with the CMS Detector

Nancy Marinelli

(Representing the CMS Collaboration)

Institute of Accelerating Systems and Applications (IASA), Athens, Greece

Abstract. The perspectives for B physics analyses with CMS are reviewed here and some of the related trigger and tracking issues are discussed.

1. INTRODUCTION

Although the current theoretically-predicted $b\bar{b}$ cross-section is uncertain, the expected value at LHC is about five orders of magnitude larger than that reachable with 2 TeV centre-of-mass energy at the Tevatron [1]. At the LHC design luminosity of $10^{34}\,\mathrm{cm}^{-2}\mathrm{s}^{-1}$, it would translate into 10^5 $b\bar{b}$ pairs/s and at the initial luminosity of $2 \times 10^{33}\,\mathrm{cm}^{-2}\mathrm{s}^{-1}$ into about 10^4 $b\bar{b}$ pairs/s. The LHC collider thus provides a unique opportunity for B physics analyses.

The amount of storage resources available at LHC is very small compared to the amount of data produced. At CMS, the initial 40 MHz p-p interaction rate will be brought down to 100 kHz by the Level 1 Trigger to finally reach about 100 Hz stored on tape. In this scenario, the trigger strategy for B physics studies is a major challenge.

The B physics program with the CMS detector is intended to cover rare B decays, CP violation probing and B_s^0-$\bar{B}_s^0$ mixing. A benchmark channel for each of these items was chosen: $B_s^0 \to \mu^+\mu^-$, $B_s^0 \to J/\psi\,\phi \to \mu^+\mu^-K^+K^-$ and $B_s^0 \to D_s^+\pi^- \to K^+K^-\pi^+\pi^-$. These channels were analyzed to develop B decay triggering strategies and to evaluate the physics outcome.

2. CMS TRIGGER STRATEGY FOR B PHYSICS

A detailed description of the CMS Trigger design and of the overall CMS Data Acquisition System is given in Refs. [2–4]. The features relevant to B physics selection are mentioned here. The CMS trigger is divided into a Level 1 Trigger and High-Level Triggers (HLT). Coarse information from the calorimeters and the muon detectors are exploited at Level 1 to select basic physics objects, subsequently refined (e.g., muon isolation) at HLT with the information provided by the tracking. The proton-proton interaction rate at LHC, 40 MHz, is brought down to 100 kHz at Level 1 and finally, after the HLT, to about 100 Hz for final data storage. In the first LHC phase at low luminosity, the CMS DAQ system is planned to be staged so that it will be able to handle at most 50 kHz Level 1 Trigger rate. However, the actual Level 1 Trigger rate so far considered for all HLT studies is only 16 kHz. A safety factor of 1/3 was applied to account for possible surprises arising from the LHC beam conditions and the CMS detector as well as uncertainties in the simulation of the basic physics processes.

The available Level 1 Trigger rate must be allocated to the various physics objects (electrons/photons, muons, jets, τ jets) to cover the widest possible range of physics for discovery. For the low luminosity period, a prototype "democratic" allocation of about 4 kHz was made for the four categories and the p_T thresholds were optimized to fulfill this requirement.

The selection of B physics processes can be triggered at Level 1 by the presence in the event of single or double muons produced in the B semileptonic decays; the Level 1 Trigger bandwidth allocated to the muons includes the single and di-muon triggers, the thresholds of which, $p_T^\mu > 14$ GeV/c and $p_T^\mu > 3$ GeV/c, yield a rate of 2.7 kHz and 0.9 kHz, respectively.

Particles originating from B hadron decays are characterized by a relatively soft p_T spectrum and muons falling in the low p_T range can well be used for triggering. Muons are preferred to electrons because of their substantially lower p_T thresholds ($p_T^{el} > 29$ GeV/c and $p_T^{el} > 17$ GeV/c).

At HLT, a total of 30 Hz is allocated to muons (single + double). The muon isolation criteria bring further down the trigger rate and reject the majority of the B hadron content; for the chosen single (double) muon threshold $p_T > 19$ GeV/c ($p_T > 7$ GeV/c) a 25 Hz (5 Hz) rate is achieved. Figure 1 shows the total muon rate at high

CP722, *B Physics at Hadron Machines*, edited by M. Paulini and S. Erhan
© 2004 American Institute of Physics 0-7354-0203-5/04/$22.00

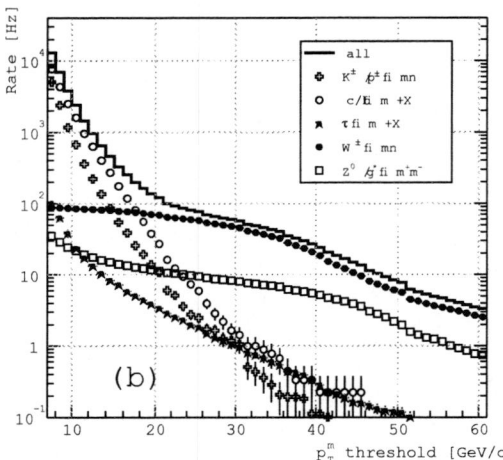

FIGURE 1. Contributions to the High-Level Trigger rate at high luminosity from all sources of muons after requiring the muon to be isolated.

luminosity as a function of the single muon p_T threshold after applying the isolation criteria. The contributions from different sources of muons to the total rate are also shown. At the working point chosen, the dominant contribution comes from the W leptonic decays. Only a small fraction (5 Hz for the single muon) is accounted for by inclusive b and c decays.

The affordable rate for final data storage on tape is only hundred events per second and they have to include all physics processes relevant for discovery. Event containing B hadrons cannot be accepted inclusively for subsequent offline selection, since their amount would be too little to explore processes with branching ratios well below 10^{-4}. A selection at HLT, involving fast tracking, is henceforth needed to identify interesting exclusive B decays.

The HLT selection will be performed in an Event Filter Farm, where each single processor is devoted to the online data access, reconstruction and analysis of a single event. The CPU time constraints today (1 GHz Pentium III processor) are about 300 ms/event. The primary task to be accomplished at HLT is then rejecting the event as fast as possible, while carefully retaining those with interesting physics content.

2.1. Tracking at High Level Trigger

The reconstruction of charged particles starts from the inner pixel detector which provides up to three accurate three-dimensional points. Track seeds are built out of pixel hit pairs compatible with a track originating from the primary interaction vertex and subsequently propagated outward in the silicon strip tracker. The full recon-

struction is highly time consuming because of the large number of hit combinations. Although tracking at HLT is required to be robust, it does not need to be as accurate as for the offline reconstruction and substantial computing time can be saved by simplifying the reconstruction procedure.

In order to speed up the tracking procedure, the concepts of "regional" and "partial or conditional" tracking were developed with the aim of reducing the number of track seeds and the number of computational steps per seed, respectively. With the regional tracking, tracks are reconstructed only in limited regions of interest in the tracker identified by the presence of a Level 1 Trigger object such as a muon. With the partial and/or conditional tracking, the track reconstruction can be stopped if, for instance, the measured p_T is below a certain threshold and no longer considered for further reconstruction, or when a certain optimal number of hits is already reconstructed. Figures 2 and 3 show the p_T and the transverse parameter resolution as a function of the number of hits reconstructed per track.

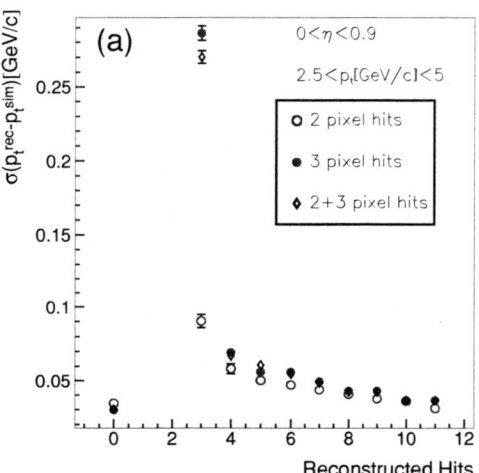

FIGURE 2. The p_T resolution for partial track reconstruction as a function of the number of reconstructed hits. The markers at zero on the horizontal axis show the resolution achieved with the full tracker reconstruction.

The results are compared with those obtained with the full reconstruction. The asymptotic value of the resolution is almost reached by reconstructing five or six hits per track while reducing by about half the reconstruction time.

3. EXCLUSIVE B DECAY CHANNELS

3.1. $B_s^0 \to \mu^+\mu^-$

The decay channel $B_s^0 \to \mu^+\mu^-$ is a Flavour-Changing-Neutral-Current process described at loop

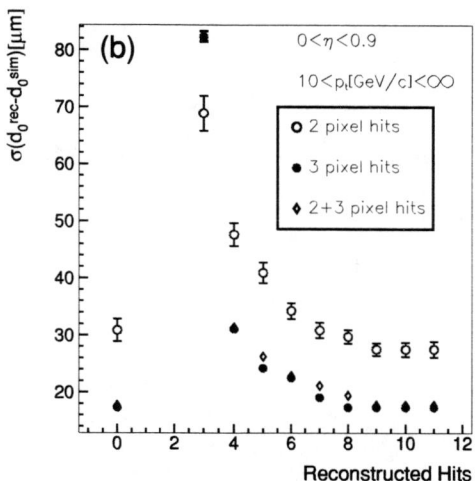

FIGURE 3. The resolution of the transverse impact parameter for partial track reconstruction as a function of the number of reconstructed hits. The markers at zero on the horizontal axis show the resolution achieved with the full tracker reconstruction.

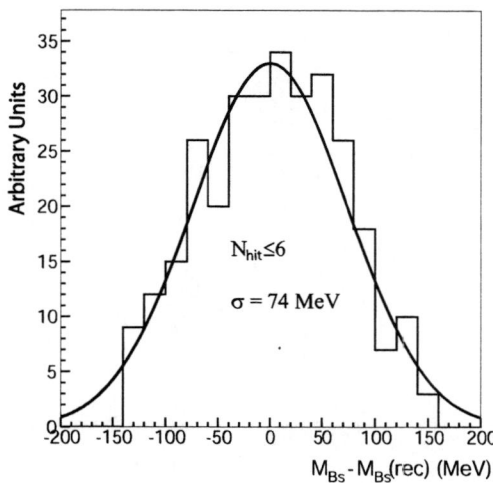

FIGURE 4. Decay channel $B_s^0 \to \mu^+\mu^-$: B_s^0 mass resolution with the HLT selection and reconstruction.

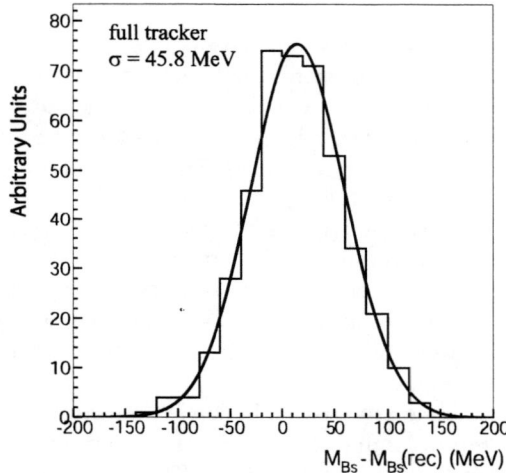

FIGURE 5. Decay channel $B_s^0 \to \mu^+\mu^-$: B_s^0 mass resolution with the offline reconstruction.

level by the Standard Model (SM). The experimental signature is unique and clean. Unfortunately, the SM predicts a tiny branching ratio ($\mathcal{O}$ (10^{-9})). It is unlikely that $B_s^0 \to \mu^+\mu^-$ will be observable before the LHC starts, unless a drastic enhancement of the branching ratio, due to contributions from new physics, shows up.

Events are triggered at Level 1 by two opposite charge muons with $p_T > 3$ GeV/c. Pixels seeds are built out of pixels hit pairs compatible with $p_T > 4$ GeV/c and transverse impact parameter below 1 mm. The seeds so obtained are used for a first reconstruction of the primary vertex. Hit pairs are then filtered with the vertex constraint and the region of interest around the Level 1 di-muon direction and used for partial/conditional track reconstruction, *i.e.*, the reconstruction is stopped if either the track p_T is lower than 4 GeV/c at 5σ, or six hits are reconstructed along the track, or the relative p_T resolution is below 2%. When two tracks with opposite charge are found, the invariant mass is calculated and required to be within 150 MeV/c^2 around the B_s^0 mass. Figures 4 and 5 show the B_s^0 mass resolution achievable at HLT and offline, respectively.

The combinatorial background is suppressed by constraining the two tracks to a common secondary vertex and imposing suitable cuts on the vertex fit χ^2 and on the transverse impact parameter. The HLT selection efficiency is about 33% while the Level 1-plus-HLT efficiency is about 5% with a yield of 47 events with 10 fb^{-1}. The average time required for reconstructing a $B_s^0 \to \mu^+\mu^-$ event is 240 ms.

The HLT selection is identical to the offline procedure described in Ref. [5]. It does, however, apply much looser cuts. The events selected by the offline analysis (seven signal events with one background event with 10 fb^{-1}) are then expected to pass the HLT selection. In such conditions a 5σ observation would be possible with about 15 fb^{-1}.

3.2. $B_S^0 \to J/\psi\,\phi \to \mu^+\mu^-\,K^+K^-$

The decay channel $B_s^0 \to J/\psi\,\phi \to \mu^+\mu^-K^+K^-$ is the gold-plated mode for probing *CP* violation. It allows the value of the *CP* violation weak phase, $\phi_s = -2\lambda^2\eta$, to be measured. This value is predicted to be extremely small, $\mathcal{O}$ (0.03), by the Standard Model. An enhancement of ϕ_s would imply contributions from new physics.

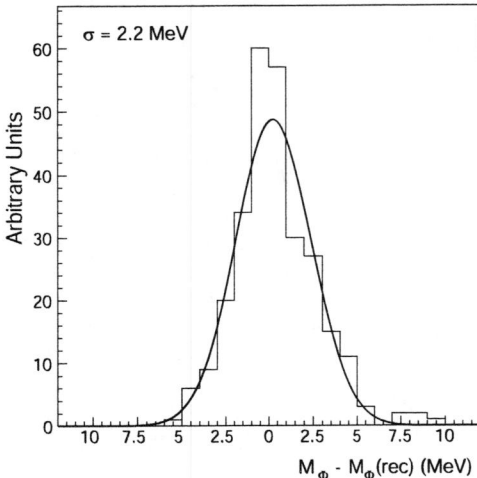

FIGURE 6. Decay channel $B_s^0 \rightarrow J/\psi\phi \rightarrow \mu^+\mu^-K^+K^-$: J/ψ mass resolution achieved with the HLT selection.

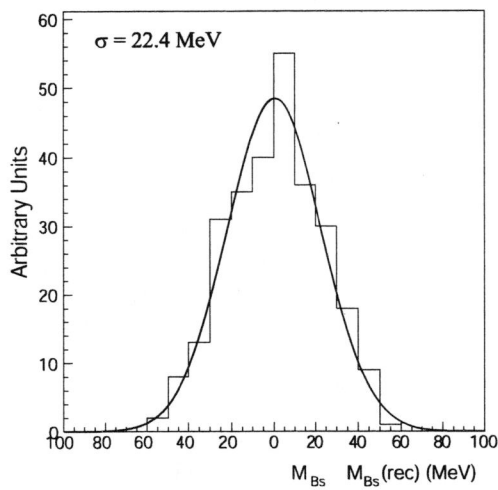

FIGURE 7. Decay channel $B_s^0 \rightarrow J/\psi\phi \rightarrow \mu^+\mu^-K^+K^-$: B_s^0 mass resolution achieved with the HLT selection.

As for the previous channel, $B_s^0 \rightarrow J/\psi\phi \rightarrow \mu^+\mu^-K^+K^-$ is triggered at Level 1 by the presence of a di-muon. The reconstruction is, in this case, more complicated and more time consuming. It can be divided into two steps, referred to as HLT step 1 and HLT step 2 in the following.

At the HLT step 1, muons from the J/ψ are reconstructed as described in the previous section. Slightly tighter cuts, however, are applied to the di-muon invariant mass and on the secondary vertex to keep the background level under control. The J/ψ mass resolution is shown in Fig. 6.

The di-muon reconstruction gives an inclusive rate of 15 Hz at low luminosity. The contribution from J/ψ produced in B decays is about 90%. The HLT step 1 takes on average 260 ms per event. The HLT step 2, concerning the ϕ and B_s^0 reconstruction, is more time demanding. First, the ϕ is reconstructed out of charged particle tracks in a cone around the J/ψ. Partial tracking is used for their reconstruction. Oppositely-charged particle track pairs with invariant mass within 10 MeV/c^2 of the ϕ are retained as candidates. Finally, the invariant mass of the J/ψ and the ϕ is calculated and required to be within 60 MeV/c^2 of the B_s^0 mass. The B_s^0 mass resolution so obtained is shown in Fig. 7.

The combined HLT-step-1 and HLT-step-2 efficiency is about 8.7%. A yield of about 84K events is expected with 10 fb^{-1} at low luminosity with a background rate smaller than 2 Hz. The total average reconstruction time is about 800 ms. The $B_s^0 \rightarrow J/\psi\phi \rightarrow \mu^+\mu^-K^+K^-$ offline reconstruction is expected to select 60% to 70% of the HLT signal. This translates to about 200k signal events with 30 fb^{-1}. Past studies [6], involving the time-dependent angular analysis of the decay products, predict a relative uncertainty on $\Delta\Gamma_s$ of about 15% and an uncer-

tainty of about 0.0025 on ϕ_s for x_s=20.

3.3. $B_s^0 \rightarrow D_s^+\pi^- \rightarrow \phi\pi^+\pi^- \rightarrow K^+K^-\pi^+\pi^-$

The Standard Model prediction for Δm_s in the B_s^0-$\bar{B}_s^0$ system is $14.8 \leq \Delta m_s \leq 25.9$ ps^{-1} at 99% CL. The present experimental 95% CL lower limit is $\Delta m_s \geq 14.4$ ps^{-1}. The study of the decay channel $B_s^0 \rightarrow D_s^+\pi^- \rightarrow K^+K^-\pi^+\pi^-$ at LHC allows the whole predicted Δm_s range to be spanned.

A fully hadronic decay channel such as $B_s^0 \rightarrow D_s^+\pi^- \rightarrow K^+K^-\pi^+\pi^-$ can be triggered at Level 1 by the muon produced in the semileptonic decay of the other B hadron present in the event. For this channel, the muon also serves for tagging the CP state of the B_s^0 at production time. Alternatively, a combination of a low-p_T muon and a low-E_T jet can be used with different threshold scenarios and trigger rates.

The $B_s^0 \rightarrow D_s^+\pi^- \rightarrow K^+K^-\pi^+\pi^-$ selection at HLT starts with the primary vertex reconstruction as for the $B_s^0 \rightarrow \mu^+\mu^-$. Since the Level 1 Trigger muon direction has no correlation with the B_s^0 direction, no region of interest can be used for applying regional tracking. Hence all tracks with p_T above 0.7 GeV/c have to be reconstructed. To reduce the reconstruction time, only three hits, two in the pixel detector and one in the innermost layer of the silicon strip detector, are used for creating the seed tracks. Also, seeds are requested to be compatible with the primary vertex along the z-direction.

Topological cuts are applied to select the ϕ, D_s^+ and B_s^0 candidates as well as cuts are applied to their invariant masses and transverse momenta. The three mass distributions after the HLT selection are shown in Figs. 8, 9 and 10. The HLT selection efficiency is about 9% and the average execution time is 250 ms per event.

The number of selected signal events depends on the HLT bandwidth available for this channel which almost likely cannot exceed 5 Hz. With this assumption, about 300 events are expected with 20 fb^{-1} which would allow the exploration of the Δm_s range up to $\Delta m_s \sim$ 20 ps^{-1}. About 1000 events would be needed to extend the sensitivity to the upper limit allowed by the Standard Model ($\Delta m_s \leq 26$ ps^{-1}). More details concerning the $B_s^0 \to D_s^+ \pi^- \to \phi \pi^+ \pi^- \to K^+ K^- \pi^+ \pi^-$ study are presented in Ref. [7].

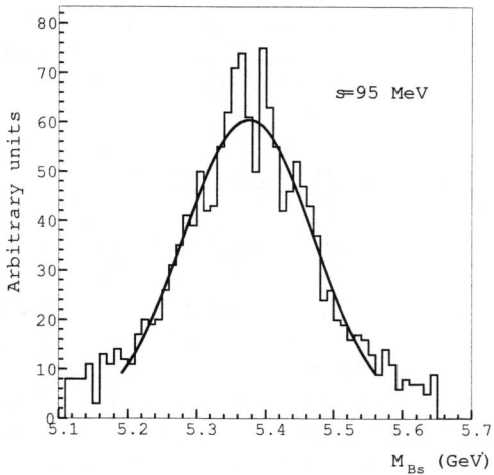

FIGURE 10. Invariant mass distribution of the B_s^0 in the decay channel $B_s^0 \to D_s^+ \pi^- \to K^+ K^- \pi^+ \pi^-$.

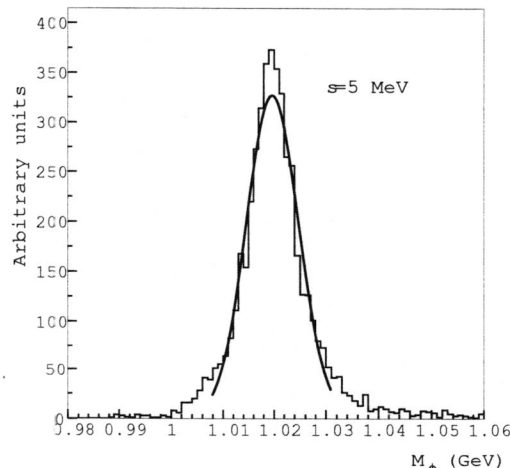

FIGURE 8. Invariant mass distribution of the ϕ in the decay channel $B_s^0 \to D_s^+ \pi^- \to K^+ K^- \pi^+ \pi^-$.

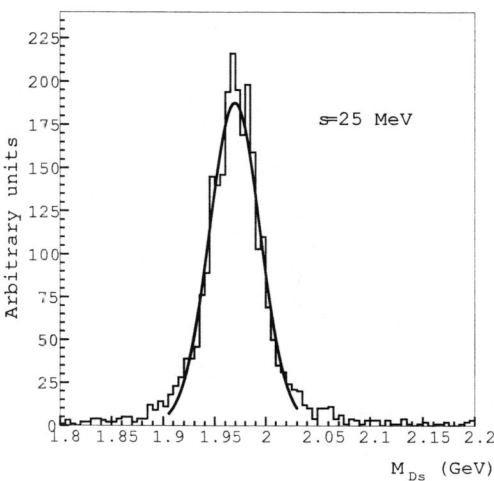

FIGURE 9. Invariant mass distribution of the D_s^+ in the decay channel $B_s^0 \to D_s^+ \pi^- \to K^+ K^- \pi^+ \pi^-$.

4. HIGH-LEVEL TRIGGER BANDWIDTH FOR EXCLUSIVE *B* DECAY CHANNELS

The present prototype HLT table does not include explicitly the *B* physics item since the allocation of *B* physics triggers is likely going to be a rather dynamic process. The rate of events selected by reconstructing online exclusive *B* decays is few Hz and an optimization of the reconstruction strategies might reduce it further. The main limitation comes from the Level 1 Trigger rate actually sustainable by the HLT reconstruction algorithms. The present average HLT processing time, including all Level 1 physics objects, is about 300 ms/event. This was used to get an estimate of the computing power needed in the low luminosity period to process the Level 1 output at 50 kHz.

Any improvement either of the processor performance (Moore's law would suggest 300 ms/event might become 30 ms/event in 2007) and of the software performance (refined algorithms as well as optimized software) would open more space for *B* physics triggers.

Another point to bear in mind is that the Level 1 Trigger thresholds were designed in order to account for a factor 1/3 of the first-phase design Level 1 output rate. The safety factor might turn out to be pessimistic: either 30 kHz become available or the time for processing each event at HLT would be three times larger than nowadays expected.

The bandwidth which can be allocated for *B* physics is likewise dependent on the actual value of the luminosity. The nominal low luminosity is 2×10^{33} cm^{-2} s^{-1}, however, at the LHC start up it is likely to expect a somehow lower value. Such an occurrence would imply a lower HLT output and larger trigger rate might be allocated to the exclusive B physics processes.

5. CONCLUSIONS

The huge $b\bar{b}$ statistics at LHC render possible the first observation of the $B_s^0 \to \mu^+\mu^-$ decay and very accurate measurements such as the extraction of the weak mixing phase ϕ_s from $B_s^0 \to J/\psi\,\phi \to \mu^+\mu^- K^+K^-$ and Δm_s from $B_s^0 \to D_s^+\pi^- \to \phi\pi^+\pi^- \to K^+K^-\pi^+\pi^-$.

Although CMS was not specifically designed for B physics studies, it can surely support a competitive program. The powerful tracking device can be exploited at HLT by designing suitably fast algorithms.

The bandwidth allocated to B physics triggers crucially depends on the technological evolution of commercial processors on one side and on the LHC beam conditions on the other.

ACKNOWLEDGMENTS

I wish to acknowledge the financial support provided through the European Community's Human Potential Programme under contract HPRN-CT-2002-00326, (PRSATLHC).

REFERENCES

1. Anikeev, K., et al., arXiv:hep-ph/0201071 (2001).
2. CMS Collaboration, The Trigger and Data Acquisition Project, Volume I: The Level-1 Trigger, Tech. Rep. CERN/LHCC 2000-038, CMS TDR 6.1 (2000).
3. CMS Collaboration, The Trigger and Data Acquisition Project, Volume II: Data Acquisition & High-Level Trigger, Tech. Rep. CERN/LHCC 2002-026, CMS TDR 6.2 (2002).
4. Konecki, M., in these proceedings (2004).
5. Nikitenko, A., Starodumov, A., and Stepanov, N., arXiv:hep-ph/9907256 (1999).
6. Standard Model Physics (and more) at the LHC, Tech. Rep. CERN 2000-004 (2000).
7. Giassi, A., Palla, F., and Starodumov, A., Study for a High Level Trigger for the Decay Channel $B_s \to D_s\pi \to \phi\pi\pi \to KK\pi\pi$", Tech. Rep. CMS NOTE 2002/045 (2002).

LHCb Physics Performance

Ulrich Uwer

(On behalf of the LHCb Collaboration)

University of Heidelberg, Physics Institute, 69120 Heidelberg, Germany

Abstract. The LHCb detector has recently undergone a reoptimization with respect to its original design. In the following, the expected physics potential of the reoptimized detector is discussed. The expectations are based on a Monte Carlo simulation of the detector performance including the simulation of the trigger. LHCb will offer an excellent opportunity to determine precisely the CKM parameters. The manifold of different *CP* measurements will over-constrain the unitarity triangles and will allow to spot new physics, if present.

1. INTRODUCTION

The LHCb experiment [1, 2] is one of the four experiments at the Large Hadron Collider (LHC) currently being built at CERN. It is a dedicated experiment to exploit the physics potential of the large amount of $b\bar{b}$ quark pairs produced with a cross section of about 500 μb in pp collisions of LHC. At a luminosity of 2×10^{32} cm^{-2}s^{-1} foreseen for the LHCb interaction point[1], 10^{12} $b\bar{b}$ pairs are expected per year (10^7 s) of LHC running.

The *B* meson system provides an excellent laboratory to study quark mixing and the related *CP* asymmetries. Large *CP* violation had been predicted by the Standard Model and, for the first time, was measured in 2001 for various *B* decays modes, such as $B^0 \to J/\psi K_s^0$ and $B^0 \to \eta_c K_s^0$ [3, 4], generated by $b \to c + W^-$ and $\bar{b} \to \bar{c} + W^+$ tree diagrams. For these *B* decays (*CP* eigenstates) the Standard Model predicts asymmetries proportional to $\sin \phi_{\rm d}$, where $\phi_{\rm d}$ is the phase of the B^0-$\bar{B}^0$ mixing. In this case, where only Standard Model box diagrams contribute to mixing, $\phi_{\rm d}$ probes the phase of the CKM matrix element V_{td} and thus the angle β of the unitarity triangle:

$$\phi_{\rm d} = -2\mathrm{arg}V_{td} = 2\beta \ .$$

First *CP* measurements are also available for the decay channel $B^0 \to \pi^+ \pi^-$. The theoretical interpretation of the extracted asymmetries however is not simple: in addition to the $\bar{b} \to \bar{u} W^+$ tree process, penguin processes are expected to contribute significantly to the decay. Without an exact theoretical determination of the penguin contri-

bution a meaningful extraction of the weak phase is difficult.

Constraints on the unitarity triangle can also be obtained from the determination of the sides through the measurement of $|V_{cb}|$, $|V_{ub}|$ and $|V_{td}|$. The precision of the extraction of the values $|V_{cb}|$, $|V_{ub}|$ and $|V_{td}|$ from the data, however, is also limited by theoretical calculations.

Precise *CP* measurements for many different channels and in particular measurements for B_s^0 meson decays are needed to test the Standard Model predictions and to over-constrain the unitarity triangle further. Different sensitivity of *B* decays to weak phases and also to potential new phases will allow to detect and to filter the effect of new physics. If present, new physics should contribute to B-$\bar{B}$ mixing and to other decays generated by loop diagrams. An example of such an over-constraining measurement is the measurement of the unitarity angle γ in three different decays as described below.

2. LHCb DETECTOR AND SIMULATION

The LHCb detector is a single-arm forward spectrometer designed to trigger and reconstruct b hadron decays in the high-multiplicity environment of LHC. The detector features excellent vertexing and particle identification, excellent tracking capabilities and good momentum resolution. The LHCb detector design has recently undergone a reoptimization with respect to its original design with the primary goal to reduce material in order to limit multiple scattering, as well as electromagnetic and hadronic interactions of B decay particles.

To study the detector and physics performance of the LHCb experiment, large Monte Carlo event samples are

---

[1] The luminosity at the LHCb interaction point is tunable. A value of 2×10^{32} cm^{-2}s^{-1} has been found as a good compromise to limit "busy" events containing more than one *pp* interaction [2].

CP722, *B Physics at Hadron Machines,* edited by M. Paulini and S. Erhan
© 2004 American Institute of Physics 0-7354-0203-5/04/$22.00

generated and passed through the detector simulation. Trigger algorithms [5] are applied and events are reconstructed including full pattern recognition. For the studies presented here about 30M minimum-bias events (mainly used for trigger studies) and about 10M inclusive $b\bar{b}$ events (used for background studies) have been generated and analyzed. In addition, signal-event samples of particular B meson decays have been produced.

2.1. Event Generation

Minimum bias proton-proton interaction at $\sqrt{s} = 14$ TeV are generated using the PYTHIA 6.2 program [6]. The PYTHIA parameters have been tuned based on available published data from UA5 and CDF [7]. Signal event samples are obtained by filtering a large minimum-bias data-set.

The bunch-crossing properties foreseen for the LHCb interaction point are taken into account. The luminous region is assumed to be a Gaussian ellipsoid with a width along the beam direction (z) of $\sigma_z = 5$ cm. Taking the bunch pattern of the machine and the predicted luminosity into account, multiple interaction per crossing ("pile-up") are simulated assuming a Poisson probability.

2.2. Detector Simulation

Generated particles are traced through the detector using the GEANT 3 package. The detector description includes the active detector components, the front-end electronics as well as all passive elements. The simulation of the subdetector response considers the resolutions and efficiencies as measured in test beam experiments. Also the effect of "spill-over", i.e., the effect of preceding or subsequent bunch crossings on the signals of the simulated interaction is taken into account[2]. Before being analyzed the data pass the Level-0 and Level-1 triggers: the trigger cuts are tuned to reach the maximum efficiency at a limited Level-0 (Level-1) output rate of 1.1 MHz (40 kHz), respectively [5]. The simulated events are reconstructed including a full pattern recognition without using any generator information.

3. SELECTION OF B DECAYS

The main challenge in the offline selection of specific B final states is to maintain high efficiency for the signal while providing a large rejection factor for combi-

---

[2] Simulated are two preceding and one subsequent bunch crossing.

TABLE 1. Efficiencies (trigger and total), annual event yields and background-to-signal ratios for different B decay channels. The yields assume a luminosity of 2×10^{32}cm^{-2}s^{-1} and a running time of 10^7s.

	$\varepsilon_{\mathrm{trig}}$	$\varepsilon_{\mathrm{tot}}$	yield	B/S
$B^0 \to \pi^+\pi^-$	34%	0.69%	26k	< 0.7
$B^0 \to K^+\pi^-$	33%	0.94%	135k	0.16
$B_s^0 \to \pi^+K^-$	37%	0.55%	5.3k	<1.3
$B_s^0 \to K^+K^-$	31%	0.99%	37k	0.3
$B_s^0 \to D_s^-\pi^+$	31%	0.34%	80k	0.3
$B_s^0 \to D_s^\mp K^\pm$	30%	0.27%	5.4k	<1.0
$B^0 \to J/\psi K_s^{0*}$	61%	1.39%	216k	0.8
$B_s^0 \to J/\psi\phi^\dagger$	64%	1.67%	100k	< 0.3
$B^0 \to K^{*0}\gamma$	38%	0.16%	35k	<0.7
$B_s^0 \to \phi\gamma$	34%	0.22%	9.3k	<2.4

* $J/\psi \to \mu^+\mu^-$
† $J/\psi \to \mu^+\mu^-$, $\phi \to K^+K^-$

natorial background. The selection of B decays exploits the long life-time and the large mass of B mesons. The decay particles should have large transverse momentum and a common vertex which is significantly different from the primary vertex of the interaction. In case that unstable daughter particles are produced, their invariant mass and their momentum are reconstructed first (e.g. D_s^- in the $K^+K^-\pi^-$ final state) and cuts on mass and momentum are applied. The invariant mass of the identified B decay particles is calculated and should agree within ± 50 GeV/c^2 with the mass of the B meson considered. Good K/π separation is necessary to distinguish the different B final states. Details on the selection of the different final states can be found in [2].

The most important source of combinatorial background for any exclusive B decay selection are other $b\bar{b}$ events. Tracks from inclusive $b\bar{b}$ events are displaced from the primary vertex and, after minimum p_T requirement, have a much larger probability to form fake secondary vertices. For many B decay selections, the simulated inclusive B sample of 10M events is yet not large enough to obtain a precise estimate of the background level. In these cases, only upper limits of the background estimate are given. Table 1 summarizes for a set of B decay channels the estimated total efficiency, the estimated trigger efficiency, the expected annual event yield (for 10^7 s) and the B/S ratio. The total efficiency includes the trigger efficiency.

4. FLAVOUR TAGGING

Flavour tagging, i.e., the identification of the initial flavour of reconstructed B^0 and B_s^0 mesons, is necessary

TABLE 2. Efficiencies and mistag rate w of different tagging algorithms.

Tag	ε_{Tag}	w	ε_{eff}
Muon	11%	35%	1.0%
Electron	5%	36%	0.4%
Kaon	17%	31%	2.4%
Q_{Vtx}	24%	40%	1.0%
Fragm. K (B_s^0)	18%	33%	2.1%

to study decays involving *CP* asymmetries and flavour oscillations. The statistical uncertainty on the measured *CP* asymmetry is directly related to the effective tagging efficiency ε_{eff}, also know as "$\varepsilon\mathcal{D}^2$" which is defined as

$$\varepsilon_{\text{eff}} = \varepsilon_{\text{tag}}(1 - 2w)^2,$$

where ε_{tag} is the tagging efficiency and w is the wrong tag fraction.

The opposite-side tagging algorithms determine the flavour of the b hadron accompanying the B meson under study. The algorithms use the charge of the lepton of semi-leptonic B decays and the charge of the kaon from the $b \to c \to s$ decay chain. Also used, is the "vertex-charge" of the B decay products forming the secondary vertex. To reduce the mistag probability cuts on p_T and the primary vertex separation of the tagging particle are applied.

Same side tagging algorithms determine directly the flavour of the signal B meson exploiting correlations in the fragmentation decay chain. The method is used to tag B_s^0 mesons. If a B_s^0 ($\bar{B}_s^0$) is produced in the fragmentation of a $b\bar{b}$ quark pair, an extra $\bar{s}$ (s) quark is available to form a kaon which is charged in about 50% of the cases and emerges from the primary vertex.

The tagging efficiencies and the mistag probability of the different tagging algorithms are listed in Table 2. The combined tagging efficiency for B mesons depends slightly on the final state[3]. For B (B_s^0) the effective tagging efficiency ε_{eff} is determined to be 3.9% (5.9%).

5. PHYSICS SENSITIVITY

The LHCb sensitivity to *CP* observables and other physics parameters have been accessed using "toy" Monte Carlo studies. For these studies event samples for defined values of the measurable physics parameters have been generated. The expected event statistics and the characteristics of the event sample (e.g. signal resolu-

---

[3] Differences result from the trigger: high multiplicity hadronic B decays have a higher chance to be triggered through the semi-leptonic decay of the accompanying (tagging) B.

tion, efficiency, purity, tagging efficiency) are taken from studies with fully-simulated events. Perfect knowledge of the signal and background parameters is assumed. The background-over-signal ratios, B/S, are taken from the full Monte Carlo simulation. In the studies involving time-dependent analysis, the proper-time distribution of the samples are fitted to extract an estimate of the statistical uncertainty. Where possible, the mistag probability is determined as in data, i.e., from an independent control sample. Maximum likelihood fits are used to extract the physics parameters from the generated distributions. The sensitivity of the physics parameters has been estimated using many Monte Carlo "toy" samples.

The time-dependent *CP* asymmetry for a B meson decaying into a *CP* eigenstate $f = \bar{f}$ is given by:

$$\mathcal{A}_f^{CP}(t) = \frac{\Gamma_{\bar{B} \to f}(t) - \Gamma_{B \to f}(t)}{\Gamma_{\bar{B} \to f}(t) + \Gamma_{B \to f}(t)}$$

$$= \frac{\mathcal{A}_f^{\text{dir}}\cos(\Delta mt) + \mathcal{A}_f^{\text{mix}}\sin(\Delta mt)}{\cosh(\Delta\Gamma t/2) - \mathcal{A}_f^{\Delta}\sinh(\Delta\Gamma t/2)} . \quad (1)$$

The quantity $\mathcal{A}_f^{\text{dir}}$ parameterizes direct *CP* violation while $\mathcal{A}_f^{\text{mix}}$ is related to mixing-induced *CP* violation. The relative decay width difference $\Delta\Gamma/\Gamma$ is expected to be 10% for B_s^0 while it can be neglected for B^0 mesons. In the latter case, the *CP* asymmetry reduces to

$$\mathcal{A}_f^{CP}(t) = \mathcal{A}_f^{\text{dir}}\cos(\Delta mt) + \mathcal{A}_f^{\text{mix}}\sin(\Delta mt) .$$

5.1. *CP* Violation in $B^0 \to J/\psi K_S^0$ Decays

For the decay $B^0 \to J/\psi K_S^0$ the Standard Model expectations are $\mathcal{A}^{\text{dir}} = 0$ and $\mathcal{A}^{\text{mix}} = \sin\phi_d = \sin 2\beta$. A binned maximum likelihood fit of the function $\mathcal{A}^{CP}(t)(1 - 2w)$ to the *CP* asymmetry of the "toy" event sample is performed. The mistag rate w is fixed to the value obtained from a control sample of $B^0 \to J/\psi K^{*0}$ events. The statistical uncertainty of w is propagated to the *CP* results. For an annual yield of 216k $B^0 \to J/\psi(\mu\mu)K_S^0$ events the average statistical uncertainty obtained for $\mathcal{A}^{\text{mix}}$ and thus $\sin 2\beta$ is 0.022 [8].

5.2. B_s^0 Mixing Phase ϕ_S and Width Difference $\Delta\Gamma_S$

The channel $B_s^0 \to J/\psi\phi$ is the SU(3) analogue of $B^0 \to J/\psi K_S^0$. It can be used to determine the phase ϕ_s due to B_s^0-$\bar{B}_s^0$ oscillation, which, in the Standard Model, should be small, of the order of 0.04. The observation of a large *CP* asymmetry would be a striking signal for new physics. Compared to $B^0 \to J/\psi K^0_S$, this channel presents the additional challenge that the J/ψ and ϕ are both vector mesons. Therefore, there are three distinct amplitudes contributing to the decay: two CP even and one CP odd. The two CP components can be disentangled on a statistical basis using the transversity angle, θ_{tr}, defined as the angle between the positive lepton and the ϕ decay plane in the J/ψ rest frame. Furthermore, it is expected that the two CP components have a non-negligible relative decay width difference, $\Delta\Gamma_s$. The large oscillation frequency Δm_s requires an excellent proper-time resolution to measure the time-dependent CP asymmetry.

"Toy" Monte Carlo event samples are generated for likely values of the physics parameters R_T (fraction of CP odd decays), $\sin\phi_s$, Δm_s and $\Delta\Gamma_s$. A likelihood fit to the $B^0_s \to J/\psi(\mu\mu)\phi(KK)$ events is used to extract the parameters R_T, $\sin\phi_s$, $\Delta\Gamma_s/\Gamma_s$, and $1/\Gamma_s$ from the generated events. The oscillation frequency Δm_s and the mistag probability w is determined from a fit simultaneously performed on generated $B^0_s \to D^-_s \pi^+$ events. The average error on $\sin\phi_s$ and $\Delta\Gamma_s/\Gamma_s$ for different settings of the physics parameters is calculated from 1000 event samples each corresponding to one year of LHCb datataking. Depending on the physics parameters, one finds for the errors $0.050 < \sigma(\sin\phi_s) < 0.088$ and $0.015 < \sigma(\Delta\Gamma_s/\Gamma_s) < 0.019$ [9].

5.3. Δm_S with $B^0_S \to D^-_S \pi^+$ and γ with $B^0_S \to D^{\mp}_S K^{\pm}$

A single tree diagram contributes to the flavour specific decay $B^0_s \to D^-_s \pi^+$. In this case, the flavour asymmetry can be defined as:

$$\mathscr{A}^{flav} = \frac{\Gamma_{\bar{B}\to f} - \Gamma_{B\to f}}{\Gamma_{\bar{B}\to f} + \Gamma_{B\to f}} = -\mathscr{D}\frac{\cos(\Delta m_s t)}{\cosh(\Delta\Gamma_s t)} ,$$

where $\mathscr{D}$ is a dilution factor due to mistagging and experimental resolution. The measurement of the asymmetry allows thus the determination of the B^0_s oscillation frequency and, optionally, the decay width difference.

The decay $B^0_s \to D^{\mp}_s K^{\pm}$ can proceed through two tree diagrams where the strong phase difference between the two diagrams is denoted by $\Delta_{T1/T2}$ in the following. The interference of the tree diagrams provides a sensitivity to the weak phase $\gamma + \phi_s$. The phase can be accessed through the measurement of the $\bar{B}^0_s$-B^0_s asymmetries of the two time dependent decay rates for $B^0_s \to D^-_s K^+$ and $B^0_s \to D^+_s K^-$.

In this study [10], $B^0_s \to D^-_s \pi^+$ and $B^0_s \to D^{\mp}_s K^{\pm}$ events are generated for different settings of the physics parameters $\gamma + \phi_s$, Δm_s and $\Delta_{T1/T2}$. The parameter $\Delta\Gamma_s/\Gamma_s$ is fixed to 10%. Since the topology of the two

decays are similar, the same mistag rate, $w = 32\%$, is assumed. The full Monte Carlo simulation results in a proper-time resolution of both event samples of $\sigma = 33$ fs (core resolution: 69%).

The expected statistical precision on Δm_s varies with its absolute value: $0.009 \text{ ps}^{-1} < \sigma(\Delta m_s) < 0.016 \text{ ps}^{-1}$ for $15 \text{ ps}^{-1} < \Delta m_s < 30 \text{ ps}^{-1}$ (one year's data). The highest Δm_s value for which oscillation can be observed (5σ significance) within the first year of running is estimated to $\Delta m_s = 68 \text{ ps}^{-1}$. Depending on the physics parameters, the expected error on $(\gamma + \phi_s)$ varies between $12°$ and $18°$ for one LHC year.

5.4. γ with $B^0 \to \pi^+\pi^-$ and $B^0_S \to K^+K^-$

In the Standard Model, the observable CP violation in the decays $B^0 \to \pi^+\pi^-$ and $B^0_s \to K^+K^-$ are both sensitive to the angle γ of the unitarity triangle. The observed time-dependent CP asymmetry of both decays can be parameterized according to Eq. (1) and allows the determination of the direct CP violation $\mathscr{A}^{dir}$ and the CP component $\mathscr{A}^{mix}$ from mixing.

For both decays, there is in addition to the $b \to u$ tree amplitudes also a large contributions from $b \to d(s)$ penguin diagrams. The observable CP asymmetry is therefore not only a function of the weak phase but also exhibits a large strong phase dependence. The hadronic contributions are usually described by parameters d and θ representing the magnitude and phase of the penguin-to-tree amplitude ratio of the decay transition. The CP asymmetry coefficients $\mathscr{A}^{dir}$ and $\mathscr{A}^{mix}$ for the decays $B^0 \to \pi^+\pi^-$ and $B^0_s \to K^+K^-$ are expressed as functions of the angle γ, the mixing phases and the hadronic parameters for the two decay channels [11]. In the limit of exact U-spin asymmetry of the strong interaction, the hadronic parameters d and θ for the two decay channels are the same. Under this assumption, the measurement of the four asymmetry coefficients allows the simultaneous determination of ϕ_d and γ, provided that ϕ_s is determined elsewhere (e.g. from $B^0_s \to J/\psi\phi$) or is negligibly small, as predicted in the Standard Model. Moreover, ϕ_d will be accurately known from measurements of the CP asymmetry of $B^0 \to J/\psi K^0_S$.

The sensitivity on $\mathscr{A}^{dir}$ and $\mathscr{A}^{mix}$ is estimated from "toy" Monte Carlo samples [12]. Signal events are generated for different settings of the physics parameters d, θ, ϕ_s, $\Delta\Gamma_s$ and Δm_s. As both decays, $B^0 \to \pi^+\pi^-$ and $B^0_s \to K^+K^-$, are topologically similar, the same tagging efficiency and mistag probability are used for the generation. The asymmetry coefficients $\mathscr{A}^{dir}$ and $\mathscr{A}^{mix}$ and their uncertainties are determined by maximizing an unbinned extended likelihood. The fit is performed on the signal samples (expected are selected samples of 26k

$B^0 \to \pi^+\pi^-$ and 37k $B_s^0 \to K^+K^-$ events) and simultaneously on samples of $B^0 \to K^+\pi^-$ and $B_s^0 \to \pi^+K^-$ events to determine the mistag probability w. Other parameters of the likelihood are the charge asymmetries $\mathscr{A}_{\pi K}$ and $\mathscr{A}_{K\pi}$, $\Delta\Gamma_s$, Δm_s, Γ_s, the two signal yields, the mean and the resolution of the B meson mass distribution and 6 parameters describing the background.

The probability density functions of the four asymmetry coefficients and of the weak phases ϕ_s and ϕ_d can be propagated into probability density functions for the quantities γ, d and θ. The sensitivity on γ is explored for different values of the physics parameters γ, d, θ, and ϕ_s. For a value of $\Delta m_s = 20\,\mathrm{ps}^{-1}$, $\Delta\Gamma_s/\Gamma_s = 0.1$, $d = 0.3$ and $\theta = 160°$, one obtains a γ sensitivity between $4°$ and $6°$ for γ between $55°$ and $105°$. Once more, it should be mentioned that this method assumes U-spin symmetry.

5.5. γ with $B \to \bar{D}^0 K^{*0}$, $D^0 K^{*0}$

The method relies on the measurement of six time-integrated decay rates [13, 14]:

$$\begin{aligned}
\Gamma_+ &= \Gamma(B \to D^0 K^{*0}) & \bar{\Gamma}_+ &= \Gamma(\bar{B} \to D^0 \bar{K}^{*0}) \\
\Gamma_- &= \Gamma(B \to \bar{D}^0 K^{*0}) & \bar{\Gamma}_- &= \Gamma(\bar{B} \to \bar{D}^0 \bar{K}^{*0}) \\
\Gamma_{CP} &= \Gamma(B \to D_{CP}^0 K^{*0}) & \bar{\Gamma}_{CP} &= \Gamma(\bar{B} \to D_{CP}^0 \bar{K}^{*0}).
\end{aligned}$$

As the decay amplitude of $B \to D_{CP}^0 K^{*0}$ is a linear composition of the amplitudes of $B \to D^0 K^{*0}$ and $B \to \bar{D}^0 K^{*0}$, the amplitudes fulfill a triangle relation and one obtains for the rates the relations,

$$\Gamma_+ = \bar{\Gamma}_- \equiv g_1 \,, \quad \Gamma_- = \bar{\Gamma}_+ \equiv g_2 \,,$$

$$\Gamma_{CP} = \frac{g_1 + g_2}{2} + \sqrt{g_1 g_2}\cos(\Delta + \gamma) \,, \qquad (2)$$

$$\bar{\Gamma}_{CP} = \frac{g_1 + g_2}{2} + \sqrt{g_1 g_2}\cos(\Delta - \gamma) \,,$$

which allow the determination of the CKM angle γ as well as the strong phase difference Δ between the two tree diagrams of the process.

As shown in Table 1, 3400 $B \to \bar{D}^0 K^{*0}$ (+ c.c.) events are expected after one year of data-taking. The branching ratio of $B \to D^0 K^{*0}$ is suppressed compared to that for $B \to \bar{D}^0 K^{*0}$. With the same total efficiency an annual yield of 490 $B \to D^0 K^{*0}$ (+ c.c.) events are expected [2]. Using the efficiencies given in Table 1, the expected number of reconstructed $B \to D_{CP}^0 K^{*0}$ (+ c.c.) events can be calculated according to Eq. (3). Assuming $\gamma = 65°$ and $\Delta = 0°$, an annual event sample of 590 $B \to D_{CP}^0 K^{*0}$ (+ c.c.) events is expected.

Using "toy" Monte Carlo event samples, the γ sensitivity is estimated for different values of γ and Δ [15]. Within the range of γ considered ($55° < \gamma < 105°$), the expected γ precision for one LHC year varies between $7°$ and $9°$ and remains unchanged for $-20° < \Delta < 20°$.

6. CONCLUSION

The reoptimized LHCb experiment can trigger and reconstruct many different b hadron final states, in particular B_s^0 states, with high statistics. The detector provides an excellent decay time resolution necessary for CP violation studies in the B_s^0 system. Excellent particle identification and mass resolution allows the efficient suppression of background.

LHCb will offer an excellent opportunity to determine precisely the CKM parameters. The manifold of different CP measurements will over-constrain the unitarity triangles and will spot new physics, if present. This has been illustrated by the measurements of the angle γ in three different ways. If there is no new physics affecting the B meson system via loop processes, all three methods should yield measurements consistent with $\gamma \approx 65°$. If present, new physics could contribute to the oscillation diagrams and/or the penguin diagrams. In this case, the measurements will give different γ values.

REFERENCES

1. Amato, S., et al., [LHCb Collaboration], Technical Proposal, Tech. Rep. CERN-LHCC/98-04 (1998).
2. Antunes Nobrega, R., et al., [LHCb Collaboration], LHCb Reoptimized Detector - Design and Performance, Tech. Rep. CERN-LHCC/2003-030 (2003).
3. Aubert, B., et al., [BaBar Collaboration], *Phys. Rev. Lett.* **87**, 091801 (2001).
4. Abe, K., et al., [Belle Collaboration], *Phys. Rev. Lett.* **87**, 091802 (2001).
5. Antunes Nobrega, R., et al., [LHCb Collaboration], Trigger Technical Design Report, Tech. Rep. CERN-LHCC/2003-031 (2003).
6. Sjostrand, T., et al., *Comput. Phys. Commun.* **135**, 238–259 (2001).
7. Bartalini P., et al., Tech. Rep. LHCb/1999-23 (1999), Contribution to the 1999 Workshop on Standard Model Physics at the LHC, CERN 2000-4, 2000.
8. Amato, S., et al., [LHCb Collaboration], The LHCb Sensitivity to $\sin 2\beta$ from $B^0 \to J/\psi(\mu^+\mu^-)K_s^0$, Tech. Rep. LHCb/2003-119 (2003).
9. Raven G., Sensitivity Studies of χ and $\Delta\Gamma$ with $B_s^0 \to J/\psi(\mu\mu)\phi(KK)$, Tech. Rep. LHCb/2003-119 (2003).
10. Hierck R., van Hunen J., and Merk M., The Sensitivity for Δm_s and $\gamma + \phi_s$ from $B_s^0 \to D_s^-\pi^+$ and $B_s^0 \to D_s^\mp K^\pm$ Decays, Tech. Rep. LHCb/2003-103 (2003).
11. Fleischer, R., *Phys. Lett.* **B459**, 306–320 (1999).
12. Vagnoni V., et al., *CP* Sensitivity with $B_{(s)}^0 \to h^+h^-$ Decays at LHCb, Tech. Rep. LHCb/2003-124 (2003).
13. Gronau, M., and Wyler, D., *Phys. Lett.* **B265**, 172–176 (1991).
14. Dunietz, I., *Phys. Lett.* **B270**, 75–80 (1991).
15. Akiba K., and Gandelmann M., γ Sensitivity with $B^0 \to D^0 K^{*0}$, Tech. Rep. LHCb/2003-105 (2003).

Physics Reach of the BTeV Experiment

Brad Cox

(Representing the BTeV Collaboration)

University of Virginia, Department of Physics, Charlottesville, VA 22901, U.S.A.

Abstract. The BTeV experiment represents "potentially the best quark flavor experiment into the next decade" according to a recent declaration of the Particle Physics Project Prioritization Panel (P5). The capabilities of this experiment will be discussed in this paper for measurements in the b and c quark systems of CP violation effects, the phases of the CKM triangle, rare decays, and, finally, the detection of new physics outside the Standard Model such as SUSY and/or the presence of extra dimensions.

BTeV Detector Layout

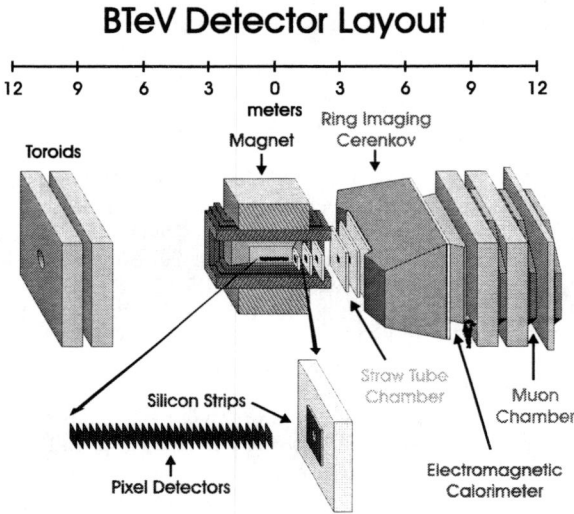

FIGURE 1. The BTeV Spectrometer.

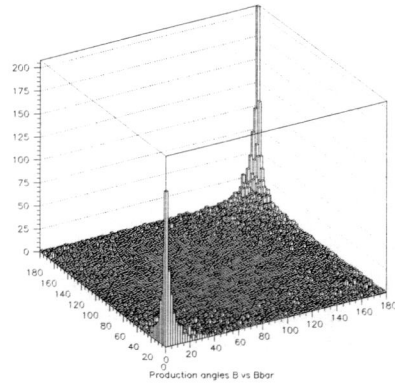

FIGURE 2. Forward Production of b and $\bar{b}$ quarks.

1. INTRODUCTION

The BTeV experiment is the culmination of the efforts of many physicists toward the development of a dedicated experiment to study b and c quark production and decays in the forward direction in a hadron collider environment. This effort started in Snowmass 86 [1] where the idea of a forward hadron collider spectrometer specifically designed for study of beauty physics was first advanced. The BTeV experiment [2] and spectrometer shown in Fig. 1 is the result of the evolution of this effort over a decade and a half.

As first recognized in Ref. [1], the main motivation for a forward collider B experiment such as BTeV is the forward nature of B production shown in Fig. 2.

Not only is B production highly peaked in the beam directions and the b and $\bar{b}$ directions highly correlated, but the higher momentum of forward produced b quarks leads to lower multiple scattering and longer decay paths (see Fig. 3) relative to centrally produced B hadrons. This results in a significantly better ability of a forward detector to separate B decays from the rest of the interaction productions.

2. PHYSICS OBJECTIVES

The major objectives of the BTeV experiment are comprehensive studies of b and c quark production, weak decays and mixing with especial attention given to precision measurement of CP violation in the B system. BTeV will measure the α, β and γ and χ phases of the CKM triangle shown in Fig. 4 and, therefore, ρ and η to determine the CP violating phase of the CKM matrix. Using

CP722, *B Physics at Hadron Machines,* edited by M. Paulini and S. Erhan

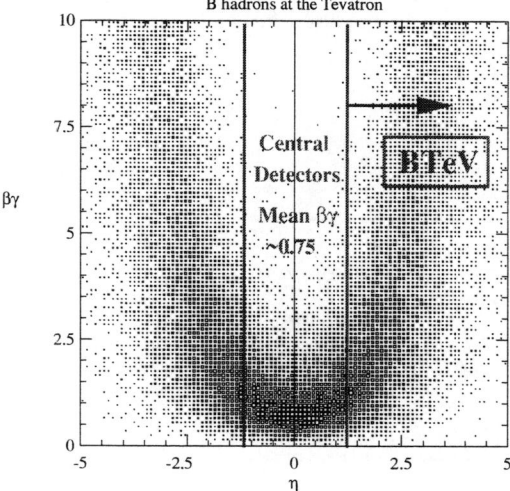

FIGURE 3. Correlation of b production pseudorapidity with $\beta\gamma$ (and therefore B decay length).

this ensemble of measurements, searches for effects of new physics will be performed.

To accomplish this, the BTeV experiment will make measurements of several B decays, some of which are given in Table 1. By these measurements, BTeV will determine the combinations of CKM matrix elements shown below which give the α, β, γ and χ phases. Depending on the magnitude of χ, it may be possible to measure it also.

$$\beta = arg(-\frac{V_{tb}V_{td}^*}{V_{cb}V_{cd}^*}) \qquad \gamma = arg(-\frac{V_{ub}^*V_{ud}}{V_{cb}^*V_{cd}})$$

$$\chi = arg(-\frac{V_{cb}V_{cs}^*}{V_{tb}V_{ts}^*}) \qquad \chi' = arg(-\frac{V_{ud}^*V_{us}}{V_{cd}^*V_{cs}})$$

α can either be calculated from β and γ assuming that $\alpha = \pi - \beta - \gamma$ or from the difficult measurement of $B^0 \to \rho\pi$ as indicated in Table 1. Note that the necessary B_s^0 measurements to determine χ and γ cannot be performed in any venue other than a hadron collider experiment.

TABLE 1. Primary B decay modes for BTeV determination of CKM angles.

Decay	CKM Triangle Angle
$B^0 \to J/\psi K_S^0$	$\sin(2\beta)$
$B^0 \to \rho\pi \to \pi^+\pi^-\pi^0$	$\sin(2\alpha)$
$B_s^0 \to D^\pm K^\mp$	$\sin(\gamma)$
$B^- \to D^0 K^-$ (and c.c.)	$\sin(\gamma)$
$B_s^0 \to J/\psi\eta'$	$\sin(2\chi)$

TABLE 2. BTeV Trigger Efficiencies.

Decay	Efficiency [%]
$B^0 \to \pi^+\pi^-$	63
$B_s^0 \to D_s K$	74
$B^- \to D^0 K^-$	70
$B^- \to K_S^0 \pi^-$	27
$B^0 \to K^+\pi^-$	63
$B^0 \to J/\psi K_S^0$	50
$B_s^0 \to J/\psi K^*$	68
$B^0 \to K^*\gamma$	40

3. *B* EVENT YIELDS

The Tevatron operating at 2 TeV and a luminosity of $10^{32}\text{cm}^{-2}\text{s}^{-1}$ produces at least 10^{11} $b\bar{b}$ pairs per 10^7 s of running. Since the BTeV trigger is based on the observation of secondary vertex tracks well separated from the primary vertex in the BTeV pixel silicon vertex detector and depends only on the lifetime of the B hadrons, the BTeV experiment is, therefore, quite catholic in its collection of various types of B decays. This trigger gives access to B^0, B^+, B_s^0, B_c and b baryon states, some of which are inaccessible to any other experiment. The trigger, which operates at the lowest level at the basic 2.5 MHz crossing rate, must be able to separate the events with B decays with good efficiency from the minimum bias backgrounds which have a cross section approximately 500 times as large as the B cross section. Moreover, at design luminosity, because of the present Tevatron 396 ns bunch separation, there are are several interactions per crossing accompanying the B event that must be dealt with. Table 2 gives the efficiencies of the trigger for various B decay modes for a trigger requirement of at least two tracks detached by more than 6 σ from the primary vertex under the condition of two minimum bias events plus a B event per crossing.

The other factor that must be considered in calculating event yield is the tagging efficiency. The BTeV experiment will use several methods for tagging the particle-antiparticle nature of the decaying B including opposite-side lepton and $K^\pm$ charge from the "other B" decay, as well as same-side jet charge and/or π^+ or $K^\pm$ charge from same-side B^0 or B_s^0 decays, respectively. The combination of these techniques yields an $\varepsilon\mathcal{D}^2$ flavor tagging efficiency of 10% for B^0 decays and 13% for B_s^0 decays (for two minimum bias interactions plus a B decay event per beam crossing). The difference in the two efficiencies is due to same-side tagging.

Given that the present design of the Tevatron collider has a crossing every 396 ns, the number of minimum bias interactions per crossing accompanying the B event will exceed 6 at a luminosity of 2×10^{32} cm^{-2}s^{-1}. At present, studies are underway with 4 and 6 interactions

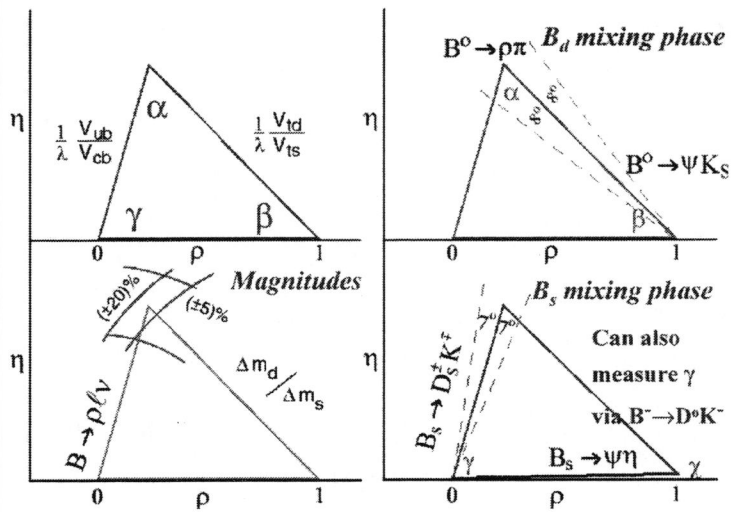

FIGURE 4. Various methods of extraction of the CKM triangle angles.

per crossing accompanying the B event. These indicate only a modest loss ($\approx 10\%$) of trigger and flavor tagging efficiencies relative to the benchmark of two minimum bias interactions per crossing. The yields of various B decays and the attendant accuracy with which CKM triangle phases and other B decay parameters can be determined are shown in Table 3.

In addition to the many B modes that are accessible to BTeV, the experiment will collect a copious number of charm decays because of its lifetime based trigger. As an example, the expectation is that greater than 10^8 $D^0 \to K\pi$ decays will be accumulated. Various physics topics such as D^0-$\bar{D}^0$ mixing can be searched for at a level much better than the present limit of $\Delta M_D \leq 10^{-1}$. It also may be possible to look for CP violation at the level of 10^{-3}.

4. NEW PHYSICS IN BTEV

As well as the precision measurement of CP violation in the b and c sectors, the unprecedented sensitivity of BTeV will allow searches in B decays for such new physics effects as supersymmetry and extra dimensions that may contribute to the various loop diagrams contributing to decays. These effects will show up in deviations from Standard Model expectations.

4.1. Supersymmetry Effects

As an example, the precision measurements of the CKM triangle angles possible in BTeV will allow a check of the Standard Model prediction of Ref. [3] that the χ, β,

and γ CKM angles should be related by

$$sin(\chi) = \lambda^2 \frac{sin(\beta)sin(\gamma)}{sin(\beta + \gamma)}$$

where $\lambda = 0.2205 \pm 0.0018$ is the Weinberg angle. As indicated in Table 3, BTeV expects to measure $sin(2\chi)$ with a error of ≈ 0.024 using $B_s^0 \to J/\psi\eta'$. Since χ is expected to be small (of order 0.03), a significant amount of $B_s^0 \to J/\psi\eta'$ data will needed to test this equality precisely.

In addition, SUSY may affect the rate of mixing and the level of direct CP violation in certain B decays. Hinchliffe and Kersting [4] calculate that the rate of mixing in $B_s^0 \to J/\psi\eta$ decay may be as much as ten times the level predicted in the Standard Model due to minimal supersymmetry contributions. They also state that the level of direct CP violation in this decay may be enhanced by minimal supersymmetry contributions whereas direct CP in $B_s^0 \to J/\psi\eta$ is almost zero in the Standard Model.

Other examples of SUSY effects in B decays are Dalitz plot interference effects [5] and polarization effects [6] in $B \to K^*\mu^+\mu^-$ decays. Curves are shown in Fig. 5 for forward-backward asymmetries of the leptons as a function of the square of the dilepton mass according to the Standard Model and according to various SUSY models. A typical error bar that can be achieved by BTeV in 10^7 seconds of operation is also shown. The expected BTeV yield of $B \to (K, K^*)\mu^+\mu^-$ is given in Table 4.

Finally, less specific minimal SUSY predictions such as those of Nir [7] predict that there will be new phases in decays such as $B^0 \to J/\psi K_S^0$, $B^0 \to \phi K_S^0$, and $D^0 \to K^-\pi^+$ in addition to the Standard Model phase β.

TABLE 3. BTeV Event yields and parameter errors in 10^7s of running.

Decay	$\mathcal{B}(\times 10^{-6})$	# of Events	S/B	Parameter	Error/Value
$B^0 \to \pi^+\pi^-$	4.5	14,600	3	Asymmetry	0.030
$B^0 \to K^+K^-$	17	18,900	6.6	Asymmetry	0.020
$B_s^0 \to D_s K^-$	300	7,500	7	$\gamma - 2\chi$	8°
$B^0 \to J/\psi K_S^0; K_S^0 \to \ell^+\ell^-$	445	168,000	10	$\sin(2\beta)$	0.017
$B^0 \to J/\psi K^0; K^0 \to \pi\ell\nu$	7	250	2.3	$\cos(2\beta)$	≈ 0.5
$B_s^0 \to D_s \pi^-$	3000	59,000	3	x_s	(75)
$B^- \to D^0(K^+\pi^-)K^-$	0.17	170	1		
$B^- \to D^0(K^+K^-)K^-$	1.1	1,000	≥ 10	γ	13°
$B^- \to K_S^0 \pi^-$	12.1	4,600	1		$\leq 4^o$ +
$B^0 \to K^+\pi^-$	18.8	62,100	20	γ	theory errors
$B^0 \to \rho^+\pi^-$	28	5,400	4.1		
$B^0 \to \rho^0\pi^0$	5	780	0.3	α	$\approx 4^o$
$B_s^0 \to J/\psi\eta$	330	2,800	15		
$B_s^0 \to J/\psi\eta'$	670	9,800	30	$\sin(2\chi)$	0.024

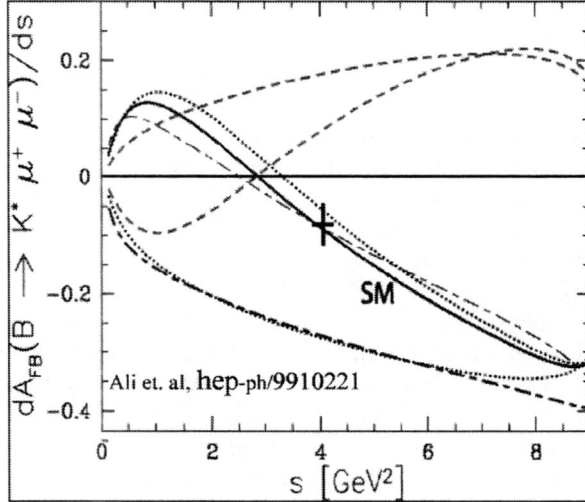

FIGURE 5. Standard Model and SUSY predictions for forward-backward muon asymmetries in $B \to K^*\mu^+\mu^-$ decays; also shown is a typical error bar for a 10^7s BTeV run.

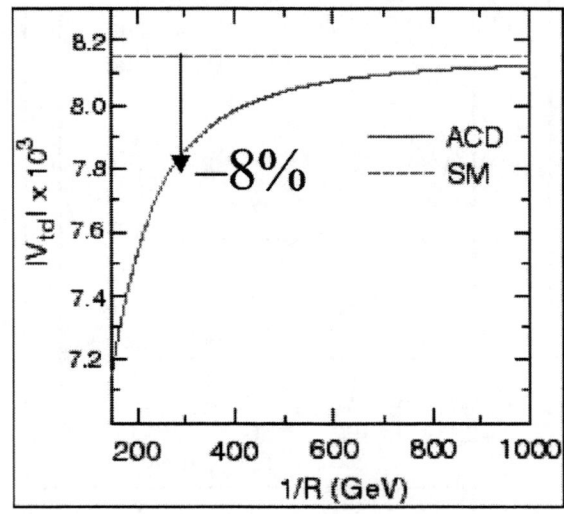

FIGURE 6. Effect on V_{td} of one extra dimension as a function of compactification scale.

TABLE 4. BTeV rare $B \to (K, K^*)\mu^+\mu^-$ yields 10^7s.

Decay	$\mathcal{B}(\times 10^{-6})$	# of Events	S/B
$B^0 \to K^*\mu^+\mu^-$	1.5	2,530	11
$B_s^0 \to K^-\mu^+\mu^-$	0.4	1,470	3.2
$b \to s\mu^+\mu^-$	5.7	4,140	0.13

4.2. Extra Dimensions

There have been many studies which try to estimate the effect of extra dimensions in B decays. The magnitude of these effects depend on the compactification scale of the extra dimensions. As an example of the effects that might arise from the presence of one extra dimension, Buras and collaborators [8] have calculated the deviation of the values of V_{td}, γ and $\mathcal{B}(B_s^0 \to \mu\mu)$ from the Standard Model expectations as a function of compactification scale $\frac{1}{R}$ using the model of Appelquist, Cheng and Dobrescu [9]. The forms of these deviations are shown in Figs. 6, 7 and 8. The magnitudes of the deviations, as noted in the figures, are -8%, -10%, and $+72\%$, respectively, for V_{td}, γ and $\mathcal{B}(B_s^0 \to \mu\mu)$ at a compactification scale of 250 GeV.

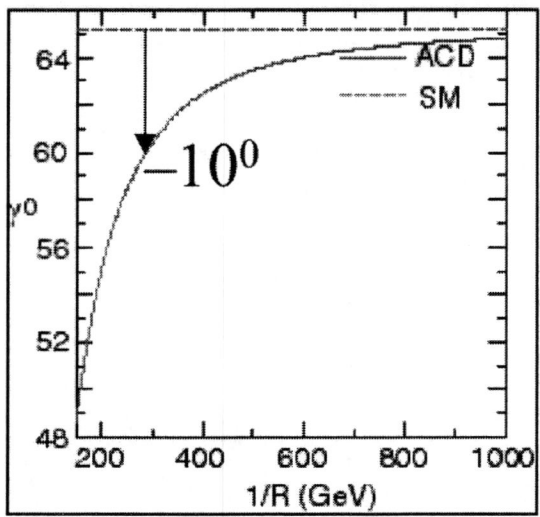

FIGURE 7. Effect on γ of one extra dimension as a function of compactification scale.

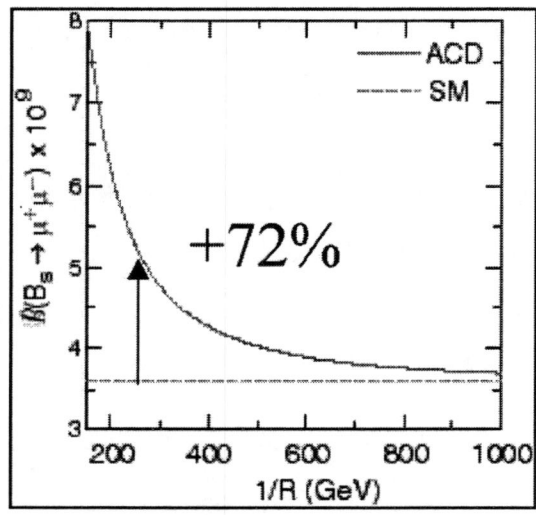

FIGURE 8. Effect on $\mathscr{B}(B_s^0 \to \mu\mu)$ of one extra dimension as a function of compactification scale.

5. COMPARISON WITH OTHER *B* EXPERIMENTS

Finally, comparisons of the sensitivity of BTeV to present e^+e^- *B* factories and the LHCb hadron collider *B* experiment at the CERN LHC can be made. Table 5 below gives relative yields, both untagged and tagged, for several important modes for both BTeV and e^+e^- *B* factories. Note also that the e^+e^- *B* factories which operate mostly at the $\Upsilon(4S)$ cannot do B_s^0, B_c, or Λ_b.

The most analogous experiment to BTeV is the LHCb experiment at the LHC. LHCb has an undeniable advantage due to a factor between 2.5 and 5 in $b\bar{b}$ cross sec-

tion at LHC CM energy of 14 TeV relative to the Tevatron CM energy of 2 TeV. The LHCb experiment will also have fewer interactions per crossing to deal with than BTeV at the same average luminosity albeit with a much closer bunch spacing (25 ns for LHCb vs. 396 ns for BTeV). However, BTeV at 2 TeV will have a lower total cross section by a factor of 1.6 than will LHCb at 14 TeV. But the basic advantages that BTeV enjoys relative to LHCb are due to the design of the experiment. The detached vertex trigger of BTeV allows BTeV to collect five times as many *B*'s per unit time as LHCb's specialized triggers with a catholic acceptance of all types of *B* decays. A second important advantage that BTeV has compared to LHCb is a much better electromagnetic calorimeter which will be crucial in measurements of modes such as $B^0 \to \rho\pi$. Finally, the fact that BTeV's vertex detector is in the magnetic field allows rejection of low energy multiple scattered tracks that otherwise might confound any trigger or subsequent event reconstruction. Table 6 below compares the resulting yields of BTeV and LHCb for particular important modes.

6. SUMMARY

By independent evaluation, the P5 committee and HEPAP have determined that the BTeV experiment has the potential to be the ultimate hadronic flavor experiment. The experiment not only will make perhaps the most precise measurements of the phases of the quark CKM matrix ever but also has the potential to detect new physics such as SUSY or extra dimensions through observations of deviations from Standard Model expectations for *B* decay rates, decay product polarizations, or *B* Dalitz plot interference effects in several modes. As Ref. [10] has noted, the importance of these BTeV measurements is not only that "indirect" searches for new physics effects in *B* decays may be first in detecting new physics but that the measurements of the effects of the new physics on *B* decay phenomena as *CP* violation or FCNC will be essential in understanding the new physics.

ACKNOWLEDGMENTS

We gratefully acknowledge the Fermi National Accelerator Laboratory, the U.S. Department of Energy, the National Science Foundation, and the Istituto Nazionale di Fisica Nucleare of Italy for their support.

TABLE 5. Comparison of BTeV Event yields (10^7s) to present B factories (500 fb^{-1}).

Decay	BTeV (10^7s)			e^+e^- B Factory (500 fb^{-1})		
	Yield	Tagged Events	S/B	Yield	Tagged Events	S/B
$B^0 \to \pi^+\pi^-$	14,175	1,420	-	223	57	-
$B^- \to D^0 K^-$	176	176	-	5	5	-
$B_s^0 \to J/\psi\eta'$	12,650	1,645	≥ 15	-	-	-
$B^- \to \phi K^-$	11,000	11,000	≥ 10	700	700	4
$B^0 \to \phi K_S^0$	2,000	200	5.2	250	75	4
$B^0 \to K^*\mu^+\mu^-$	2530	2530	11	≈ 50	≈ 50	3
$B_s^0 \to \mu^+\mu^-$	6	0.7	≥ 15	0	0	-
$B^0 \to \mu^+\mu^-$	1	0.1	≥ 10	0	0	-
$D^{*+} \to \pi^+ D^0, D^0 \to K\pi$	$\approx 10^8$	$\approx 10^8$	large	8×10^5	8×10^5	large

TABLE 6. Comparison of BTeV Event yields to LHCb (10^7s).

Decay	$\mathcal{B}$	BTeV (10^7s)		LHCb (10^7s)	
		Yield	S/B	Yield	S/B
$B_s^0 \to J/\psi\eta'$	1.0×10^{-4}	12,650	≥ 15	-	-
$B^0 \to \rho^+\pi^-$	2.8×10^{-5}	5,400	4.1	2,140	0.8
$B^0 \to \rho^0\pi^0$	0.5×10^{-5}	776	0.3	880	not known

REFERENCES

1. Cox, B., Gilman, F., and Gottschalk, T., "Heavy Flavors," in *Physics of the Superconducting Supercollider*, edited by R. Donaldson and J. Marx, American Physical Society - Division of Particles and Fields, Washington, DC, 1986, pp. 33–44.
2. BTeV Collaboration, BTeV Experiment Proposal and TDR, Tech. Rep. BTeV-doc-316 (2003), URL: http://www-btev.fnal.gov/public/hep/general/proposal/index.sh%tml.
3. Silva, J. P., and Wolfenstein, L., *Phys. Rev.* **D55**, 5331–5333 (1997).
4. Hinchliffe, I., and Kersting, N., *Phys. Rev.* **D63**, 015003 (2001).
5. Greub, C., Ioannisian, A., and Wyler, D., *Phys. Lett.* **B346**, 149–158 (1995).
6. Ali, A., Ball, P., Handoko, L. T., and Hiller, G., *Phys. Rev.* **D61**, 074024 (2000).
7. Nir, Y., arXiv:hep-ph/9911321 (1999).
8. Buras, A. J., Spranger, M., and Weiler, A., *Nucl. Phys.* **B660**, 225–268 (2003).
9. Appelquist, T., Cheng, H.-C., and Dobrescu, B. A., *Phys. Rev.* **D64**, 035002 (2001).
10. Masiero, A., and Vives, O., *Ann. Rev. Nucl. Part. Sci.* **51**, 161–187 (2001).

ATLAS *B* Physics Performance Update

Paula Eerola

(Representing the ATLAS Collaboration)

Lund University, Experimental High Energy Physics, SE-22100 Lund, Sweden

Abstract. An update of the *B* physics performance of the ATLAS experiment is presented. After a short review of *B* production and decays at LHC, the main features of the ATLAS detector are described, and the present construction status is shown. Physics topics which are presented are: precision measurements, rare decays, and *B* production. It is shown that despite of recent changes in the trigger strategy and initial detector layout, ATLAS retains an excellent potential for measurements of $\sin 2\beta$ and rare dimuon decays. A large spectrum of *B* physics studies is also feasible - here we present examples of precision measurements with B_s^0 mesons, B_c mesons, and *B* production.

1. INTRODUCTION

1.1. *B* Production and Decays at the LHC

The Large Hadron Collider, LHC, is a proton-proton collider designed to operate at a center-of-mass energy of 14 TeV and at high luminosity, starting in the second half of 2007. The high collision energy and luminosity of LHC gives the possibility of exploring a new high-energy frontier at the TeV scale. The design luminosity of the LHC is $\mathscr{L} = 1 \times 10^{34}\,\mathrm{cm^{-2}s^{-1}}$. The initial luminosity, however, will be $\mathscr{L} = 2 \times 10^{33}\,\mathrm{cm^{-2}s^{-1}}$, and it is this initial running period which is most feasible for *B* physics studies in view of trigger rates, pile-up, event reconstruction, *etc.* Some parts of the *B* physics program can be continued even at the design luminosity, such as rare *B* decays to dimuons.

The LHC allows for high-precision measurements of *B* hadrons, in particular of the B_s^0, B_c, and *b* baryons. LHC is thus complementary to the present asymmetric e^+e^- *B* factories. The statistics at LHC will be much larger than at the Tevatron, which gives the possibility for testing the origin of *CP* violation by over-constraining the Unitarity Triangle and searching for rare decays mediated by loop diagrams in the Standard Model. This gives the possibility to observe deviations from the Standard Model and to disentangle any new physics components from the Standard Model ones.

For the proton-proton operation of LHC, there will be two general-purpose experiments, ATLAS and CMS, and an experiment designed for *B* physics, LHCb, operating at a reduced luminosity $\mathscr{L} = 2 \times 10^{32}\,\mathrm{cm^{-2}s^{-1}}$. The general-purpose experiments are, however, in many ways complementary to LHCb for *B* physics studies.

ATLAS/CMS cover the central rapidity region while LHCb is a forward spectrometer. ATLAS/CMS triggers are based on lepton signatures while the LHCb trigger is a mixture of lepton and hadron triggers. Hadron triggers make it possible for LHCb to collect a large sample of purely hadronic decays, but the advantage of the lepton triggers is a clean sample and the possibility to continue data-taking even at the highest LHC luminosity.

The total cross-section at the LHC is about 100 mb, and the total $b\bar{b}$ cross-section is about 500 μb. The $b\bar{b}$ cross-section is about 100 μb within the ATLAS (CMS) acceptance requiring that at least one of the *b* quarks has $|\eta(b)| < 2.5$ and $p_T(b) > 10$ GeV/c[1]. The sample sizes for a year of operation are comparable in ATLAS (CMS) and LHCb, the difference being that the data samples of the general-purpose detectors are dominated by dimuon events, while about half of the LHCb event sample is collected with the hadronic trigger.

1.2. The ATLAS Detector

1.2.1. Overview

The ATLAS experiment has been designed to maximize the discovery potential for new physics and to have the capability of high-accuracy measurements. ATLAS accommodates, nevertheless, features which make it possible to incorporate an ambitious *B* physics program in particular in the first years of running during the lower

---

[1] η is pseudorapidity, and p_T is transverse momentum with respect to the beam direction.

CP722, *B Physics at Hadron Machines*, edited by M. Paulini and S. Erhan
© 2004 American Institute of Physics 0-7354-0203-5/04/\$22.00

luminosity operation. For a full review of the ATLAS detector and physics performance, see Refs. [1, 2].

ATLAS has a tracking system covering the pseudo-rapidity range $|\eta| < 2.5$. The tracking system comprises pixel and silicon strip detector layers, and a transition radiation tracker (TRT). The pixel layer closest to the beampipe provides high-accuracy measurements of the primary and secondary vertices. Electrons can be identified for transverse momenta down to 0.5 GeV/c by using the e/π separation provided by the TRT. For more details on the tracking system, see Ref. [3]. The LAr electromagnetic calorimeters can be used for electron and photon identification and measurements starting from about $E_T = 2$ GeV. High-precision muon momentum measurements are achieved by combining information from the muon spectrometer and the inner tracking system. The muon system is embedded in an air-core toroid magnet, while the inner detector is surrounded by a smaller central solenoid, providing a 2 T magnetic field.

ATLAS has a flexible and efficient multi-level trigger system. The system consists of three levels, reducing the trigger rates from 40 MHz to $\mathcal{O}(200)$ Hz. The multi-level scheme allows for selecting events with multiple trigger objects with low p_T or E_T-thresholds. This is particularly crucial for the B physics program, where a significant reduction of the event rate has to be achieved on-line, due to the large total cross-section and high luminosity.

1.2.2. Construction Status

All subdetector parts of ATLAS are under construction. Installation activities in LHC Point 1 have started: parts of the underground cavern (UX15) were delivered to ATLAS in April 2003, and the cavern hand-over to ATLAS took place in May 2003. The barrel calorimeter system is the first one to be installed in the cavern. The installation will start in March 2004, and the foreseen mechanical completion date is November 2004. The installation schedule thereafter is: barrel toroid 06/2004 - 03/2005, end-cap calorimeters and barrel muon detectors 03/2005 - 04/2006, end-cap large muon detectors 09/2005 - 03/2006, inner detector 10/2005 - 06/2006, end-cap toroids 01/2006 - 07/2006, and end-cap small muon detectors 04-05/2006. The final work in the cavern will take place June-August 2006, and the global commissioning of ATLAS is foreseen to take 60 days from August to October 2006. 30 days are reserved for cosmic tests between November and December 2006 [2]. The construction status in the cavern in December 2003 is shown in Fig. 1.

[2] The schedule is obtained from M. Kotamäki, ATLAS Schedule and Milestones Manager, private communication.

FIGURE 1. Construction status in the ATLAS cavern on December 9, 2003.

1.2.3. Initial Detector and Recent Design Changes

Some parts of ATLAS subdetectors are foreseen to be installed at a later stage: the middle pixel barrel layer, the second layer of end-cap pixel disks, and the TRT C-wheels (the most outward wheels at $|z| = 2818$-3363 mm). Compared to the full detector, the effect on the proper-time resolution of B decays is small, since the closest layer will have the largest significance for the proper-time resolution. Furthermore, the fact that there are fewer measurement points is compensated by the reduced amount of material. The mass resolutions are not affected by the staging. It should be pointed out, however, that there are other factors which have deteriorated the proper-time resolution compared to what was presented in the ATLAS Physics Technical Design Report in 1999 [1, 2]. The closest pixel layer is now located at an average radius of 5.5 cm instead of 4.3 cm, the material in the beam pipe has increased, and the pixel length in z will be 40 μm instead of 300 μm. We are in the process of evaluating the effects of the worse proper-time resolution and re-optimizing the cuts for the Δm_s measurement.

The B physics trigger strategy has recently been revised due to changes in the planning for the LHC initial luminosity, staging of some subdetector parts, and tight funding constraints. The trigger scenario which we assumed in previous studies was based on a single muon trigger at Level-1, with $p_T > 6$ GeV/c and $|\eta| < 2.4$, followed by muon confirmation at the Level-2 and a track search over the whole inner detector.

The new trigger scenario is based on multi-object triggers. The Level-1 B physics trigger has to consist of at least two muons (dimuon trigger), a muon and a low-E_T jet (muon-jet trigger), or a muon and a low-E_T electromagnetic cluster (muon-electron trigger). The trigger is foreseen to be flexible so that the baseline is the dimuon trigger at the nominal initial luminosity, and further triggers are added and/or thresholds are lowered later in the beam-coast or for lower luminosity fills. The new Level-2 B physics trigger strategy is based on a track search within a Region of Interest given by the Level-1 calorimeter trigger object (electron or jet), instead of the full scan of the inner detector. Results are promising so far, and a strong reduction in processing requirements compared to the previous scenario can be achieved at the expense of some loss in efficiency. For more details see Ref. [4].

2. PRECISION MEASUREMENTS

2.1. Angles of the Unitarity Triangle

When neutral B and $\bar{B}$ mesons decay into the same final CP eigenstate, CP violation can in general have contributions both from direct CP violation and from interference between direct and mixed decays. In this case, the time-dependent CP asymmetry can be expressed as follows:

$$
\begin{aligned}
\mathscr{A}_{CP}(t) &= \frac{\Gamma(B_q^0(t) \to f) - \Gamma(\bar{B}_q^0(t) \to f)}{\Gamma(B_q^0(t) \to f) + \Gamma(\bar{B}_q^0(t) \to f)} \quad (1) \\
&= A_{CP}^{dir}(B_q^0 \to f)\cos(\Delta m_q t) \\
&+ A_{CP}^{int}(B_q^0 \to f)\sin(\Delta m_q t),
\end{aligned}
$$

where A_{CP}^{dir} is the direct CP violation amplitude and A_{CP}^{int} is the mixing-induced CP violation amplitude. Subscript q indicates a d- or an s-quark, and Δm_q is the mass difference of the B_q^0 eigenstates. Here it is assumed that there is no width difference of the B_q^0 eigenstates, which is valid for B^0 mesons. For B_s^0 however, the width difference can be sizable, 15% or so, and the formula has to be modified accordingly.

If the decay of a neutral B meson is dominated by a single CKM amplitude, the mixing-induced CP violation amplitude is $A_{CP}^{int} = \Lambda \sin(\phi_M - \phi_D)$. The mixing phase ϕ_M is $+2\beta$ in case of B^0 mesons and $-2\delta\gamma$ for B_s^0 mesons, and Λ is the CP eigenvalue of the final state. The CP violating weak decay phase ϕ_D is zero for dominant $\bar{b} \to \bar{c}c\bar{s}(\bar{d})$ amplitudes, and -2γ for dominant $\bar{b} \to \bar{u}u\bar{d}(\bar{s})$ amplitudes.

The angle β can be measured cleanly using the decays $B^0 \to J/\psi K_S^0$. This decay is dominated by the decay amplitude $\bar{b} \to \bar{c}c\bar{s}$, since the penguins are expected to be small and moreover, they have the same weak phase as the tree-level decays. Therefore, the direct CP viola-

tion contribution is very small and can be neglected, *i.e.* $A_{CP}^{dir} \simeq 0$, and the mixing-induced CP violation amplitude is $A_{CP}^{int} = -\sin 2\beta$.

ATLAS will be able to make a very precise measurement of the CP violation parameter $\sin 2\beta$. The event samples are:

- $B^0 \to J/\psi K_S^0 + \bar{B} \to X$, $J/\psi \to \mu^+\mu^-$, $p_T(\mu_1) > 5$ GeV/c, $p_T(\mu_2) > 6$ GeV/c,
- $B^0 \to J/\psi K_S^0 + \bar{B} \to \mu X$, $J/\psi \to e^+e^-$, $p_T(e) > 1$ GeV/c, $p_T(\mu) > 6$ GeV/c.

In our study of this channel, the K_S^0 was required to decay into charged pions. The tagging methods used were same-side jet-charge tagging and lepton (electron and muon) tagging for the dimuon sample, and muon tagging for the dielectron sample (a muon is always included in the triggered data sample).

The sensitivity of the $\sin 2\beta$ measurement was estimated with a maximum likelihood fit. The detector performance parameters were first determined from a full detector simulation and reconstruction of the signal events and the background. The time resolution function was approximated with a Gaussian with a width $\sigma = 69$ fs, and the wrong-tag fractions were assumed to be $w = 0.24$ for the lepton tags, and $w = 0.41$ for the jet-charge tags. The decay-time distribution of the background was determined from the simulation and parametrized as $C \exp(-\Gamma_0 t)$, where C is the level of the background and Γ_0 is the average decay rate. The direct CP violation term A_{CP}^{dir} was assumed to be zero. The parameter $\sin 2\beta$ was thus the only free parameter in the fit. More details of the likelihood fit can be found in Ref. [5].

The expected event yields and the sensitivities to $\sin 2\beta$ are summarized in Table 1. Results are given both for the new trigger scheme, assuming only a dimuon trigger with p_T thresholds of 5 GeV/c and 6 GeV/c for the muons, and the old full trigger scheme, with lower dimuon trigger thresholds (3 GeV/c and 6 GeV/c) combined with the dielectron sample which has been triggered at Level-1 with a single muon. The new trigger scheme is able to collect dielectrons with a single muon, but the efficiency will be lower than in case of the old trigger scheme. The trigger efficiency for the dielectron sample is still under study. The sensitivity to $\sin 2\beta$ with the new trigger scheme, assuming conservatively that only the dimuon sample is available, is 0.016, while the sensitivity with the old full trigger is 0.010. The systematic uncertainty was estimated to be 0.005.

The angle α can be measured with the decay $B^0 \to \pi^+\pi^-$. This decay has, however, contributions from both penguin and tree decay graphs. Including the penguin contributions, the CP violation amplitudes are: $A_{CP}^{dir} = 2\frac{A_P}{A_T}\sin(\delta)\sin(\alpha)$ and

244

TABLE 1. ATLAS statistical sensitivity to sin2β with 30 fb^{-1}. Here, $\mu6 = (p_T(\mu) > 6$ GeV/$c)$, $\mu5 = (p_T(\mu) > 5$ GeV/$c)$, $\mu3 = (p_T(\mu) > 3$ GeV/$c)$, $e1 = (p_T(e) > 1$ GeV/$c)$.

	$J/\psi(\mu6\mu5)$	$J/\psi(\mu6\mu3)$	$J/\psi(e1e1)$
Number of reconstructed events	250,000	490,000	15,000
Signal/background ratio	32	28	16
$\Delta\sin2\beta$, lepton tag	0.030	0.023	0.018
$\Delta\sin2\beta$, jet-charge tag	0.019	0.015	-
$\Delta\sin2\beta$, combined tags	0.016	0.013	0.018
$\Delta\sin2\beta$(stat.), combined results	0.016	0.010	

$A_{CP}^{int} = -\sin(2\alpha) - 2\frac{A_P}{A_T}\cos(\delta)\cos(2\alpha)\sin(\alpha)$, where A_P/A_T is the ratio of the penguin and tree amplitudes, and δ is the strong phase difference between the amplitudes. If the ratio A_P/A_T can be predicted accurately with the help of additional branching ratio measurements, α and δ can be extracted from the measured asymmetry.

Other two-body final states $B_{d,s}(\Lambda_b) \to h^+h^-$ make it difficult to obtain a clean signal without charged-hadron identification. ATLAS can use a maximum likelihood fit to extract the CP asymmetry parameters A_{CP}^{dir} and A_{CP}^{int} for all the decay modes (both signal and backgrounds) together with the ratio A_P/A_T and the strong phase δ. The precision on A_{CP}^{int} does not depend on the actual values of α, A_P/A_T and δ, but the resulting sensitivities on α and γ do. Due to changes in the trigger, particle identification and decay-time resolution, the old results in Ref. [2] are largely obsolete by now. A new analysis is in progress. It can be expected that ATLAS alone will not have a good sensitivity in this channel, but can contribute to the overall LHC combined measurement.

2.2. B_s^0 Decays

ATLAS can perform precise measurements of the parameters of the B_s^0-$\bar{B}_s^0$ system: the width Γ_s, the width difference $\Delta\Gamma_s$ and the mass difference Δm_s of the mass eigenstates. The experimental determination of these parameters will be valuable input for determining parameters of the CKM matrix: the ratio $\Delta m_s/\Delta m_d$ is related to $|V_{ts}/V_{td}|^2$. ATLAS can also probe the B_s^0 mixing phase $\phi_M = -2\delta\gamma = -2\lambda^2\eta$. In the Standard Model, this phase is expected to be negligibly small (0.024-0.054 radians, see Ref. [6]), so a measurable effect would be a sign of physics beyond the Standard Model.

The parameter Δm_s will be measured from flavour-specific final states $B_s^0 \to D_s^+\pi$ and $B_s^0 \to D_s^+a_1$. Already with an integrated luminosity of 10 fb^{-1} the sensitivity of the Δm_s measurement will reach 36 ps^{-1}, covering fully the Standard Model allowed range (14.3 ps^{-1} to

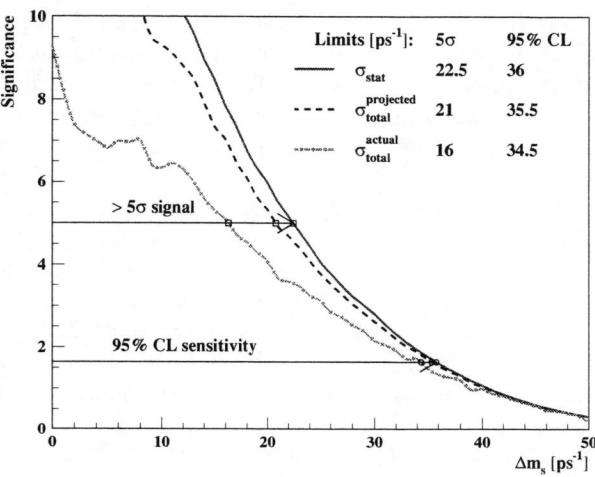

FIGURE 2. The significance of the oscillation signal versus Δm_s for 10 fb^{-1}. The solid line shows the significance including only statistical errors, the dotted line the significance including also systematic errors (present-day knowledge), and the dashed line including projected systematic errors (systematic errors as we expect them to be at the time of the LHC start-up, see [7]).

26 ps^{-1}) [7, 8]. The measurement reach is shown in Fig. 2.

The parameters $\Delta\Gamma_s$, Γ_s and ϕ_M can be determined by angular analyses of the final state $B_s^0 \to J/\psi\phi$. The parameter $\Delta\Gamma_s$ can be determined with a relative error of 12% (stat) with 30 fb^{-1}. The sensitivity to ϕ_M depends on the actual value of the mixing parameter Δm_s. For Δm_s in the range (14-27) ps^{-1}, the sensitivity achieved with the analysis of the decay mode $B_s^0 \to J/\psi\phi$ is between 0.08 and 0.15 radians.

The decay mode $B_s^0 \to J/\psi\eta$ is analogous to the mode $B_s^0 \to J/\psi\phi$, and it can thus be used as a cross-check for measurements of Γ_s and ϕ_M. Since the final state $J/\psi\eta$ is a CP eigenstate, an angular analysis is not needed. The reconstruction is, however, hampered by the small expected branching fraction and the low efficiency of η reconstruction. In ATLAS, about 10,000 signal events can be reconstructed with an integrated luminosity of 30 fb^{-1}, with a signal-to-background ratio of about 1:1 [9]. ATLAS can thus measure the branching fraction of this decay mode, and contribute to the lifetime measurement of the B_s^0 (Γ_s). The sensitivity to the weak mixing phase ϕ_M is in the range 0.27-0.31 radians for $\Delta m_s = (14 - 21)\,\mathrm{ps}^{-1}$.

2.3. B_c Properties

The expected large production rates at the LHC will allow for precision measurements of the properties of the B_c mesons, such as mass, lifetime, production mechanism, decay properties, *etc.* The B_c meson is particularly interesting from a QCD point of view, being a particle composed of two heavy quarks with explicit (non-hidden) flavour.

Studies of the experimental aspects of B_c mesons require a dedicated event generator, since producing events through normal b quark fragmentation (*e.g.* through the standard PYTHIA generator [10]) is far too ineffective. Two dedicated Monte Carlo generators have recently been interfaced with PYTHIA, a fragmentation approximation model, and a full matrix-element calculation for the subprocess $gg \to B_c(B_c^*) + b + \bar{c}$ based on an extended helicity approach (grouping of the 36 contributing Feynman diagrams into gauge-invariant subgroups) [11]. The cross-sections for B_c and B_c^* obtained with the full matrix-element calculation are shown in Fig. 3a.

Assuming a branching fraction of $B(b \to B_c) = 10^{-3}$, and the old ATLAS trigger scenario, the number of events in the $B_c \to J/\psi(\mu\mu)\pi$ decay mode would be 5 600 for an integrated luminosity of 20 fb^{-1}. This mode can be used for mass and lifetime measurements, since the final state can be fully reconstructed. The decay mode $B_c \to J/\psi(\mu\mu)\mu\nu$ is interesting, for example for determining the term $|V_{cb}|$ of the CKM matrix. The ATLAS trigger is less effective for hadronic decay modes due to the muon trigger requirement: for example, in the decay mode $B_c \to B_s^0\pi$ the number of events would be about 100 per 20 fb^{-1}.

$B_c \to J/\psi(\mu\mu)\pi$ events were generated with the BCVEGPY simulation program interfaced with PYTHIA, and passed through the full ATLAS detector simulation and event reconstruction (initial layout). The mass resolution was found to be 74 MeV/c^2, as shown in Fig. 3b. More detailed studies of B_c mesons are in progress.

3. RARE DECAYS

The flavour-changing neutral currents $b \to s$ and $b \to d$ occur only at loop level in the Standard Model, and the branching ratios are of the order of 10^{-5} or less. These decays, so called rare decays, constitute thus an excellent probe of new physics effects, originating from new particles in the loop, tree-level decays, or modified couplings, which typically enhance the branching fractions of these channels.

The dimuon decays $B_{s,d} \to \mu^+\mu^-$ are self-triggering modes with a clear signature. The dimuon trigger allows for triggering these events even in high-luminosity data taking. The branching fractions in the Standard Model are tiny, $\mathcal{B}(B_s^0 \to \mu^+\mu^-) = 3.5 \cdot 10^{-9}$ and $\mathcal{B}(B^0 \to \mu^+\mu^-) = 1.5 \cdot 10^{-10}$ [12], so these channels are ideal for searches for effects beyond the Standard Model.

After one year of data taking at high luminosity (integrated luminosity of 100 fb^{-1}), ATLAS is expected to observe a $B_s^0 \to \mu^+\mu^-$ signal with a 4.3σ significance. The event rates are summarized in Table 2. There is an indication of possible improvement of background conditions with another vertex-fit procedure, but this work is in progress.

The decay mode $B^0 \to K^{*0}\mu^+\mu^-$ ($B^0 \to \rho^0\mu^+\mu^-$) is sensitive to the CKM matrix element $|V_{ts}|$ ($|V_{td}|$). Combining these decay modes one can obtain an independent measurement of the involved matrix elements, with a reduced dependence on the form factors. ATLAS is expected to measure the ratio of the branching fractions $\mathcal{B}(B^0 \to K^{*0}\mu^+\mu^-)/\mathcal{B}(B^0 \to \rho^0\mu^+\mu^-)$ with a statistical precision of 14% with an integrated luminosity of 30 fb^{-1}. The event rates are shown in Table 2.

The shape of the forward-backward asymmetry of the two muons in the decay $B^0 \to K^{*0}\mu^+\mu^-$ is sensitive to new physics effects. The forward and backward hemispheres are defined according to the angle between the lepton ℓ^+ and the B meson direction in the rest frame of the lepton pair. The Standard Model predicts an asymmetry which is positive at low $\hat{s}$ ($\hat{s} = q^2/M_B^2$), zero at $\hat{s} \simeq 0.14$, and negative at high $\hat{s}$ apart from the resonance region. In the Minimal Supersymmetric Standard Model (MSSM), the shape of the forward-backward asymmetry is sensitive to the Wilson coefficient $C_{7\gamma}$. For $C_{7\gamma} > 0$ the shape of the asymmetry is similar to that in the Standard Model, but for $C_{7\gamma} < 0$ the asymmetry is negative at low $\hat{s}$ and can thus be easily distinguished from the shape predicted by the Standard Model.

In the lowest mass region $\hat{s} = \hat{s}_{min}$ - 0.14 ($\hat{s}_{min} = 4m_\ell^2/M_B^2$, where m_ℓ is the lepton mass), the average forward-backward asymmetry A_{FB} is expected to be +10% in the Standard Model. The statistical sensitivity for the forward-backward asymmetry, ΔA_{FB}, is expected to be $\pm 5\%$ in ATLAS with the $B^0 \to K^{*0}\mu^+\mu^-$ sam-

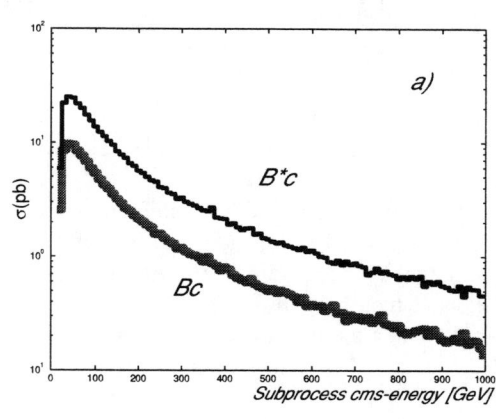
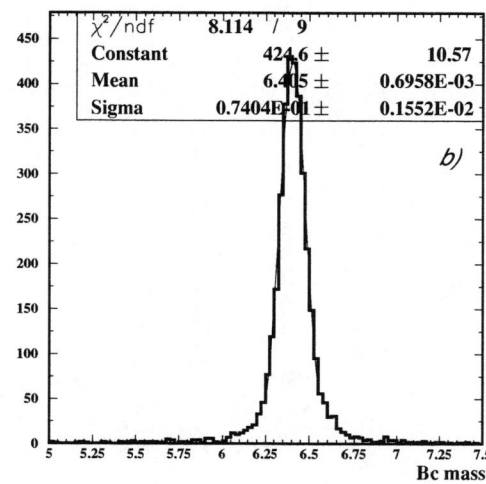

FIGURE 3. a) The total production cross-sections of B_c and B_c^* versus subprocess centre-of-mass energy, from Ref. [11]. b) The reconstructed $B_c \to J/\psi\,\pi$ mass distribution.

TABLE 2. Event rates of rare decays for ATLAS.

Channel	Expected $\mathscr{B}$ (SM)	Integrated luminosity (fb^{-1})	N(signal)	N(background)
$B_s^0 \to \mu^+\mu^-$	$3.5 \cdot 10^{-9}$	100	92	660
$B^0 \to \mu^+\mu^-$	$1.5 \cdot 10^{-10}$	100	14	660
$B^0 \to K^{*0}\mu^+\mu^-$	$1.5 \cdot 10^{-6}$	30	2,000	290
$B^0 \to \rho^0\mu^+\mu^-$	$1 \cdot 10^{-7}$	30	220	950
$B_s^0 \to \phi\mu^+\mu^-$	$1 \cdot 10^{-6}$	30	410	140

ple collected with 30 fb^{-1}. In the MSSM, the average A_{FB} is expected to be $(-17\pm0.5)\%$ with $C_{7\gamma}<0$, so the expected measurement accuracy can clearly distinguish the two cases. The distributions are shown in Fig. 4.

ATLAS can also study the decay mode $B^0 \to K^{*0}\gamma$, which has an expected branching ratio $\mathscr{B} = (4.2 \pm 0.4) \cdot 10^{-5}$. The expected statistics is 10,500 events per 30 fb^{-1}, with a signal significance of 5.7σ. The resolution for the B mass measurement is 57 MeV/c^2.

4. *B* PRODUCTION

The LHC will allow for probing strong interactions in a new region both in energy and in fractional momentum (Bjorken x). The high $b\bar{b}$ cross-section will enable measurements of the single-b cross-section, $b\bar{b}$ correlations and multiple heavy quark production well beyond the Tevatron reach. The event samples available are exclusive B decays (low-p_T region), semi-inclusive $B \to J/\psi X$ decays, and b-tagged jets at very high p_T.

There has been a long-standing discrepancy between the beauty cross-sections measured by CDF and DØ and NLO QCD calculations [13–15]. The calculated beauty

cross-section in the central region was roughly a factor of two below the measured values. The Tevatron results have been statistics limited, in particular when it comes to $b\bar{b}$ correlations. At LHC, the statistics will be sufficient for using exclusive channels instead of b-jets for correlation studies. In particular the kinematic region in which the b quarks are close to each other in azimuthal angle is sensitive to higher-order QCD contributions.

ATLAS can measure $b\bar{b}$ production correlations using exclusive B decays and semileptonic decays to muons. In the simulation study, the azimuthal angle $\Delta\phi$ was measured from the angular difference between the J/ψ and muon in the following decay modes:

- $\bar{b} \to B^0 \to J/\psi(\mu\mu)K_S^0 + b \to \bar{B} \to \mu X$,
- $\bar{b} \to B_s^0 \to J/\psi(\mu\mu)\phi + b \to \bar{B} \to \mu X$,

The numbers of events were 48,000 and 32,000, respectively, per 30 fb^{-1}. No degradation of efficiency was observed at low angular differences. The contribution from $K \to \mu\nu$ background was, however, found to be significant at low angular differences in case of the B_s^0 decays due to the hadronization correlation of B_s^0 and kaons.

An example of a specific measurement with b baryons is the measurement of Λ_b polarization. The angular dis-

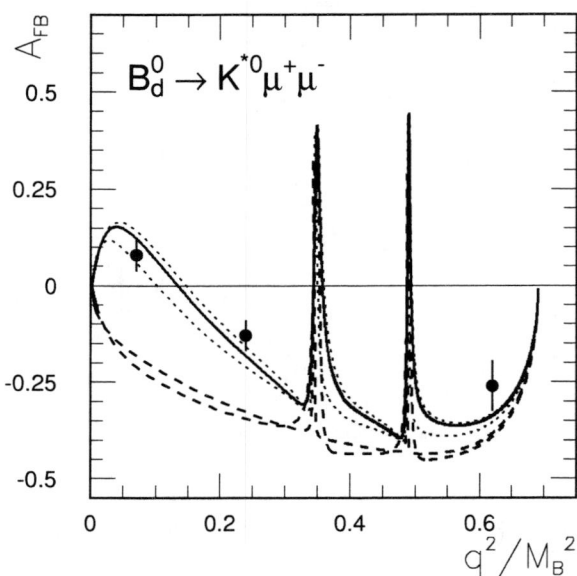

FIGURE 4. The forward-backward asymmetry A_{FB} in decays $B^0 \to K^{*0} \mu^+ \mu^-$. The points show the simulated experimental values, averaged over three different regions of $\hat{s}$, with statistics corresponding to 30 fb^{-1}. The solid line shows the expected Standard Model distribution, the dotted lines the MSSM distributions with $C_{7\gamma} > 0$, and the dashed lines the MSSM distributions with $C_{7\gamma} < 0$.

tribution of the $\Lambda_b \to J/\psi (\mu\mu) \Lambda (p\pi)$ depends on five measurable decay angles, helicity amplitudes and the polarization P_b. A combined fit with 75,000 events (statistics with 30 fb^{-1}) gives a statistical sensitivity of 0.016 to the polarization measurement. More details of the analysis can be found in Ref. [2].

5. SUMMARY

ATLAS is preparing a multi-thematic physics program for a thorough study of B hadron production and decays. The large samples of all types of B hadrons allow for high-precision measurements, searches for evidence of physics beyond the Standard Model, and measurements of beauty production. In particular, ATLAS will be able to make a very precise measurement of the angle β. ATLAS will also perform precise measurements of the properties of B_s^0 and B_c mesons and b baryons.

Rare decays of the type $B \to \mu\mu(X)$ have a favourable experimental signature, allowing ATLAS and CMS to trigger these events also at the highest LHC luminosity. These experiments will thus be the first ones which can probe the decays $B_{s,d} \to \mu\mu$.

All the ATLAS subdetectors are now under construction, and the aim is to commission the detector in 2006 and be ready for regular data taking in 2007. The new ATLAS B physics trigger strategy aims at maximizing the physics potential within the tight constraints.

ACKNOWLEDGMENTS

This overview is based on studies performed within the ATLAS Collaboration, and the author wishes to thank colleagues for providing her with the material. The studies have made use of the physics analysis framework and tools which are the result of collaboration-wide efforts.

REFERENCES

1. ATLAS Collaboration, ATLAS Detector and Physics Performance Technical Design Report, Tech. Rep. CERN/LHCC/99-14, ATLAS TDR 14, Vol I (1999).
2. ATLAS Collaboration, ATLAS Detector and Physics Performance Technical Design Report, Tech. Rep. CERN/LHCC/99-15, ATLAS TDR 15, Vol II (1999).
3. Moser, H.-G., in these proceedings (2004).
4. George, S., in these proceedings (2004).
5. Smizanska, M., ATLAS: B Physics Reach, Tech. Rep. ATLAS Scientific Note SN-ATLAS-2003-033, to be published in Eur. Phys. J. Direct (2003).
6. Ali, A., and London, D., *Eur. Phys. J.* **C9**, 687–703 (1999).
7. Epp, B., Ghete, V. M., and Nairz, A., *Eur. Phys. J. direct* **C4**, N3 (2002).
8. Epp, B., Ghete, V. M., and Nairz, A., *ECONF* **C0304052**, WG205 (2003).
9. Driouichi, C., Eerola, P., Melcher, M., Ohlsson-Malek, F., and Viret, S., *Eur. Phys. J. direct* **C4**, N2 (2002).
10. Sjostrand, T., et al., *Comput. Phys. Commun.* **135**, 238–259 (2001).
11. Chang, C.-H., Driouichi, C., Eerola, P., and Wu, X. G., arXiv:hep-ph/0309120 (2003).
12. Ali, A., *J. Phys.* **G18**, 1605–1626 (1992).
13. Abe, F., et al., [CDF Collaboration], *Phys. Rev. Lett.* **75**, 1451–1455 (1995).
14. Abe, F., et al., [CDF Collaboration], *Phys. Rev.* **D53**, 1051–1065 (1996).
15. Abachi, S., et al., [D0 Collaboration], *Phys. Lett.* **B370**, 239–248 (1996).

THE FUTURE OF FLAVOR PHYSICS

Flavor Prospects

Edward Witten

School of Natural Sciences, Institute for Advanced Study, Princeton, NJ 08540, U.S.A.

Abstract. I will briefly offer a few theoretical reflections on the problem of quark and lepton flavor, focusing on the central role of fermion chirality.

1. INTRODUCTION

First of all, the quarks and leptons as we see them have an odd-looking set of Standard Model quantum numbers:

$$\begin{pmatrix} u \\ d \end{pmatrix}_{1/3} \oplus \bar{u}_{-4/3} \oplus \bar{d}_{2/3} \oplus \begin{pmatrix} \nu \\ e^- \end{pmatrix}_{-1} \oplus e_2^+. \quad (1)$$

When one first sees this structure, it looks like a bizarre hodge-podge. The fractional electric charges and the parity violation in the gauge couplings are perhaps the biggest surprises. The parity violation, in particular, may look ugly the first time that one sees it.

The quantum numbers have, however, been at least partly demystified by a series of familiar insights:

- Naturalness – the chiral structure of the fermions explains why the quarks and leptons that we see are "light" compared to the scale of gravity or GUT's. There may be many fermions with parity-conserving couplings, but they received bare masses at the GUT or Planck scale, and we do not know about them. We see the stragglers, the ones whose chiral quantum numbers prevent them from getting mass in the absence of electroweak symmetry breaking.

- Anomaly Cancellation – the strange fractional charges cannot be changed at random or the theory would be inconsistent because of "triangle anomalies." The actual fermion quantum numbers seen in the Standard Model are one of the simpler anomaly-free chiral representations of the gauge group $SU(3) \times SU(2) \times U(1)$.

- Grand Unified Theories (GUT's) – most ambitious of all, the grand unified $SU(5)$ theory, and its refinements, potentially explain the structure of a generation of quarks and leptons by unifying the pieces in a (relatively) simple representation of a simple gauge group.

To whatever extent we do or do not understand the structure of a single generation, we also have to face the fact that nature has presented us with three of them, which we could call the electron, muon, and tau families. Why did nature repeat the structure in this way? "Who ordered that?", Rabi asked about the muon. Apart from understanding why there are three flavors, we also want to understand the "flavor structure" of the masses and interactions.

In fact, most measured particle physics parameters – both Standard Model parameters and beyond – are flavor parameters. The flavor parameters in the Standard Model are the quark and lepton masses and the CKM mixing matrix. The most striking thing to explain about these parameters is the extreme smallness of some quark and lepton masses – the fact that the electron is more than 100,000 times lighter than the W boson, for example.

In physics beyond the Standard Model, we have the neutrino mass differences and the PMNS neutrino mixing matrix. Here the big news, perhaps not always emphasized as much as it might be, is that neutrino masses are about right for GUT's! The order of magnitude of neutrino masses that was suggested over twenty years ago based on the GUT see-saw mechanism has turned out to be about right. Tentatively assuming that this is on the right track, the most striking thing to explain about neutrino parameters is that neutrinos have large mixing angles, in marked contrast to the smallness of the quark mixing angles.

Most attempts to explain flavor begin by asking why nature has repeated herself, why there are three flavors. One then tries to understand the mass and mixing parameters. Numerous approaches have been tried. These include approaches based on:

- Symmetries
- Compositeness
- Unification
- Extra Dimensions

CP722, *B Physics at Hadron Machines,* edited by M. Paulini and S. Erhan

I will make a few remarks about each.

1.1. Symmetries

A simple idea to account for why there are several families, is to assume the existence of a new "horizontal" symmetry group such as $SO(3)$ or possibly a finite group with the three families in a three-dimensional representation of this group. One then of course needs a suitable method of breaking the horizontal symmetry. In many attempts, including the Froggatt-Nielsen model, in some unbroken-symmetry limit, the CKM matrix is diagonal. Its off-diagonal terms are then related to small symmetry-breaking effects.

1.2. Compositeness

Here, the idea is to get the observed quarks and leptons as bound states from some more elementary constituents – with three generations because of different internal wave functions. I'd say that the chirality has made this largely unworkable, at least until now. The problem is that as the electron, for example, is known to be pointlike down to very short distances, the mass scale of the interactions that generate the quarks and leptons as composites is much higher than the mass scale of the observed fermions. Such dynamics is at least unusual. Some possibilities are known. In practice, they always involve generating the observed chiral fermions from other primordial chiral fermions, that in practice, turn out to be more numerous than the fermions that we are trying to generate. Thus, the naive goal of generating the observed chiral fermions from a more simple structure is not attained, at least in any obvious way.

1.3. Gauge Unification

Another approach to flavor involves gauge unification. Grand Unified Theories or GUT's explained a single family in terms of the gauge structure of a GUT representation such as the $\bar{5} + 10$ of $SU(5)$, or the 16 of $SO(10)$. This elegant structure is one of the prime achievements of GUT's.

It is natural to try to do better. Can we find a larger gauge group with a larger irreducible representation that contains several families of the Standard Model?

This is an interesting program, and several Lie groups such as $SO(16)$ and E_8 have some of the right properties. Like most models based on compositeness, this program goes wrong because of *chirality* – the families that one generates come with antifamilies of the opposite hand-

edness. This is another important lesson in how fundamental the handedness of the fermions is.

What goes wrong? Ideally, we'd like to find a "beyond GUT" gauge group G with an anomaly-free simple representation R leading in terms of the Standard Model to n families and m anti-families where

$$n - m = 3. \qquad (2)$$

Then the Higgs mechanism, with families and antifamilies pairing up with large bare masses, could reduce us to the real world, in which apparently $n = 3, m = 0$.

But this doesn't work. When we try to work out an actual model, we get $n - m = 0$. For example, we could take the gauge group to be $SO(18)$, with R being the spinor representation. In this example, we find $n = m = 8$.

Or we could use the group E_8. This group is worthy of describing nature as it is the biggest and most splendid of the exceptional simple Lie groups. If nature were described by $SU(5)$, we would always wonder, "Why not $SU(6)$?" Likewise, $SO(18)$ would raise the question, "Why not $SO(19)$ or $SO(22)$?" But if nature is described by E_8, there is no such question: it is the finest of its type, the end of the line.

Moreover, the Standard Model gauge group and chiral fermions can be neatly embedded in E_8. But when we do this, we get $n = m = 4$.

So we learn another lesson in how the handedness of the observed fermions constrains efforts to improve upon the Standard Model. This handedness is really fundamental in our present understanding of nature. As I have tried to convey, it is characterized by all of the following:

- It is odd at first sight.
- It accounts for the "lightness" of the observed fermions.
- It helps separate the sheep from the goats in terms of theories beyond the Standard Model.

As I will explain more fully later, the handedness of the fermions may well mean that flavor is even more fundamental as it first appears.

1.4. Extra Dimensions

Another idea about flavor is to use extra dimensions. For example, the (heterotic) string models of the mid-1980's are for the flavor problem somewhat similar to the GUT models that I just mentioned, except that the unification occurs in ten dimensions, and the starting point is more constrained – the starting gauge group has to be $SO(32)$ or E_8. When one tries to count generations and antigenerations, one starts in a sense with $n = m = \infty$, because of the Kaluza-Klein harmonics. The infinity is regularized by the bare masses, and one is left with a finite remainder that depends on the topology of the extra dimensions.

In the original model, the number of light generations – 3 in the real world – comes out to be $\chi/2$, where χ is the most elementary topological invariant, the "Euler characteristic" of the extra dimensions.

In the process, E_8 is revived. Four-dimensional E_8 doesn't work because of the fermion chirality, but ten-dimensional E_8 is (almost) forced on us by string theory, and leads in a simple way to the right Standard Model gauge group and chiral fermions.

From this point of view, topology (of the extra dimensions) plays the role for flavor that we might have assigned to symmetries. The number of generations comes from topology (the Euler characteristic) and more subtle aspects of topology (and complex geometry) control the fermion masses and CKM matrix.

I have described the oldest models, but nowadays, a much wider range of models of flavor are derived from extra dimensions. In string theory, there are many new types of models, frequently "dual" to the original ones in different regions of their parameter space, but possibly giving important new insights or more relevant to nature, and sometimes quantitatively quite different. There are also numerous "bottom-up" models, often but not always incorporating aspects of the string models.

2. LESSONS

One lesson we might learn from this little survey of models is that while there may be a "theory of flavor" at accelerator energies, it may also be that the origin of flavor is at much higher energies. Furthermore, understanding flavor is inseparable from understanding a larger unification of particle forces, possibly including gravity.

Indeed, fermion chirality is an important hint in this direction. We might draw a contrast between flavor and another notorious problem in particle physics – the "gauge hierarchy" problem, which is the question of why the mass scale of observed particle physics is so tiny compared to the mass scale of gravity and, possibly, of Grand Unification.

Technically, the problem arises because the Higgs boson φ can have a bare mass $m^2\varphi^2$ and loop corrections will naturally renormalize m^2 up to an amount of order $\alpha = e^2/\hbar c$ times the scale at which the ultraviolet divergences are somehow cut off.

There are numerous proposals for what this cutoff may be – among them Supersymmetry, technicolor, models with large extra dimensions or low-scale unification with gravity, and "little Higgs" models. But all models that give any sort of rational explanation for the lightness of the elementary particle scale do this with a mechanism that can be probed at accelerator energies.

For example, in the case of Supersymmetry, the cut-off involves a cancellation between the known particles and their superpartners, and is only sufficiently effective if the superpartners weigh no more than a few hundred GeV, or a few TeV if one stretches things.

So we expect that the LHC, or possibly Fermilab, can reveal a mechanism, or at least some of the ingredients of a mechanism, that stabilizes the scale of electroweak symmetry breaking and hence the particle bare masses. If instead the LHC would discover only a Standard Model Higgs boson and no further mechanism accounting for the stability of the elementary particle mass scale, this would sharply contradict our way of thinking.

I don't think we have the same expectation that the origin of flavor is accessible in an equally direct way in accelerator experiments. In some models it is, but in some perfectly plausible models (like the string models I mentioned) flavor originates at much higher energies, and then penetrates down to the energies at which we make our observations.

The difference arises because of chirality, which can keep fermions light compared to the mass scale at which their structure emerges. Fermion masses are "protected." By contrast, the Higgs boson is highly non-chiral, and "anything" can contribute to its mass. The Higgs boson is in need of protection.

Part of the fascination of flavor is precisely that the chiral and CKM and PMNS structure may thus originate from the GUT or string scale. But how can we learn more? More clues about flavor are emerging from current experiments, including those reported at this meeting, on CKM physics, CP violation, neutrino masses and mixing, and studies of rare or forbidden K and charm and B decays.

We may have the chance to learn many more clues about flavor physics when we can probe the mechanism of electroweak symmetry breaking at accelerators. Many of the models lead to dozens of new flavor observables, and not just the few additional unmeasured flavor observables that we currently know about.

For example, with Supersymmetry, there would be dozens of new flavor observables involving the superpartners. Not only is the number of new observables large, but there is quite a bit of mystery about them. When it comes to flavor (and CP, and baryon number), the supersymmetric models, despite their other virtues, definitely have the potential to spoil some of the merits of the Standard Model. Why don't the superpartners mediate flavor changing neutral current processes at a level that would contradict what we see? It is a conundrum, with some possible solutions, none of which appears perfect. Yet many clues (involving the successes of SUSY-GUT's and gauge coupling unification as well as the hierarchy problem) suggest that Supersymmetry may well be there at TeV energies. If so, one of these days we will be able to study the superpartners in detail, and our limited supply of clues about the flavor problem will get considerably longer.

The whole story of the superpartners will be extremely rich and elaborate, almost surely requiring lepton colliders as well as hadron colliders for its elucidation.

Most other attempts at the gauge hierarchy problem lead to some of the same issues. They generate lots of new flavor-dependent observables to study at accelerators, and like Supersymmetry they have the potential to generate unwanted flavor-changing (and baryon-number changing) processes.

We may also gain more information about flavor if we are lucky enough to discover proton decay, whose occurrence at a rate not too far from the present observational bounds is suggested by SUSY-GUT's and related models. Does the proton decay to e or μ? Is the outgoing lepton left-handed or right-handed? Is the hadronic part of the final state strange or non-strange? There is another mixing matrix here analogous to the CKM matrix, but involving GUT interactions instead of weak interactions. It would be wonderful to be able to measure it, at least partly.

In short, we might find a theory of flavor at accelerator energies. The resulting discoveries would certainly be very rich. Alternatively, flavor might be even more fundamental if – as fermion chirality suggests – it originates at the GUT scale. In this case, we can hope to obtain numerous new clues about *how* flavor emerges, especially if we have the good fortune to discover Supersymmetry or another mechanism accounting for the scale of particle bare masses.

Footprints of New Physics in the *B* System

Yuval Grossman

Department of Physics, Technion–Israel Institute of Technology, Technion City, 32000 Haifa, Israel [1]
and
Stanford Linear Accelerator Center, Stanford University, Stanford, CA 94309, U.S.A.
and
Santa Cruz Institute for Particle Physics, University of California, Santa Cruz, CA 95064, U.S.A.

Abstract. In the first part of the talk, the flavor physics input to models beyond the Standard Model is described. In the second part of the talk, we discuss several observables that are sensitive to new physics. We explain what type of new physics can produce deviations from the Standard Model predictions in each of these observables.

1. INTRODUCTION

The success of the Standard Model (SM) can be seen as a proof that it is an effective low energy description of Nature. Yet, there are many reasons to believe that the SM has to be extended. A partial list includes the hierarchy problem, the strong *CP* problem, baryogenesis, gauge coupling unification, the flavor puzzle, neutrino masses, and gravity. We are therefore interested in probing the more fundamental theory. One way to go is to search for new particles that can be produced in yet unreached energies. Another way to look for new physics is to search for indirect effects of heavy unknown particles. In this talk, we explain how flavor physics is used to probe such indirect signals of physics beyond the SM.

2. NEW PHYSICS AND FLAVOR

In general, flavor bounds provide strong constraints on new physics models. This fact is called "the new physics flavor problem". The problem is actually the mismatch between the new physics scale that is required in order to solve the hierarchy problem and the one that is needed in order to satisfy the experimental bounds from flavor physics [1]. Here we explain what is the new physics flavor problem and discuss ways to solve it.

In order to understand what is the new physics flavor problem, let us first recall the hierarchy problem [2]. In order to prevent the Higgs mass from getting a large radiative correction, new physics must appear at a scale that is a loop factor above the weak scale

$$\Lambda \lesssim 4\pi m_W \sim 1\,\text{TeV}. \tag{1}$$

Here, and in what follows, Λ represents the new physics scale. Note that such TeV new physics can be directly probed in collider searches.

While the SM scalar sector is unnatural, its flavor sector is impressively successful[2]. This success is linked to the fact that the SM flavor structure is special. First, the charged current interactions are universal. (In the mass basis, this is manifest through the unitarity of the CKM matrix.) Second, Flavor-Changing-Neutral-Currents (FCNC's) are highly suppressed: they are absent at the tree level and at the one loop level they are further suppressed by the GIM mechanism. These special features are important in order to explain the observed pattern of weak decays. Thus, any extension of the SM must conserve these successful features.

Consider a generic new physics model, that is, a model where the only suppression of FCNC's processes is due to the large masses of the particles that mediate them. Naturally, these masses are of the order of the new physics scale, Λ. Flavor physics, in particular measurements of meson mixing and *CP* violation, put severe constraints on Λ.

---

[1] Permanent address.

[2] The flavor structure of the SM is interesting since the quark masses and mixing angles exhibit hierarchy. These hierarchies are not explained within the SM, and this fact is usually called "the SM flavor puzzle". This puzzle is different from the new physics flavor problem that we are discussing here.

CP722, *B Physics at Hadron Machines*, edited by M. Paulini and S. Erhan
© 2004 American Institute of Physics 0-7354-0203-5/04/$22.00

In order to find these bounds we take an effective field theory approach. At the weak scale we write all the non-renormalizable operators that are consistent with the gauge symmetry of the SM. In particular, flavor-changing four Fermi operators of the form (the Dirac structure is suppressed)

$$\frac{q_1 \bar{q}_2 q_3 \bar{q}_4}{\Lambda^2}, \qquad (2)$$

are allowed. Here q_i can be any quark flavor as long as the electric charges of the four fields in Eq. (2) sum up to zero[3]. The strongest bounds are obtained from meson mixing and CP violation measurements:

- K physics: K-$\bar{K}$ mixing and CP violation in K decays imply

$$\frac{s\bar{d}s\bar{d}}{\Lambda^2} \quad \Rightarrow \quad \Lambda \gtrsim 10^4 \text{ TeV}. \qquad (3)$$

- D physics: D-$\bar{D}$ mixing implies

$$\frac{c\bar{u}c\bar{u}}{\Lambda^2} \quad \Rightarrow \quad \Lambda \gtrsim 10^3 \text{ TeV}. \qquad (4)$$

- B physics: B-$\bar{B}$ mixing and CP violation in B decays imply

$$\frac{b\bar{d}b\bar{d}}{\Lambda^2} \quad \Rightarrow \quad \Lambda \gtrsim 10^3 \text{ TeV}. \qquad (5)$$

Note that the bound from kaon data is the strongest.

There is tension between the new physics scale that is required in order to solve the hierarchy problem, Eq. (1), and the one that is needed in order not to contradict the flavor bounds, Eqs. (3)–(5). The hierarchy problem can be solved with new physics at a scale $\Lambda \sim 1$ TeV. Flavor bounds, on the other hand, require $\Lambda > 10^4$ TeV. This tension implies that any TeV scale new physics cannot have a generic flavor structure. This is the new physics flavor problem.

Flavor physics has been mainly an input to model building, not an output. The flavor predictions of any new physics model are not a consequence of its generic structure but rather of the special structure that is imposed to satisfy the severe existing flavor bounds.

Any viable TeV new physics model has to solve the new physics flavor problem. We now describe several ways to do so that have been used in various models.

(i) Minimal Flavor Violation (MFV) models [3]. In such models the new physics is flavor blind. That is, the only source of flavor violation are the Yukawa couplings.

This is not to say that flavor violation arises only from W exchange diagrams via the CKM matrix elements. Other flavor contributions exist, but they are related to the Yukawa interactions. Examples of such models are gauge mediated Supersymmetry breaking models [4] and models with universal extra dimensions [5]. In general, MFV models predict small effects in flavor physics.

(ii) Models with flavor suppression mainly in the first two generations. The hierarchy problem is connected mainly to the third generation since its couplings to the Higgs field are the largest. Flavor bounds, however, are most severe in processes that involve only the first two generations. Therefore, one way to ameliorate the new physics flavor problem is to keep the effective scale of the new physics in the third generation low, while having the effective new physics of the first two generations at a higher scale. Examples of such models include Supersymmetric models with the first two generations of quarks heavy [6] and Randall-Sundrum models with bulk quarks [7]. In general, such models predict large effects in the B and B_s^0 systems, and smaller effects in K and D mixings and decays.

(iii) Flavor suppression mainly in the up sector. Since the flavor bounds are stronger in the down sector, one way to go in order to avoid them is to have new flavor physics mainly in the up sector. Examples of such models are Supersymmetric models with alignment [8] and models with discrete symmetries [9]. In general such models predict large effects in charm physics and small effects in B, B_s^0 and K mixing and decays.

(iv) Generic flavor suppression. In many models some mechanism that suppresses flavor violation for all the quarks is implemented. Examples of such models are Supersymmetric models with spontaneously broken flavor symmetry [10] and models of split fermions in flat extra dimension [11]. In general, such models can be tested with flavor physics.

3. PROBING NEW PHYSICS WITH FLAVOR

Any TeV new physics model has to deal with the flavor bounds. Depending on the mechanism that is used to deal with flavor, the prediction of where deviation from the SM can be expected varies. It is important, however, that in many cases large effects are expected. Thus, we hope that we will be able to find such signals.

Generally, it is easier to search for new physics effects where they are relatively large. Namely, in processes that are suppressed in the SM, in particular in meson mixing, loop mediated decays, and CKM suppressed amplitudes. It is indeed a major part of the B factories program to study such processes. Below we give several examples

---

[3] We emphasize that there is no exact symmetry that can forbid such operators. This is in contrast to operators that violate baryon or lepton number that can be eliminated by imposing symmetries like $U(1)_{B-L}$ or R-parity.

for ways to search for new physics.

Before proceeding we emphasize the following point: *At present there is no significant deviation from the SM predictions in the flavor sector*. In the following we give examples of deviations from the SM predictions that are below the 3σ level. In particular, we choose the following possible tests of the SM: Global fit, $\mathscr{A}_{CP}(B \to J/\psi K_S^0)$ vs $\mathscr{A}_{CP}(B \to \phi K_S^0)$, $B \to K\pi$ decays, and polarization in $B \to VV$ decays. There are many more possible tests. Our choice of examples here is partially biased toward cases where the present experimental ranges deviate by more than one standard deviation from the SM predictions. While, as emphasized above, one should not consider these as significant indications for new physics, it should be interesting to follow future improvements in these measurements. Furthermore, it is an instructive exercise to think what one would learn if the central value of these measurements turn out to be correct. As we will see, this would not only indicate new physics, but actually probe the nature of the new physics.

3.1. Global Fit

One way to test the SM is to make many measurements that determine the sides and angles of the unitarity triangle, namely, to over-constrain it [12]. Another way to put it is that one tries to measure ρ and η in many possible ways. λ, A, ρ and η are the Wolfenstein parameters. We emphasize that this is not the only way to look for new physics. It is just one among many possible ways to look for new physics.

The global fit is done using measurements of (or bounds on) $|V_{cb}|$, $|V_{ub}/V_{cb}|$, ε_K, B-$\bar{B}$ mixing, B_s^0 mixing, and $\mathscr{A}_{CP}(B \to J/\psi K_S^0)$. The fit is very good, as can be seen in Fig. 1. Clearly, there is no indication for new physics from the global fit. There are many more measurements that at present have very little impact on the fit. In the future, such measurements can be included, and then discrepancies may show up.

3.2. *CP* Asymmetries in $b \to s\bar{q}q$ Modes

The time dependent *CP* asymmetry in B decays into a *CP* eigenstate, f_{CP}, is given by [13]

$$\mathscr{A}_{CP}(B \to f_{CP}) \equiv$$
$$\frac{\Gamma(\bar{B}(t) \to f_{CP}) - \Gamma(B(t) \to f_{CP})}{\Gamma(\bar{B}(t) \to f_{CP}) + \Gamma(B(t) \to f_{CP})} =$$
$$-\frac{(1 - |\lambda|^2)\cos(\Delta m_B t) - 2Im\lambda \sin(\Delta m_B t)}{1 + |\lambda|^2} \equiv$$
$$S \sin(\Delta m_B t) - C \cos(\Delta m_B t). \qquad (6)$$

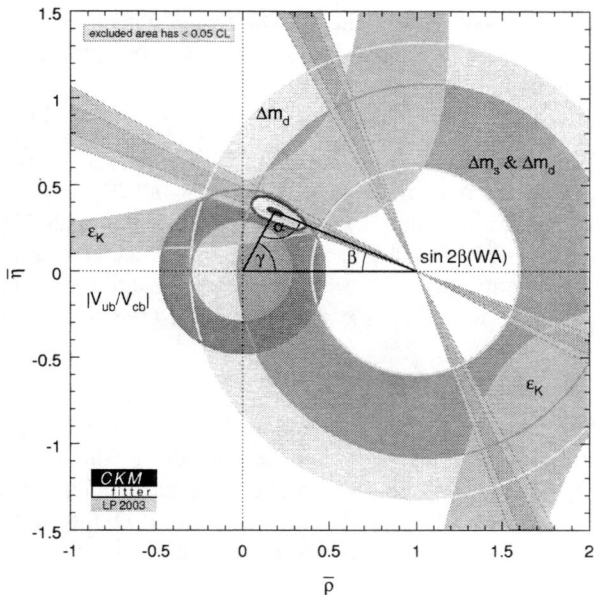

FIGURE 1. Global fit to the unitarity triangle [12]. The fit is based on the measurements of $|V_{cb}|$, $|V_{ub}/V_{cb}|$, ε_K, B-$\bar{B}$ mixing, and $\mathscr{A}_{CP}(B \to J/\psi K_S^0)$ and the bound on B_s^0 mixing.

Here $\Delta m_B \equiv m_H - m_L$ and the last line defines S and C. Furthermore,

$$\lambda \equiv \left(\frac{q}{p}\right)\left(\frac{\bar{A}}{A}\right), \qquad (7)$$

where $\bar{A} \equiv A(\bar{B} \to f_{CP})$ and $A \equiv A(B \to f_{CP})$. The neutral B meson mass eigenstates are defined in terms of flavor eigenstates as

$$|B_{L,H}\rangle = p|B\rangle \pm q|\bar{B}\rangle. \qquad (8)$$

In the $|\lambda| = 1$ limit, which is a very good approximation in many cases, Eq. (6) reduces to the simple form

$$\mathscr{A}_{CP}(B \to f_{CP}) = Im\lambda \sin(\Delta m_B t). \qquad (9)$$

In that case $Im\lambda$ is just the sine of the phase between the mixing amplitude and twice the decay amplitude.

In the SM, the mixing amplitude is[4]

$$\arg(A_{mix}) = 2\beta. \qquad (10)$$

The phase of the decay amplitude depends on the decay mode. $B \to J/\psi K_S^0$ is mediated by the tree level quark decay $b \to c\bar{c}s$ which has a real amplitude, namely,

$$\arg(A_{b \to c\bar{c}s}) = 0, \qquad (11)$$

---

[4] Here, and in what follows, we use the standard parameterization of the CKM matrix. The results, of course, do not depend on the parameterization we choose.

and therefore $Im\lambda = \sin 2\beta$. The penguin $b \to s\bar{q}q$ ($q = u, d, s$) decay amplitude is also real to a good approximation, namely,

$$\arg(A_{b \to s\bar{q}q}) = 0. \qquad (12)$$

We learn that also in that case $Im\lambda = \sin 2\beta$. In particular, the $B \to \phi K_S^0$, $B \to \eta' K_S^0$, $B \to \pi^0 K_S^0$, and $B \to K^+ K^- K_S^0$ are examples of decays that are dominated by the $b \to s\bar{q}q$ transition. They are of particular interest since their CP asymmetries have been measured. We conclude that to first approximation the SM predicts

$$S_{\psi K_S^0} = -S_{K^+ K^- K_S^0} = S_{\phi K_S^0} = S_{\eta' K_S^0} = S_{\pi^0 K_S^0} \qquad (13)$$

Furthermore, for all these modes the SM predicts $|S| = \sin 2\beta$. Note that in order to violate the predictions of Eq. (13), new physics has to affect the decay amplitudes. New physics in the mixing amplitude shifts all the modes by the same amount, leaving Eq. (13) unaffected.

3.2.1. SU(3) Analysis

In order to probe new physics we need to know the theoretical uncertainties in the predictions of Eq. (13). Theoretical estimates are that they are less than $\mathscr{O}(1\%)$ for $S_{\psi K_S^0}$, of $\mathscr{O}(5\%)$ for $S_{\phi K_S^0}$ and $S_{\eta' K_S^0}$, $\mathscr{O}(10\%)$ for $S_{\pi^0 K_S^0}$ and $\mathscr{O}(20\%)$ for $S_{K^+ K^- K_S^0}$ [14, 17–19]. Here we like to show one way to bound these theoretical uncertainties.

The SM amplitude for $b \to s\bar{q}q$ ($q = u, d, s$) penguin dominant decay modes can be written as follows:

$$A_f \equiv A(B^0 \to f) = V_{cb}^* V_{cs} a_f^c + V_{ub}^* V_{us} a_f^u. \qquad (14)$$

The second term is CKM-suppressed compared to the first one since

$$\mathscr{I}m\left(\frac{V_{ub}^* V_{us}}{V_{cb}^* V_{cs}}\right) = \left|\frac{V_{ub}^* V_{us}}{V_{cb}^* V_{cs}}\right| \sin\gamma = \mathscr{O}(\lambda^2), \qquad (15)$$

where $\lambda = 0.22$ is the Wolfenstein parameter. For final states with zero strangeness, f', we write the amplitudes as

$$A_{f'} \equiv A(B^0 \to f') = V_{cb}^* V_{cd} b_{f'}^c + V_{ub}^* V_{ud} b_{f'}^u. \qquad (16)$$

Here neither term is CKM suppressed compared to the other. We use SU(3) flavor symmetry to relate the $a_f^{u,c}$ amplitudes to sums of $b_{f'}^{u,c}$.

It is convenient to define

$$\xi_f \equiv \frac{V_{ub}^* V_{us} a_f^u}{V_{cb}^* V_{cs} a_f^c}, \qquad (17)$$

such that we expect $|\xi_f| \ll 1$. Then we rewrite the amplitude of Eq. (14) as

$$A_f = V_{cb}^* V_{cs} a_f^c (1 + \xi_f). \qquad (18)$$

Finite ξ_f result in deviations from the leading order result, which, to first order in $|\xi_f|$ read

$$-\eta_f S_f - \sin 2\beta = 2\cos 2\beta \sin\gamma \cos\delta_f |\xi_f|, \qquad (19)$$

$$C_f = -2\sin\gamma \sin\delta_f |\xi_f|. \qquad (20)$$

where η_f is the CP of the final state and $\delta_f = \arg(a_f^u/a_f^c)$.

We discuss here a way to estimate ξ_f using SU(3) (or equivalently U-spin) [15, 17–19]. The basic idea is to relate $b \to s$ to $b \to d$ penguin amplitudes. In the later the tree amplitude is enhanced and thus there is larger sensitivity to it. Then, using SU(3), the tree amplitude in the $b \to d$ decay is related to the one in $b \to s$ decay.

The crucial question, when thinking of the deviation of $-\eta_f S_f$ from $\sin 2\beta$, is the size of a_f^u/a_f^c. While a_f^c is dominated by the contribution of $b \to s\bar{q}q$ gluonic penguin diagrams, a_f^u gets contributions from both penguin diagrams and $b \to u\bar{u}s$ tree diagrams. For the penguin contributions, it is clear that $|a_f^u/a_f^c| \sim 1$. (The a_f^c term comes from the charm penguin minus the top penguin, while the up penguin minus the top penguin contributes to a_f^u.) Thus our main concern is the possibility that the tree contributions might yield $|a_f^u/a_f^c|$ significantly larger than one.

We first provide a simple explanation of the method. Let us assume that the decays to final strange states, f, are dominated by the a_f^c terms and that those to final states with zero strangeness, f', are dominated by the $b_{f'}^u$ terms. Thus we can estimate $|a_f^c|$ and $|b_{f'}^u|$ from the measured branching ratios (or the upper bounds on them). Then the SU(3) relations give upper bounds on certain sums of the $b_{f'}^c$ and a_f^u amplitudes from the extracted values of a_f^c and $b_{f'}^u$, respectively. This then gives a bound on $|a_f^u/a_f^c|$, and consequently on $|\xi_f|$.

Actually, the assumptions made in the previous paragraph can be avoided entirely [18, 19]. The SU(3) relations actually provide an upper bound on

$$\hat{\xi}_f \equiv \left| \frac{V_{us}}{V_{ud}} \times \frac{V_{cb}^* V_{cd} a_f^c + V_{ub}^* V_{ud} a_f^u}{V_{cb}^* V_{cs} a_f^c + V_{ub}^* V_{us} a_f^u} \right|$$

$$= \left| \frac{\xi_f + (V_{us} V_{cd})/(V_{ud} V_{cs})}{1 + \xi_f} \right|. \qquad (21)$$

If the bound on $\hat{\xi}_f$ is less than unity, then it gives a bound on $|\xi_f|$.

In general we can write

$$a_f^q = \sum_{f'} x_{f'} b_{f'}^q, \qquad (22)$$

where $q = u, c$ and $x_{f'}$ are Clebsch-Gordon coefficients, which are calculated using group theory properties of

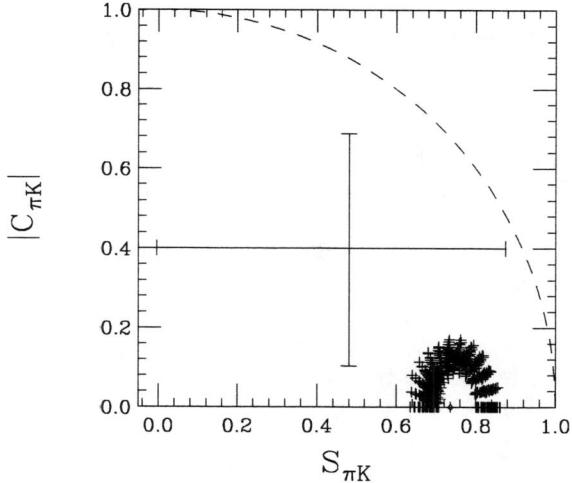

FIGURE 2. Points in the $S_{\pi K}-|C_{\pi K}|$ plane allowed by the SU(3) relations. The small plotted point denotes the pure-penguin value $S_{\pi K} = \sin 2\beta$, $C_{\pi K} = 0$. The point with large error bars denotes the current experimental value. The dashed arc denotes the boundary of allowed values: $S_{\pi K}^2 + C_{\pi K}^2 \leq 1$.

SU(3). Then, using the relevant measured rates, we get

$$\hat{\xi}_f \leq \lambda \sum_{f'} |x_{f'}| \sqrt{\frac{\mathscr{B}(f')}{\mathscr{B}(f)}}. \qquad (23)$$

These bounds are exact in the SU(3) limit.

The SU(3) relations have worked out in details for several modes [18, 19]. Using the tables in [18] relations for many other modes can be found. Such relation will be important once the asymmetries in those modes will be measured. Here we present only the simplest bound, that is for $\pi^0 K_S^0$. The SU(3) relation reads

$$a(\pi^0 K^0) = b(\pi^0\pi^0) + b(K^+K^-)/\sqrt{2}. \qquad (24)$$

The available experimental data is

$$\mathscr{B}(B^0 \to \pi^0 K^0) = (11.92 \pm 1.44) \times 10^{-6},$$
$$\mathscr{B}(B^0 \to \pi^0\pi^0) = (1.89 \pm 0.46) \times 10^{-6},$$
$$\mathscr{B}(B^0 \to K^+K^-) < 0.6 \times 10^{-6}. \qquad (25)$$

which lead to

$$\hat{\xi}_{\pi K} < 0.13, \quad |S_{\pi K} - \sin 2\beta| < 0.19, \quad |C_{\pi K}| < 0.26 \qquad (26)$$

We expect $\mathscr{B}(B^0 \to K^+K^-)$ to be much smaller than the present bound. If this is indeed the case we will be able to neglect it and we will get not only a bound on $\hat{\xi}_{\pi K}$, but an actual estimate. Note also that the bounds in Eq. (26) are correlated. This can be seen in Fig. 2 where the allowed value for $S_{\pi K}$ and $C_{\pi K}$ are plotted.

3.2.2. Status of Penguin Decays

The data do not show a clear picture yet. Using the most recent results [20], the world averages of the asymmetries are[5]

$$S_{\psi K_S^0} = +0.73 \pm 0.05,$$
$$S_{\eta' K_S^0} = +0.27 \pm 0.21,$$
$$S_{\phi K_S^0} = -0.15 \pm 0.70,$$
$$S_{\pi K_S^0} = 0.48^{+0.38}_{-0.47} \pm 0.11,$$
$$-S_{K^+K^- K_S^0} = +0.51 \pm 0.26^{+0.18}_{-0.00}. \qquad (27)$$

In particular, both $S_{\phi K_S^0}$ and $S_{\eta' K_S^0}$ are more then one standard deviation away from $S_{\psi K_S^0}$. (Since the theoretical errors on $S_{K^+K^- K_S^0}$ are large and due to the brief nature of this talk, we do not discuss this mode any further.)

Assuming that these anomalies are confirmed in the future, we ask what can explain them. We have to look for new physics that can generate a situation where all the asymmetries above are different. The one loop processes in the SM are expected to receive large new physics effects. Moreover, we expect the shift from $\sin 2\beta$ to be different in the those modes since the ratio of the SM and new physics hadronic matrix elements is in general different. On the contrary, $B \to J/\psi K_S^0$ is a CKM favored tree level decay in the SM and thus we do not expect new physics to have significant effects. We conclude that new physics in the $b \to s\bar{q}q$ decay amplitude generally gives different asymmetries in all the modes [22].

It is interesting to ask what we would learn if it turns out that $S_{\psi K_S^0} \neq S_{\phi K_S^0}$ but $S_{\eta' K_S^0}$ and $S_{\pi^0 K_S^0}$ are consistent with $S_{\psi K_S^0}$. Such a situation can be the result of new parity conserving penguin diagrams [23, 24]. To understand this point note that $B \to \phi K_S^0$ is parity conserving while $B \to \eta' K_S^0$ is parity violating. Thus, parity conserving new physics in $b \to s$ penguins only affects $B \to \phi K_S^0$. While generically new physics models are not parity conserving, there are models that are approximately parity conserving. Supersymmetric $SU(2)_L \times SU(R) \times$ Parity models provide an example of such an approximate parity conserving new physics framework [23, 24].

---

[5] We use the PDG prescription of inflating the errors when combining measurements that are in disagreement [21]. Simply combining the errors there is one change in (27), $S_{\phi K_S^0} = -0.15 \pm 0.33$.

3.3. $B \to K\pi$

Consider the four $B \to K\pi$ decays and the underlying quark transitions that mediate them:

$$
\begin{aligned}
B^+ &\to K^0\pi^+ & b &\to d\bar{d}s, \\
B^+ &\to K^+\pi^0 & b &\to d\bar{d}s \quad \text{or} \quad b \to u\bar{u}s, \\
B^0 &\to K^+\pi^- & b &\to u\bar{u}s, \\
B^0 &\to K^0\pi^0 & b &\to d\bar{d}s \quad \text{or} \quad b \to u\bar{u}s.
\end{aligned} \tag{28}
$$

In the SM these modes can be used to measure γ. Moreover, there are many SM relations between these modes that can be used to look for new physics [25].

There are three main types of diagrams that contribute to these decays. The strong penguin diagram (P), the tree diagram (T) and the Electro-Weak (EW) penguin diagram (P_{EW}). It is important to understand the relative magnitudes of these amplitudes. Due to the ratio between the strong and electroweak coupling constants, $P \gg P_{EW}$. The relation between P and T is not as simple. On the one hand, P is a loop amplitude while T is a tree amplitude. On the other hand, the CKM factors in T are $O(\lambda^2) \sim 0.05$ smaller than in P. Thus, it is not clear which amplitude is dominant. Experimentally, it turns out that $P \gg T$. Thus, to first approximation all the four decay rates in Eq. (28) are mediated by the strong penguin amplitude and therefore have the same rate (up to Clebsch-Gordon coefficients). Yet, there are corrections to this expectation due to the sub-leading T and P_{EW} amplitudes.

Due to the hierarchy of amplitudes, there are many approximate relations between the four $B \to K\pi$ decay modes. Let us consider one particular relation, called the Lipkin sum rule [26]. As we explain below the Lipkin sum rule is interesting since the correction to the pure P limit is only second order in the small amplitudes.

The crucial ingredient that is used in order to get useful relations is isospin. Penguin diagrams are pure $\Delta I = 0$ amplitudes, while T and P_{EW} have both $\Delta I = 0$ and $\Delta I = 1$ parts. The Lipkin sum rule, which is based only on isospin, reads [26]

$$
\begin{aligned}
R_L &\equiv \frac{2\Gamma(B^+ \to K^+\pi^0) + 2\Gamma(B^0 \to K^0\pi^0)}{\Gamma(B^+ \to K^0\pi^+) + \Gamma(B^0 \to K^+\pi^-)} \\
&= 1 + O\left(\frac{P_{EW} + T}{P}\right)^2.
\end{aligned} \tag{29}
$$

Experimentally the ratio was found to be [27]

$$
R_L = 1.24 \pm 0.10. \tag{30}
$$

Using theoretical estimates [28] that

$$
\frac{P_{EW}}{P} \sim \frac{T}{P} \sim 0.1, \tag{31}
$$

we expect

$$
R_L = 1 + O(10^{-2}). \tag{32}
$$

We learn that the observed deviation of R_L from 1 is an $O(2\sigma)$ effect.

What can explain $R_L - 1 \gg 10^{-2}$? First, note that any new $\Delta I = 0$ amplitude cannot significantly modify the Lipkin sum rule since it modifies only P. From the measurement of the four $B \to K\pi$ decay rates we roughly know the value of P. This tells us that new physics cannot modify P in a significant way. What is needed in order to explain $R_L - 1 \gg 10^{-2}$ are new "Trojan penguins", P_{NP}, which are isospin breaking ($\Delta I = 1$) amplitudes. They modify the Lipkin sum rule as follows

$$
R_L = 1 + O\left(\frac{P_{NP}}{P}\right)^2. \tag{33}
$$

In order to reproduce the observed central value a large effect is needed, $P_{NP} \approx P/2$ [29]. In many models there are strong bounds on P_{NP} from $b \to s\ell^+\ell^-$. Leptophobic Z' is an example of a viable model that can accommodate significant Trojan penguins amplitude [30].

3.4. Polarization in $B \to VV$ Decays

Consider B decays into light vectors, in particular,

$$
B \to \rho\rho, \qquad B \to \phi K^*, \qquad B \to \rho K^*. \tag{34}
$$

Due to the left handed nature of the weak interaction, in the $m_B \to \infty$ limit we expect [24, 31]

$$
\frac{R_T}{R_0} = O\left(\frac{1}{m_B^2}\right), \qquad \frac{R_\perp}{R_\parallel} = 1 + O\left(\frac{1}{m_B}\right), \tag{35}
$$

where R_0 (R_T, $R_\perp$, $R_\parallel$) is the longitudinal (transverse, perpendicular, parallel) polarization fraction. Recall that $R_T = R_\perp + R_\parallel$ and $R_0 + R_T = 1$.

To understand the above power counting consider for simplicity the pure penguin $B \to \phi K^*$ decays. It is convenient to work in the helicity basis ($\mathscr{A}_-$, $\mathscr{A}_+$ and $\mathscr{A}_0$), which is related to the transversity basis via

$$
\mathscr{A}_{\parallel,\perp} = \frac{\mathscr{A}_+ \pm \mathscr{A}_-}{\sqrt{2}}, \tag{36}
$$

and the longitudinal amplitude is the same in the two bases. We consider the factorizable helicity amplitudes, namely, those contributions which can be written in terms of products of decay constants and form factors. In the SM they are proportional to

$$
\mathscr{A}_0 \propto \frac{f_\phi m_B^3}{m_{K^*}}\left[\left(1 + \frac{m_{K^*}}{m_B}\right)A_1 - \left(1 - \frac{m_{K^*}}{m_B}\right)A_2\right] \tag{37}
$$

$$
\mathscr{A}_\pm \propto f_\phi m_\phi m_B\left[\left(1 + \frac{m_{K^*}}{m_B}\right)A_1 \pm \left(1 - \frac{m_{K^*}}{m_B}\right)V\right],
$$

260

where terms of order $1/m_B^2$ were neglected. The $A_{1,2}$ and V are the $B \to K^*$ form factors, which are all equal in the $m_B \to \infty$ limit [32]. Thus, to leading order in α_s [33]

$$\frac{A_2}{A_1} \sim \frac{V}{A_1} = 1 + \mathcal{O}\left(\frac{1}{m_B}\right). \quad (38)$$

Using Eqs. (37) and (38) we see that the helicity amplitudes exhibit the following hierarchy [24, 31]

$$\frac{\mathscr{A}_+}{\mathscr{A}_0} \sim \mathcal{O}\left(\frac{1}{m_B}\right), \qquad \frac{\mathscr{A}_-}{\mathscr{A}_0} \sim \mathcal{O}\left(\frac{1}{m_B^2}\right). \quad (39)$$

Using Eq. (36) the relations in Eq. (35) immediately follow.

An intuitive understanding of these relations can be obtained by considering the helicities of the $q\bar{q}$ pair that make the vector meson. In the valence quark approximation, when they are both right-handed (left-handed) the vector meson has positive (negative) helicity. When they have opposite helicities the vector meson is longitudinally polarized. In the $m_B \to \infty$ limit the light quarks are ultra relativistic and their helicities are determined by the chiralities of the weak decay operators. Since the weak interaction involves only left-handed b decays, the three outgoing light fermions do not have the same helicities. For example, the leading operator generates decays of the form

$$\bar{b} \to \bar{s}_R s_L \bar{s}_R. \quad (40)$$

(The spectator quark does not have preferred helicity.) Since the ϕ is made from an s quark and an $\bar{s}$ antiquark, in this limit it has longitudinal helicity. For finite m_B each helicity flip reduces the amplitude by a factor of $1/m_B$. To get positive helicities one spin flip, that of the s quark, is required. To get negative helicities, spin flips of the two antiquarks are needed.

The relations in (35) receive factorizable as well as non-factorizable corrections. Some of these corrections have been calculated, with the result that they do not significantly modify the leading-order results [31]. Still, in order to get a clearer picture, more accurate determinations of the corrections are needed.

Observation of $R_\perp \gg R_\parallel$ would signal the presence of right-handed chirality effective operators in B decays [23, 24]. The hierarchy between $\mathscr{A}_+$ and $\mathscr{A}_-$ generated by the opposite chirality operator, $\tilde{Q}_i$, (obtained from Q_i via a parity transformation) is flipped compared to the hierarchy generated by the SM operator. Such right-handed chirality operators lead to an enhancement of R_T and therefore can also upset the first relation in Eq.(35).

The polarization data are as follows [27]. The longitudinal fraction has been measured in several modes

$$R_0(B^0 \to \phi K^{*0}) = 0.58 \pm 0.10,$$
$$R_0(B^+ \to \phi K^{*+}) = 0.46 \pm 0.12,$$

$$R_0(B^+ \to \rho^0 K^{*+}) = 0.96 \pm 0.16,$$
$$R_0(B^+ \to \rho^+ \rho^0) = 0.96 \pm 0.07,$$
$$R_0(B^0 \to \rho^+ \rho^-) = 0.99 \pm 0.08. \quad (41)$$

There is only one measurement of the perpendicular polarization [34]

$$R_\perp(B^0 \to \phi K^{*0}) = 0.41 \pm 0.11. \quad (42)$$

Using $R_0 + R_\perp + R_\parallel = 1$ we extract

$$R_\parallel(B^0 \to \phi K^{*0}) = 0.01 \pm 0.15. \quad (43)$$

We see that in $B \to \rho\rho$ and $B \to K^*\rho$ the SM prediction $R_T/R_0 \ll 1$ is confirmed, although $R_T/R_0 \gg 1/m_B^2$ remains a possibility. Since in these modes R_T is very small, the second SM prediction, $R_\perp \approx R_\parallel$, cannot be tested yet.

The situation is different in $B \to \phi K^*$. First, the data favor $R_T/R_0 = \mathcal{O}(1)$, which is not a small number. Second, one also finds that $R_\perp/R_\parallel \gg 1$. Both of these results are in disagreement with the SM predictions in Eq. (35).

It is interesting that the preliminary data indicate that the SM predictions do not hold in $B \to \phi K^*$. This is a pure penguin $b \to s\bar{s}s$ decay. The decays where the SM predictions appear to hold, $B \to K^*\rho$ and particularly $B \to \rho\rho$, on the other hand, have significant tree contributions. It is thus important to obtain polarization measurements in other modes, especially the pure penguin $b \to s d\bar{d}$ decay $B^+ \to K^{*0}\rho^+$.

With more precise polarization data it may therefore be possible to determine whether or not there are new right-handed currents, and if so whether or not they are only present in $b \to s\bar{s}s$ decays.

4. CONCLUSIONS

The main goal of high energy physics is to find the theory that extends the SM into shorter distances. Flavor physics is a very good tool for such a mission. Depending on the mechanism for suppressing flavor changing processes, different patterns of deviation from the SM are expected to be found. In some cases almost no deviations are expected, while in other we expect deviations in specific classes of processes. While there is no signal for such new physics yet, there are intriguing results. More data is needed in order to look further for fundamental physics using low energy flavor changing processes.

ACKNOWLEDGMENTS

The work of YG is supported by the Department of Energy, contract DE-AC03-76SF00515 and by the Department of Energy under grant No. DE-FG03-92ER40689.

REFERENCES

1. For a discussion for supersymmetric models see, for example, Z. Ligeti and Y. Nir, Nucl. Phys. Proc. Suppl. **111**, 82 (2002) [hep-ph/0202117]; Y. Grossman, Y. Nir and R. Rattazzi, Adv. Ser. Direct. High Energy Phys. **15**, 755 (1998) [hep-ph/9701231].

2. See, for example, M. Schmaltz, Nucl. Phys. Proc. Suppl. **117**, 40 (2003) [hep-ph/0210415]; G. F. Giudice, hep-ph/0311344.

3. For a review see, for example, A. J. Buras, hep-ph/0307203.

4. For a review see, Y. Shadmi and Y. Shirman, Rev. Mod. Phys. **72**, 25 (2000) [hep-th/9907225].

5. Flavor aspects of universal extra dimensions are discussed in A. J. Buras, M. Spranger and A. Weiler, Nucl. Phys. B **660**, 225 (2003) [hep-ph/0212143].

6. M. Dine, R. G. Leigh and A. Kagan, Phys. Rev. D **48**, 4269 (1993) [hep-ph/9304299]; A. Pomarol and D. Tommasini, Nucl. Phys. B **466**, 3 (1996) [hep-ph/9507462]; G. R. Dvali and A. Pomarol, Phys. Rev. Lett. **77**, 3728 (1996) [hep-ph/9607383]; A. G. Cohen, D. B. Kaplan and A. E. Nelson, Phys. Lett. B **388**, 588 (1996) [hep-ph/9607394].

7. T. Gherghetta and A. Pomarol, Nucl. Phys. B **586**, 141 (2000) [hep-ph/0003129]; S. J. Huber and Q. Shafi, Phys. Lett. B **498**, 256 (2001) [hep-ph/0010195]; S. J. Huber, Nucl. Phys. B **666**, 269 (2003) [hep-ph/0303183]; K. Agashe, A. Delgado, M. J. May and R. Sundrum, hep-ph/0308036; G. Burdman, hep-ph/0310144.

8. Y. Nir and N. Seiberg, Phys. Lett. B **309**, 337 (1993) [hep-ph/9304307]; M. Leurer, Y. Nir and N. Seiberg, Nucl. Phys. B **420**, 468 (1994) [hep-ph/9310320].

9. For one of the early models see S. Pakvasa and H. Sugawara, Phys. Lett. B **73**, 61 (1978).

10. M. Leurer, Y. Nir and N. Seiberg, Nucl. Phys. B **398**, 319 (1993) [hep-ph/9212278].

11. N. Arkani-Hamed and M. Schmaltz, Phys. Rev. D **61**, 033005 (2000) [hep-ph/9903417]; Y. Grossman and G. Perez, Phys. Rev. D **67**, 015011 (2003) [hep-ph/0210053].

12. A. Hocker, H. Lacker, S. Laplace and F. Le Diberder, Eur. Phys. J. C **21**, 225 (2001) [hep-ph/0104062]. Recent fits can be found in the CKMfitter home page at ckmfitter.in2p3.fr.

13. For a review, notation and formalism, see Y. Nir, *Lectures at XXVII SLAC Summer Institute on Particle Physics*, hep-ph/9911321; G.C. Branco, L. Lavoura and J.P. Silva, *"CP violation," Oxford, UK: Clarendon (1999)*; K. Anikeev *et al.*, hep-ph/0201071.

14. D. London and A. Soni, Phys. Lett. B **407**, 61 (1997) [hep-ph/9704277]; M. Beneke and M. Neubert, Nucl. Phys. B **651**, 225 (2003) [hep-ph/0210085];

15. M. J. Savage and M. B. Wise, Phys. Rev. D **39**, 3346 (1989) [Erratum-ibid. D **40**, 3127 (1989)]; A. S. Dighe, Phys. Rev. D **54**, 2067 (1996) [arXiv:hep-ph/9509287];

16. B. Grinstein and R. F. Lebed, Phys. Rev. D **53**, 6344 (1996) [arXiv:hep-ph/9602218]; A. S. Dighe, M. Gronau and J. L. Rosner, Phys. Rev. D **57**, 1783 (1998) [arXiv:hep-ph/9709223]; N. G. Deshpande, X. G. He and J. Q. Shi, Phys. Rev. D **62**, 034018 (2000) [arXiv:hep-ph/0002260].

17. Y. Grossman, G. Isidori and M. P. Worah, Phys. Rev. D **58**, 057504 (1998) [arXiv:hep-ph/9708305].

18. Y. Grossman, Z. Ligeti, Y. Nir and H. Quinn, Phys. Rev. D **68**, 015004 (2003) [arXiv:hep-ph/0303171].

19. M. Gronau, Y. Grossman and J. L. Rosner, arXiv:hep-ph/0310020.

20. T. E. Browder, arXiv:hep-ex/0312024.

21. K. Hagiwara *et al.*, Particle Data Group, Phys. Rev. D **66**, 010001 (2002).

22. See, for example, Y. Grossman and M. P. Worah, Phys. Lett. B **395**, 241 (1997) [hep-ph/9612269]; A. Kagan, in proceedings of the 7th International Symposium on Heavy Flavor Physics, Santa Barbara, CA July 1997, hep-ph/9806266; R. Fleischer and T. Mannel, Phys. Lett. B **511**, 240 (2001) [hep-ph/0103121]; G. Hiller, Phys. Rev. D **66**, 071502 (2002) [hep-ph/0207356]; A. Datta, Phys. Rev. D **66**, 071702 (2002) [hep-ph/0208016]; M. Raidal, Phys. Rev. Lett. **89**, 231803 (2002) [hep-ph/0208091]; R. Harnik, D. T. Larson, H. Murayama and A. Pierce, hep-ph/0212180; C. W. Chiang and J. L. Rosner, Phys. Rev. D **68**, 014007 (2003) [hep-ph/0302094]; G. L. Kane, P. Ko, H. b. Wang, C. Kolda, J. h. Park and L. T. Wang, Phys. Rev. Lett. **90**, 141803 (2003) [hep-ph/0304239]; A. K. Giri and R. Mohanta, Phys. Rev. D **68**, 014020 (2003) [hep-ph/0306041]; J. F. Cheng, C. S. Huang and X. h. Wu, hep-ph/0306086; R. Arnowitt, B. Dutta and B. Hu, hep-ph/0307152.

23. A. L. Kagan, lecture at SLAC Summer Institute, August 2002, www.slac.stanford.edu/gen/meeting/ssi/2002/kagan1.html#lecture2.

24. A. L. Kagan, talk at first workshop on the discovery potential of an asymmetric B factory at 10^{36} luminosity, May 2003, www.slac.stanford.edu/BFROOT/www/Organization/1036_Study_Group/0303Workshop/index.html

25. For recent reviews, see: R. Fleischer, Phys. Rept. **370**, 537 (2002) [hep-ph/0207108]; J. L. Rosner, hep-ph/0304200; M. Gronau, hep-ph/0306308.

26. H. J. Lipkin, hep-ph/9809347; Phys. Lett. B **445**, 403 (1999) [hep-ph/9810351]; M. Gronau and J. L. Rosner, Phys. Rev. D **59**, 113002 (1999) [hep-ph/9809384].

27. J. Fry, talk presented at the 21st International Symposium On Lepton And Photon Interactions At High Energies (LP03) 11-16 Aug 2003, Batavia, Illinois, conferences.fnal.gov/lp2003/program/papers/fry.ps

28. M. Beneke, G. Buchalla, M. Neubert and C. T. Sachrajda, Nucl. Phys. B **606**, 245 (2001) [hep-ph/0104110].

29. M. Gronau and J. L. Rosner, hep-ph/0307095; A. J. Buras, R. Fleischer, S. Recksiegel and F. Schwab, hep-ph/0309012.

30. Y. Grossman, M. Neubert and A. L. Kagan, JHEP **9910**, 029 (1999) [hep-ph/9909297]; K. Leroux and D. London, Phys. Lett. B **526**, 97 (2002) [hep-ph/0111246].

31. A. L. Kagan, in preparation.

32. J. Charles, A. Le Yaouanc, L. Oliver, O. Pene and J. C. Raynal, Phys. Rev. D **60**, 014001 (1999) [hep-ph/9812358]; C. W. Bauer, S. Fleming, D. Pirjol and I. W. Stewart, Phys. Rev. D **63**, 114020 (2001) [hep-ph/0011336].

33. M. Beneke and T. Feldmann, Nucl. Phys. B **592**, 3 (2001) [hep-ph/0008255]; G. Burdman and G. Hiller, Phys. Rev. D **63**, 113008 (2001) [hep-ph/0011266].

34. [Belle Collaboration], hep-ex/0307014.

Ultra-Rare B Decays

Benjamín Grinstein

University of California, San Diego, Department of Physics, La Jolla, CA 92093-0319, U.S.A.

Abstract. A good place to look for deviations from the Standard Model is in decay modes of B mesons, like purely leptonic decays $B \to \ell \nu$, for which a very long Standard Model lifetime is due to an accidental suppression of the decay amplitude. For other rare decay modes involving no hadrons in the final state (e.g., $B \to \gamma \ell^+ \ell^-$, $B \to \gamma \ell \nu_\ell$ and $B \to \nu \bar{\nu} \gamma$) new results on QCD factorization in exclusive processes show that all the decay rates are given in terms of a single universal form factor. Hence, trustworthy relations between different processes can be used to test the Standard Model of electroweak interactions. Sometimes, surprisingly, a large energy expansion may allow computation when a hadron is in the final state. An example is $B \to \pi \ell^+ \ell^-$ which can be used to settle the ambiguity in α from a measurement of $\sin 2\alpha$ from CP asymmetries.

1. INTRODUCTION

For the purposes of this talk, I will define "Ultra-Rare" decays as decays that have a small branching fraction, smaller than 10^{-4}, are not the standard "Rare decays," namely, radiative $b \to s\gamma$ and $b \to s\ell^+\ell^-$, and are not purely hadronic. Table 1 shows a partial list of processes and a rough estimate of their branching fraction in the Standard Model (SM).

2. TREE-LEVEL DECAYS

Borrowing from the classic result for the pion lifetime we obtain,

$$\Gamma(B_q^+ \to \ell^+ \nu_\ell) = \frac{G_F^2 m_{B_q} m_\ell^2 f_{B_q}^2}{8\pi} |V_{qb}|^2 \left(1 - \frac{m_\ell^2}{m_{B_q}^2}\right)^2.$$

(1)

The $V - A$ vertex of the electroweak interactions gives rise to the famous helicity suppression, a factor of m_ℓ^2. The suppression, which accounts for the wide range in BR's for $\ell = e, \mu, \tau$ in Table 1, is due to the vector nature of the interaction, not to the precise combination of V and A currents. The small numbers for the $\ell = e$ and μ modes are good news. There is no helicity suppression for new physics that couples quarks to leptons through a scalar vertex, as would result from exchange of a scalar field. Such new physics could be hiding under the unsuppressed rate for $B \to X_u \ell \nu$, but the new coupling could be so large as to dominate $B \to \ell \nu$ by orders of magnitude! Table 2 shows experimental bounds: there is plenty of room for improvement (and for new physics to show up).

TABLE 1. Estimated Standard Model branching ratios of Ultra-Rare B decays.

Decay	SM Prediction
$B^+ \to e^+ \nu_e$	7×10^{-12}
$B^+ \to \mu^+ \nu_\mu$	1×10^{-6}
$B^+ \to \tau^+ \nu_\tau$	7×10^{-5}
$B_c^+ \to e^+ \nu_e$	3×10^{-9}
$B_c^+ \to \mu^+ \nu_\mu$	1×10^{-4}
$B_c^+ \to \tau^+ \nu_\tau$	3×10^{-2}
$B^0 \to e^+ e^-$	7×10^{-12}
$B^0 \to \mu^+ \mu^-$	1×10^{-10}
$B^0 \to \tau^+ \tau^-$	3×10^{-8}
$B^+ \to \ell^+ \nu \gamma$	$10^{-7} - 10^{-5}$
$B_s^0 \to \ell^+ \ell^- \gamma (e, \mu)$	$(2-5) \times 10^{-9}$
$B^0 \to \ell^+ \ell^- \gamma$	$(3-6) \times 10^{-10}$
$B_s^0 \to \nu \bar{\nu} \gamma$	7×10^{-8}
$B_s^0 \to e^+ e^- e^+ e^-$	4×10^{-10}
$B_s^0 \to e^+ e^- \mu^+ \mu^-$	1×10^{-10}
$B_s^0 \to \mu^+ \mu^- \mu^+ \mu^-$	4×10^{-11}
$B_s^0 \to \gamma\gamma$	—
$B^0 \to X_s \nu \bar{\nu}$	4×10^{-5}
$B^0 \to K \sum_i \nu_i \bar{\nu}_i$	$(2-9) \times 10^{-6}$
$B^0 \to K^* \sum_i \nu_i \bar{\nu}_i$	$(2-20) \times 10^{-6}$

SUSY with R-Parity Violation. As an example of the kind of new physics, these processes are sensitive to consider supersymmetric extensions of the Standard Model with a superpotential that breaks R-parity:

$$W = \frac{1}{2} \lambda_{ijk} L_i L_j E_k^c + \lambda'_{ijk} L_i Q_j D_k^c + \frac{1}{2} \lambda''_{ijk} U_i^c D_j^c D_k^c. \quad (2)$$

It is customary to set $\lambda''_{ijk} = 0$ for proton stability. Fig. 1 shows contributions to $B^- \to \ell^- \nu$ from slepton

CP722, *B Physics at Hadron Machines*, edited by M. Paulini and S. Erhan
© 2004 American Institute of Physics 0-7354-0203-5/04/$22.00

TABLE 2. Experimental bounds.

Decay	Upper Bound
$B^+ \to e^+ \nu_e$	1.5×10^{-5} [5]
$B^+ \to \mu^+ \nu_\mu$	6.5×10^{-6} [6, 7]
$B^+ \to \tau^+ \nu_\tau$	4.1×10^{-4} [8, 9]

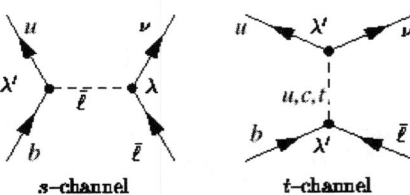

FIGURE 1. Sparticle exchange contribution to $B^- \to \ell^- \nu$.

(s-channel) and squark (t-channel) exchange, which depend on $\lambda\lambda'$ and on $(\lambda')^2$, respectively. The experimental bounds from B and D purely leptonic decays impose restrictions on these couplings for fixed sparticle masses. Typical upper bounds [1–4] on any one combination $\lambda\lambda'$ are in the range 2×10^{-2} to 7×10^{-5}.

3. DECAYS $B \to \gamma \ell^+ \ell^-$

Ward Identities. Until recently the theory of these decays was in disarray. It is now appreciated [10, 11] that some form factors are related by ward identities, while some calculations using *ad-hoc* hadronic models included more independent form factors than necessary, giving inconsistent results. Exactly the same issues arise in closely related modes, like $B \to K^* \gamma$.

The derivation of the Ward identities is simple. Consider the amplitude for $B \to f \gamma$ through an interaction described by the local operator $\mathcal{O}(x)$:

$$\langle \gamma(q,\varepsilon) f | \mathcal{O}(0) | B(v) \rangle$$
$$= -ie\varepsilon_\mu^* \int \mathrm{d}^4 x e^{iq\cdot x} \langle f | \mathrm{T} j_\mu^{\mathrm{e.m.}}(x) \, \mathcal{O}(0) | B(v) \rangle. \quad (3)$$

Here the electromagnetic current $j_\mu^{\mathrm{e.m.}} = +\frac{2}{3}(\bar{u}\gamma_\mu u + \bar{c}\gamma_\mu c) - \frac{1}{3}(\bar{d}\gamma_\mu d + \bar{s}\gamma_\mu s + \bar{b}\gamma_\mu b)$ couples to the photon of polarization vector ε_μ. The Ward identity is obtained by considering a derivative on the matrix element on the right hand side of Eq. (3). Acting on the time order product it gives nothing from the divergence of the conserved current plus a term from the derivative of the step function for the $x^0 = t$ coordinate:

$$-iq_\mu \int \mathrm{d}^4 x e^{iq\cdot x} \langle f | \mathrm{T} j_\mu^{\mathrm{e.m.}}(x) \, \mathcal{O}(0) | B(v) \rangle$$
$$= \int \mathrm{d}^3 x e^{-i\vec{q}\cdot\vec{x}} \langle f | [j_0^{\mathrm{e.m.}}(\vec{x}), \mathcal{O}(\vec{0})] | B(v) \rangle. \quad (4)$$

The commutator is non-vanishing, $[j_0^{\mathrm{e.m.}}(\vec{x}), \mathcal{O}(\vec{0})] \neq 0$ only if $\mathcal{O}$ carries electric charge. This is the case in, e.g., $B^+ \to \gamma e^+ \nu$, $B \to \pi(\rho)\gamma e^+ \nu$ or $B \to \bar{D}^{(*)} \gamma e^+ \nu$.

As an example of an application consider axial current transitions, $\mathcal{O} = \bar{b}\gamma_\nu\gamma_5 q$, and take a non-hadronic final state,

$$-iq_\mu \int \mathrm{d}^4 x e^{iq\cdot x} \langle 0 | \mathrm{T} j_\mu^{\mathrm{e.m.}}(x) \, (\bar{b}\gamma_\nu\gamma_5 q)(0) | B(v) \rangle$$
$$= (Q_b - Q_q) \langle 0 | \bar{b}\gamma_\nu\gamma_5 q | B(v) \rangle$$
$$= (Q_b - Q_q) f_B m_B v_\mu.$$

The non-local matrix element can be parametrized in terms of five form-factors $f_i(q^2, v \cdot q)$

$$-i \int \mathrm{d}^4 x e^{iq\cdot x} \langle 0 | \mathrm{T} j_\mu^{\mathrm{e.m.}}(x) \, (\bar{b}\gamma_\nu\gamma_5 q)(0) | B(v) \rangle$$
$$= f_1 g_{\mu\nu} + f_2 v_\mu v_\nu + f_3 q_\mu q_\nu + f_4 q_\mu v_\nu + f_5 v_\mu q_\nu. \quad (5)$$

The Ward identity then implies only three are independent:

$$(v \cdot q) f_2 + q^2 f_4 = (Q_b - Q_q) f_B m_B, \quad (6)$$
$$f_1 + q^2 f_3 + (v \cdot q) f_5 = 0. \quad (7)$$

Finally, since $q^2 = 0$ for an on-shell photon, the matrix element that appears in the amplitude for $B \to \gamma \ell \nu_\ell$ is given in terms of two form factors

$$\langle \gamma(q,\varepsilon) | \bar{b}\gamma_\mu\gamma_5 q | B(v) \rangle = -f_5 [(v \cdot q)\varepsilon_\mu^* - (v \cdot \varepsilon^*) q_\mu]$$
$$+ (v \cdot \varepsilon^*) v_\mu \frac{1}{v \cdot q} (Q_b - Q_q) f_B m_B. \quad (8)$$

Models of Form Factors. Four independent form factors are needed to describe the amplitude for $B \to \gamma \ell^+ \ell^-$:

$$\langle \gamma(q) | \bar{q}\gamma_\mu\gamma_5 b | \bar{B}(p) \rangle = ie\varepsilon^{*\alpha}(q)[g_{\mu\alpha}(p \cdot q) - p_\alpha q_\mu] \frac{F_A}{M_B}$$

$$\langle \gamma(q) | \bar{q}\gamma_\mu b | \bar{B}(p) \rangle = e\varepsilon^{*\alpha}(q)\varepsilon_{\mu\alpha\rho\sigma} p^\rho q^\sigma \frac{F_V}{M_B}$$

$$\langle \gamma(q) | \bar{q}\sigma_{\mu\nu}\gamma_5 b | \bar{B}(p) \rangle k^\nu = e\varepsilon^{*\alpha}(q)[g_{\mu\alpha} p \cdot q - p_\alpha q_\mu] F_{TA}$$

$$\langle \gamma(q) | \bar{q}\sigma_{\mu\nu} b | \bar{B}(p) \rangle k^\nu = ie\varepsilon^{*\alpha}(q)\varepsilon_{\mu\alpha\rho\sigma} p^\rho q^\sigma F_{TV}.$$

Here, $k = p - q$. The form factors F_A, F_V, F_{TA} and F_{TV} depend only on one variable which can be taken to be the photon energy E in the B restframe. A large energy effective theory [12] (LEET) can be used to show [13] that at large E all form factors are given in terms of one universal function,

$$F_V \simeq F_A \simeq F_{TA} \simeq F_{TV} \simeq \zeta_\perp^\gamma(E, M_B) \propto \frac{f_B M_B}{E}. \quad (9)$$

The proportionality constant in the last step is determined by

$$R \equiv \int \mathrm{d}x \, \frac{\phi_B(x)}{x}, \quad (10)$$

264

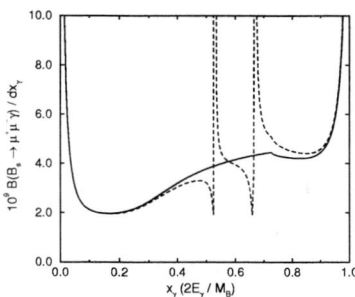

FIGURE 2. Photon spectrum in $B \to \gamma\mu\mu$ in a light front model [10].

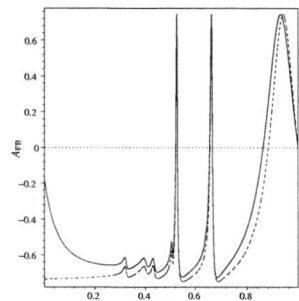

FIGURE 3. Photon spectrum in the forward-backward asymmetry in $B \to \gamma\mu\mu$. The dashed line is as predicted by LEET at lowest order, the solid line is a LEET inspired model [14].

which is however not known.

The light-front quark model [10] and a LEET inspired modification [14] have been introduced to capture these features in calculations of the $B \to \gamma\ell^+\ell^-$ amplitude, as shown in Figs. 2 and 3.

New Physics. Interest in very rare decay arises from the expectation that they are good probes of new physics. To evaluate this, one may either look at specific extensions of the Standard Model of electroweak interactions or, more generally, augment the Standard Model with non-renormalizable interactions parameterizing the effects of new physics. The most general effective Hamiltonian of four fermion interactions that may contribute to $B \to \gamma\ell^+\ell^-$ is [15]

$$
\mathcal{H}_{eff} = \frac{G\alpha}{\sqrt{2}\pi} V_{tq} V_{tb}^* \left\{ C_{SL} \bar{q} i\sigma_{\mu\nu} \frac{q^\nu}{q^2} b_L \bar{\ell}\gamma^\mu \ell \right.
$$

$$
+ C_{LL}^{tot} \bar{q}\gamma_\mu b_L \bar{\ell}\gamma^\mu \ell_L + C_{LR}^{tot} \bar{q}\gamma_\mu b_L \bar{\ell}\gamma^\mu \ell_R + C_{RL} \bar{q}\gamma_\mu b_R \bar{\ell}\gamma^\mu \ell_L
$$

$$
+ C_{RR} \bar{q}\gamma_\mu b_R \bar{\ell}\gamma^\mu \ell_R + C_{LRLR} \bar{q} b_R \bar{\ell}\ell_R + C_{RLLR} \bar{q} b_L \bar{\ell}\ell_R
$$

$$
+ C_{LRRL} \bar{q} b_R \bar{\ell}\ell_L + C_{RLRL} \bar{q} b_L \bar{\ell}\ell_L + C_T \bar{q}\sigma_{\mu\nu} b \bar{\ell}\sigma^{\mu\nu} \ell
$$

$$
\left. + C_{BR} \bar{q}_L i\sigma_{\mu\nu} \frac{q^\nu}{q^2} b_R \bar{\ell}\gamma^\mu \ell + iC_{TE} \varepsilon^{\mu\nu\alpha\beta} \bar{q}\sigma_{\mu\nu} b \bar{\ell}\sigma_{\alpha\beta} \ell \right\}
$$

$$(11)$$

In the Standard Model only three of these terms are present. In standard notation [16], $C_{BR} = -2m_b C_7$, $C_{LL}^{tot} = C_9 - C_{10}$ and $C_{LR}^{tot} = C_9 + C_{10}$. Fig. 4 shows [15] the effect of turning on and varying the individual coefficients C_X on the branching fraction for $B \to \mu^+\mu^-\gamma$.

4. FACTORIZATION AND SYMMETRY

From the QCD point of view, all three processes $B \to \gamma\ell\nu_\ell$, $B \to \gamma\gamma$ and $B \to \gamma\ell^+\ell^-$ are basically the same. Fig. 5 shows a Feynman diagram describing all

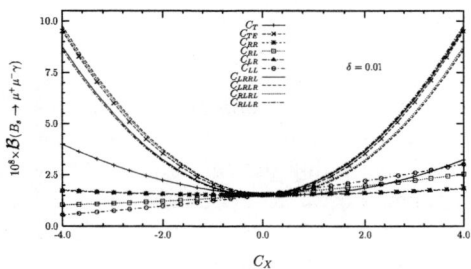

FIGURE 4. Effect of New Physics, as parametrized by the coefficients C_X of the effective Hamiltonian in (3), on $\mathcal{B}(B \to \mu^+\mu^-\gamma)$ [15].

three processes. However, the nature of the interaction represented by the gray blob is different in each of these processes. To show [17–21] that the three amplitudes are related, one must show factorization and use either heavy quark or LEET symmetry to relate all form factors, as was seen in the previous section. For example, Fig. 6 shows the ratio of the spectra,

$$
R_q(E_\gamma) = \frac{d\Gamma(\bar{B}_q \to \gamma e^+ e^-)/dE_\gamma}{d\Gamma(B^+ \to \gamma e^+ \nu)/dE_\gamma} \tag{12}
$$

as a function of the photon energy fraction $x_\gamma = 2E_\gamma/M_B$.

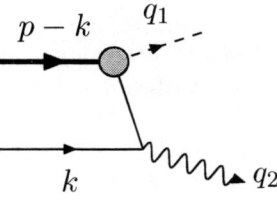

FIGURE 5. Feynman Diagram for $B \to \gamma\ell\nu_\ell$, $B \to \gamma\gamma$ and $B \to \gamma\ell^+\ell^-$ (understood as dressed with gluons).

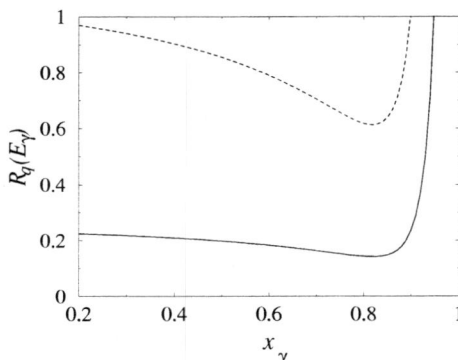

FIGURE 6. Ratio of the spectra $R_q(E_\gamma)$, Eq. (12), as a function of the photon energy fraction $x_\gamma = 2E_\gamma/M_B$. Solid line is $R_d \times 10^{-4}$, dashed line is $R_s \times 10^{-3}$ [20]. Long distance effects are not included.

$B \to \nu\bar{\nu}\gamma$. While admittedly experimentally "challenging," the lack of long distance contamination makes this process appealing. If measured we would have direct, untainted access to the underlying universal form factor in $B \to \gamma\ell\nu_\ell$, $B \to \gamma\gamma$ and $B \to \gamma\ell^+\ell^-$ among others. In the Standard Model, the decay is described by the effective Hamiltonian

$$\mathcal{H}_{\text{eff}} = \frac{\alpha G_F}{\sqrt{2}\sin^2\Theta_W}V_{tb}V_{ts}^*C^{\text{SM}}\sum_{l=e,\mu,\tau}[\bar{\nu}_{Ll}\gamma_\mu\nu_{Ll}][\bar{s}_L\gamma^\mu b_L].$$
(13)

The coefficient C^{SM} characterizes the short distance effects giving rise to the interaction. It is generated by the box diagram in Fig. 7 and a Z-penguin, which give

$$C^{\text{SM}} = \frac{x_t}{8}\left[\frac{x_t+2}{x_t-1}+\frac{3(x_t-2)}{(x_t-1)^2}\ln x_t\right],$$
(14)

where $x_t = (m_t/m_W)^2$.

In extensions [22, 23] of the Standard Model with only $(V-A)\times(V-A)$ couplings, the decay rate is characterized in terms of one coefficient $C = C^{\text{SM}}+C^{\text{new}}$:

$$\frac{d\Gamma}{dx} = \left|\frac{\alpha G_F V_{tb}V_{ts}^*C}{\sqrt{2}\sin^2\Theta_W}\right|^2 x(1-x)^3\left[|F_A(x)|^2+|F_V(x)|^2\right],$$
(15)

where $x = 2E_\gamma/M_B$. Models with additional generations or additional Higgs doublets are examples of extensions of the Standard Model preserving the $(V-A)\times(V-A)$ structure. Fig. 8 shows the ratio of total rate for $B \to \nu\bar{\nu}\gamma$ in a type-II two Higgs doublet model to the corresponding rate in the Standard Model, as a function of the Higgs mass and the mixing angle $\tan\beta$ characterizing the relative sizes of the expectation values of the two Higgs doublets.

Go off-shell: $B \to \gamma^*\ell^+\ell^-$. The unknown universal form factor for all processes above can be computed if

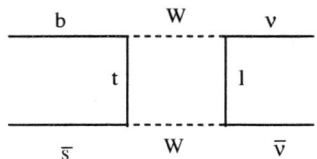

FIGURE 7. Feynman diagram for $B \to \nu\bar{\nu}\gamma$ in the Standard Model.

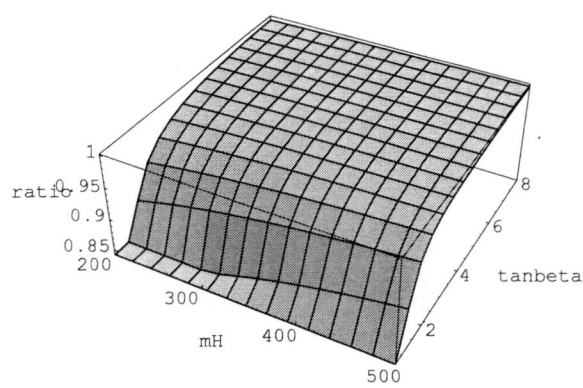

FIGURE 8. Ratio [24] of total rate for $B \to \nu\bar{\nu}\gamma$ in a type-II two Higgs doublet model to total rate in the Standard Model, as a function of the Higgs mass and the ratio of expectation values $\tan\beta$.

the photon is very off-shell. The intermediate light quark propagator in Fig. 5 goes off-shell and the rate can be computed (invoking local duality at large virtuality). We will see how this idea has been exploited in $B \to \pi\gamma^*$ below, but to my knowledge no one has applied it in $B \to \gamma^*\ell^+\ell^-$. Dincer and Sehgal [24] have estimated the decay rate for $B \to \ell^+\ell^-\ell^+\ell^-$ retaining only the Double-Dalitz conversion, and using the leading LEET relation Eq. (9) with a model to determine the proportionality constant in the last step in Eq. (9). Their results for the branching fractions are entered in Table 1. That the rates for $B \to \gamma^*\ell^+\ell^-$ are suppressed relative to the on-shell $B \to \gamma\ell^+\ell^-$ should not surprise the reader: there is an additional cost of phase space and electromagnetic coupling constant (perhaps surprising is that the suppression is far less than $\alpha/4\pi \sim 10^{-3}$).

5. $B \to \pi\ell^+\ell^-$ AND RESOLVING AMBIGUITIES IN α

Things get more complicated, but also more interesting when hadrons are involved in the final state. When one replaces the lepton pair by a quark pair, there are additional strong interactions which must be accounted for. The amplitude for $B \to \pi \ell^+ \ell^-$ is dominated by short distance transitions $b \to d\ell^+\ell^-$ and, to lesser extent, $b \to d\gamma$, which arise from virtual W and top-quark exchange. At long distances these are described by the following local operators:

$$\mathcal{O}_7 = \frac{e}{16\pi^2} m_b (\bar{d}_L \sigma^{\mu\nu} b_R) F_{\mu\nu}, \qquad (16)$$

$$\mathcal{O}_9 = \frac{e^2}{16\pi^2} (\bar{d}_L \gamma^\mu b_L) \bar{e} \gamma_\mu e, \qquad (17)$$

$$\mathcal{O}_{10} = \frac{e^2}{16\pi^2} (\bar{d}_L \gamma^\mu b_L) \bar{e} \gamma_\mu \gamma_5 e. \qquad (18)$$

Computation of the amplitude for $B \to \pi \ell^+ \ell^-$ requires knowledge of the same form factors that arise in the semileptonic decay, $B \to \pi \ell \nu$:

$$\langle \pi(p') | \bar{d} \gamma^\mu b | B(p) \rangle = (p + p')^\mu f_+ + (p - p')^\mu f_- \quad (19)$$

To calculate the contribution from the short distance $b \to d\gamma$ an additional form factor, not directly measured in the semileptonic decay, is also required:

$$\langle \pi(p') | \bar{d} \sigma^{\mu\nu} b | B(p) \rangle = 2ih(p^\nu p'^\mu - p^\mu p'^\nu) \quad (20)$$

Since this is a small contribution to the total amplitude, and there exist several methods for inferring this, we will not discuss it here further.

As is often the case for exclusive decays, there is long distance contamination to the amplitude. The interference between the long distance amplitude ($\propto V_{ub}V_{ud}^*$) and the short distance amplitude ($\propto V_{tb}V_{td}^*$) depends on $\cos\alpha$. While the branching fraction is small, of order of a few times 10^{-9} in the restricted region of phase space useful for the analysis below, even a rough measurement of the lepton mass spectrum may be enough to differentiate among the values of $\cos\alpha$ allowed by a $\sin 2\alpha$ determination through CP asymmetries. Remarkably the long distance contributions are computable [25] to leading order in a large energy expansion.

Fig. 9 is a schematic representation of the long distance contribution in which the photon is emitted from the light quark in the incoming B meson, while Fig. 10 shows the contribution with an emission from a quark in the outgoing pion. Formally, the hadronic part of the amplitude,

$$H^\mu = \langle \pi | \int d^4 x \, e^{iq \cdot x} \, T(j_{\rm em}^\mu(x) \mathcal{H}_{\rm eff}'(0)) | B \rangle \quad (21)$$

does not admit such a decomposition. However, when the pion recoils at large energy this relation factors,

$$H^\mu = \kappa \int d^4 x \, e^{iq \cdot x} \big[\langle \pi | \, T(j_{\rm em}^\mu(x) j_\lambda(0)) | 0 \rangle \tfrac{1}{2} f_B p_B^\lambda$$

$$+ \langle \pi | j_\lambda(0) | 0 \rangle \langle 0 | T(j_{\rm em}^\mu(x) J^\lambda(0)) | B \rangle \big], \quad (22)$$

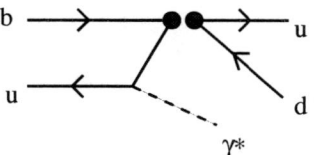

FIGURE 9. Schematic representation of long-distance contribution to $B \to \pi \ell^+ \ell^-$, with γ^* emitted from initial quark.

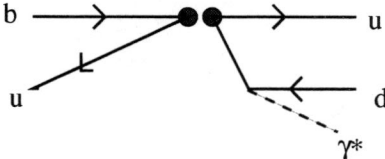

FIGURE 10. As in Fig. 9 but with γ^* emitted from final quark.

giving meaning to the picture described above. The first line in Eq. (22) corresponds to γ^* emission from the pion. The factor of $f_B p_B^\lambda$ is just the amplitude for the B meson to be annihilated by the axial-current. The first line can be computed exactly using a Ward identity (giving $-e\kappa f_\pi f_B p_B^\mu$). The second line involves the same form factor as in the $B \to \ell^+\ell^-\ell^+\ell^-$ amplitude of the previous section. This can be computed in an expansion in $\Lambda_{\rm QCD} m_b/q^2$. The first correction in this expansion is computable too. One obtains:

$$H^\mu = -\tfrac{4}{3} e\kappa f_\pi f_B p_B^\mu (1 + \tfrac{2}{3} \bar{\Lambda} m_b/q^2), \quad (23)$$

where $\bar{\Lambda} = m_B - m_b \sim \Lambda_{\rm QCD}$ is a standard HQET parameter.

Denoting by C_i the coefficients of the short distance interactions Eqs. (16)–(18) and putting it all together the rate is

$$\frac{d\Gamma}{dq^2} = |V_{tb}V_{td}^*|^2 \frac{G_F^2 \alpha^2 m_B^3}{3 \times 2^9 \pi^5} \frac{(m_B^2 - q^2)^3}{m_B^6} \Big[|C_{10} f_+|^2$$

$$+ \Big| C_9 f_+ + 2m_b C_7 h$$

$$- \frac{16\pi^2}{3} \frac{V_{ub}V_{ud}^*}{V_{tb}V_{td}^*} \frac{c(m_b) f_\pi f_B}{q^2} \big(1 + \frac{2\bar{\Lambda} m_b}{3q^2}\big) \Big|^2 \Big].$$

$$(24)$$

Figure 11 shows the rate as a function of the lepton invariant mass q^2 for $\cos\alpha = 0$ (solid line), $\cos\alpha = -1$ (dashed) and $\cos\alpha = 1$ (dotted). Only the region below the J/ψ threshold is shown to avoid complications due to charm. The shaded region indicates the size of the parametric error, from the sub-leading terms in our combined expansion in $1/E_\pi$ and $1/q^2$. The $1/q^2$ expansion becomes unmanageable below a couple of GeV/c^2. It

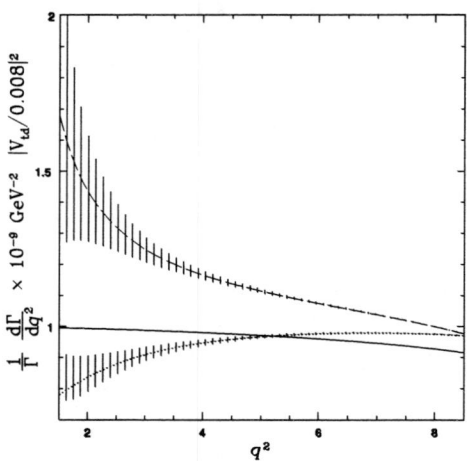

FIGURE 11. Differential Decay Rate for $B \to \pi \ell^+ \ell^-$ as a function of the lepton pair invariant mass q^2 [25]. The top (dashed), middle (solid) and lower (dotted) lines are drawn for $\cos \alpha = -1, 0$ and 1, respectively. The shaded region shows the uncertainty in the calculation as given by the next order terms in the combined expansions in $1/E_\pi$ and $1/q^2$.

may be possible to extend the prediction to lower values of q^2 by relating the $q^2 \to 0$ limit of the form factor to the on-shell universal form factor for $B \to \gamma \ell \nu_\ell$, $B \to \gamma \gamma$ and $B \to \gamma \ell^+ \ell^-$ discussed above. In any case, in the region shown, the very distinct interference behavior of the different signs of $\cos \alpha$ should make it possible to unravel the ambiguity in α left behind by a determination of $\sin 2\alpha$ from *CP* asymmetries.

6. CONCLUSIONS

There is a multitude of ultra-rare processes which can be studied in ultra-luminous colliders or B factories. In summary, there are three important lessons that we wish to emphasize:

- Modes which have very small branching fractions in the Standard Model because of an accidental suppression are very welcome. It is in such modes that new physics could most dramatically show up.

- While we cannot compute form factors in QCD reliably, many processes, including $B \to \gamma \gamma$, $B \to \gamma \ell \nu_\ell$, $B \to \gamma \ell^+ \ell^-$ and $B \to \nu \bar{\nu} \gamma$, are determined in terms of the same universal form factor. *Caveat emptor:* this statement is true of the leading amplitudes in a large energy expansion.

- Large energy expansion may allow computation (or relation to other observables) of ultra-rare processes even with a hadron in the final state. An example is $B \to \pi \ell^+ \ell^-$ which can be used to settle the am-

biguity in α from a measurement of $\sin 2\alpha$ from *CP* asymmetries.

ACKNOWLEDGMENTS

This work is funded in part by the Department of Energy under contract DE-FG03-97ER40546.

REFERENCES

1. Baek, S., and Kim, Y. G., *Phys. Rev.* **D60**, 077701 (1999).
2. Dreiner, H. K., Polesello, G., and Thormeier, M., *Phys. Rev.* **D65**, 115006 (2002).
3. Akeroyd, A. G., and Recksiegel, S., *Phys. Lett.* **B541**, 121–128 (2002).
4. Akeroyd, A. G., and Recksiegel, S., *Phys. Lett.* **B554**, 38–44 (2003).
5. Artuso, M., et al., [CLEO Collaboration], *Phys. Rev. Lett.* **75**, 785–789 (1995).
6. Abe, K., et al., [Belle Collaboration] (2001), prepared for 20th International Symposium on Lepton and Photon Interactions at High Energies (LP 01), Rome, Italy, 23-28 Jul 2001.
7. Aubert, B., et al., [BaBar Collaboration], arXiv:hep-ex/0307047 (2003).
8. Aubert, B., et al., [BaBar Collaboration], arXiv:hep-ex/0303034 (2003).
9. Abreu, P., et al., [DELPHI Collaboration], *Phys. Lett.* **B496**, 43–58 (2000).
10. Geng, C. Q., Lih, C. C., and Zhang, W.-M., *Phys. Rev.* **D62**, 074017 (2000).
11. Grinstein, B., and Pirjol, D., *Phys. Rev.* **D62**, 093002 (2000).
12. Dugan, M. J., and Grinstein, B., *Phys. Lett.* **B255**, 583–588 (1991).
13. Charles, J., Le Yaouanc, A., Oliver, L., Pene, O., and Raynal, J. C., *Phys. Rev.* **D60**, 014001 (1999).
14. Kruger, F., and Melikhov, D., *Phys. Rev.* **D67**, 034002 (2003).
15. Aliev, T. M., Ozpineci, A., and Savci, M., *Phys. Lett.* **B520**, 69–77 (2001).
16. Grinstein, B., Savage, M. J., and Wise, M. B., *Nucl. Phys.* **B319**, 271–290 (1989).
17. Korchemsky, G. P., Pirjol, D., and Yan, T.-M., *Phys. Rev.* **D61**, 114510 (2000).
18. Li, H.-n., *Phys. Rev.* **D64**, 014019 (2001).
19. Lunghi, E., Pirjol, D., and Wyler, D., *Nucl. Phys.* **B649**, 349–364 (2003).
20. Descotes-Genon, S., and Sachrajda, C. T., *Nucl. Phys.* **B650**, 356–390 (2003).
21. Bosch, S. W., arXiv:hep-ph/0308319 (2003).
22. Barakat, T., *Nuovo Cim.* **110A**, 631–635 (1997).
23. Dincer, Y., arXiv:hep-ph/0204183 (2002).
24. Dincer, Y., and Sehgal, L. M., *Phys. Lett.* **B556**, 169–176 (2003).
25. Grinstein, B., Nolte, D. R., and Rothstein, I. Z., *Phys. Rev. Lett.* **84**, 4545–4548 (2000).

B Physics in the LHC Era

Robert N. Cahn

Lawrence Berkeley National Laboratory, 1 Cyclotron Rd., Berkeley CA 94720, U.S.A.

Abstract. The nature of B physics will be transformed by the LHC. The nature of that transformation will depend on what we find there. B physics will no longer have a compelling role in discovering New Physics, but it will remain a central part of our exploration of fundamental interactions.

1. B PHYSICS IN A HISTORICAL CONTEXT

Is B physics the K physics of the 21st century? A case can be made that the K meson taught us most of what we know about particle physics. From it, we learned about strangeness and about mixing of neutral particles. The τ-θ puzzle led to the discovery of parity violation. From strangeness came $SU(3)$ and quarks. Of course the K meson brought us CP violation and the absence of neutral weak currents led to the postulate of charm. More recently, in K decay we saw direct CP violation.

The results so far in B physics are impressive, especially the confirmation at the few-percent-level of the CKM prediction for $\sin 2\beta$. It is too soon to know whether B physics will bring any revolutions of the sort due to the K meson.

The agenda for fundamental physics today includes the perennial favorite electroweak symmetry breaking, together with grand unification/extra dimensions, the baryon-antibaryon asymmetry, dark matter, and dark energy. B physics has a plausible connection to the first three. That connection relies on the hope that deviations from Standard Model predictions discovered in B physics, will demonstrate the need to extend the Standard Model in some more or less well-defined way.

2. VIRTUAL ATTACK ON FUNDAMENTAL PROBLEMS

Two B physics processes that are especially susceptible to modification by New Physics are mixing and penguin decays. These are illustrated in Fig. 1.

So far, no flaws in the Standard Model have been revealed. Still, there are some intriguing results that might portend contradictions. Let us briefly review some of the interesting measurements from BaBar, Belle, and CLEO

and consider what might be possible with much larger data samples, either at high luminosity e^+e^- machines or at LHCb and BTeV.

2.1. $J/\psi K_S$

This is truly the Golden Mode, the ideal both experimentally and theoretically. The statistical errors reported by BaBar and Belle are 0.067 and 0.057 with 80 fb^{-1} and 140 fb^{-1}. The systematic errors are correspondingly 0.034 and 0.028. With 0.5 ab^{-1} we might get a combined error of 0.03. This is within reach in a few years at both of these experiments. With BTeV or LHCb we could imagine an error of 0.017. Even better results might be achieved at a Super-B-factory.

2.2. ϕK_S

The CP eigenstate ϕK_S is even more inviting theoretically, since the decay is likely pure penguin. Unfortunately, the rates for this process are much less than for $J/\psi K_S$. As a consequence, the statistical errors at present are about 0.43 in $\sin 2\beta$, that is about six times as great as they are in $J/\psi K_S$. Indeed, even with BTeV or LHCb, the statistical errors would be about 0.14 for a year's running. A Super-B-factory would do better, reaching perhaps 0.05, that is, comparable to the error today in $J/\psi K_S$.

2.3. B_S Mixing

The mixing of B_s mesons is completely analogous to that of B^0 mesons except that everywhere the d quark is replaced by an s quark. Thus for the ratio of the

CP722, B Physics at Hadron Machines, edited by M. Paulini and S. Erhan

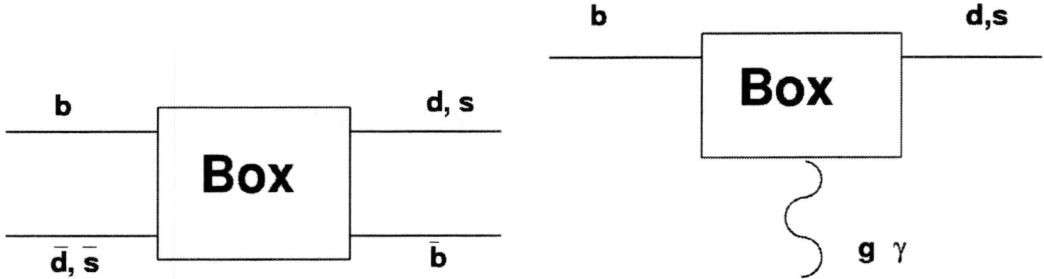

FIGURE 1. Both mixing and penguin decays are sensitive to New Physics, which may lie inside the box.

$x = \Delta m / \Gamma$ parameters, we find

$$\frac{x_s}{x_d} = \frac{m_{B_s} \eta_{B_s} B_{B_s} f_{B_s}^2}{m_{B_d} \eta_{B_d} B_{B_d} f_{B_d}^2} |V_{ts}/V_{td}|^2. \qquad (1)$$

Here, $\eta_{B_{d,s}}$ is a known QCD correction, $B_{B_{d,s}}$ is a non-perturbative matrix element, and $f_{B_{d,s}}$ is the decay constant for the $B_{d,s}$. While the $B_{B_{d,s}}$ and $f_{B_{d,s}}$ are not known to high precision, the ratios of the values for d and s have a smaller uncertainty. Thus, the awaited measurement of x_s at Fermilab will give an important new constraint on the unitarity triangle, the most demanding one since the measurement of $\sin 2\beta$. We anticipate knowing $|V_{td}/V_{ts}|$ to 10%.

Earlier hopes that this measurement might be in hand already, have turned out to be far too optimistic. At this conference, we heard [1] that to achieve 5σ significance will take approximately ≈ 1.7 fb^{-1} if $\Delta m_s = 18$ ps^{-1} or ≈ 3.2 fb^{-1} if $\Delta m_s = 24$ ps^{-1}. There is some danger that the measurement of Δm_s will turn out to be LHC Era physics.

2.4. $B_S \to J/\psi \phi$

This process is the B_s analog of the Golden Mode $B^0 \to J/\psi K$ with the spectator d quark turned into s. It is thus the hadron-collider Golden Mode and the asymmetry should be very small since both mixing and decay include only second and third generation quarks. We know that Standard Model CP violation requires three generations, so all CP violation here is suppressed. BTeV expects to be able to measure $\sin 2\chi$, the analog of $\sin 2\beta$, to ± 0.024, a very impressive result.

2.5. $B^0 \to \pi\pi, \pi\rho, \rho\rho$

In early discussions of the CP program for B physics, $B^0 \to \pi\pi$ seemed like the natural means of determining the unitarity triangle angle α. That hope has dimmed as

the branching ratios turned out to be small. Because the decay involves not only the tree process $b \to u\bar{u}d$, but the $b \to d$ penguin, it is necessary to make a number of measurements in addition to the obvious time dependence in $B^0 \to \pi^+\pi^-$. The most daunting of these are the measurements of the separate branching fractions $B^0 \to \pi^0\pi^0$ and $\bar{B}^0 \to \pi^0\pi^0$. Only recently has the combined branching fraction been measured at $(2.1 \pm 0.6 \pm 0.3) \times 10^{-6}$ [2] and $(1.7 \pm 0.6 \pm 0.2) \times 10^{-6}$ [3]. The great difficulty of this measurement is one of the strongest arguments for a high-luminosity B-factory.

There are alternatives to the $\pi\pi$ mode for measuring α. The $\rho\pi$ channel has already yielded interesting results on CP asymmetries. In the long run, a full time-dependent Dalitz plot analysis has the potential to measure α. Recently, the $\rho\rho$ channel has become an interesting candidate. Higher decay rates make this more accessible than $\pi\pi$ and even $\rho\pi$. There are complications here from having three partial waves, but this may not be too big of a problem. Certainly, understanding backgrounds from other channels will be critical.

2.6. $b \to s\gamma$, $b \to s\ell\ell$

Theorists want results for $b \to s\gamma$ and experimenters want to measure $B \to K^*\gamma$. The inclusive measurement requires making a cut to include only events with gammas above some energy. Further reductions in background can be made by requiring a lepton from the other B meson.

In $b \to s\ell\ell$, there are electroweak contributions, not just electromagnetic ones. Again we have the conflict between the experimenters' desire for exclusive final states and the theorists' desire for the inclusive measurement. Of particular interest is the measurement of the angular distribution of the leptons. Interference between axial and vector couplings can produce an asymmetry that is sensitive to new physics contributions.

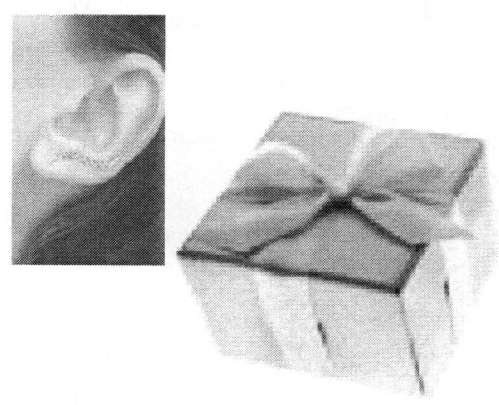

FIGURE 2. To sense New Physics inside the Boxes of B mixing and penguin diagrams, we can only shake the Boxes and hope to distinguish some new tones above those dictated by the Standard Model.

3. ROLE OF B PHYSICS IN LHC ERA

The dream of B physics today is to use high statistics data from the two B-factories to find an unambiguous discrepancy with the Standard Model. This will be ascribed to new particles in those Boxes appearing in mixing and in penguin diagrams. We would next see the announcement in the New York Times of the discovery of supersymmetry.

This dream is likely to remain a dream in the pre-LHC era. We can see from the statistical limitations of all of the measurements described above, except for the Golden Mode $B^0 \to J/\psi K_S^0$ (and the anticipated result for Δm_s) that a five-σ discrepancy is quite unlikely, unless the extraordinary central value found by Belle for $B^0 \to \phi K_S^0$ holds up. Truly precise measurements will come only with LHCb and BTeV, and a Super-B-factory, should one be built.

That means that the precision data will arrive at the same time as LHC opens up the energy frontier. The New Physics Box will be available to those doing B physics and those looking at the energy frontier. However, while the frontiersmen will be allowed to open the Box, the B physics community will be allowed only to shake it, hoping to interpret the faint rumblings inside as signals of the real contents.

4. HISTORY OF VIRTUAL DISCOVERIES

No matter how loudly we insist that deviations from the Standard Model would constitute the discovery of New

FIGURE 3. At the LHC, ATLAS and CMS will open the Box to inspect the New Physics directly.

Physics, history argues otherwise. There may be some dispute as to who really deserves credit for discovering the W boson, but the names proposed would probably include members of UA 1 and UA 2, who found the W in its leptonic decays at the $Sp\bar{p}S$ at CERN in 1983. Advocates of virtual discovery ought to point to the scientists in Figure 4. More recent examples of virtual discover-

FIGURE 4. The discoverers of the W boson through its virtual appearance. Rutherford discovered β decay. Fermi identified β decay as probably the consequence of an interaction analogous to electromagnetism.

ies include that of the Z by the Gargamelle experiment, which first established the existence of neutral (flavor-conserving) weak currents and that of charm by Ben Lee and Mary K. Galliard, who predicted the charm quark mass from the absence of neutral flavor-changing currents.

5. WHAT WILL WE FIND AT THE LHC?

The future of B physics depends quite directly on what is found at the LHC. In the most favorable scenario a whole new spectroscopy is found. This might or might not be supersymmetry. This would be the particle physics jackpot. All of those goodies that the LHC would have revealed to be inside the Box would necessarily make

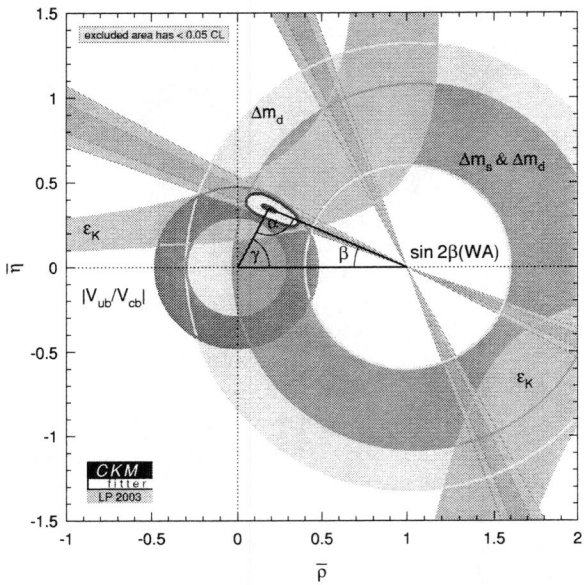

FIGURE 5. A contemporary unitarity triangle from the CKM Fitter team.

some sound, perhaps muffled, that would, in principle, be perceptible in *B* physics. This would indeed be a great challenge: to verify, at least semi-quantitatively, the virtual effects of the new particles on the behavior of *B* mesons.

Another possibility is that we will find a single, orthodox Higgs boson. This would make most of us happy for about 24 hours. After that, gloom would set in while we looked for other work.

Yet another grim possibility is that no sharp phenomena will appear, perhaps only the strong interactions among longitudinal gauge bosons. This would make life at the LHC tough and would not benefit *B* physics either.

The first significant running at LHC may seal the future of *B* physics, indeed the future of all high energy physics.

The consensus of the theoretical community is that there will be a new spectroscopy. Indeed, you need to wonder whether the young theorists being trained today are aware that supersymmetry has not yet been established.

If *B* physics is to meet the challenge of observing virtual effects from New Physics discovered at the LHC, unprecedented precision will be needed.

6. WHAT LIMITS PRECISION TESTS OF THE STANDARD MODEL?

Consider a contemporary plot of the Unitarity Triangle, shown in Fig. 5.

The big swaths associated with the various measurements are not wide because the experiments are imprecise but rather because there are limits to our theoretical understanding. Because the theoretical uncertainties arise from non-perturbative quantities, we are likely to depend ultimately on lattice gauge calculations. The increasing precision of these calculations has been discussed at this conference [4].

It is not enough to do better calculations: we must know how good they really are.

7. THEORY MUST BECOME AS RIGOROUS AS EXPERIMENT

In evaluating whether a measurement conforms to the expectations of the Standard Model, we need to take into account both statistical and systematic errors. In many instances, however, it is the theoretical error that dominates. It follows then that we must have an equivalent understanding of the theoretical errors. Without an understanding of errors, the theoretical results are really not of much use. In particular, the lattice gauge calculations need to incorporate methodology that gives meaningful error estimation.

If we are going to demand such things, and we must, then we need to accept that the lattice gauge community will need resources appropriate to the task. The evaluation of the appropriate level of investment needs to consider the lattice gauge effort as an integral part of the experimental program. Without it, the experimental results will not be nearly so worthwhile. Proposals for lattice gauge calculations in support of the *B* physics program need to include milestones and standards for evaluating the progress made.

8. *B* PHYSICS ISN'T JUST AN ATTACK ON THE STANDARD MODEL

The Standard Model is an astonishing achievement. We assert that it describes all of physics up to an extraordinary mass scale (well, setting aside those nasty problems with dark matter and dark energy). The verification of this model is an enterprise worthy of real effort. The verifications of QED and more recently of the electroweak model in the *W-Z* sector are among the great achievements of physical science. The analogous verification of the weak decay component of the electroweak model is an equally compelling task. The role of *B* physics in this program is central. It is imperative that we continue the exploration of the Unitarity Triangle, even when it is no longer the best means of seeking New Physics.

QCD is part of the Standard Model and *B* physics provides an excellent perspective on QCD. In the absence of a complete solution to QCD, we shall always be reduced to approximations. The heavy quark approximation provides a platform for exploring QCD. Already much progress has been made here, but much more will be possible with better measurements and better lattice gauge calculations.

Finally, we need to remember that the greatest discoveries are often those that are not anticipated. We have had recent reminders of this in the discovery of the $D_s(2317)^*, D_s(2458)$ and the $X(3872)$.

ACKNOWLEDGMENTS

This work was supported in part by the Director, Office of Science, Office of High Energy and Nuclear Physics, of the U.S. Department of Energy under Contract DE-AC0376SF00098.

REFERENCES

1. Miao, T., in these proceedings (2004).
2. Aubert, B., et al., [BaBar Collaboration], *Phys. Rev. Lett.* **91**, 241801 (2003).
3. Abe, K., et al., [Belle Collaboration], *Phys. Rev. Lett.* **91**, 261801 (2003).
4. el Khadra, A., in these proceedings (2004).

PANEL DISCUSSION: MODELS AND METHODS: CAN THEORY MEET THE *B* PHYSICS CHALLENGE?

Models and Methods:
Can Theory Meet the *B* Physics Challenge?

Joel N. Butler

Fermi National Accelerator Laboratory, Batavia, IL 60510, U.S.A.

Abstract. Theory and experiment work together in the effort to obtain insight from *B* physics. At Beauty 2003, we held a panel discussion among eminent theorists working in this area. Below we present the motivation for this panel, the charge to the panelists, and some brief comments concerning the discussions and main conclusions.

1. THE MOTIVATION

Theory and experiment work together in the effort to extract knowledge from *B* physics. Throughout the Beauty 2003 conference, we have heard the latest experimental results concerning *B* decays, mixing, and *CP* violation and about the expectations for results from ongoing experiments and from new experiments planned to start around 2007-2009. We have also heard many theoretical talks, ranging from speculations on what a theory of flavor might look like to how Standard Model (SM) parameters can be disentangled from hadronic complications, to how "new physics" can be detected against "Standard Model backgrounds."

The emphasis of the next rounds of *B* experiments is on "new physics" beyond the Standard Model (BSM). This physics could appear for the first time in *B* decays, through to failure of certain processes to conform to SM expectations. Perhaps more likely, *B* physics will contribute to the interpretation of new physics discovered elsewhere, for example at the LHC or the Tevatron. It is expected that there will be many interpretations for the new phenomena and they may have definite, but different, implications for the pattern of *B* decays, mixing, and *CP* violation. By studying these phenomena in *B* decays with enough statistics, it may be possible to eliminate some of these interpretations and to pin down the parameters of others.

Since much of the ability of *B* physics to contribute to the development of insight into new physics depends on theoretical input, it is fair to ask whether theory is up to the task. We therefore posed two questions:

- Will theory provide enough knowledge and control of models to help use the data to extract "detailed information" about "New Physics" from the measurements that it will be possible to make?

- Will theory be able to master the hadronic difficulties to allow us to cleanly separate Standard Model predictions of weak parameters from hadronic effects
    - to permit us to perform all the consistency checks provided by the Standard Model to look for deviations that could indicate "New Physics", and
    - to be able to use a much larger set of *B* decays to address these issue than the "theoretically pristine" subset that we are now restricted to.

2. CHARGE TO THE PANEL

Based on the above considerations, we assembled a panel of six experts in various aspects of *B* physics. Their "charge" was

At this conference, we have seen the latest results and projections of experimental sensitivities from future experiments to a variety of *B* decays. The stated goal of all experiments now is not only to measure Standard Model parameters but to find deviations in *CP*-violating or rare *B* decays that would signify New Physics or, if New Physics were found elsewhere to help in its interpretation. We ask the panel to discuss, how they think *B* physics will unfold over the next decade, whether the projected advance of theoretical models and tools are adequate to maximize the knowledge gained from the expected level of precision of experimental results, and what are the prospects for progress towards a theory of flavor.

CP722, *B Physics at Hadron Machines,* edited by M. Paulini and S. Erhan

3. CONDUCT AND COMPOSITION OF THE PANEL

The conduct of the panel was as follows: each of the six panelists was asked to give a brief statement, lasting about 10 minutes, addressing the charge. This was followed by a discussion among the panelists and finally by an open session where the audience asked questions and engaged in debate with the panel members.

The panelists, listed in the order in which they gave their initial presentations, were:

- Lincoln Wolfenstein
- Gino Isidori
- Ulrich Nierste
- Amarjit Soni
- David London
- Robert Cahn

4. SOME OBSERVATIONS

Several of the panelists have provided written summaries of their remarks and of their views of what transpired during the subsequent discussions. Here, I make a few general observations extracted from my own notes.

An important observation, in this panel and in the talks, was that some tests of flavor symmetry suggest a scale for new flavor physics $> 10^2$ TeV while the Electroweak sector indicates New Physics at the ~ 1 TeV scale, that nevertheless could have strong impact on flavor physics. Any New Physics scenario that does contribute to CP violation, rare decays or mixing must include a mechanism to respect the constraints already imposed by existing measurements and limits. A second point is that decays involving Flavor Changing Neutral Currents (FCNC) are excellent places to look for contributions from new physics. There are many examples where theoretical uncertainties are less than 10%. Looking at specific decays where the Standard Model predicts (near)zero CP asymmetry, (near) zero branching fraction, or near-zero mixing (as in charm) are especially promising. It is possible to measure CKM parameters from tree-level diagrams only and then to compare them with measurements using FCNC processes. Differences would signify New Physics. A third important observation was that the panelists felt that B physics' main role would be to help with the elucidation of New Physics discovered elsewhere by direct observation. The panelists agreed that it was unlikely that the first New Physics observations would produce a clear and unambiguous picture and that precision measurements in B, charm, and kaon decays may play an important role in clarifying the situation or at least in eliminating false explanations. It

was suggested that the Electroweak and flavor physics community would do well to talk to one another more to help speed up the process of considering new results in both areas together. A fourth general point of agreement was that theory errors would improve over this period both through progress in lattice gauge calculations, and in analytic approaches and model building but they might not be accurate enough to permit a firm conclusion that a result violated the Standard Model unless the deviation was very large. Even if they did, it might be very difficult to identify the nature of the New Physics. The speakers stressed that experiments should take maximum advantage of the existence of the subset of theoretically clean decays and do an excellent job of studying them. Fifth, there was agreement that the study of both the kaon system and the B system are important, and that, in the B system, B_s^0 studies would be especially significant. Sixth, we were cautioned that even if physics beyond the Standard Model turns out to have large parameters, their manifestation could wind up being small so that it would be good to plan experiments that look for deviations from SM expectations at the order of $\sim 10^{-3}$. This "guess" is obtained by analogy with CP violation in K decays, which is small even though, in some sense CP-violating parameters in the Standard Model are large. Finally, there seemed to be agreement that with the advent of the LHC, the HEP community finally would begin to see the New Physics and that, one way or another, B, kaon and charm physics were likely to play an important role in elucidating or constraining its interpretation.

Finally, in behalf of the conference organizers and for myself, I would like to thank the six panelists for their participation in this event, for their many insights, and for the candor, humor, and grace with which they were willing to discuss this complex topic.

ACKNOWLEDGMENTS

This work received partial support from Fermilab which is operated by University Research Association Inc. under Contract No. DE-AC02-76CH03000 with the United States Department of Energy.

Models and Methods:
Can Theory Meet the *B* Physics Challenge?

Lincoln Wolfenstein

Carnegie Mellon University, Department of Physics, Pittsburgh, PA 15213, U.S.A.

Abstract. This is the write-up of my contribution to the Panel Discussion on "Models and Methods: Can Theory Meet the *B* Physics Challenge?" chaired by Joel Butler at the Beauty 2003 Conference.

1. INTRODUCTION

For 25 years after 1964, all *CP* violation could be explained by one parameter, a term in the kaon mass matrix that violated *CP* and was *CPT* invariant. This could be due to a superweak interaction with a strength more than 10 billion times weaker than the normal weak interaction. Later in the 1990's, *CP* violation in the kaon decay amplitude was established from the measurement of the parameter ε'_K. However, this was only 4 parts per million and so might be due to some new very weak physics.

It became clear in 1983 with the measurement of the *B* lifetime that the CKM model predicted large *CP* violation in the *B* system. The first verification of this qualitative prediction came with the measurement of $\sin 2\beta$. However, this could be attributed to *CP* violation in B^0-$\bar{B}^0$ mixing and so alternatively a very weak $\Delta B = 2$ interaction. Thus I believe the next crucial step is the equivalent of the ε'_K experiment for the *B* system.

2. CURRENT SITUATION

Within the Standard Model, this means showing that γ is not zero. This can be done by measuring the *S* parameter in a decay like $B^0 \rightarrow \pi^+\pi^-$ and showing that it does not equal $\sin 2\beta$. Alternatively since there is evidence for a significantly strong final state phase, finding a difference in the decay rates of *B* and $\bar{B}$ to final states f and $\bar{f}$, respectively, or, equivalently the *C* parameter in the time-dependent decay to a *CP* eigenstate.

The next step is NOT a precise determination of γ in addition to β, or equivalently, my parameters ρ and η. No one cares what the exact values are. The goal is to find quantitative checks of the CKM model or, hopefully, signals of new physics. Of course, this can be done by precise determinations of the parameters by different methods and seeing if they agree. In particular, it is important to see if the determination of ρ and η based on the phase and magnitude of the mixing is consistent with the value of γ determined from decays.

The real challenge is to evaluate the theoretical uncertainties. One needs to identify those measurements for which these are small. Of course, there is the possibility that the new physics is relatively large as suggested by some preliminary results discussed in these proceedings and we can proceed to the next step of trying to delineate the new physics. For the moment, the theorists must patiently await more accurate experimental results and the experimentalists must await patiently cleaner theoretical predictions.

CP722, *B Physics at Hadron Machines,* edited by M. Paulini and S. Erhan

Models and Methods:
Can Theory Meet the *B* Physics Challenge?

Ulrich Nierste

Fermi National Accelerator Laboratory, Batavia, IL 60510, U.S.A.

Abstract. The *B* physics experiments of the next generation, BTeV and LHCb, will perform measurements with an unprecedented accuracy. Theory predictions must control hadronic uncertainties with the same precision to extract the desired short-distance information successfully. I argue that this is indeed possible, discuss those theoretical methods in which hadronic uncertainties are under control and list hadronically clean observables.

The target of *B* physics is short-distance physics associated with the electroweak and even higher scales: *CP* violation, CKM elements and the search for new physics. Ideally, one wants to gain enough experimental information to disentangle Standard Model (SM) and new physics and to quantify the magnitudes and phases of the CKM elements and the parameters of the new theory precisely. Short-distance physics couples to quarks, but in experiments we encounter hadrons. The theorist's task is to relate the hadronic amplitudes to the quark-level transition. This step involves non-perturbative QCD. It is therefore natural to ask whether theory can keep up with the increasing precision of future *B* physics experiments like BTeV and LHCb.

It is clear that hadronic models are not an acceptable tool for the extraction of fundamental parameters, because the uncertainties of model calculations are uncontrollable. Unfortunately, models are still used to estimate hadronic matrix elements, often in the disguise of 'plausible dynamical assumptions' or similar paraphrases. Yet the only reliable methods to deal with non-perturbative QCD are those which

 (i) are based on (approximate) symmetries of the QCD Lagrangian or

 (ii) involve a systematic expansion in a small parameter.

In these cases, one can assess the uncertainty of the calculation from the size of the symmetry breaking parameter or the expansion parameter. In certain cases even corrections to the symmetry limit can be computed to first order in the symmetry breaking parameter (often the first-order corrections are simply zero) or sub-leading terms of the expansion can be computed as well.

In *B* physics the *CP* symmetry of the strong interaction turns out to be most useful. From the searches for electric dipole moments, we know that QCD obeys *CP* to an accuracy of at least one part in a billion. This allows us to relate the matrix elements of *B* and *B̄* decays to each other. In certain cases, we can define quantities from which the hadronic physics drops out and the measurement is directly related to the desired short-distance physics. The most prominent example is the *CP* asymmetry $\mathcal{A}_{CP}(B^0 \to J/\psi K_S^0) = \sin 2\beta$ [1, 2]. This cancellation of hadronic elements from *CP* asymmetries usually fails when different operators with different weak phases interfere. We find this situation in K^0-$\bar{K}^0$ mixing or in the 'penguin pollution' in $B^0 \to \pi^+\pi^-$. Actually, there also is a penguin pollution in $\mathcal{A}_{CP}(B^0 \to J/\psi K_S^0)$, but it is suppressed by two powers of the Wolfenstein parameter $\lambda = 0.22$ and a loop factor. It yields only a correction of 1% or less. Our second sharpest tool is the isospin symmetry of QCD (relating *u* and *d* quarks), which is excellently fulfilled in *B* decays and, *e.g.*, used in the determination of α from $B \to \pi\pi$ decays [3]. Isospin symmetry can be enlarged to SU(3)$_F$, which, however, is broken substantially, because the strange quark mass is much larger then the up and down quarks masses. Especially the U-spin subgroup of SU(3)$_F$, which transforms *d* and *s* quarks into each other, is widely used. The SU(3)$_F$ breaking corrections are not always easy to estimate, but usually believed to be below 30%.

The large *b* quark mass opens the possibility to expand matrix elements in $\Lambda_{QCD}/m_b \sim 0.1$. To this end different observables require different theoretical methods. For inclusive decay rates one can employ an operator product expansion, the *Heavy Quark Expansion* (HQE) [4–8]. The HQE is used to determine $|V_{cb}|$ and $|V_{ub}|$ from inclusive semi-leptonic *B* decays. The leading term in the HQE of these decay rates is given by the calculable *b* quark decay rate and corrections of order Λ_{QCD}/m_b are absent, so that one can extract $|V_{cb}|$

CP722, *B Physics at Hadron Machines,* edited by M. Paulini and S. Erhan
© 2004 American Institute of Physics 0-7354-0203-5/04/$22.00

TABLE 1. Purity classification of theoretical methods. QCDF/SCET is not rated yet.

Rating:	method:	example:
*****	CP or isospin symmetry of QCD	$\gamma - 2\beta_s$ from $B_s^0 \to D_s^\pm K^\mp$
****	CP or isospin symmetry of QCD plus $\mathcal{O}(\lambda^2)$-suppressed penguin	β from $B \to J/\psi K_S^0$
***	HQE	$|V_{cb}|$ from incl. decays
***	HQET	$|V_{cb}|$ from $B \to D^* \ell \nu_\ell$
**	four-quark matrix elements from unquenched lattice QCD	B^0-$\bar{B}^0$ mixing
*	$SU(3)_F$ symmetry	γ from $B_s^0 \to K^+ K^-$ and $B^0 \to \pi^+ \pi^-$

with a small uncertainty. Heavy Quark Effective Theory (HQET) is another framework employing a systematic expansion in Λ_{QCD}/m_b [9–15]. It combines features (i) and (ii), since it exploits extra symmetries occurring in the limit $m_b \to \infty$ to constrain *e.g.* form factors. A prime application of HQET is the determination of $|V_{cb}|$ from $B \to D^* \ell \nu_\ell$ decays at zero recoil, because here Λ_{QCD}/m_b corrections vanish as well [16]. A more recent development is the application of heavy quark methods to hadronic two-body decays in the framework of *QCD Factorization* (QCDF) [17, 18]. A formulation of this concept in the language of effective field theories is the *Soft Collinear Effective Theory* (SCET) [19, 20]. With the help of QCDF/SCET one can express a large class of hadronic decays in terms of a few hadronic parameters and *e.g.* extract the CKM phase γ from $B \to K\pi$ decays [21, 22]. At present some conceptual issues and the calculability and size of the Λ_{QCD}/m_b corrections are unclear, so that the accuracy of QCDF/SCET calculations is hard to assess. But before the start of BTeV and LHCb we can expect clarifications from the confrontation of the predictions with more precise data. A common feature of all the described heavy quark methods is that possible terms of the form $\exp(-\kappa m_b/\Lambda_{QCD})/m_b^n$ are missed. This issue has been discussed at length for the case of HQE, where it has been speculated that such exponential terms turn into damped oscillating terms proportional to $\sin(\kappa m_b/\Lambda_{QCD})/m_b^n$ [23] ("violation of quark-hadron duality"). While operator product expansions are widely used in many areas of QCD, no such terms have been observed so far. Finally unquenched lattice computations are solely QCD-based, too. Applications to *b* physics either use effective field theories like HQET or simulations with dynamical *b* quarks. In the latter case one must take the *b* mass lighter than in Nature and finally extrapolate to the physical mass using again information from HQET. Also light quark masses cannot be simulated with their actual values and chiral extrapolations are needed. Unquenched calculations are a new topic in lattice QCD and we don't know yet whether all systematic errors are sufficiently under control to take the challenge from BTeV and LHCb. I sum-

marize this discussion with my purity classification in Table 1. There are many more five and four star methods: for example the CP analyses in $B^\pm \to K^\pm D^0$, $B \to \pi\pi$, $B_s^0 \to \psi\phi, \psi\eta$ and $B_s^0 \to \phi\phi, \phi\eta, \eta\eta$ yielding the CP phases γ, α and β_s, or the determination of $|V_{td}/V_{ub}|$ from $\mathcal{B}(B^0 \to \ell^+ \ell^-)/\mathcal{B}(B^+ \to \ell^+ \nu_\ell)$.

It should be stressed that also the methods with three or less stars are suited to probe and possibly falsify the Standard Model. *b* physics provides us with a plethora of observables yielding redundant information on the short-distance physics of interest. New physics couples to quarks, which hadronize in many different ways. The corresponding rates are differently affected by hadronic physics. For example, parity-conserving new physics in $b \to s\bar{s}s$ transitions will possibly be seen first in $B^0 \to \phi K_S^0$ and then be confirmed through an angular analysis of $B \to \phi K^*$.

One point, however, must be stressed: In the presence of new physics, interference effects between the Standard Model amplitude and the new amplitude introduces hadronic uncertainties. This implies that we can falsify the SM from clean observables, but we cannot necessarily determine the parameters of the new theory cleanly. A prominent example for this situation is $\mathcal{A}_{CP}(B^0 \to \phi K_S^0)$. The best strategy for analyzing hints of new physics is the study of observables which are zero or very small in the Standard Model. Consider the situation that you'll find Δm_{B_s} off from the SM prediction by 10%. Did you find new physics or did the lattice people compute $f_{B_s}^2 B_{B_s}$ incorrectly? If the new contribution to B_s^0-$\bar{B}_s^0$ mixing does not come from the CKM mechanism, it will also affect the CP asymmetry $\mathcal{A}_{CP}(B_s^0 \to \psi\phi)$, which in this example can be enhanced by a factor of 3 compared to its small SM value. Important "near zero predictions" of the SM are

- certain CP asymmetries,
- certain rare decays,
- FCNC's in the charm system.

As an example, I mention the *CP* asymmetry in flavor-specific decays (meaning $\bar{B} \not\to f$ and $B \not\to \bar{f}$) [24, 25]:

$$
\begin{aligned}
\mathscr{A}_{\mathrm{fs}} &= \frac{\Gamma(\bar{B}(t) \to f) - \Gamma(B(t) \to \bar{f})}{\Gamma(\bar{B}(t) \to f) + \Gamma(B(t) \to \bar{f})} \\
&= -(5.0 \pm 1.1) \times 10^{-4},
\end{aligned}
$$

with *e.g.* $f = X\ell^{-}\bar{\nu}_{\ell}$. $\mathscr{A}_{\mathrm{fs}}$ is GIM suppressed in the SM and new physics can give an $\mathcal{O}(1)$ contribution.

In conclusion, the answer to the question posed to the panelists is definitely yes! There are many four- and five-star observables, which theory can predict with hadronic uncertainties below 1%. Yet once Standard Model contributions and new physics effects are found to interfere, this cleanliness can be lost and one may have to pursue other avenues to quantify the parameters of the new theory precisely. To this end "near zero predictions" of the Standard Model are useful, because they can be dominated by new physics.

ACKNOWLEDGMENTS

I thank the organizers of the conference for the opportunity to give this presentation and for financial support.

REFERENCES

1. Carter, A. B., and Sanda, A. I., *Phys. Rev.* **D23**, 1567 (1981).
2. Bigi, I. I. Y., and Sanda, A. I., *Nucl. Phys.* **B193**, 85 (1981).
3. Gronau, M., and London, D., *Phys. Rev. Lett.* **65**, 3381–3384 (1990).
4. Khoze, V. A., and Shifman, M. A., *Sov. Phys. Usp.* **26**, 387 (1983).
5. Shifman, M. A., and Voloshin, M. B., *Sov. J. Nucl. Phys.* **41**, 120 (1985).
6. Shifman, M. A., and Voloshin, M. B., *Sov. Phys. JETP* **64**, 698 (1986).
7. Shifman, M. A., and Voloshin, M. B., *Zh. Eksp. Teor. Fiz.* **91**, 1180 (1986).
8. Bigi, I. I. Y., Uraltsev, N. G., and Vainshtein, A. I., *Phys. Lett.* **B293**, 430–436 (1992), [Erratum-ibid. **B297** (1992) 477].
9. Eichten, E., and Hill, B., *Phys. Lett.* **B234**, 511 (1990).
10. Grinstein, B., *Nucl. Phys.* **B339**, 253–268 (1990).
11. Georgi, H., *Phys. Lett.* **B240**, 447–450 (1990).
12. Eichten, E., and Hill, B., *Phys. Lett.* **B243**, 427–431 (1990).
13. Falk, A. F., Grinstein, B., and Luke, M. E., *Nucl. Phys.* **B357**, 185–207 (1991).
14. Georgi, H., Grinstein, B., and Wise, M. B., *Phys. Lett.* **B252**, 456–460 (1990).
15. Manohar, A. V., and Wise, M. B., *Cambridge Monogr. Part. Phys. Nucl. Phys. Cosmol.* **10**, 1–191 (2000).
16. Luke, M. E., *Phys. Lett.* **B252**, 447–455 (1990).
17. Beneke, M., Buchalla, G., Neubert, M., and Sachrajda, C. T., *Phys. Rev. Lett.* **83**, 1914–1917 (1999).
18. Beneke, M., Buchalla, G., Neubert, M., and Sachrajda, C. T., *Nucl. Phys.* **B591**, 313–418 (2000).
19. Bauer, C. W., Fleming, S., and Luke, M. E., *Phys. Rev.* **D63**, 014006 (2001).
20. Bauer, C. W., Fleming, S., Pirjol, D., and Stewart, I. W., *Phys. Rev.* **D63**, 114020 (2001).
21. Beneke, M., Buchalla, G., Neubert, M., and Sachrajda, C. T., *Nucl. Phys.* **B606**, 245–321 (2001).
22. Beneke, M., and Neubert, M., *Nucl. Phys.* **B675**, 333–415 (2003).
23. Chibisov, B., Dikeman, R. D., Shifman, M. A., and Uraltsev, N., *Int. J. Mod. Phys.* **A12**, 2075–2133 (1997).
24. Ciuchini, M., Franco, E., Lubicz, V., Mescia, F., and Tarantino, C., *JHEP* **08**, 031 (2003).
25. Beneke, M., Buchalla, G., Lenz, A., and Nierste, U., *Phys. Lett.* **B576**, 173–183 (2003).

Models and Methods:
Can Theory Meet the *B* Physics Challenge?

David London

Physics Department, McGill University, 3600 University St., Montréal QC, Canada H3A 2T8
and
Lab. René J.-A. Lévesque, Université de Montréal, C.P. 6128, succ. centre-ville, Montréal, QC, Canada H3C 3J7

Abstract. I am skeptical about the likelihood of significant theoretical progress over the next few years in reducing the hadronic uncertainty in *B* decays. While it is not difficult to establish that new physics is present in the *B* system, its identification will almost certainly require direct observation at high-energy colliders. On the other hand, *B* experiments will be necessary to probe the *CP* nature of the new physics couplings. Thus, *B* physics experiments and high-energy experiments are complementary, and I hope that experimentalists in these two areas will coordinate their efforts.

In setting the stage for this panel discussion, Joel Butler sent out an email asking several questions:

At this conference, we have seen the latest projections of experimental sensitivities to a variety of B decays. The stated goal of all experiments is to find deviations in CP-violating and/or rare B decays that would signify "new physics." We pose to this panel the following questions:

1. To what extent can the various models of new physics be distinguished by this projected statistical precision? Where else should we look?

2. The channels discussed by experimenters are generally those for which theoretical difficulties are either known to be small or where some theoretically clean CP-violating amplitude can be isolated by measuring a related set of decays. Many more decays could be used, and statistical precision could be improved, if hadronic effects could be disentangled from the weak interaction effects. To what extent will theory be able to accomplish this disentanglement on the relevant timescale?"

The two questions are clearly related to one another. Let me begin with the second. To be honest, I am skeptical that theory will be able to "meet the challenge" of hadronic uncertainties. At present, there are two approaches on the market which address hadronic physics: analytical methods and lattice calculations. I address these in turn.

Over the past ten years, there has been a great deal of progress in analytical methods: heavy-quark effective theory (HQET), QCD factorization, perturbative QCD, soft-collinear effective theory. All of these techniques

are based on an expansion in terms of $1/m_b$, and some beautiful results have been obtained in the $m_b \to \infty$ limit. Unfortunately, m_b is obviously not infinite, and one has to deal with $1/m_b$ corrections.

Now, the problem is that, though the corrections are $\mathcal{O}(1/m_b)$, we don't know what goes in the numerator. This is very process-dependent, but one can easily have corrections at the level of 1 GeV/5 GeV ~ 20%. That is, these $1/m_b$ corrections need not be small! Thus, the uncertainty in all analytical methods is of the order of 20%. Even the higher-order corrections, which are $\mathcal{O}(1/m_b^2)$, are ~ 5%.

In the coming years, the experimental error on measurements of *B* processes will be reduced considerably. Will we see a sufficient improvement in these analytical methods and will the theoretical uncertainty be reduced to below the experimental error? This is not impossible. After all, HQET has allowed us to obtain V_{cb} with a precision of better than 5%. However, I would not assume that such an improvement will occur, particularly over a short timescale of about five years. The case of V_{cb} looks to me exceptional, in that the leading-order $1/m_b$ corrections were shown to vanish [1]. Other processes receive nonzero corrections of $\mathcal{O}(1/m_b)$, and these look to be very difficult to get under control. Note: I am not claiming that this will not happen, and I am certainly not suggesting that these subjects are not worth working on. However, from an experimental point of view, I would not count on such improvements being made in analytical methods.

The second approach to hadronic physics in *B* decays is lattice calculations. This is extremely promising, and may eventually allow us to reliably compute hadronic

CP722, *B Physics at Hadron Machines*, edited by M. Paulini and S. Erhan

quantities. However, here too, I am quite cautious about the prospects for obtaining the necessary precision, especially in the short term. In this case, the problem is one of error estimation. There are two events which have coloured my opinion.

The first is the computation of the B^0 decay constant f_{B_d}. The lattice value of this quantity has changed enormously over the past ten years. In 1996, lattice calculations gave $f_{B_d} = 170^{+55}_{-50}$ MeV [2]. Since that time the central value has increased steadily, until we now have $f_{B_d} = 230 \pm 28$ MeV [3]. Now, the two values do not really disagree, but the central values have clearly changed significantly. In particular, the 1996 value does not include the 2003 central value within 1σ.

Second, the importance of chiral logs in unquenched calculations was recently noted [4]. This has had a large effect on computations of decay constants. Prior to this, the value for $\xi \equiv f_{B_s}/f_{B_d}$ was taken to be $\xi = 1.15 \pm 0.05$; subsequently it was found to be $\xi = 1.32 \pm 0.10$ [4]. The decay constants for D mesons changed similarly: in 2000, the lattice value $f_{D_s}/f_D = 1.11 \pm 0.02$ was given [5], while in 2003 we have $f_{D_s}/f_D = 1.22 \pm 0.04$ [6]. Today's central values are clearly well outside the errors quoted only a few years ago.

In both cases, we understand the source of the changes. These are due to scaling violation, unquenching and chiral logs. Still, the fact remains that there have been (at least) two unanticipated theoretical developments that have led to large changes in the central values of the lattice calculation of decay constants. Furthermore, in both cases the modified values lie outside of the errors given previously. This is somewhat troubling: how do we know that there won't be further modifications? More importantly, it does make one wonder about the error estimation – are the systematics sufficiently well understood? Can we count on the precision of future calculations?

In addition, the computation of decay constants is relatively easy; what we really want are B matrix elements, which are very hard to calculate. I am fairly confident that, given time, lattice calculations will settle down, and we will end up with reliable computations of the various hadronic quantities we need in B decays. However, I am skeptical that, in the short term, we can be sufficiently confident about the given errors to use lattice results to claim the discovery of new physics in a particular B process. Again, I am not suggesting that lattice calculations are not worth doing. Rather I am saying that experimentalists need to be cautious about quoted errors, at least over the next few years.

I am therefore unconvinced at present that theory will be able to deal with the hadronic uncertainties on the timescale of, say, five years, *i.e.* of this generation of B experiments. I would be very hesitant to use such theoretical input in the search for new physics.

Now, if theory is unable to "meet the challenge" of dealing with hadronic physics, what are we to do? This brings me to the first question posed by Joel Butler. The goal of B physics is to find evidence for physics beyond the Standard Model (SM), and to identify it. In my opinion, it is not that difficult to find evidence of new physics – there are numerous signals of new physics in *CP*-violating processes in the B system [7]. These have varying degrees of theoretical input. However, by performing a systematic search for new effects, it should be possible to establish with certainty that such new physics is present, even without precise knowledge of hadronic quantities.

On the other hand, it is very difficult, if not impossible, to identify the physics beyond the SM through measurements of B decays. After all, all we will have are indirect effects of such new physics. (And note that the precise knowledge of hadronic physics would not help in this respect.) To identify the new physics, we will have to turn to high-energy experiments which specifically look for new physics at the weak scale. That is, we will have to produce the new particles directly, probably at a high-energy collider like the LHC. The lesson here is that B physics experiments cannot do it all alone: although evidence for new physics can be found in B decays, it will require high-energy colliders to definitively identify it.

On the other hand, high-energy colliders cannot do it all alone either. Although they will be able to produce the new particles directly, they will be unable to probe the *CP* nature of their couplings. That is the domain of B physics. Thus, B physics experiments and high-energy experiments are complementary.

I hope that experimentalists in the two areas – B physics and high-energy measurements – will be able to coordinate their efforts. This is the only way, we will end up with a complete description of the physics beyond the SM.

ACKNOWLEDGMENTS

This work was supported by NSERC of Canada.

REFERENCES

1. Luke, M. E., *Phys. Lett.* **B252**, 447–455 (1990).
2. Wittig, H., arXiv:hep-ph/9606371 (1996).
3. For example, see the CKMfitter group (2003), URL: http://www.slac.stanford.edu/~laplace/ckmfitter.html.
4. Kronfeld, A. S., *eConf* **C020620**, FRBT05 (2002).
5. Becirevic, D., *Nucl. Phys. Proc. Suppl.* **94**, 337–341 (2001).
6. Becirevic, D., arXiv:hep-ph/0310072 (2003).
7. London, D., in these proceedings (2004).

List of Participants

Marco Adinolfi	University of Oxford	m.adinolfi1@physics.ox.ac.uk
Valeri Andreev	UCLA	Valeri.Andreev@cern.ch
Marina Artuso	Syracuse University	artuso@physics.syr.edu
David Asner	University of Pittsburgh	asner@lepp.cornell.edu
Paul Avery	University of Florida	avery@phys.ufl.edu
Roy Briere	Carnegie Mellon University	rbriere@andrew.cmu.edu
Joel Butler	Fermilab	butler@fnal.gov
Robert Cahn	LBNL	rncahn@lbl.gov
Olivier Callot	LAL-Orsay and CERN	Olivier.Callot@cern.ch
Mario Calvetti	INFN	calvetti@fi.infn.it
Junegone Chay	Korea University	chay@korea.ac.kr
Chunhui Chen	University of Pennsylvania	cchen23@hep.upenn.edu
David Cinabro	Wayne State University	cinabro@physics.wayne.edu
Brad Cox	University of Virginia	cox@uvahed.phys.virginia.edu
Alakabha Datta	University of Toronto	datta@physics.utoronto.ca
Giuseppe Della-Ricca	Universita' & INFN - Trieste	dellaricca@ts.infn.it
Karl Ecklund	Cornell University	kme@mail.lepp.cornell.edu
Paula Eerola	Lund University	paula.eerola@hep.lu.se
Aida El-Khadra	University of Illinois Urbana	axk@uiuc.edu
Jürgen Engelfried	Universidad de San Luis Potosi	jurgen@ifisica.uaslp.mx
Samim Erhan	UCLA	Samim.Erhan@cern.ch
Tom Ferguson	Carnegie Mellon University	ferguson@cmphys.phys.cmu.edu
Sean Fleming	Carnegie Mellon University	spf@andrew.cmu.edu
Simon George	RHUL	S.George@rhul.ac.uk
Lawrence Gibbons	Cornell University	lkg@mail.lepp.cornell.edu
Karen Gibson	Carnegie Mellon University	krg@andrew.cmu.edu
Frederick Gilman	Carnegie Mellon University	gilman@andrew.cmu.edu
Gavril Giurgiu	Carnegie Mellon University	ggiurgiu@fnal.gov
Benjamin Grinstein	UC San Diego	ben@ben-lee.ucsd.edu
Yuval Grossman	Technion/SLAC	yuval@Slac.Stanford.edu
Robert Harr	Wayne State University	harr@physics.wayne.edu
Carsten Hast	SLAC	hast@slac.stanford.edu
Takeo Higuchi	KEK	takeo.higuchi@kek.jp
Richard Holman	Carnegie Mellon University	rh4a@andrew.cmu.phys
Gino Isidori	INFN - Frascati	Gino.Isidori@lnf.infn.it
Vivek Jain	Brookhaven National Lab	vj@bnl.gov
Hidekazu Kakuno	Tokyo Institute of Technology	kakuno@hp.phys.titech.ac.jp
Min Jeong Kim	Carnegie Mellon University	mjkim@fnal.gov
Marcin Konecki	University of Basel	marcin.konecki@cern.ch
Robert Kroeger	University of Mississippi	kroeger@relativity.phy.olemiss.edu
Adam Leibovich	University of Pittsburgh	akl2@pitt.edu
Ling-Fong Li	Carnegie Mellon University	lfli@andrew.cmu.edu
Cheng-Ju Lin	Fermilab	cjl@fnal.gov

David London	Universite de Montreal	london@lps.umontreal.ca
Frederic Machefert	LAL - CNRS Orsay	frederic.machefert@in2p3.fr
Krishnamurthy Mahalaxmi	University of Tennesee	kmaha@slac.stanford.edu
Petar Maksimovic	Johns Hopkins University	petar@jhu.edu
Nancy Marinelli	IASA-University of Athens	nancy.marinelli@cern.ch
Daniel Marlow	Princeton University	marlow@Princeton.edu
Andreas Meyer	Hamburg University	andreas.meyer@desy.de
Ting Miao	Fermilab	tmiao@fnal.gov
Kenkichi Miyabayashi	Nara Women's University	miyabaya@phys.nara-wu.ac.jp
Hans-Günther Moser	Max-Planck-Insitut Munich	moser@mppmu.mpg.de
Homer Neal	Yale University	homer.neal@yale.edu
Ulrich Nierste	Fermilab	nierste@fnal.gov
Fabrizio Parodi	Genova University & INFN	Fabrizio.Parodi@ge.infn.it
Manfred Paulini	Carnegie Mellon University	paulini@cmu.edu
Gerhard Raven	NIKHEF/VU Amsterdam	Gerhard.Raven@nikhef.nl
Jonathan Rosner	University of Chicago	rosner@hep.uchicago.edu
Ira Rothstein	Carnegie Mellon University	izr@andrew.cmu.edu
James Russ	Carnegie Mellon University	russ@cmphys.phys.cmu.edu
Peter Schlein	UCLA	Peter.Schlein@cern.ch
Lalit Sehgal	RWTH Aachen	sehgal@physik.rwth-aachen.de
Dhiren Shah	The M.S.University of Baroda	dhiren70_shah@indiatimes.com
Paul Shepard	University of Pittsburgh	shepard@pitt.edu
Amarjit Soni	Brookhaven National Lab	soni@bnl.gov
Iain Stewart	MIT	iains@mit.edu
Sheldon Stone	University of Syracuse	stone@physics.syr.edu
Nikolai Terentiev	Carnegie Mellon University	teren@fnal.gov
Christophe Thiebaux	LLR Ecole Polytechnique	thiebaux@in2p3.fr
Vivek Tiwari	Carnegie Mellon University	vtiwari@cmu.edu
Karim Trabelsi	University of Hawaii	karim@phys.hawaii.edu
Ulrich Uwer	University of Heidelberg	uwer@PhysI.uni-heidelberg.de
Marc Verderi	LLR - Ecole Polytechnique	verderi@in2p3.fr
Helmut Vogel	Carnegie Mellon University	helmut.vogel@cmu.edu
Eckhard Von Toerne	Kansas State University	evt@phys.ksu.edu
Jianchun Wang	Syracuse University	jwang@phy.syr.edu
Michael Wang	Fermilab	mwang@fnal.gov
Yili Wang	Iowa State University	yiliwa@iastate.edu
Alan Watson	University of Birmingham	alan.watson@cern.ch
Edward Witten	IAS Princeton	witten@ias.edu
Lincoln Wolfenstein	Carnegie Mellon University	lincolnw@andrew.cmu.edu
Feng Wu	Carnegie Mellon University	fwu@andrew.cmu.edu
Thomas Ziegler	Princeton University	tziegler@princeton.edu
Daria Zieminska	Indiana University	daria@indiana.edu
Martin zur Nedden	Humboldt University of Berlin	nedden@mail.desy.de

Scientific Program

Morning Session: CHAIR: Fred Gilman (Carnegie Mellon)

Recent Results

9.00 - 9.10	Welcome by Lincoln Wolfenstein [5]
	Welcome by MCS Dean Richard McCullough [5]
9.10 - 9.30	Recent Results from CLEO [17+3]
	Karl Ecklund (Cornell University)
9.30 - 9.55	Recent B Physics Results from DØ [20+5]
	Vivek Jain (BNL)
9.55 - 10.20	Recent Heavy Flavour Results from CDF [20+5]
	Robert Harr (Wayne State University)

10.20 - 10.45: COFFEE BREAK

Recent Results and Grid

10.45 - 11.10	Recent Results from Belle [20+5]
	Takeo Higuchi (KEK)
11.10 - 11.30	Recent Results from BaBar (ϕK_S^0) [17+3]
	Krishnamurthy Mahalaxmi (University of Tennessee)
11.30 - 12.00	CKM Matrix Elements & Fits [25+5]
	Hidekazu Kakuno (Tokyo Institute of Technology)
12.00 - 12.30	Grid Computing in High Energy Physics [25+5]
	Paul Avery (University of Florida Gainesville)

12.30 - 2.00: LUNCH (Luncheon served)

Afternoon Session: CHAIR: Bob Cahn (LBNL)

Flavour Physics and *CP* Violation

2.00 - 2.45	Flavour Physics in the Grande Scheme [35+10]
	Edward Witten (IAS Princeton)
2.45 - 3.10	*CP* Violation at Belle [20+5]
	Kenkichi Miyabayashi (Nara University)
3.10 - 3.35	*CP* Violation Prospects at the Tevatron [20+5]
	Petar Maksimovic (Johns Hopkins University)

3.35 - 4.05: COFFEE BREAK

CP Violation continued

4.05 - 4.50	Status and Future of Gamma [35+10]
	Jonathan Rosner (University of Chicago)
4.50 - 5.30	Towards alpha and gamma at the *B* Factories [35+5]
	Alan Watson (University of Birmingham)

Morning Session: CHAIR: Paul Shepard (University of Pittsburgh)

Lattice and Charm

9.00 - 9.30	Lattice QCD now and in > 5 Years [25+5]
	Aida El-Khadra (University of Illinois Urbana-Champaign)
9.30 - 9.50	$CLEO_C$ Status and Prospects [17+3]
	David Asner (University of Pittsburgh)
9.50 - 10.15	Review of Recent Results in Charm Physics [20+5]
	Jürgen Engelfried (Universidad Autonoma de San Luis Potosi)

10.15 - 10.45: COFFEE BREAK

B Production and Spectroscopy

10.45 - 11.10	Final $b\bar{b}$ Cross Section from HERA-B [20+5]
	Martin zur Nedden (Humboldt University Berlin)
11.10 - 11.30	Hadronic Beauty Production at HERA [17+3]
	Andreas Meyer (University of Hamburg)
11.30 - 12.00	Heavy Flavour Production and Cross Sections at the Tevatron [25+5]
	Chunhui Chen (University of Pennsylvania)
12.00 - 12.30	B Reconstruction and Spectroscopy at DØ: Present & Future [25+5]
	Eckhard von Toerne (Kansas State University)

12.30 - 2.00: LUNCH (Luncheon served)

Afternoon Session: CHAIR: Marina Artuso (Syracuse University)

Charm at B Factories

2.00 - 2.20	Review of New D_S States [17+3]
	Jianchun Wang (Syracuse University)
2.20 - 2.45	B Decays to Charm and Charmonium at BaBar [20+5]
	Christophe Thiebaux (LLR - Ecole Polytechnique Palaiseau)
2.45 - 3.05	Charm Physics at Belle [17+3]
	Karim Trabelsi (University of Hawaii)

Detector and Hardware Updates

3.05 - 3.30	Status of Atlas Inner Detector [20+5]
	Hans-Günther Moser (MPI Munich)
3.30 - 4.00	CMS Detector Update [25+5]
	Homer Neal (Yale University)

4.00 - 4.30: COFFEE BREAK

Afternoon Session: (continued) CHAIR: Tom Ferguson (Carnegie Mellon)

Detector and Hardware Updates continued

4.30 - 5.00	BTeV Detector Update [25+5]
	David Cinabro (Wayne State University)
5.00 - 5.25	Re-optimized LHCb Detector & Tracking Performance [20+5]
	Gerhard Raven (NIKEHF)
5.25 - 5.45	Status of LHCb RICH & Hadron Particle Identification [17+3]
	Marco Adinolfi (Oxford)
5.45 - 6.05	Status of LHCb Calorimeter & Muon system plus Lepton Identification
	Frederic Machefert (LAL Orsay)
6.05 - 6.30	Atlas Software for B Physics [20+5]
	Fabrizio Parodi (Universita di Genova/INFN)

Thursday October 16, 2003

Morning Session: CHAIR: Jonathan Rosner (University of Chicago)

V_{cb}, V_{ub} and Factorization

9.00 - 9.25	Semileptonic BR's and Moments [20+5]
	Giuseppe Della Ricca (Universita' & INFN - Trieste)
9.25 - 9.45	Review of V_{cb} [17+3]
	Marina Artuso (Syracuse University)
9.45 - 10.05	Review of V_{ub} [17+3]
	Lawrence Gibbons (Cornell University)
10.05 - 10.35	QCD, Factorization and SCET [25+5]
	Iain Stewart (MIT)

10.35 - 11.00: COFFEE BREAK

B Lifetimes and Mixing

11.00 - 11.20	B Mixing and Lifetimes at the B Factories: Present & Future [17+3]
	Dan Marlow (Princeton)
11.20 - 11.50	B Lifetime Measurements at the Tevatron [25+5]
	Daria Zieminska (Indiana University)
11.50 - 12.20	B Tagging and Mixing at the Tevatron [25+5]
	Ting Miao (Fermilab)

Afternoon Program:

12.30 pm: Tour to Fallingwater (Boxed lunch provided)

Evening: Dinner at Seven Springs en route back

Morning Session: CHAIR: Linclon Wolfenstein (Carnegie Mellon)

	Rare Decays
9.00 - 9.30	Theory of Radiative and Rare Decays [25+5] Gino Isidori (INFN Frascati)
9.30 - 9.55	Radiative and Rare Decays at BaBar [20+5] Carsten Hast (SLAC)
9.55 - 10.15	Radiative Decays at Belle: Present & Future [17+3] Tom Ziegler (Princeton)
10.15 - 10.40	Spectroscopy and Rare Decays at CDF: Present and Future [20+5] Cheng-Ju Stephen Lin (Fermilab)

10.40 - 11.10: COFFEE BREAK

	Beyond the Standard Model
11.10 - 11.55	Footprints of New Physics in the B System [35+10] Yuval Grossman (Technion and SLAC)
11.55 - 12.25	CP Violation beyond the Standard Model [25+5] David London (Universite de Montreal)

12.25 - 2.00: LUNCH (Luncheon served)

Afternoon Session: CHAIR: Sheldon Stone (Syracuse University)

	B **Trigger at Future Experiments**
2.00 - 2.30	Atlas B Physics Trigger [25+5] Simon George (Royal Holloway College London)
2.30 - 3.00	Online Event Selection at the CMS Experiment [25+5] Marcin Konecki (University of Basel)
3.00 - 3.20	LHCb Trigger: Implementation and Performance [17+3] Oliver Callot (LAL Orsay and CERN)
3.20 - 3.50	BTeV Trigger [25+5] Michael Wang (Fermilab)

3.50 - 4.20: COFFEE BREAK

	Future B **Physics**
4.20 - 4.50	Physics of Super-Rare Decays [25+5] Benjamin Grinstein (UC San Diego)
4.50 - 5.35	B Physics in the LHC Era [35+10] Bob Cahn (LBNL)

Evening program:

6.30 pm: Conference Banquet at Carnegie Museum of Art and Natural History
(Cocktail Reception and Dinner)

Morning Session: CHAIR: Samim Erhan (UCLA)

B Physics at Future Hadron Machines

9.00 - 9.30	CMS B Physics Update [25+5]
	Nancy Marinelli (IASA/Athens)
9.30 - 9.55	LHCb Physics Performance [20+5]
	Ulrich Uwer (University of Heidelberg)
9.55 - 10.25	Physics of BTeV [25+5]
	Brad Cox (University of Virginia)
10.25 - 10.55	Atlas B Physics Performance Update [25+5]
	Paula Eerola (University of Lund)

10.55 - 11.15: COFFEE BREAK

Panel Discussion

"Models and Methods: Can Theory meet the B Physics Challenge?"
Organized and chaired by Joel Butler (Fermilab)

Panel members: Bob Cahn (LBNL), Gino Isidori (INFN Frascati), David London (Universite de Montreal), Amarjit Soni (BNL), Uli Nierste (Fermilab), Lincoln Wolfenstein (Carnegie Mellon)

11.15 - 1.15	Introduction [5]
	Opening Statements by Panelists [10 min each]
	Panel discussion: [30]
	How might the new physics unfold?
	What progress will be made towards a theory of flavour?
	Additional questions from the audience [25]

End of Conference

293

AUTHOR INDEX

A

Adinolfi, M., 119
Artuso, M., 151
Asner, D., 82
Avery, P., 131

B

Butler, J. N., 277

C

Cahn, R. N., 269
Callot, O., 214
Chen, C., 67
Cinabro, D., 109
Cox, B., 236

D

Della Ricca, G., 146

E

Ecklund, K. M., 3
Eerola, P., 242
Engelfried, J., 79

G

George, S., 201
Gibbons, L., 156
Grinstein, B., 263
Grossman, Y., 255

H

Harr, R., 11

I

Isidori, G., 172

J

Jain, V., 7

K

Konecki, M., 207
Krishnamurthy, M., 16

L

Lin, C.-J. S., 193
London, D., 50, 283

M

Machefert, F., 123
Maksimović, P., 30
Mantry, S., 141
Marinelli, N., 225
Meyer, A. B., 63
Miao, T., 172
Miyabayashi, K., 23
Moser, H.-G., 99

N

Neal, H., 104
Nierste, U., 280

P

Parodi, F., 127
Pirjol, D., 141

R

Raven, G., 114
Rosner, J. L., 35

S

Stewart, I. W., 141

T

Trabelsi, K., 92

U

Uwer, U., 231

V

von Toerne, E., 72

W

Wang, J., 87
Wang, M. H. L. S., 218

Watson, A., 42
Witten, E., 251
Wolfenstein, L., 279

Z

Ziegler, T., 188
Zieminska, D., 167
zur Nedden, M., 57